ERGONOMIC WORKPLACE DESIGN
FOR HEALTH, WELLNESS, AND PRODUCTIVITY

Human Factors and Ergonomics Series

ERGONOMIC WORKPLACE DESIGN
FOR HEALTH, WELLNESS, AND PRODUCTIVITY

edited by

Alan Hedge

CRC Press
Taylor & Francis Group
Boca Raton London New York

CRC Press is an imprint of the
Taylor & Francis Group, an **informa** business

First published in paperback 2024

First published 2017 by CRC Press
2385 NW Executive Center Drive, Suite 320, Boca Raton FL 33431

and by CRC Press
4 Park Square, Milton Park, Abingdon, Oxon, OX14 4RN

CRC Press is an imprint of Taylor & Francis Group, LLC

© 2017, 2024 Taylor & Francis Group, LLC

Library of Congress Cataloging-in-Publication Data

Names: Hedge, Alan, editor.
Title: Ergonomic workplace design for health, wellness, and productivity / editor, Alan Hedge.
Description: Boca Raton : Taylor & Francis, a CRC title, part of the Taylor & Francis imprint, a member of the Taylor & Francis Group, the academic division of T&F Informa, plc, [2016] | Series: Human factors and ergonomics | Includes bibliographical references and index.
Identifiers: LCCN 2016002569 | ISBN 9781466598430 (alk. paper)
Subjects: LCSH: Human engineering. | Industrial hygiene. | Industrial safety. | Labor productivity.
Classification: LCC TA166 .E6931324 2016 | DDC 620.8/2--dc23
LC record available at http://lccn.loc.gov/2016002569

ISBN: 978-1-4665-9843-0 (hbk)
ISBN: 978-1-03-291981-2 (pbk)
ISBN: 978-1-315-37400-0 (ebk)

DOI: 10.1201/9781315374000

Visit the Taylor & Francis Web site at
http://www.taylorandfrancis.com

and the CRC Press Web site at
http://www.crcpress.com

Contents

SECTION I   The Physical Environment

SECTION II   Ergonomic Design Issues in Workplaces

SECTION III   Emerging Ergonomic Workplace Design Issues

Preface

Every day billions of people around the world go to work. Work is fundamental to human societies. Work partly defines us as individuals, and certain professions can serve as status symbols. Many people spend years in education systems training for a work career. Losing one's job can be a significant stressor, as can retirement from work. Anything that improves the conditions of work has an enormous impact on the well-being of vast numbers of people. Ergonomics is the science of work, and it is a valuable discipline that focuses on improving the ability of people to perform work. Ergonomics adopts a systems approach to designing effective work, and that requires consideration of relevant cognitive, physical, and organizational factors. Indeed, in the International Ergonomics Association's description of ergonomics, it describes these three sets of factors. Yet there is also a crucial fourth factor, namely, the environment. As this book will demonstrate, the ergonomic design of the environment is an essential, yet all too often overlooked, component of the work systems design process.

All human work, whether physical, mental, or both, occurs somewhere, and the design of the work environment obviously plays a critical role in the ability of a person to perform their work. Work performance can suffer if the environmental conditions are suboptimal, such as workplaces that are too cold or too hot, where the lighting is too bright or too dim, where it is too noisy, where the air is polluted, or where the work setting is vibrating or in motion. Also, a suboptimal spatial layout of a workplace can detrimentally affect work postures, which in turn impacts health, wellness, and task performance.

Although early humans were most likely nomadic, where possible they inhabited places and natural structures such as caves, which offered protection against elements and predators, and which served as congregation places. Eventually, some 5000 years ago, developments in agriculture allowed communities to settle in specific locations starting the processes of urbanization and civilization as we now know it. At that time, it is likely that a majority of workers did most of their work outdoors, with activities such as hunting and fishing, agriculture, road building, construction, and fighting battles.

The industrial revolution that began around 1750 marked the acceleration in the movement of work from outdoors in fields to indoors in factories. In developed countries today, a majority of workers perform their work inside some kind of designed structure, such as a building or a vehicle. How well the designed environment supports their work plays a significant role in factors such as the risks of work-related injuries, accidents, and productivity.

Although the designed environment plays an obvious role in impacting human behavior, this often gets overlooked, even in the ergonomic analysis of work. For example, task analysis methods typically focus on the work content and the physical actions involved in performing work, and cognitive task analysis, workload measurement, and error analysis methods focus on the mental processes involved in completing the tasks, yet such methods typically neglect the consideration of the physical environment design changes that either positively or negatively impact the work processes. We all know from personal experience how critical the design of the environment is to the successful performance of work. If you use an iPad, you may have struggled to read the screen in bright sunlight because the ambient lighting overwhelms the luminance of the screen, or, conversely, you may have struggled to read a printed menu in a dimly lit restaurant where the lighting is insufficient for easy legibility of the text. You may be an adroit typist, but if you are using a laptop while riding on a bus that is driving along a bumpy highway, you will have experienced how difficult it is to maintain adequate performance and to minimize errors because the environment is not supporting your ability to do work. You may have experienced feelings of drowsiness when sitting in a crowded meeting in an inadequately ventilated room, and this occurs because of an accumulation of carbon dioxide. Your manual dexterity and cognitive abilities are substantially impaired by exposure to very cold conditions, and your energy levels may be set by hot and humid conditions. Environmental conditions, such as the thermal environment, the luminous environment, the acoustic environment, and the vibration, all impact our comfort, health, and performance. Quite simply, we are animals with biological systems that are adapted to a relatively narrow range of environmental conditions, and if we are to be successful when inside human-designed enclosures, ranging from submarines to spacecraft, from cars to buildings, then we must pay close attention to optimizing these environmental conditions to maximize our ability to perform work efficiently and effectively.

This book provides a good overview of these environmental requirements. But just knowing the environmental conditions by itself is not sufficient to ensure that our performance is optimized. Our capabilities are limited by our chronobiology—there are times of the day when we expect to be able to sleep and other times when we are alert. Unfortunately, in our 24/7 societies, there are many jobs that require people to work at those times of the day when our bodies are least prepared for this. In addition, our capabilities are also limited by factors such as our size, reach distances, and strength, and so the physical arrangement of tools and other work artifacts is critical if we are to demonstrate maximum performance ability while minimizing the risks of errors, accidents, and injuries. To illustrate these issues and other related considerations, this book also presents workplace design considerations for a wide variety of workplace settings. In most of the settings that are described, the ergonomics considerations focus on physical design issues, and one fact that remains invariant is that whenever we can position a person so that they can perform their work while in a neutral posture, whether sitting or standing, then we will maximize their physical capabilities and their endurance and minimize the possibilities of developing work-related injuries.

This book contains the latest information from internationally recognized ergonomics experts. In Section I, the first seven chapters of the book, the physical environmental conditions necessary for optimal health, wellness, and productivity are presented. In Chapter 1, Hedge describes the basic computer workstation design requirements for a healthy posture. We are homiotherms, and in Chapter 2, Parsons presents the thermal environment requirements for comfort, health, and performance. In many indoor settings, the thermal conditions are linked with ventilation, and in Chapter 3, Wargocki provides a comprehensive review of the optimal indoor air quality requirements. Noise can be stressful and can interfere with work performance, and in Chapter 4, Oseland and Hodsman tell us the requirements for the design of a successful acoustic environment. In many work settings, the worker is in motion or using tools that vibrate, and in Chapter 5, Burgess-Limerick describes what is acceptable and what will interfere with our ability to work as well our health and well-being. In Chapter 6, Figueiro and Rea summarize the lighting conditions essential for optimal visual performance in indoor workplaces. For optimal health, our bodies need to be synchronized with the environment, yet in our 24/7 world many people have to work at times when our body is not at its best, and Puttonen discusses this important topic of shift work in Chapter 7.

In Section II, there are eight chapters that present the application of ergonomics in different workplaces. Perhaps one of the commonest workplaces in the modern world, the office looks innocuous, but in Chapter 8, Vink et al. discuss a range of ergonomic issues with various types of office designs. Especially in the United States, healthcare is provided 24 hours each day, every day, and it is a sector that now is in transition as new healthcare information technologies permeate many aspects of medical care, and in Chapter 9, Springer describes a selection of these issues. Likewise, many control center operations involve 24/7 working, and their heavy emphasis on computing technology presents unique challenges, as shown by Papic in Chapter 10. Our education systems are critical in providing a future workforce with the necessary knowledge and skills for success, and in Chapter 11, Straker and Howie review the important contributions that ergonomics makes to the design of school settings. Universities are the pinnacle of many education systems, and in addition to teaching students, they are typically large institutions that fulfill a variety of research and other functions. In Chapter 12, Nou shows the importance of ergonomics in a variety of these settings. In Chapter 13, Burt describes the value of ergonomics in the design of laboratories and laboratory equipment that is used many research settings, from universities to biotechnology and pharmaceutical companies.

For many people, hotels provide temporary vacation accommodation and/or a temporary workplace; hospitality settings are complex environments that present ergonomists with a variety of challenges, and in Chapter 14, Punnett et al. discuss a wide range of these issues. In Chapter 15, Robertson and Maynard systematically look at the ergonomic design challenges presented by the growth in teleworking, where the residence also becomes the workplace for at least part of the working week.

The final six chapters in Section III address emerging ergonomic design issues. In Chapter 16, Peacock et al. examine some of the issues of transportation systems, including the role of vehicles as modern workplaces. Many organizations are experimenting with replacing more traditional office designs. In Chapter 17, Brand describes the drivers for these new ways of working (WOW) and presents alternative workplace design strategies, and in Chapter 18, McAtamney et al. summarize a range of ergonomic design considerations associated with new WOW settings. The green building movement has transformed the construction industry worldwide, and ergonomic designs can play a valuable role in the creation of sustainable buildings, as described by Dorsey in Chapter 19. Innovation is the lifeblood of most organizations, and in Chapter 20, Yoon and Chung outline a number of important elements for designing 3C workplaces that can foster connectedness, collaboration, and creativity, and present recent work on this topic using social sensing technology. Finally, in Chapter 21, Hedge and Pazell discuss the benefits of ergonomics and wellness programs, which are traditionally separated in organizations with ergonomics being a part of Health and Safety and wellness a part of the Human Resources, and they discuss the importance of new initiatives aimed at a total systems approach to the design of workplaces to promote employee health, wellness, and productivity.

Acknowledgments

The topic of workplace design often gets little attention in the human factors and ergonomics world. This book will hopefully serve to raise the profile of workplace ergonomic design. But this book probably would not have materialized without the foresight of Gavriel Salvendy of Purdue University and Tsinghua University, for it is he who enthusiastically suggested that I compile this book. But also, I could not have completed this task without the willingness and dedication of all the contributors. I also acknowledge all of the ergonomics practitioners who help those who have been injured because of the inadequate design of their workplaces. I hope that this book will serve as a stimulus that triggers greater interest in the importance of workplace design in the human factors and ergonomics community, for every day what we do directly affects the health, wellness, and productivity of millions of workers around the world.

Editor

Alan Hedge is a professor in the Department of Design and Environmental Analysis, Cornell University, where he also directs the Cornell Human Factors and Ergonomics laboratory. His research and teaching activities focus on ergonomic designs that promote health, comfort, and productivity, especially in healthcare and office workplaces.

He is a fellow of the Human Factors and Ergonomics Society (HFES), and he was awarded the 2003 Alexander Williams Jr. Design Award and the 2009 Oliver Hansen Outreach Award by the HFES. He is also a fellow of the International Ergonomics Association, a chartered ergonomist in the UK, and a certified professional ergonomist.

Contributors

Christine Aickin
Workability Pty Ltd.
Sydney, Australia

W. Gary Allread
Institute for Ergonomics
Ohio State University
Columbus, Ohio

Iris Bakker
Levenswerken
Boskoop, The Netherlands

Jay L. Brand
School of Education
Andrews University
Berrien Springs, Michigan

Robin Burgess-Limerick
Minerals Industry Safety and Health Centre
University of Queensland
Brisbane, Australia

Cynthia M. Burt
Environmental Health
University of California, Los Angeles
Los Angeles, California

David Caple
David Caple and Associates Pty Ltd.
Ivanhoe, Australia

Carlo Caponecchia
School of Aviation
University of New South Wales
Sydney, Australia

Susan S. E. Chung
American Society of Interior Designers
Washington, DC

Julie Dorsey
Department of Occupational Therapy
Ithaca College
Ithaca, New York

Mariana G. Figueiro
Lighting Research Center
Rensselaer Polytechnic Institute
Troy, New York

Liesbeth Groenesteijn
Charly Green
Bilthoven, The Netherlands

Alan Hedge
Department of Design and Environmental
    Analysis
Cornell University
Ithaca, New York

Paige Hodsman
Saint-Gobain Ecophon
Tadley, United Kingdom

Erin Howie
School of Physiotherapy and Exercise Science
Curtin University
Perth, Australia

Martin Mackey
Ageing, Work and Health Research Unit and
    Clinical and Rehabilitation Sciences FRG
University of Sydney
Sydney, Australia

Wayne S. Maynard
Liberty Mutual Research Institute for Safety
Hopkinton, Massachusetts

Lynn McAtamney
Atune Health Centres
Newcastle, Australia

Danny S. Nou
Occupational Biomechanics
University of California, Davis
Davis, California

Nigel Oseland
Workplace Unlimited
Berkhamsted, United Kingdom

Matko Papic
Evans Consoles Corporation
Calgary, Alberta, Canada

Ken Parsons
Loughborough Design School
Loughborough University
Loughborough, United Kingdom

Sara Pazell
Viva Health at Work
Brisbane, Australia

Brian Peacock
Singapore Institute of Management University
Singapore, Singapore

Chui Yoon Ping
Singapore Institute of Management University
Singapore, Singapore

Laura Punnett
Department of Work Environment and Center
    for Women and Work
University of Massachusetts Lowell
Lowell, Massachusetts

Sampsa Puttonen
Finnish Institute of Occupational Health
Helsinki, Finland

Mark S. Rea
Lighting Research Center
Rensselaer Polytechnic Institute
Troy, New York

Michelle M. Robertson
Liberty Mutual Research Institute for Safety
Hopkinton, Massachusetts

Noor Nahar Sheikh
University of Massachusetts Lowell
Lowell, Massachusetts

Tim Springer
Human Environment Research Organization
    (HERO) Inc.
Chicago, Illinois

Leon Straker
School of Physiotherapy and Exercise Science
Curtin University
Perth, Australia

Peter Vink
Industrial Design Engineering
Delft University of Technology
Delft, The Netherlands

Pamela Vossenas
Worker Safety and Health Program
UNITE HERE! International Union
New York, New York

Pawel Wargocki
Department of Civil Engineering
Technical University of Denmark
Lyngby, Denmark

So-Yeon Yoon
Department of Design and Environmental
    Analysis
Cornell University
Ithaca, New York

Section I

The Physical Environment

1 Introduction to Workplace Ergonomics and Issues of Health and Productivity in Computer Work Settings

Alan Hedge

CONTENTS

1.1 INTRODUCTION

In the United States, the number of white-collar or no-collar office workers has grown from ~18% of employees in 1900 to ~60% of employees in 2010 (Cenedella 2010). The microcomputer revolution of the 1970s and the 1980s has led to the vast majority of U.S. office workers using a computer for some part of their work activities. This is especially true for the office workplace, where the vast majority of these workers use a computer at their workplace and also likely use a computer elsewhere when away from their office, such as at home. The technology shift from paper to computer that began in the 1980s marked the beginning of a major change in the emphasis in the practice of ergonomics because a growing number of office workers using computers began to develop work-related musculoskeletal injuries, and the culprit was the poor design of their workspaces. The ergonomic redesign of office workspaces emerged as both a means of rehabilitating workers who had become injured and also a means of preventing workers from becoming injured. This focus on the physical design of workspaces also forms the basis of technical standards and, in 1998, saw the release of the first U.S. computer workstation design standard (American National Standards Institute [ANSI] Human Factors Society [HFS] 100 1988). Since that time, considerable research has been conducted to investigate how to optimize the design of different computer components, such as keyboards, mice, trackballs, touchpads, voice recognition systems, and computer screens. More recently, a revised standard was promulgated that gives designers greater guidance on how to optimize the physical layout of a computer workspace (ANSI HFES 100 2007). As technology continues to develop so does research continue to be conducted on the optimal arrangements of computer technologies to maximize worker productivity and comfort and minimize their risks of injuries. What we know is that the concept of *neutral posture working* has emerged as being of fundamental importance to much of this work, and it provides a basis for the physical ergonomics design of modern computer workplaces. The knowledge of neutral posture working is of such value because the capabilities of the human body change quite slowly, whereas the technologies that are used to perform work can change very rapidly. We also know that our ability to perform tasks is dependent on the level of comfort that we are experiencing. For example, if one has a toothache, a headache, or a backache, then this impairs both physical and cognitive capabilities. In short, we know that "pain distracts the brain." Yet when we look at workers in many of our designed settings, we find a high prevalence of those who are experiencing frequent discomfort and often musculoskeletal pain, and for these individuals, it is impossible for them to perform their work at an optimum level. There is also abundant evidence that placing individuals in work settings that have been designed to promote a neutral posture while performing a task can eliminate pain and discomfort. Consequently, the principles of neutral posture working are summarized in the ANSI HFES 100 ergonomic standards (2007), and they serve as the goal of ergonomic interventions that focus on redesigning individual workspaces. The fundamental requirements for neutral posture working with a computer are summarized in the following sections.

1.2 NEUTRAL POSTURE WORKING

Every articulating joint of the body has a normal range of motion. Working with the body positioned in a neutral posture means that no parts of the body are bent, twisted, or otherwise contorted away from a normal, relaxed, and comfortable position. For specific body segments, this means that a neutral posture conforms to the following guidelines (note that these positions are not absolute and a task may require intermittent excursions beyond them, but sustained postures outside of the neutral posture can cause discomfort and injuries):

- Neck—The neck is balanced and aligned with the top of the spine with minimal forward flexion or backward extension (dorsiflexion), and not laterally bent or twisted.
- Back—The whole spine is erect in a normal S shape with no part of the spine being uncomfortably flexed or extended and with no segment being laterally bent or twisted. If the spine

is in an S shape but in a reclined posture, then this should be supported by a suitable back support, such as an ergonomic chair back.

- Shoulders—The shoulders are relaxed and symmetrical; neither shoulder should be elevated, hunched or twisted.
- Upper arms—The upper arms are relaxed by the side of the body with minimal abduction or no adduction, as close to vertical as possible with minimal forward extension or backward flexion.
- Elbows/Forearms—The elbows/forearms are close to horizontal, not flexed, and forearms, not twisted into the extremes of pronation or supination.
- Wrists/Hands—The wrists/hands are straight and level, not laterally bent, extended upward or flexed downward, or twisted into the extremes of pronation or supination.
- Thighs—When seated, the thighs should be close to horizontal or slightly declined, well supported without uncomfortable compression, and when standing, these should be vertically aligned and not twisted.
- Knees—The popliteal angle behind the knee should be 90° or greater; otherwise, the blood flow to the lower legs is impeded. When standing, these should not be uncomfortably bent.
- Lower legs—When seated, the lower legs should be close to vertical or slightly angled so that the feet lie ahead of the knees. They must be free from uncomfortable compression. When standing, these should be vertically aligned and not twisted.
- Ankles/Feet—The feet can be flat on the floor beneath the lower legs or if the flower legs are outstretched then the feet should be on an inclined foot support.

These neutral posture guidelines also form the basis for posture targeting methods, such as the rapid upper limb assessment method (McAtamney and Corlett 1993) and the rapid entire body assessment method (Hignett and McAtamney 2000). Several field studies have confirmed the importance of neutral posture working for computer workers in offices and demonstrated how this results in a very substantial decrease in the prevalence of work-related upper body musculoskeletal symptoms (Rudakewych et al. 2001; Hedge et al. 2002, 2011; Hedge 2013; Hedge and Puleio 2014).

Figures 1.1 and 1.2 show the examples of a person in a neutral posture for sitting and standing computer use. Note that in these figures, the keyboard is placed on a height-adjustable downward-tilting platform that can also accommodate a mouse (not shown) and that has been adjusted so the hands are relatively leveled with the fingertips resting on the keytops, but the computer screen has limited height adjustability and ideally should be placed a little higher than shown to minimize any forward neck flexion. The person is positioned centered on their input devices and computer screen.

1.3 ERGONOMIC GUIDELINES FOR ARRANGING A COMPUTER WORKSTATION

Today many workers sit or stand to use a computer to perform their work tasks. Creating a good ergonomic working arrangement for safe computer use is important to maximize worker performance and minimize the risks of musculoskeletal injury. There is a wide variety of workplace settings in which computers are used, but for office workplaces, the following ergonomic considerations are important.

1.3.1 How Will the Computer Be Used?

To answer the question, how will the computer be used? requires knowledge about the characteristics of the user or the users and also the daily duration of their computer use. If only one person is using the computer, then the workspace arrangement can be optimized for that person's size and shape, and the features such as a height-adjustable chair may be unnecessary if the person has

FIGURE 1.1 Seated neutral posture for a computer worker.

a chair that fits their body dimensions. However, in many situations where the furniture is being bought for large numbers of workers, it is advisable to buy ergonomic products that provide adjustability to fit any worker from the dimensions of a 5th percentile woman to a 95th percentile man. Providing products with a suitable range of adjustability and easy and quick adjustments is essential if the same product is going to be used by several people, such as with shift work in say a hospital. If the workspace arrangement does not fit the anthropometrics of the worker to support neutral posture working, then s/he will adjust their body to the work tools and most likely end up in a deviated, nonneutral work posture that impedes their productivity and increases injury risks.

Consideration of how long each person will be using the computer is important. If it is a few minutes in total each day, then the ergonomic issues may not be a high priority. If it is for a few minutes at a time but there is a high frequency of use, as with say computer cart use by a hospital nurse, then quick and easy adjustments are a priority. If it is to be used by a person for more than one hour per day, then it is advisable that an ergonomic workspace arrangement be created. If it is more than four hours each day, then this definitely requires an ergonomic workspace arrangement.

1.3.2 WHAT KIND OF COMPUTER WILL BE USED?

There are at least three different types of computers that a worker could use, a desktop, a laptop, and a tablet, and each has different needs for the design of the workspace.

1.3.2.1 Desktop Computer

Most ergonomic guidelines for computer workstation arrangements assume that the worker will be using a desktop system where the computer screen is separate from the keyboard/mouse and the central processing unit (CPU). The critical considerations for desktop computer use are the

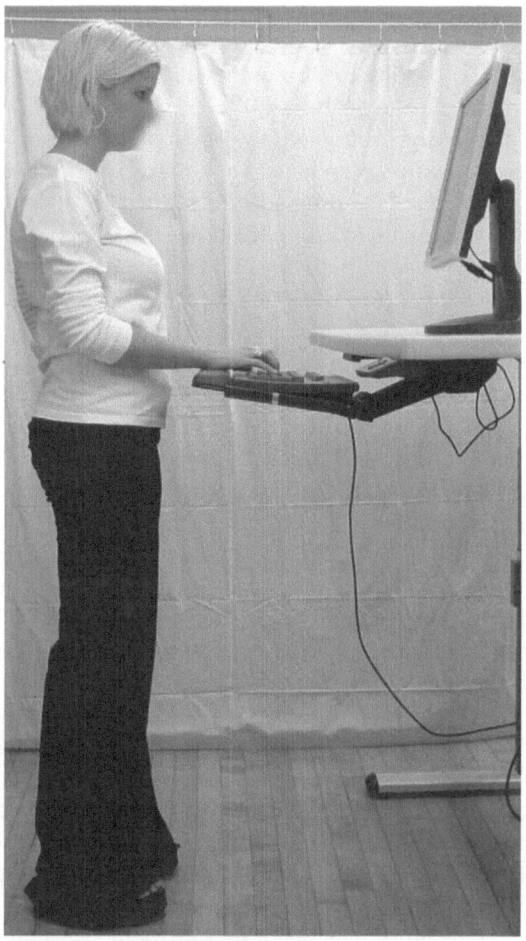

FIGURE 1.2 Standing neutral posture for a computer worker.

positioning of the keyboard and the mouse, the position of the computer screen, and whether the worker will be sitting or standing. If the worker also has to frequently access the CPU, then placing this close so that the worker can reach the CPU while still in a neutral posture is important.

1. Keyboard—If most of the work being done involves typing text, then the worker should be centered on the alphanumeric keyboard. Hedge (2004) summarizes the ergonomic considerations with computer keyboards and Kroemer (2001) provides an excellent annotated bibliography of the keyboard literature from 1878 to 1999. Rempel et al. (2008) have shown that the wrist extension during typing can elevate carpal tunnel pressure. Most modern keyboards are fairly flat and asymmetrical—the alphanumeric keyboard is to the left and a numeric keypad, to the right. If the outer edges of such a keyboard are used as landmarks for centering the keyboard and the monitor, the worker's hands, especially the right hand, will be ulnar-deviated because the alphanumeric keys will be to the left of the user's midline. Positioning such a keyboard so that the center of the alphanumeric keys, the H key, is centered on the midline of the user will reduce the hand deviation. However, if most of the entry work that is being done involves using the number pad, then aligning this with the right hand with the arm relaxed by the side of the body will reduce hand deviation. If the person is left-handed, then a left-handed keyboard or a separate number pad can be

used to align with their left hands. Placing the keyboard on an adjustable height and angle downward-tilting platform allows the keyboard to be positioned slightly below elbow level and the downward-tilting keytops can be used with the hands in a flat, neutral posture (Hedge et al. 1999; Simoneau and Marklin 2001).

2. Ergonomic and alternative keyboards—Many ergonomic keyboards are keyboards where the alphanumeric keys typically are split at an angle, split into two halves of the keyboard, dished, or otherwise arranged. The rationale for most split keyboards, whether they are fixed split angle or adjustable split angle, is to reduce ulnar deviation, and it can be traced back to the Crandall New Model typewriter of 1886. Contrary to expectations, even with a conventional flat keyboard, ulnar deviation is often not extreme and from studies of intracarpal tunnel pressure (Honan et al. 1995), ulnar deviation appears to be less important than wrist extension during typing. When typing, Baker et al. (2015) conducted a randomized crossover trial that tested fixed split-angle or standard flat keyboards for five months with 77 symptomatic computer operators in their workplace, and found no significant changes in discomfort with the fixed split-angle keyboard and a comfort preference for the flat keyboard. Some keyboards are completely split, and each half of the keyboard can even be mounted on chair arms. Hedge and Shaw (1996) studied a chair-mounted split keyboard and found that this design significantly reduced ulnar deviation, but did not reduce wrist extension and typing speed was slower, although the accuracy was unaffected. Muss and Hedge (1999) studied a vertical split keyboard used with or without articulating forearm supports, and found that the vertical keyboard significantly improved the proportion of typing movements performed in a neutral zone of wrist motion (71% for flexion/extension; 78% for radial/ulnar movements) compared with the conventional keyboard (44% and 25%, respectively), but the typing performance was slightly slower for the vertical keyboard. For a nontouch typist, such alternative designs can significantly impair the typing performance. Split designs typically focus on reducing the ulnar deviation of the hand, but research studies suggest that vertical hand posture (wrist extension) is more important (Hedge et al. 1999). There is no consistent research evidence that most of the alternative keyboard designs currently available really produce any substantial postural, performance, and usability benefits. Other keyboard designs have been developed such as chordic keyboards, which reduce the number of keys so that different letters are generated by the combined pressing of keys, like playing chords on a piano. People can memorize around 59 different chords, but even after 10 hours of practice typing, the speed is only around 14 words per minute, which is much slower than an average typist who types around 40 words per minute (Kroemer 1992). Ting and Hedge (2001) found that the typing speed was only ~9 words per minute for a hybrid chordic keyboard and game controller. Typing on a flat multitouch keyboard is significantly slower than on a conventional keyboard, and, even though there was significantly less wrist extension, the multitouch keyboard is judged to be less comfortable (Thom-Santelli and Hedge 2005). Typing on a laser-projected keyboard is also slower; ~17 words per minute and 8.6% errors compared with a conventional keyboard where the typing speed was ~40 words per minute and the error rate was 5.3% (Wang and Hedge 2008). For most people, a conventional flat keyboard design will work without substantially increasing injury risks if it is positioned so that the hands are in a neutral posture.

3. Mouse—Computer mice are available in many different shapes and sizes. Whatever the design of the mouse, it is important that it is used with the hand in a neutral posture as much as possible. Research suggests that 15° of wrist extension is a limit above which there is a rapid rise in intracarpal pressure that can cause median nerve compression (Honan et al. 1995). A study of 100 mouse users showed that 97% of mouse users use this with their hand in more than 15° of wrist extension (Lee et al. 2008). When sitting, the optimal position for a convex mouse is when this is on a keyboard platform 1–2″ (25–50 mm)

above the thighs that is movable over the numeric keypad so that it is in line with the right hand (if the person is left-handed, then the same vertical position to the left side of the alphanumeric keyboard works well). In this position the hand will be in a more neutral posture (Damann and Kroemer 1995). Compared with a more conventional convex mouse design, the use of vertical mouse designs can actually increase wrist extension deviation, which is undesirable (Hedge et al. 2010; Feathers et al. 2013), and can slow performance (Gustafsson and Hagberg 2003). Although a slanted mouse design can put the hand into a neutral posture, this also can slow performance (Hedge et al. 2010). Other cursor control input devices are available that typically center their control location on the keyboard, which is a position that has been shown to allow right- or left-hand use and to reduce wrist deviation (Dennerlein and Johnson 2006). One such device, the Rollermouse, has been shown to yield performance comparable to a conventional computer mouse (Bohan et al. 2003). However, a disadvantage of this central location for a cursor control device is that the worker has to reach over this input device to access the keyboard keys for typing.

4. Phone—Keeping the phone in close proximity is recommended for anyone who frequently uses it. If the person is right-handed, then positioning the phone to the right side of the user and within the zone of comfortable reach for the work surface is recommended, and vice versa if the user is left-handed. For very frequent phone, use a wireless/Bluetooth headset or speakerphone, or a shoulder cradle to reduce the lateral bending of the neck, if the phone is a landline.

1.3.2.2 Laptop/Notebook Computer

Originally designed as mobile computers to be frequently used for short periods of computer work, laptops/notebook computers typically have a keyboard with an integrated pointing device, usually a touchpad that is connected to the computer screen. The guidelines for laptop use are more difficult because often the laptop design is inherently problematic—when the screen is at a comfortable height and the distance from the user the keyboard is not, and vice versa. If a laptop has a separate screen and a keyboard, then that can be arranged as described in the following section for a tablet computer. If the keyboard and the screen are connected as one unit, then for sustained laptop use, or where the laptop is replacing a desktop, it is recommended that the worker be provided with the following:

1. Laptop riser—A laptop riser is used to elevate the screen to a comfortable viewing height and then provide a wireless keyboard and mouse so that the position of the display can be adjusted independent of the position of the input devices, and placing these on a height-adjustable downward tilting keyboard/mouse platform is preferable. This reduces neck flexion and also improves typing performance (Berkhout et al. 2004; Asundi et al. 2012).
2. External screen—If a laptop riser is not available, then an external computer screen or a docking station that connects to an external computer screen can be used.

1.3.3.3 Tablet Computer

Originally called slate computers, tablets are mobile computers designed to be frequently used for short periods of information consumption, compared with information creation work on desktops and laptops. Typically, a tablet has a capacitive screen design that can display a virtual touch screen keyboard, although tablets also typically support third-party external physical keyboards and mice. Compared with larger, heavier tablets, the performance seems to be comparable for smaller to medium tablets, and these are rated as more usable and less fatiguing, especially tablet designs with a ledge or handle-shape on the back and a rubberized textured surface (Pereira et al. 2013). Holding a tablet in one hand for more than 10 min can result in a high level of fatigue (Chau and Wells 2015). Intensive use of a poorly positioned tablet typically results in extreme neck flexion, and this can increase injury risks, and the resulting neck discomfort has been called *iPad neck* (Young et al. 2012).

There are now numerous products that allow a tablet to be supported on a work surface in a position that minimizes neck flexion so that an external keyboard and mouse can be used while the person is sitting in a more neutral posture.

1.3.3 What Chair Will Be Used?

Although sitting for prolonged periods in static postures can be detrimental to health (Buckley et al. 2015), the chair is an antigravity device that reduces the workload on the body, and it is important that a worker has the ability to sit in a comfortable chair for at least a part of their workday. If only one person is using this chair, it can be at a fixed height provided that it is comfortable to sit on and has a good backrest that provides lumbar support. If, however, more than one person will be using the chair or a single chair model is being purchased for many different workers, then the chair must have certain ergonomic features. Table 1.1 summarizes the requirements for an ergonomic chair from the ANSI HFES 100 standard (2007).

In addition to the list of requirements, the ANSI HFES 100 (2007) standard also lists a number of recommendations for the design of ergonomic chairs, and these are summarized in Table 1.2.

As Helander (2003) notes, users cannot easily perceive many of the ergonomics chair features that are designed to relieve sitting discomfort because the differences in pressure due to different body postures cannot be sensed by the spine; small changes in angle cannot be sensed by the joints; and many of the chair controls are hidden from view beneath the seat pan. However, users can perceive esthetic features, and their ratings of chair comfort and choice of chair tend to be based on esthetics rather than on ergonomic features.

Even if a chair has all of the required and recommended features, there is no guarantee that the worker will use these and correctly adjust the chair for themselves (Vink et al. 2007). Helander et al. (1995) investigated how people adjusted their chair for 26 chairs with a total of 24 different types of control arrangements and found that, although the chair with the greatest number of adjustability controls was judged to be the most comfortable, it took significantly greater time to adjust and this requires more training. This issue of control complexity is further discussed by Vink in Chapter 8, where he also reports that a majority of office workers may not know how to correctly adjust their chairs. Simple controls and, where possible, automated controls improve usability.

When sitting in the chair, the seat pan should be at least 1″ wider than a user's hips and thighs on either side. The seat pan should not be too long for a user's legs; otherwise, it may either compress behind the knees or prevent the user from fully leaning back against the lumbar support. Most ergonomic chairs have a seat pan with a waterfall front that prevents the seat from compression behind the knees. The seat pan should also be contoured to allow even weight distribution, and it should be comfortable to sit on. If there is insufficient hip room, this can encourage a forward-flexed posture on the seat pan, and this posture may create thigh compression problems. If the seat pan is made from low-density foam, then continuous use may cause it to become permanently deformed and then it will not provide adequate-cushioned support. Insufficient cushioning and inappropriate contouring can cause discomfort, imbalance, and hip and back fatigues. For preference, the seat pan height should be easily adjustable while sitting on the chair. Some chairs have a mechanical height-adjustment (spinning) mechanism that may also be acceptable. The height of the seat pan should be aligned level with the front of their knees or be slightly below the level when feet are stable on the ground.

Chairs can be covered in a variety of upholstery materials, each of which has benefits and concerns. Vinyl and vinyl-like coverings are easy to clean and spill resistant, but they do not breathe and if the chair begins to heat up under the thighs, uncomfortable amounts of moisture can accumulate. Cloth upholstery is the most common covering, but this is less resistant to spills and more difficult to clean. A cloth-covered seat pan can also become warm and moisture laden, and cloth-covered foam seat pans can be a significant source of dust mite allergen (O'Reilly et al. 1998). Mesh chair seats can focus compressive forces under the hips and the thighs, and curved mesh chair backs can

TABLE 1.1

Chair Requirements Summarized from ANSI HFES 100

Item	Requirements	Yes/No
Chair	Shall have a lumbar support	
	Shall have a backrest that reclines	
	Shall have a seat pan that adjusts for height and tilt	
	Shall support at least one of the two other seated reference postures in addition to the upright sitting posture	
	Shall provide support to the user's back and thighs in the chosen reference postures	
Seat pan and backrest adjustments	Shall be height adjustable	
	Shall have a user adjustment for tilt	
Backrest	Shall not constrain the user's torso to a position forward of vertical	
	Shall not force a torso–thigh angle less than 90°	
	Shall allow adjusting the angle between the backrest and the seat pan to an angle of 90° or greater	
	Shall allow the user to recline to at least 15° from the vertical	
Armrests	Shall provide sufficient clearance to allow the user to sit or stand without interference	
	Shall not cause the user to violate any of the following postural guidelines:	
	• Elbow angles between 70° and 135°	
	• Shoulder abduction angles less than 20°	
	• Shoulder flexion angles less than 25°	
	• Wrist flexion angles less than 30°	
	• Wrist extension angles less than 30°	
	• Torso-to-thigh angles equal to or greater than 90°	
Seat height	Shall be adjustable by the user over a minimum range of 11.4 cm (4.5″) within the recommended range of 38–56 cm (15–22″)	
	Manufacturer shall provide information to show which of the three seated postures the chair will accommodate	
Depth and front edge of the seat pan	Shall, if nonadjustable, be no greater than 43 cm (16.9″)	
	Shall include 43 cm (16.9″) if adjustable	
Seat pan width	Shall be at least 45 cm (17.7″) wide	
Seat pan angle	Shall have a user-adjustable range of at least 4°, which includes a reclined position of 3°	
Seat pan–backrest angle	Shall be able to achieve a position that is vertical or to the rear of vertical	
	Shall have an adjustment range of 15° or more within the range of 90° and 120° relative to horizontal if the backrest is adjustable	

Source: Human Factors and Ergonomics Society, ANSI HFES 100, Human Factors Engineering of Computer Workstations, Santa Monica, California, 2007. With permission.

give better support than flat mesh designs (Agarwal and Hedge 2006). Some chair mesh materials can stretch with time, and mesh can accumulate dust and then become abrasive for clothing. When selecting a chair covering, think about cleaning and maintenance issues and plan appropriately.

Contrary to widely held belief, research shows that the best seated posture is a reclined posture of 100°–110° and not the erect 90° posture that is often portrayed as being an ergonomic sitting posture (Andersson and Ortengren 1974; Andersson et al. 1975; Grandjean and Kroemer 1997; Wilke et al. 1999; Gscheidle and Reed 2004). In a slightly reclined posture, the chair back begins to support some of the body weight, and this reduces the activity of the back muscles (Park et al. 2000) and reduces spinal compression (Leivseth and Drerup 1997). In this recommended posture, the chair starts to work for the body, and there are significant decreases in postural muscle activity and in

TABLE 1.2
Chair Recommendations from ANSI HFES 100

Item	Recommendations	Yes/No
Chair	Should be adjustable to provide clearance under the work surface	
	Should provide information to the user as to the recommended use and adjustment of the chair	
Casters	Should be appropriate for the type of flooring at the workstation	
Seat pan and backrest adjustments	Should be wide enough to accommodate the clothed hip width of a 95th percentile female	
	Should be of sufficient depth to allow the user's back to be supported by the backrest without contact between the back of the user's knee and the front edge of the seat pan	
	Should have a tilt lock or stop position that the user can select while seated, if a tilt lock is provided (a stop limits the motion in one direction, whereas a lock limits the movement in two directions)	
	Should have a rounded front edge	
Backrest	Should allow the user to control the resistance necessary to recline the backrest	
	Should provide support to the lumbar and thoracic regions of the back	
	Should have a means of adjusting the backrest tension	
Armrests	Should be adjustable in height	
	Should allow adjustment of the clearance width between the armrests	
	Should be detachable	
	Should adjust in height from 17 to 27 cm (6.7–10.6″) above the compressed seat pan height	
	Should be designed to evenly distribute forces over the contact area	
	Should not create excessive pressure points	
	Should not irritate or abrade the skin	
	Should be able to be detached from the chair, if necessary, to fit the workplace	
	Fixed-height armrests should be between 18 and 27 cm (7.1–10.6″) above the compressed seat pan height	
	The clearance between armrests should be at least 46 cm (18.1″)	
	The clearance between armrests should be adjustable by the user (for example, pivot or otherwise move)	
Seat pan–backrest angle	If the backrest recline angle exceeds 120° from the horizontal, the backrest should have a headrest, preferably user adjustable	
Backrest height and width	Should be at least 45 cm (17.7″) above the compressed seat height	
	If fixed, the lumbar support area of the backrest should be located between 15 and 25 cm (5.9 and 9.8″) above the compressed seat height	
	The position of the center of the lumbar support should be user adjustable between 15 and 25 cm (5.9 and 9.8″) above the compressed seat height	
	The width of the backrest should be at least 36 cm (14.2″)	

Source: Human Factors and Ergonomics Society, ANSI HFES 100, Human Factors Engineering of Computer Workstations, Santa Monica, California, 2007. With permission.

intervertebral disk pressure in the lumbar spine. Erect sitting is not relaxed sitting or a sustainable posture over a long duration, whereas reclined sitting is. Moreover, many ergonomic chairs also incorporate a dynamic chair back whereby the back of the chair moves to stay in contact with the worker's back as s/he moves. The use of a dynamic chair back results in less spinal compression that occurs in a fixed back chair (van Dieën et al. 2001).

Many chairs have cushioned lumbar supports that can be height and depth adjusted to best fit a user's shape. If the chair will be used by multiple users, then this level of adjustment can be beneficial. If the chair has a fixed height lumbar support and it feels comfortable when a user sits back against this, and that user will be the primary user of the chair, then a fixed lumbar support may be acceptable. Many chairs also have back supports that are large enough to provide midback and upper-back support, in addition to good lumbar support. As described earlier, the movement of the back while sitting helps to maintain a healthy spine. Chairs that allow for easy reclining that provide good back support in different recline postures and that have a back that tracks where the user's back is are preferable. Locking the chair backrest in one position is not generally recommended or beneficial to users.

Other useful chair features include height- and position-adjustable chair armrests, which can be helpful to aid ingress and egress from the chair. Also, the armrests can be useful for the occasional resting of the arms (e.g., when on the phone, sitting back relaxing). However, the use of chair arm-rests does not necessarily improve hand/wrist posture when typing or mousing (Barrero et al. 1999). It is not a good idea to permanently rest the forearms on armrests while you are typing or mousing because this can compress the flexor muscles, and some armrest designs, especially narrow and hard armrests, can create ulnar nerve compression at the elbow, and consequently broader, flatter, padded armrest designs are preferable. Ideally, it should be easy to move the armrests out of the way when the worker needs unimpeded access to their keyboard and mouse. Chairs with headrests can be beneficial (Monroe et al. 2001).

If chair mobility is important to help with work then the chair should have at least a five pedestal bases with casters that freely glide over the floor surface, and choosing a chair that easily swivels can also be of benefit.

1.3.4 What Work Surface Furniture Will Be Used?

For any sustained period of work, the computer must be placed on a stable working surface (nothing that bounces) with adequate room for proper arrangement of the task tools (e.g., keyboard, mouse, documents).

1.3.4.1 Fixed-Height Work Surface

Ideally, for neutral posture working, a work surface should be at a height that is around the worker's seated or standing elbow height. Table 1.3 shows these heights for a 5th and 50th percentile woman and for a 50th and 95 percentile man (note that the distributions overlap so that the standing elbow height of a 95th percentile U.S. woman is equivalent to that of a 41st percentile U.S. man, and the standing elbow height of a 5th percentile U.S. man is equivalent to that of a 54th percentile U.S. woman). Many office workers sit at work surfaces that are 30″ (762 mm) high, and Table 1.3 shows that this is higher than the seated elbow heights of all woman and almost all men. However, when a keyboard, mouse, laptop, or tablet is placed on that work surface, it is too high for sustained use in a neutral posture. Ideally, the work surface should be adjustable over a range of 5.3″ (136 mm) for seated workers, 10″ (267 mm) for standing workers, or 23.7″ (603 mm) for sit–stand working desks.

TABLE 1.3

Elbow Heights for 5th and 50th Percentile U.S. Women and 50th and 95th Percentile U.S. Men either Standing or Sitting

Posture	U.S. Women (5th Percentile)	U.S. Women (50th Percentile)	U.S. Men (50th Percentile)	U.S. Men (95th Percentile)
Sitting	23.1″ (587 mm)	23.3″ (631 mm)	25″ (635 mm)	28.4″ (723 mm)
Standing	36.3″ (923 mm)	39.6″ (1005 mm)	43.3″ (1100 mm)	46.8″ (1190 mm)

If the work surface is not height adjustable and is too high for the worker to adopt a neutral arm/wrist/hand posture, then a height- and angle-adjustable keyboard platform can be used to lower the keyboard surface to an appropriate height (see Table 1.3). It is also preferable that any such platform allows for angle adjustability so that the keyboard can decline away from the worker so that the hands can be in a neutral posture (Hedge et al. 1999).

The ANSI HFES 100 ergonomics standard (2007) provides a series of requirements for office work surfaces, and these are summarized in Table 1.4.

In addition to the list of requirements, the ANSI HFES 100 standard (2007) also lists a number of recommendations for the design of ergonomic work surfaces, and these are summarized in Table 1.5.

1.3.4.2 Height-Adjustable Sit–Stand Workstation

An average person makes 66 sit-to-stand changes per workday (Dall and Kerr 2010). To help combat the potential perils of prolonged sitting (Dunstan et al. 2011; Hamilton et al. 2007; Buckley et al. 2015), the use of a height-adjustable work surface for sitting and standing work is becoming more popular. However, there is limited evidence that sit–stand furniture has cost-effective benefits, unless other changes in work practices are also made. The evidence suggests that there may be a reduction in back discomfort (Hedge and Ray 2004), but the research for this has not used adequate comparison groups (e.g., testing people who stand for the same time at the same frequency without doing keyboard/mouse work). There is no evidence that sit–stand improves wrist posture when typing or mousing on a flat surface, but the addition of a downward-tilting keyboard platform can help (Hedge et al. 2005). Logically, the potential benefit of sit-to-stand is just the intermittent changes between sitting and standing. But standing in a static posture is even more tiring than sitting in a static posture, and prolonged standing can result in greater risks of varicose veins (Tüchsen et al. 2000, 2005), carotid artery disease (Krause et al. 2000), and back pain (Gallagher et al. 2014).

Recent research suggests that sit–stand workstations that can be quickly adjusted allow each worker to easily modify their work surface height throughout the day, and this may reduce musculoskeletal discomfort and improve work performance (Hedge and Ray 2004; Karakolis and Callaghan 2014). However, correctly adjusting the height of the work surface to support the keyboard and mouse, and the height of the computer screen is extremely important.

With posture, the need to keep the body in a neutral posture is the same for height-adjustable, split work surfaces and sit–stand work surfaces. If the surface is too low below the elbow height, the hands will be in greater wrist extension, and the neck will be in forward flexion. If the surface is too high above the elbow height, the elbow will be in sustained flexion, and the neck may be dorsiflexed. It is impossible to position a single flat work surface at an appropriate height for the five main tasks of office work—keyboarding, mousing, writing, viewing documents, and viewing the screen—because these all require different heights for an optimal arrangement. When the work surface is set for a comfortable writing height (28–30″; 71–76 cm), a negative-slope keyboard tray system serves as an effective solution that incorporates a platform for height and angle adjustment for the keyboard (Hedge et al. 1999) and a platform that places the mouse just above the level of the keytops to maintain a neutral hand posture when mousing (Damann and Kroemer 1995). Screen monitor height, distance, and tilt can be adjusted by a separate monitor arm, which reduces neck discomfort (Boothroyd and Hedge 2007). There are also split work surface designs that allow for the separate adjustment of the keyboard and mouse surface and the monitor surface.

1.3.5 Display Positioning

Aligning the body and the head with what needs to be seen in the environment is crucial to the ability to maintain a neutral posture. Just like a car driver is positioned to view the road directly ahead so a computer worker needs to be positioned so that s/he can directly view the relevant visual

TABLE 1.4
Work Surface Requirements from ANSI HFES 100

Item	Requirements	Yes/No
Controls	Shall not intrude into the leg and foot clearance spaces when not in use	
	Shall not interfere with users' typical work activities	
Adjustable surfaces	Shall use a fail-safe mechanism to prevent inadvertent movement	
	Shall use a control-locking mechanism to prevent inadvertent operation	
Pinch points	Shall be avoided by means of design or guarding	
Leg and foot clearance	Shall provide adequate leg and foot clearances in the chosen reference posture or postures	
Input device location	Shall adjust in height, or a combination of height and tilt, to allow placement of the input device within the recommended space	
Seated and standing work	Shall provide adequate leg and foot clearances	
	Shall provide adequate space for multiple input devices (e.g., keyboard and mouse)	
Sit–stand work	Shall accommodate at least one of the three seated reference postures in addition to the standing reference posture	
Monitor support surface	Shall allow users to adjust the line-of-sight (viewing) distance between their eye point and the front (first) surface of the viewable display area	
	Shall allow users to adjust the tilt and the rotation angle between their eye point and the front (first) surface of the viewable display area	
Workstation adjustments	Shall not interfere with users' work activities or pose hazards during use	
Finish of furniture and accessories	Shall have radii of at least 3 mm	
Operator clearances	Shall accommodate at least two of the three seated reference working postures (declined, upright, or reclined)	
	Shall be	
	• 52 cm (20.5″) wide	
	• 44 cm (17.3″) deep at the level of the knee	
	• 60 cm (23.6″) deep at the level of the foot	
	• Adjustable between 50 and 72 cm (19.7 and 28.3″) in height at the edge of the work surface closest to the operator	
	• Adjustable between 50 and 64 cm (19.7 and 25.2″) in height at the horizontal position of the knee	
	• At least 10 cm (3.9″) in height at the position of the foot	
Monitor support surface/ device	Manufacturer shall specify the size and weight of monitor that can be accommodated by the support surface because monitor support surfaces may not be compatible with certain-sized monitors	
	Manufacturer shall specify the range of adjustment if the support surface is adjustable	
Input-device support surface	Shall adjust in height, or a combination of height and tilt	
	Manufacturer shall provide information regarding the range of height adjustment	
	Manufacturer shall provide information regarding the tilt adjustments	
Sit–stand working postures: height adjustable surface	Shall adjust in height between 56 and 118 cm (22 and 46.5″) as measured from the floor to the surface at the front edge of the support	
	Shall comply with the clearance requirements specified when used in the seated position	

(Continued)

TABLE 1.4 (CONTINUED)
Work Surface Requirements from ANSI HFES 100

Item	Requirements	Yes/No
Sit–stand working postures: height and tilt adjustable surface	Shall accommodate seated workers by adjusting in height in some portion of the range between 56 and 72 cm (22 and 28.3″) as measured from the floor to the surface at the front edge of the support	
	Shall accommodate standing workers by providing additional height adjustability (greater than 72 cm [28.3″]) when combined with tilt as described in the equation $A + \sin(B) \times C = $ input device height	
	Shall adjust in tilt in some portion of the range between +20° and −45°, to include 0	
	Shall comply with the clearance requirements specified in Section 8.3.2.1 when used in the seated position	

Source: Human Factors and Ergonomics Society, ANSI HFES 100, Human Factors Engineering of Computer Workstations, Santa Monica, California, 2007. With permission.

information from a sitting or a standing neutral posture. When the primary visual display is a computer screen, this means that it should be positioned as follows.

The computer screen should be placed directly in front of the worker and facing them, not angled to the left or the right so that they have to twist their neck or torso to view the screen. This helps to eliminate too much neck twisting. If someone is working with a large monitor or spends most of their time working with software, like MS Word, which defaults to creating left aligned new pages, then aligning the worker's head/body to a point about 1/3 of the distance across the monitor from their left side will help to minimize lateral head movements.

While such positioning addresses the lateral position of the visual field, it is also important to address the vertical position of information. Once a screen is well positioned, use the screen scroll bars to ensure that what is being viewed most is in the center of the monitor rather than at the top or the bottom of the screen. The screen should be positioned at a comfortable viewing height that does not require tilting the head up to see items or bending the neck down to see items. When comfortably seated, a worker's eyes should be approximately in line with a point on the screen about 2–3″ (50–74 mm) below the top of the screen so that most of the central region of the screen can be viewed without any head movement. As a rule of thumb, the worker should sit back in their chair in a slight recline, at an angle of around 100°–110°, then they should hold their right arm out horizontally at shoulder level, and their middle finger should almost touch the center of the screen (see Figure 1.3). From this starting position, a worker then can make minor changes to screen height and angle to suit their viewing needs.

We see more visual field below the horizon than above this (look down a corridor and you will see more of the floor than the ceiling), so at this position, the user should comfortably be able to see more of the screen. Research shows that the center of the monitor should be about 17°–18° below horizontal (Sommerich et al. 2001) for optimal viewing, and this is where it will be if the worker follows the simple arm extension/finger pointing tip. If a user has to crane their neck forward to see the screen, then it is positioned too low. If they have to tilt their head backward to see the screen then it is too high. In either situation, repeated exposure to this posture will increase the risks of neck/shoulder pain. If a user is wearing bifocals, trifocals, or progressive lens, then the screen position and the tilt angle can be fine adjusted when they are sitting back in their chair in a reclined posture (at around 11°), until they can see the screen with the head in a neutral posture. If the text looks too small, then the user should either use a larger font or magnify the screen image in the software program rather than sitting closer to the monitor.

TABLE 1.5
Work Surface Recommendations from ANSI HFES 100

Item	Recommendations	Yes/No
Device cabling	Should be placed to avoid interference with the operation of workstation components	
	Should be placed to avoid creating hazards for people or equipment in the workstation	
Leg and foot clearances	Should not hinder the foot, the leg, or the knee in alternative or auxiliary (non-video display terminal [VDT]) work positions	
Horizontal work envelope	Should accommodate the user postural design criteria:	
	• Elbow angles between 70° and 135°	
	• Shoulder abduction angles less than 20°	
	• Shoulder flexion angles less than 25°	
	• Wrist flexion angles less than 30°	
	• Wrist extension angles less than 30°	
	• Torso-to-thigh angles equal to or greater than 90°	
	Should be at least 70 cm (27.6″) wide	
	Should locate the most commonly used objects in the primary work zone	
Monitor support surface	Should allow users with normal visual capabilities to adjust the line-of-sight (viewing) distance between their eyes and the front (first) surface of the viewable display area within the range of 50–100 cm (19.7–39.4″)	
Workstation adjustments	Should be usable by users while in the relevant reference postures	
Finish of furniture and accessories	Secondary user contact edges should have radii of at least 2 mm	
Surface gloss	Should have a matte finish that provides a specular reflectance of no more than 45 gloss units at an angle of 60° as measured with instruments and procedures that conform to ASTM D523-89 (1999) Standard Test Method for Specular Gloss (American Society for Testing and Materials 1999)	
Work surface	Should be at least 70 cm (27.6″) wide	
	Depth should allow a viewing distance of at least 50 cm (19.7″)	
	Depth should allow positioning of the monitor so that the angle between the horizontal level of the eyes and the center of the screen ranges between 15° and 25°	
	Depth should allow positioning of the entire viewing area (e.g., including the keyboard) in an arc 60° below horizontal eye level	
Monitor support surface/ device	Should be designed so as to allow placement of the viewing area of the screen at a minimum viewing distance of 50 cm (19.7″)	
	Should be designed so as to allow placement of the monitor's viewing area below the user's horizontal eye height	
	Should be stable during use	
	Should not interfere with the user's ability to adjust the height, tilt, and rotation of the monitor	
Input-device support surface	Should adjust fore and aft in the horizontal plane	
	Should adjust in side-to-side placement within the optimal area for input devices	
	Should tilt	

Source: Human Factors and Ergonomics Society, ANSI HFES 100, Human Factors Engineering of Computer Workstations, Santa Monica, California, 2007. With permission.

FIGURE 1.3   Ideal screen distance for a seated computer worker.

If the work being performed involves reading or transcribing any paper documents, then these should be placed as close to the computer screen as possible and a document holder can be used to position documents at a similar angle to the screen so that the eyes do not have to refocus when moving from the documents to the screen and vice versa. Three types of document holders can be used:

1. Screen-mounted document holder—If single sheets of a small number of sheets of paper are to be read, then a screen-mounted document holder positioned to the same side of the computer screen that is the worker's dominant eye can be used.
2. In-line document holder—This typically sits between the keyboard/keyboard tray and the computer screen and is aligned with the body midline so that all the worker has to do is look down to see the documents and raise their eyes to see the screen.
3. Freestanding document holder—This should be positioned adjacent to the same side of the screen as the dominant eye; it should be slightly tilted backward and/or curved so that it follows a curve from the side of the screen.

Finally, there are natural changes in vision that occur in most people during their early 40s, and these generally shorten the focal length of the eye and reduce its transparency so it is important to periodically have a visual health checkup by a qualified professional.

1.3.6   WHERE WILL THE COMPUTER BE USED?

The environmental conditions where the computer will be used are important. This is especially relevant for residential use or for use in nonwork settings such as hotels, airports, etc.

1.3.6.1   Lighting

The ambient lighting where work is being done should not be too bright; otherwise, this will cause veiling glare on the screen, and also ensure that the screen is free from any bright light reflections (specular glare). If glare is a problem, then this can be addressed by moving the screen location (this mostly helps specular glare), lowering the light level (this mostly helps with veiling glare), and using a good quality antiglare screen. It is important to position the computer monitor screen so that it is not backed up to a bright window or facing a bright window so that the screen looks washed out (a shade or drapes will control the window brightness). Where possible, sitting sideways to a window is recommended.

1.3.6.2   Ventilation

The computer should be used somewhere that has adequate fresh air ventilation and that has adequate heating or cooling to provide thermally comfortable working conditions.

1.3.6.3   Noise

Noise can cause stress and that tenses muscles that in turn can increase injury risks. Find a quiet place for computer work. Listening to low-volume music, preferably light classical, such as Mozart, can boost productivity (Tayyari and Smith 1987; Smith et al. 2010) and mask office noises (Schlittmeier and Hellbruck 2009). Music can be played through earbuds/headphones or a noise-cancelling headset if it is a noisy environment such as an airplane.

1.3.7   Organizing an Optimal Work Pace

Taking frequent, brief rest breaks can improve the well-being by reducing musculoskeletal discomfort and can boost work productivity (Henning et al. 1996, 1997; Galinsky et al. 2000; McLean et al. 2001; Montie et al. 2004).

The following break schedules can be beneficial.

1.3.7.1   Eye Breaks

Looking at a computer screen for a while causes some changes in how the eyes work, slows the blink rate, and exposes more of the eye surface to the air. These effects can be mitigated by briefly looking away from the screen for a minute or two to a more distant scene, preferably something more than 20 ft (6 m) in the distance, every 20 min, which lets the ciliary muscles inside the eye relax, and by rapidly blinking the eyes for a few seconds to refresh the tear film and clear dust from the eye surface.

1.3.7.2   Microbreaks

Most typing is done in bursts of activity rather than continuously. Between these bursts of activity, the hands can be rested in a relaxed, flat, straight posture. During a microbreak, which typically lasts less than 2 min, brief stretching, standing up, moving around, or doing a different work task, e.g., making a phone call, can rest the muscles involved in typing and mousing. A microbreak is not necessarily a break from work, but it is a break from the use of a particular set of muscles that is doing most of the work (e.g., the finger flexors, if a lot of typing is being done).

1.3.7.3   Rest Breaks

Brief rest breaks around every 30 min in which a worker can stand up, move around, and do something else, such as going and getting a drink of water, tea, coffee, or whatever, will allow resting of the primary work muscles and exercising of different muscles that will also improve circulation and lessen feelings of tiredness and reduce reports of low back pain (Sheahan et al. 2015).

1.3.7.4   Exercise Breaks

There are many stretching and gentle exercises that can help relieve muscle fatigue. Doing these every 1–2 hours throughout the day can help to reduce overall fatigue.

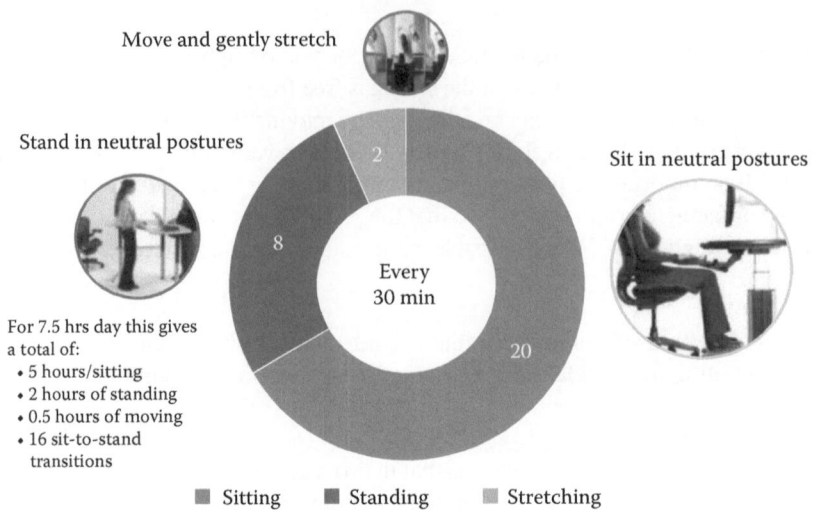

FIGURE 1.4   Optimal work pattern. (From Hedge, A., Sit-Stand Working Programs, http://ergo.human.cornell .edu/CUESitStandPrograms.html, 2015. With permission.)

Working at a computer can be hypnotic, and that often disrupts the sense of time resulting in longer duration of typing and mousing without a rest than is recommended. There are several ergonomic software programs that provide break reminders and can be installed on a computer. The best software will run in the background, and it will monitor how much typing and mousing have occurred and use this as the basis for triggering rest break reminders, and many programs also suggest simple exercises that can be done during breaks. Wearable activity trackers and a new generation of smart watches can also provide activity alerts for taking rest breaks.

With the current trend for using sit-to-stand workstations, Figure 1.4 shows a good break regime pattern to try and develop at work.

1.3.8   OTHER ERGONOMIC OFFICE PRODUCTS

Many kinds of products are labeled as being ergonomically designed, but this is often not true and some of these so-called ergonomic products can actually make matters worse in terms of injury risks and lowering productivity. There are many computer-related products that are marketed as being ergonomic, and in addition to those already discussed in this chapter, the most common ones are as follows.

1.3.8.1   "Ergonomic" Mice

Many of these mouse designs or alternative input device designs can work well to improve your hand/wrist posture. However, it is important to check that you can use these with your upper arm relaxed and as close to your body as possible and with your hand in a neutral posture. Overreaching to any ergonomic mouse defeats any benefits of this design.

1.3.8.2   Wrist Rests

Research studies have failed to demonstrate any substantial benefits with using wrist rests. Some wrist rest designs have no beneficial effects on wrist posture (Cook et al. 2004). A wrist rest, especially one that is narrow, curved/domed, and soft, may actually increase the pressure inside the carpal tunnel by compressing the undersurface of the wrist (Horie et al. 1993). The pressure applied to the underside of the carpal tunnel is transferred into the tunnel itself via the transverse carpal ligament and that intracarpal pressure can double when resting on a wrist rest compared with floating the hands over a keyboard. The best design for a wrist rest is one with a broad, flat, firm surface to support the heel of the palm on this and not compress the wrist. Resting in between bursts of typing is preferred to continuous resting while typing. Firmer rather than softer wrist rests are preferred because these will not contour to the wrist, restrict the freedom of movement of the hands, or encourage more lateral deviation during typing. The used pattern can often be seen in areas of erosion on the surface of a typical wrist rest, which shows the area of compressive forces on the wrist. The hands should be able to glide above the surface of any wrist rest during typing, rather than being in a fixed position on the rest while typing.

1.3.8.3 Wrist Support Braces/Gloves

There is no consistent research evidence that wearing wrist supports during computer use actually helps reduce the risk of hand/wrist injury. A wrist support should keep the hands flat and straight, not bent or extended. Wearing such a wrist support at night when sleeping may help relieve symptoms for those with carpal tunnel syndrome.

1.3.8.4 Forearm Supports/Resting Forearms on Chair Arms

Resting the forearms on any support while typing has the potential for restricting the circulation to the finger flexor muscles in the forearm and compressing the ulnar nerve in the elbow. Resting on chair arms while typing does not improve hand/wrist posture (Barrero et al. 1999). If the keyboard/mouse is appropriately arranged, they should be accessible with the user's arms in a neutral position (close to the body and with the upper arm hanging in a relaxed way) which does not pose any significant neck or shoulder load. If forearm supports are required, it is usually a sign of a poor ergonomic arrangement.

1.3.8.5 Footrest

If your feet cannot rest on the floor when your legs are in a comfortable position or if you want to stretch your feet out in front of you, like when driving a car, a freestanding floor-mounted support will allow you to rest your feet out in front of you in a comfortable manner. Look for a design that allows foot rocking movements to assist with the circulation to the lower legs.

Finally, before buying any ergonomic product, it is worthwhile asking the following four questions:

1. Do the product design and the manufacturer's claims make sense?
2. What research evidence can the manufacturer provide to support their claims? Be cautious with products that have not been studied by researchers.
3. Does it feel comfortable to use the product for a long period? Some ergonomic products may feel strange or slightly uncomfortable at first because they often produce a change in your posture but the changes can be beneficial in the long term. Think of some products as being like new shoes that initially may feel strange but then feel comfortable after being used for a while. If a product continues to feel uncomfortable after a reasonable trial period (say at least a week), then stop using it.
4. What do ergonomics experts say about the product? If they do not recommend it, do not use it.

1.4 CONCLUSIONS

This chapter has presented a brief summary of the recommendations for healthful ways of using computer technologies in the workplace, and it has given the supportive evidence where this is available. The following list summarizes the main recommendations from this chapter:

- Keep arms and elbows relaxed and close to body.
- Keep wrists flat and straight in relation to the forearms to use keyboard/mouse/input device.
- Use a negative-tilt keyboard tray with an upper mouse platform or downward-tilting platform adjacent to the keyboard.

- Use a stable work surface and a stable (no-bounce) keyboard tray.
- Use a good chair with a dynamic chair back and sit back on this.
- Sit at arm's length from the monitor screen.
- Position the top of the monitor casing 2–3″ (5–8 cm) above seated eye level.
- Center the monitor screen and the keyboard in front of you.
- Position the screen to be glare free or use an optical glass antiglare filter where needed.
- Use a document holder, preferably in line with the computer screen.
- Keep your feet on the floor or a stable footrest.
- Take frequent short breaks (microbreaks).

The other chapters in this book will provide more details on healthful and productive ways of working in specific workplaces.

REFERENCES

Agarwal, A., and A. Hedge. 2006. A 3D body scan method to investigate how flexible material chair backs respond to the seated body. *Proc HFES 50th Annu Meet*. 507:804–8.

Andersson, B. J., and R. Ortengren. 1974. Lumbar disc pressure and myoelectric back muscle activity during sitting: II. Studies on an office chair. *Scand J Rehabil Med*. 6(3):115–21.

Andersson, B. J., R. Ortengren, A. L. Nachemson, G. Elfström, and H. Broman. 1975. The sitting posture: An electromyographic and discometric study. *Orthop Clin North Am*. 6(1):105–20.

ANSI HFES 100. 2007. Human Factors Engineering of Computer Workstations. Human Factors and Ergonomics Society, Santa Monica, CA.

ANSI HFS 100. 1988. American National Standard for Human Factors Engineering of Visual and Display Terminal Workstations. Human Factors Ergonomics Society, Santa Monica, CA.

Asundi, K., D. Odell, A. Luce, and J. T. Dennerlein. 2012. Changes in posture through the use of simple inclines with notebook computers placed on a standard desk. *Appl Ergon*. 43(2):400–7.

Baker, N. A., K. K. Moehling, and S. Y. Park. 2015. The effect of an alternative keyboard on musculoskeletal discomfort: A randomized cross-over trial. *Work*. 50(4):677–86.

Barrero, M., A. Hedge, and T. Muss. 1999. Effects of chair arm design on wrist posture. *Proc HFES 43rd Annu Meet*. 43(1):584–8.

Berkhout, A. L., K. Hendriksson-Larséna, and P. Bongers. 2004. The effect of using a laptopstation compared to using a standard laptop PC on the cervical spine torque, perceived strain and productivity. *Appl Ergon*. 35(2):147–52.

Bohan, M., J. Slocum, D. Shaikh, and B. Chaparro. 2003. A comparison of cursor-control performance of the Rollermouse station™ and the standard mouse. *Proc HFES 47th Annu Meeti*. 1:741–5.

Boothroyd, K., and A. Hedge. 2007. Effects of an LCD arm on comfort, posture and preference in an architectural practice. *Proc HFES 51st Annu Meet*. 51(8):549–53.

Buckley, J. P., A. Hedge, T. Yates, R. J. Copeland, M. Loosemore, M. Hamer, G. Bradley, and D. W. Dunstan. 2015. The sedentary office: A growing case for change towards better health and productivity; Expert statement commissioned by Public Health England and the Active Working Community Interest Company. *Brit J Sport Med*. 49(21):1357–62.

Cenedella, M. 2010. Great news! We've become a white-collar nation. *Business Insider* (http://www.business insider.com/great-news-weve-become-a-white-collar-nation-2010-1).

Chau, L., and R. Wells. 2015. Biomechanical loading on the hand, wrist, and forearm when holding a tablet computer. *IIE Trans Occup Ergon Hum Factors*. 3:105–14.

Cook, C., R. Burgess-Limerick, and S. Papalia. 2004. The effects of wrist rests and forearm supports during keyboard and mouse use. *Int J Indust Ergon*. 33:463–72.

Dall, P. M., and A. Kerr. 2010. Frequency of the sit to stand task: An observational study of free-living adults. *Appl Ergon*. 41(1):58–61.

Damann, E. A., and K. H. E. Kroemer. 1995. Wrist posture during computer mouse usage. *Hum Fac Erg Soc P*. 39(10):625–9.

Dennerlein, J. T., and P. W. Johnson. 2006. Changes in upper extremity biomechanics across different mouse positions in a computer workstation. *Ergonomics*. 49(14):1456–9.

Dunstan, D. W., A. A. Thorp, and G. N. Healy. 2011. Prolonged sitting: Is it a distinct coronary heart disease risk factor? *Curr Opin Cardiol.* 26(5):412–9.

Feathers, D. J., K. Rollings, and A. Hedge. 2013. Computer mouse issues for college students aged 18–25. *Work.* 44(1):115–22.

Galinsky, T. L., N. G. Swanson, S. S. Sauter, J. J. Hurrell, and L. M. Schleifer. 2000. A field study of supplementary rest breaks for data-entry operators. *Ergonomics.* 43(5):622–38.

Gallagher, K. M., T. Campbell, and J. P. Gallagher. 2014. The influence of a seated break on prolonged standing induced low back pain development. *Ergonomics.* 57(4):555–62.

Grandjean, E., and K. H. E. Kroemer. 1997. *Fitting the Task to the Human: A Textbook of Occupational Ergonomics,* 5th Edition. CRC Press, Boca Raton, FL.

Gscheidle, G. M., and M. P. Reed. 2004. Sitter-selected postures in an office chair with minimal task constraints. *Hum Fac Erg Soc P.* 48(8):1086–90.

Gustafsson, E., and M. Hagberg. 2003. Computer mouse use in two different hand positions: Exposure, comfort, exertion and productivity. *Appl Ergon.* 34(2):107–13.

Hamilton M. T., D. G. Hamilton, and T. W. Zderic. 2007. Role of low energy expenditure and sitting in obesity, metabolic syndrome, type 2 diabetes, and cardiovascular disease. *Diabetes.* 56:2655–67.

Hedge, A. 2004. Keyboards. In Bainbridge, W. S. (ed), *Berkshire Encyclopedia of Human–Computer Interaction.* Berkshire Publishing Group, Great Barrington, MA. 1:401–5.

Hedge, A. 2013. Benefits of a proactive office ergonomics program. *Hum Fac Erg Soc P.* 57(1):536–40.

Hedge, A. 2015. Sit-Stand Working Programs (http://ergo.human.cornell.edu/CUESitStandPrograms.html, accessed March 18, 2016).

Hedge, A., and E. J. Ray. 2004. Effects of an electronic height-adjustable worksurface on self-assessed musculoskeletal discomfort and productivity among computer workers. *Hum Fac Erg Soc P.* 48(8):1091–5.

Hedge, A., and J. Puleio. 2014. Proactive office ergonomics really works. *Hum Fac Erg Soc P.* 59(1):482–6.

Hedge, A., and G. Shaw. 1996. Effects of a chair-mounted split keyboard on performance, posture and comfort. *Hum Fac Erg Soc P.* 40(13):624–8.

Hedge, A., D. J. Feathers, and K. A. Rollings. 2010. Ergonomic comparison of slanted and vertical computer mouse designs. *Hum Fac Erg Soc P.* 54(6):561–5.

Hedge, A., J. Jagdeo, A. Agarwal, and K. Rockey-Harris. 2005. Sitting or standing for computer work—Does a negative-tilt keyboard tray make a difference? *Hum Fac Erg Soc P.* 49(8):808–12.

Hedge, A., S. Morimoto, and D. McCrobie. 1999. Effects of keyboard tray geometry on upper body posture and comfort. *Ergonomics.* 42(10):1333–49.

Hedge, A., J. Puleio, and V. Wang. 2011. Evaluating the impact of an office ergonomics program. *Hum Fac Erg Soc P.* 55(1):594–8.

Hedge, A., M. Rudakewych, and L. Valent-Weitz. 2002. Investigating total exposure to WMSD risks: The roles of occupational and non-occupational factors. *Hum Fac Erg Soc P.* 46(15):1325–9.

Helander, M. 2003. Forget about ergonomics in chair design? Focus on aesthetics and comfort! *Ergonomics.* 46(13–14):1306–19.

Helander, M. G., L. Zhang, and D. Michel. 1995. Ergonomics of ergonomic chairs: A study of adjustability features. *Ergonomics.* 38(10):2007–78.

Henning, R. A., E. A. Callaghan, A. M. Ortega, G. V. Kissel, J. I. Guttman, and H. A. Braun. 1996. Continuous feedback to promote self-management of rest breaks during computer use. *Int J Indust Ergon.* 18:71–82.

Henning, R. A., P. Jacques, G. V. Kissel, A. B. Sullivan, and S. M. Alteras-Webb. 1997. Frequent short rest breaks from computer work: Effects on productivity and well-being at two field sites. *Ergonomics.* 40(1):78–91.

Hignett, S., and L. McAtamney. 2000. Rapid entire body assessment (REBA). *Appl Ergon.* 31(2):201–5.

Honan, M., E. Serina, R. Tal, and D. Rempel. 1995. Wrist postures while typing on a standard and split keyboard. *Hum Fac Erg Soc P.* 39(5):366–8.

Horie, S., A. Hargens, and D. Rempel. 1993. Effect of keyboard wrist rest in preventing carpal tunnel syndrome. *Proc Amer Pub Health Assoc Annu Meet.* p. 319. (Abstract).

Karakolis, T., and J. P. Callaghan. 2014. The impact of sit–stand office workstations on worker discomfort and productivity: A review. *Appl Ergon.* 45(3):799–806.

Krause, N., J. W. Lynch, G. A. Kaplan, R. D. Cohen, R. Salonen, and J. T. Salonen. 2000. Standing at work and progression of carotid atherosclerosis. *Scand J Work Environ Health.* 26(3):227–36.

Kroemer, K. H. E. 1992. Performance on a prototype keyboard with ternary chorded keys. *Appl Ergon.* 23(2):83–90.

Kroemer, K. H. E. 2001. Keyboards and keying: An annotated bibliography of the literature from 1878 to 1999. *Univ Access Inform Soc.* 1(2):99–160.

Lee, D., H. McLoone, and J. T. Dennerlein. 2008. Observed finger behaviour during computer mouse use. *Appl Ergon.* 39(1):107–13.

Leivseth, G., and B. Drerup. 1997. Spinal shrinkage during work in a sitting posture compared to work in a standing posture. *Clin Biomech.* 12(7/8):409–18.

McAtamney, L., and E. N. Corlett. 1993. RULA: A survey method for the investigation of work-related upper limb disorders. *Appl Ergon.* 24(2):91–9.

McLean, L., M. Tingley, R. N. Scott, and J. Rickards. 2001. Computer terminal work and the benefit of micro-breaks. *Appl Ergon.* 32(3):225–37.

Monroe, M. J., C. M. Sommerich, and G. A. Mirka. 2001. The influence of head, forearm and back support on myoelectric activity, performance and subjective comfort during a VDT task. *Hum Fac Erg Soc P.* 45(14):1082–6.

Montie, D. L., C. A. Putnam, A. P. Fenety, and K. M. Harman. 2004. VDT users' self-administration of micro-breaks during two hours of computer use: Microbreak frequency, duration and in-chair movement. *Bridging the Gap: Proc 35th Annu Conf ACE.* 4 pp.

Muss, T., and A. Hedge. 1999. Effects of a vertical-split keyboard on posture, comfort and performance. *Hum Fac Erg Soc P.* 43(6):496–500.

O'Reilly, J. T., P. Hagan, P. R. Gots, and A. Hedge. 1998. *Keeping Buildings Healthy: How to Monitor and Prevent Indoor Environmental Problems.* J. Wiley & Sons, New York.

Park, M., J. Kim, and J. Shin. 2000. Ergonomic design and evaluation of a new VDT workstation chair with keyboard–mouse support. *Int J Indust Ergon.* 26(5):537–48.

Pereira, A., T. Miller, Y. Huang, D. Odell, and D. Rempel. 2013. Holding a tablet computer with one hand: Effect of tablet design features on biomechanics and subjective usability among users with small hands. *Ergonomics.* 56(9):1363–75.

Rempel, D. M., P. J. Keir, and J. M. Bach. 2008. Effect of wrist posture on carpal tunnel pressure while typing. *J Orthop Res.* 26(9):1269–73.

Rudakewych, M., L. Valent-Weitz, and A. Hedge. 2001. Effects of an ergonomic intervention on musculoskeletal discomfort among office workers. *Hum Fac Erg Soc P.* 45(10):791–5.

Schlittmeier, S. J., and J. Hellbruck. 2009. Background music as noise abatement in open-plan offices: A laboratory study on performance effects and subjective preferences. *App Cog Psychol.* 23(5):684–97.

Sheahan, P. J., T. L. Diesbourg, and S. L. Fischer. 2015. The effect of rest break schedule on acute low back pain development in pain and non-pain developers during seated work. *Appl Ergon.* 53(Part A):64–70.

Simoneau, G. G., and R. W. Marklin. 2001. Effect of computer keyboard slope and height on wrist extension angle. *Hum Factors.* 43(2):287–98.

Smith, A., B. Waters, and H. Jones. 2010. Effects of prior exposure to office noise and music on aspects of working memory. *App Aspects Audit Distract.* 12(49):235–43.

Sommerich, C. M., S. M. B. Joines, and J. P. Psihogios. 2001. Effects of computer monitor viewing angle and related factors on strain, performance, and preference outcomes. *Hum Factors* 43(1):39–55.

Tayyari, F., and J. L. Smith. 1987. Effect of music on performance in human-computer interface. *Hum Fac Erg Soc P.* 31(12):1321–5.

Thom-Santelli, J., and A. Hedge. 2005. Effects of a multitouch keyboard on wrist posture, typing performance and comfort. *Hum Fac Erg Soc P.* 49(5):646–50.

Ting, A., and A. Hedge. 2001. An ergonomic evaluation of a hybrid keyboard and game controller. *Hum Fac Erg Soc P.* 45(7):677–81.

Tüchsen, F., N. Krause, H. Hannerz, H. Burr, and T. S. Kristensen. 2000. Standing at work and varicose veins. *Scand J Work Environ Health.* 26(5):414–20.

Tüchsen, F., H. Hannerz, H. Burr, and N. Krause. 2005. Prolonged standing at work and hospitalisation due to varicose veins: A 12 year prospective study of the Danish population. *Occup Environ Med.* 62(12):847–50.

van Dieën, J. H., M. P. de Looze, and V. Hermans. 2001. Effects of dynamic office chairs on trunk kinematics, trunk extensor EMG and spinal shrinkage. *Ergonomics.* 44(7):739–50.

Vink, P., R. Porcar-Seder, A. Page de Poso, and F. Krause. 2007. Office chairs are often not adjusted by end-users. *Hum Fac Erg Soc P.* 51(17):1015–9.

Wang, V., and A. Hedge. 2008. A usability evaluation of a laser projection virtual keyboard. *Hum Fac Erg Soc P.* 52(6):537–41.

Wilke, H. J., P. Neef, M. Caimi, T. Hoogland, and L. E. Claes. 1999. New in vivo measurements of pressures in the intervertebral disc in daily life. *Spine.* 24(8):755–62.

Young, J. G., M. Trudeau, D. Odell, K. Marinelli, and J. T. Dennerlein. 2012. Touch-screen tablet user configurations and case-supported tilt affect head and neck flexion angles. *Work.* 41:81–91.

2 Designing Thermal Environments for Comfort, Health, and Performance

Ken Parsons

CONTENTS

2.1 INTRODUCTION

An objective for environmental design is to provide comfort, well-being, health, and performance to people who experience the environment, and maybe stimulation, inspiration, pleasure, and excitement. Implicit in that objective is the avoidance of discomfort and dissatisfaction; avoidance of conditions detrimental to health, which can lead to illness, injury, or even death; and enhancement of performance and productivity, which includes the avoidance of conditions that reduce motivation, physical capacity, including manual dexterity, and cognitive ability, or cause distraction and time off from work. This chapter provides principles and methods for the design and assessment of thermal environments. It describes how people respond to hot, moderate, and cold environments and presents principles, techniques, and tools for measuring thermal stress and its effects, and how they can be used in practical applications.

2.2   THE SIX BASIC PARAMETERS

Heat stress, cold stress, and thermal comfort are not defined by just one factor of the thermal environment such as air temperature, but by the interaction of six factors. These are often referred to as the *six basic parameters*, as representative values of continually changing variables are usually used in assessment. The six factors are air temperature, radiant temperature, air velocity, air humidity (the environmental factors), clothing of the person, and metabolic heat generated by the person (personal factors). A specification of instruments and how to measure the environmental factors is provided in International Standards Organization (ISO) 7726 (1998). The measurement or the estimation of all six variables to which people are exposed, and hence the six basic parameters, should always be a starting point for any design and assessment of the thermal environment.

2.3   HUMAN THERMOREGULATION

People are homeotherms and attempt to maintain an internal body temperature of around 37°C (98.6°F). Human thermoregulation could be considered to be the process by which they defend that position. Thermal stress can be defined as the environment made up of the six basic parameters. It is the interaction of the parameters that provides the thermal stress on the body. Thermal strain is the response of the person to thermal stress. If environmental or other conditions tend toward reducing (cold stress) or increasing (heat stress) internal body temperature, it will elicit both behavioral and physiological responses as the body attempts to defend and maintain an optimum condition. A system diagram is shown in Figure 2.1. This is adapted and modified from Parsons (2014).

The system is stimulated by both skin temperature and internal body temperature. Physiological responses are driven by the difference between the set point core temperature (which varies around 37°C [98.6°F]) and the actual core temperature of the body (e.g., brain temperature). If there is a tendency for the body to lose heat and hence for the body temperature to fall, the posterior hypothalamus promotes vasoconstriction where blood is withdrawn from the extremities (arms, legs, hands, and feet). This reduces skin temperature and hence heat loss from the skin (shell) in an attempt to maintain the internal body temperature (core). If this response is insufficient to reverse any fall in the internal body temperature, then the body generates heat by nonshivering thermogenesis, which is an increase in muscle tone, and then shivering which can provide significant additional metabolic heat production.

If the environmental conditions, the activity, and the clothing combine to provide a tendency for the body temperature to rise, the anterior hypothalamus initiates vasodilation, where blood flows to the skin and the extremities, hence raising the skin temperature and promoting heat loss. If that is insufficient, then sweating occurs which allows heat to be lost from the skin by evaporation. There is a connection between the anterior and the posterior hypothalamus to prevent instability and cold responses and heat responses working against each other.

A powerful form of human thermoregulation is behavioral. Not only does the body continuously, automatically, and unconsciously detect, process, and respond to its thermal state, but it also consciously recognizes or feels its thermal state (hot, cold, etc.). This is mainly by feeling the skin condition. This state is consciously compared with a desired state and so comfort, or something related to it, in addition to the internal body temperature, can be regarded as a controlled variable. When a person feels too cold, then s/he will move away from the environment that is causing discomfort, turn up the heating, close the windows, add clothing layers, reduce the exposed surface area by changing posture, and so on. When a person feels too hot, they may move away, lower the thermostat or the ventilation controls, take off clothing layers, increase the exposed surface area by changing posture, open the windows, and so on. In Figure 2.1, this is stimulated by skin temperature and how wet (W) the person feels. The effectiveness of behavioral thermoregulation and which behaviors are carried out will depend upon what is possible in the environmental context

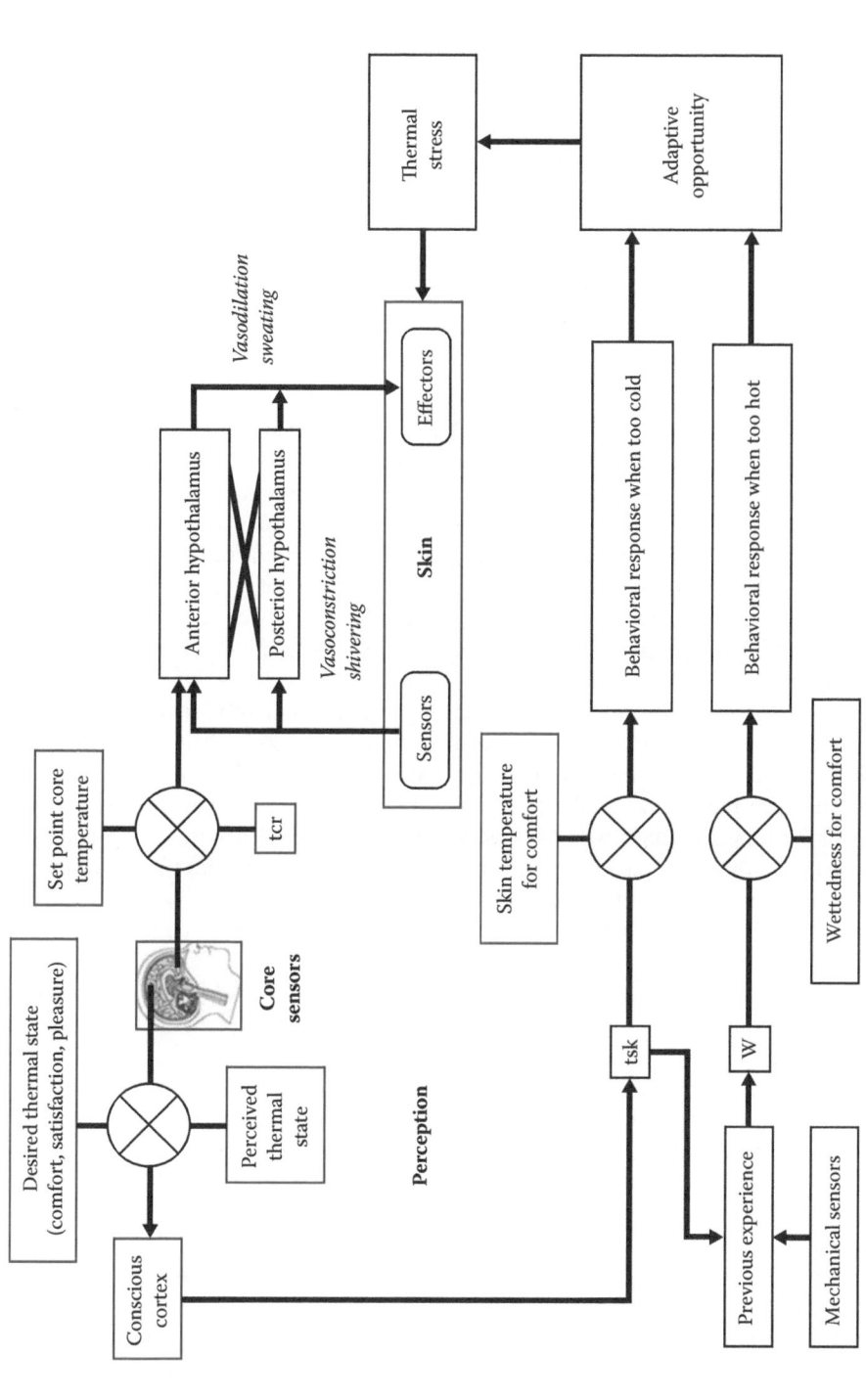

FIGURE 2.1 System of human thermoregulation including physiological and behavioral responses. (Adapted from Parsons, K. C., *Human Thermal Environments*, Third ed., Taylor & Francis, New York, 2014. With permission.)

considered. This is termed *adaptive opportunity*. A person restricted by the environment (e.g., some buildings do not allow window opening) or by the job they are doing (e.g., some jobs do not allow movement away from a workstation or adjustment of uniform) or by a physical or a mental condition (babies or children, people with disabilities who are not mobile or cannot adjust clothing) will have limited adaptive opportunity. Any appropriate behavior, however, will reduce the thermal stress (heat or cold—defined by the six basic parameters), and this will be fed back into the conscious cortex as shown in Figure 2.1. Both systems are continuous and complementary. It is reasonable to assume that behavioral responses are often a first reaction and anticipate and reduce the need for significant physiological responses.

2.4 RESPONSES RELATED TO HUMAN THERMOREGULATION

Physiological responses to heat and cold include vasoconstriction, vasodilation, sweating, and shivering. Relevant to these responses are a number of mechanisms and phenomena that can influence health and performance. These are the following:

- Cold-induced vasodilatation (CIVD)—CIVD occurs when the extremities become cold for prolonged periods, and after a bout of severe vasoconstriction, the hands suddenly become warm as vasoconstriction ceases below skin temperatures of about 12°C (53.6°F) and blood flows to the skin. This phenomenon is often experienced when playing in the snow. After the hands (or other parts) warm up, vasoconstriction begins to work again and reduces the skin temperature. So a cycle of warm and cold persists. This is known as the *hunting phenomenon* (Lewis 1930). The warming of the hands promotes comfort and manual performance but can lead to a dangerous loss of heat in cold conditions.
- Selective brain cooling (SBC)—SBC has been suggested as a method specifically designed to keep the brain temperature from overheating (Cabanac 1995). It uses the blood flow direction and the heat exchange blood vessel systems to remove heat from the brain. This is found in warm-blooded animals (e.g., cats and dogs), but it is debatable whether such a mechanism occurs in humans who do not have the physiological structures associated with this mechanism, that is, a rete mirabile, which is a network of arterioles that allow effective heat exchange and hence cooling due to breathing and panting.
- Brown fat—Brown fat is an adipose tissue enriched with mitochondria and blood vessels, providing a resource for heat production and distribution. Its role in human thermoregulation is unclear, but it is thought to be effective in babies and not adults. However, there is some dispute about this, and it may be possible that brown fat increases as a result of cold acclimatization in adults.
- Countercurrent heat exchange—Countercurrent heat exchange is a system of heat exchange to prevent heat from transferring out of the body. The cool blood returning in veins is directed away from the skin and passes the warm blood flowing down arteries in the arms and the legs. The heat from the body is therefore transferred to the venous blood and returned to the core.
- Arteriovenous anastomoses—Arteriovenous anastomoses are small blood vessels that make a direct connection between outgoing blood and incoming blood, to regulate heat loss or gain from the skin surface.
- Piloerection—Piloerection is when hairs stand more erect in cold conditions to provide an increased insulating air layer between the skin and the environment. This may be effective even in humans, by reducing heat loss, particularly when shivering, and also to provide an extra insulating air layer under the clothing.

2.5 HEAT ILLNESS

There are several stages of heat illness:

- Heat stroke—Heat stroke is an unacceptable rise in the internal body temperature leading to confusion, convulsions, collapse, and eventually death. It is caused by excessive thermal stress (combination of the six basic parameters) and is the problem most heat stress assessment and safety procedures are designed to avoid. An internal body temperature rise to 38°C (100.4°F) is often regarded as an upper safety limit with progressive symptoms often rapidly occurring above that value and becoming extreme above around 39°C (102.2°F). The problems include the denaturing of body protein and the irreversible damage to vital organs (kidneys, liver, lungs, brain) as the body temperatures rise toward 44°C (111.2°F) and above. Even at core temperatures below 38°C (100.4°F), however, there can be an unacceptable strain on the body and, in particular, vulnerable people are at risk (e.g., some elderly people and the sick), and illness and death can often occur due to heart and respiratory conditions.
- Heat syncope—Heat syncope is caused by the blood pooling in the feet and away from the brain, leading to fainting. It is often found in people standing for long periods in the heat.
- Heat cramps—Heat cramps occur due to low salt levels caused by excessive sweating or excessive drinking of water.
- Heat rashes—Heat rashes can occur in skin soaked with sweat that cannot evaporate, often trapped beneath clothing.
- Heat exhaustion—Heat exhaustion is a general incapacity caused by a prolonged exposure to heat.

2.6 ACCLIMATIZATION

Acclimatization is the progressive change in a physiological response that occurs after repeated exposure to heat. It has a number of properties but the most significant is that of the ability to sweat more and to sweat earlier at the onset of heat stress. This provides greater potential for evaporative cooling in hot conditions, and it will reduce heat illness by providing a greater capacity to tolerate heat. Fuller descriptions of this are provided by Leithead and Lind (1964), Goldman (1988), Kenney et al. (2012), and Parsons (2014).

Behavioral acclimatization refers to the behaviors that are employed to reduce thermal illness. They are important in both hot and cold conditions and are usually learned through experience and exposure by living in the hot or the cold climate. Indigenous populations learn the behaviors as part of their culture. The avoidance of heat or cold is often a major behavior for maintaining health. Parsons (2014) provides a rule of thumb mnemonic for the avoidance of problems in the heat. He recommends SHAFTS: be sensible (appropriate behavior), hydrated, acclimatized, fit, thin, and sober (avoid alcohol and inappropriate drugs).

2.7 COLD INJURY

Cold can provide significant discomfort and the strain can lead to death in vulnerable people. Extreme cold can lead to cold injury and death in all people. *Clinical hypothermia* is defined as a rectal temperature below 35°C (95°F), but significant problems occur at much higher values. Any drop in the internal body temperature will be after significant vasoconstriction leading to low and uncomfortable skin temperatures. This will also provide a strong drive for behavioral responses and distraction, a change in mood and a loss in manual dexterity and strength.

Two recognized cold injuries are nonfreezing cold injury and frostbite. Nonfreezing cold injury (often called *trench foot*) occurs at temperatures above freezing and after prolonged exposure when wet and cold. This causes damage to tissues, often in the feet. Frostbite is an irreversible damage caused by the freezing of the tissue. It usually occurs in the extremities (fingers, toes, nose, etc.) because of their exposed position and because they have a large surface area for heat loss compared to their mass or volume. When the fluid in cells freezes, it expands and damages the cell wall and the contents, leading to permanent damage.

2.8 PHYSIOLOGICAL INDICATORS OF HUMAN HEALTH, COMFORT, AND PERFORMANCE

Four physiological measures are recognized as indicators of human response to heat and cold and are often used in personal monitoring systems to monitor and preserve the health of a person working in the heat or the cold. These are internal body temperature, skin temperature, heart rate, and mass loss. These are described in ISO 9886 (2004). The internal body temperature provides an indication of body heat gain or loss and the rate of heat gain or loss. It is measured in the ear (aural—using sensors; or tympanic—often measured with an infrared device), under the tongue (with mouth closed), in the esophagus next to the heart (by swallowing a sensor on a wire), in the alimentary canal (using a radio pill), in the rectum (10 cm beyond the anal sphincter), and more (urine temperature, under the arm, on the forehead, etc.). All have specific properties and errors in measurement and will be selected depending upon context.

The skin temperature will be measured on the extremities (fingers, toes) in the cold and over the body to provide mean skin temperature, which is usually determined from a weighted average of temperatures measured at defined points on the body (see ISO 9886 [2004]).

The heart rate is a general indicator of stress and total heart rate (HR) (in beats per minute [bpm]) can be apportioned into causes such as beats due to resting, activity, psychological effects, and thermal stress (Vogt et al. 1981). Mass loss indicates sweat production, sweat evaporated, and sweat trapped in clothing, by measuring the weight of a person before and after exposure (to heat and/or exercise), with and without clothing. It provides an indication of the thermal strain and the clothing performance but is not a continuous measure that can be easily used for personal monitoring.

A personal monitoring system to preserve health will include a combination of measures and criteria for deciding when the strain is unacceptable and the person should be removed from the thermal stress. The criteria vary depending upon context. The internal body temperatures of greater than 38.0–38.5°C (100.4–101.3°F) are often taken as limits for action; however, it will depend upon other factors. The time to remove clothing after exposure to heat is important; 38.5°C (101.3°F) may be regarded as high when resting but not so high in lightly clothed people taking part in fun runs and marathons. In the cold, any fall in the internal body temperature may be regarded as unacceptable with 36°C (96.8°F) as a lower limit. A maximum heart rate related to age is often taken during exercise (e.g., 200 – age in years, bpm). The mean skin temperature will provide an indication of thermal discomfort (as it varies from 33°C [91.4°F]) and general cold stress, and the rate of convergence of mean skin temperature and internal body temperature is an indication of heat strain. Local skin temperatures (of hands and feet) can indicate discomfort and potential loss in performance and health, especially in the cold. The effects can occur at skin temperatures below 30°C (86°F), and skin temperatures below 25°C (77°F) can be uncomfortable; below 15°C (59°F) is an indicator of loss in performance, and below 6°C (42.8°F) provides an alarm criterion for potential health problems. The appropriate combination of measures and criteria will depend upon the context of the activity and must be integrated into a more comprehensive set of working practices for hot or cold environments (e.g., ISO 15743 [2008]). Subjective measures can complement physiological measurements. They can be useful and informative; however, caution must be used in extremes of hot and cold where the judgment of a person may be impaired.

2.9　THE THERMAL AUDIT

A powerful technique for use in the design and assessment of thermal environments is the thermal audit (Parsons 1992). It involves the calculation of the heat transferred into and out of the body using the body heat equation, sometimes called the *heat balance equation* as, to maintain a constant internal body temperature of around 37°C (98.6°F), the heat production and the heat inputs to the body must be balanced by the heat outputs from the body. That is, the metabolic heat production (metabolic rate minus any external work = M − W) must be balanced by the net of heat losses and gains by convection (C), radiation (R), and evaporation (E) from the body. That is, from the skin and by breathing. The amount of heat involved for each of these pathways can be calculated from the measures of the six basic parameters. This provides a rational approach to assessing a thermal environment and hence a thermal audit of heat production and loss. A full description with examples is provided by Parsons (2014). The power of the technique lies in the interpretation of the audit. To preserve comfort, the required skin temperatures and the skin wetness will have to be achieved as well as heat balance. The degree of discomfort will then depend upon the difference between heat production and loss when the body is assumed to be in this desired thermal state (Fanger 1970). An indication of heat stress can be calculated from the thermal audit using the sweat rate required to achieve heat balance and the rate of heat storage, as well as hydration state (from sweat loss, e.g., Vogt et al. [1981]). In cold conditions, how much clothing insulation is required to maintain an acceptable state can be calculated, as well as exposure times, determined by a calculation of heat loss (negative heat storage) before the skin and the core temperatures reach unacceptable levels. The thermal audit is the principle behind rational techniques for assessing thermal environments, which are described in the following.

2.10　THE THERMAL INDEX

A widely used indicator of thermal stress (the integrated effects of the six basic parameters) is the thermal index. This is a single number that represents the stress and varies with the thermal strain on a person. A thermal comfort index is a single number that varies with the level of thermal comfort or discomfort of people, a cold stress index with strain in the cold, and a heat stress index with heat strain. There have been many studies to determine thermal indices and many exist. A common approach is to use the concept of the *standard environment*. This is where the index value is defined as the temperature of a standard environment that will provide the same effect on people as the actual environment under consideration. For example, the actual environment under investigation will be defined by its air temperature, radiant temperature, humidity, and air velocity, as well as the clothing worn and the activity of the people in the environment. This will provide a state of comfort, heat strain or cold strain. The thermal index will then be the temperature of a standard environment (e.g., where air temperature = radiant temperature, still air, relative humidity = 50% with the same clothing and activity as in the actual environment) that provides the same thermal strain. The challenge in defining the index is then to identify a method for determining the equivalent thermal strain. The examples of indices defined in this way are the effective temperature (ET) (Houghton and Yagloglou 1923) and the standard effective temperature (Gagge et al. 1973).

Three types of thermal index can be identified, although definitions overlap:

1. Direct thermal indices are instruments that respond to environmental conditions in a way related to human response (i.e., to an interaction of the effects of air temperature, radiant temperature, humidity, and air velocity). Limit values can then be used to monitor the conditions and will be determined by the clothing worn and the activity level of the people exposed to the thermal conditions. An example of this is the wet bulb globe temperature (WBGT) heat stress index (ISO 7243 1989).

2. Empirical thermal indices are essentially databases of human responses to a wide range of conditions, such that they can be used to predict responses of people when exposed to thermal environments within the range of conditions included in the database. The ET index (Houghton and Yagloglou 1923) is an example of an empirical index.
3. Rational thermal index are based upon a thermal audit as described above, an example being the required sweat rated index (Vogt et al. 1981).

2.11 THERMAL MODELS

Thermal models combine a representation of the human body in terms of its mass, volume, dimensions, and thermal properties of tissues (the passive system) with a system of human thermoregulation (the controlling system) and the equations to calculate heat transfer between body parts and between the clothed body and the environment. The combined system provides a model for predicting human responses for any environment and is therefore suited for investigating human response as well as designing environments to preserve health, comfort, and performance. Models are used in computer-aided design and are powerful tools for investigating human thermal environments from prototype design to assessment. A more detailed description of thermal models is provided by Parsons (2014).

2.12 ESTIMATION OF METABOLIC HEAT PRODUCTION

People produce heat by burning food in oxygen, the total amount of heat being the integration of the heat produced in each living cell. Actually, cells produce energy and part of that energy will be used for work, but most of the energy will be produced as heat, and the body needs this heat to survive. The minimum amount of heat produced for a totally inactive person is termed the *basal metabolic heat*, and this is often taken as a total heat production of around 45 W m^{-2}. As a person becomes more active, additional heat is produced (mainly in the muscles). Human thermoregulation is essentially the mechanism by which we lose or preserve metabolic heat in order to maintain optimum conditions which includes an internal body temperature of around 37°C (98.6°F). The total amount of heat produced by a person (watts = joules per second) is the integration of the heat produced in all of the cells in his or her body. Larger people will have more cells; muscular people relatively, more muscle; and so on. To be able to provide consistent values across groups, for a given activity, the values are often given in relation to the size (surface area, mass, volume, etc.) of the person. ISO 8996 (2004) uses watts per square meter (W m^{-2}) of the body surface area. Table 2.1 provides examples for a range of simple activities. ISO 8996 (2004) provides values for a wide range of activities. An estimate of 1.8 m^2 for an average male and 1.6 m^2 for an average female is often

TABLE 2.1
Estimates of Metabolic Rate for Basic Activities

Basic Activity	Estimate of Metabolic Rate (W m^{-2})
Lying (basal minimum)	45
Sitting	58
Standing	65
Low general activity	65
Medium	100
Slow walking on a level even path	110
High	230
Very high	290
Going upstairs at normal pace	400

taken. These estimates are probably now too low for western populations due to secular trends (not to mention obesity). For specific populations around the world, the estimate of average surface area for the population of interest should be used when calculating heat production in watts.

2.13 ESTIMATION OF THE THERMAL PROPERTIES OF CLOTHING

Clothing provides a resistance to heat transfer between the body and the environment. In the heat, this can restrict heat loss and for active people wearing protective clothing and equipment, this can be dangerous if sufficient controls are not in place. The process of heat transfer through clothing worn by a person is complex and not fully understood. Three properties of clothing are often considered. These are the dry insulation, the vapor permeability, and the ventilation properties. Dry insulation is measured in clo values, where 1 clo is said to be the clothing insulation of a normal business suit and is given a clothing insulation value of 0.155 m^2 °C W^{-1}, 0 clo is a nude person, and so on. Insulation values for a range of clothing ensembles are provided by Olesen and Dukes-Dubos (1988) (see Table 2.2). Vapor permeation properties are particularly relevant for people who are sweating. A value of 0.0155 m^2 kPa W^{-1} is a typical value, where the *kilopascal* term accounts for the difference in partial vapor pressure between that on the skin and that in the air (humidity). The ventilation properties of clothing are related to the type of activity and the clothing design, in particular, the openings and the vents in clothing. The exchange (L min^{-1}) of, often saturated, air at skin temperature with outside air caused by ventilation can cause significant heat loss. A range of values from 20 to over 100 L min^{-1} is typical, from resting to running,

TABLE 2.2
Work Clothing Ensembles: Dry Thermal Insulation Values

Work Clothing		I_{cl} Clo	I_{cl} m^2 °C W^{-1}
Underpants with	Boiler suits, socks, shoes	0.79	0.110
	Shirt trousers, socks, shoes	0.75	0.115
	Shirt, boiler suits, socks, shoes	0.80	0.125
	Shirt, trousers, jacket, socks, shoes	0.85	0.135
	Shirt, trousers, smock, socks, shoes	0.90	0.140
Underwear with	Short sleeves and legs, shirt, trousers, socks, shoes	1.00	0.155
	Short sleeves and legs, shirt, trousers, jacket, socks, shoes	1.10	0.170
	Long legs and sleeves, thermojacket, socks, shoes	1.20	0.185
	Short sleeves and legs, shirt, trousers, jacket, thermojacket, socks, shoes	1.25	0.190
	Short sleeves and legs, boiler suit, thermojacket + trousers socks, shoes	1.40	0.220
	Short sleeves and legs, shirt, trousers, jacket, heavy quilted outer jacket and overalls, socks, shoes	1.55	0.225
	Short sleeves and legs, shirt, trousers, jacket, heavy quilted outer jacket and overalls, socks, shoes	1.85	0.285
	Short sleeves and legs, shirt, trousers, jacket, heavy quilted outer jacket and overalls, socks, shoes, cap, gloves	2.00	0.310
	Long sleeves and legs, thermojacket + trousers, outer thermojacket + trousers socks, shoes	2.20	0.340
	Long sleeves and legs, thermojacket + trousers, parka with heavy quilting, overalls with heavy quilting, socks, shoes, cap, gloves	2.55	0.395

Source: Olesen, B. W., and F. N. Dukes-Dubos, *Performance of Protective Clothing*, American Society for Testing Materials, Philadelphia, Pennsylvania, 1988. With permission.

depending on clothing type, wind conditions, and so on. The integration of these properties and more determines the effects of clothing. The tables of properties, derived from measurements on human subjects and specialist thermal mannequins, provide values for thermal assessment, including the use of thermal indices, models, and thermal audit. Parsons (2014) and ISO 9920 (2007) provide more details and a comprehensive range of values for garments and clothing ensembles.

2.14 HEAT STRESS

Heat stress is the combination of the six basic parameters that provides a tendency for the body temperature to rise. People respond by behaving to reduce the heat stress if they can and by vasodilation and sweating to increase heat loss. If this is not sufficient, then the body temperature will rise leading to unacceptable strain and even death. To prevent heat casualties, a comprehensive management system is required involving monitoring of the environment and the people exposed and modeling the heat transfer and making predictions of allowable exposure times usually based upon internal body temperature and dehydration limits.

A simple monitoring index that is recognized and internationally used is the WBGT index. This is the temperature of a measuring instrument that responds to the integration of air temperature, radiant temperature, air velocity, and humidity. It was first developed to monitor environments in military training camps in the United States as an improved method to avoid heat casualties (Yaglou and Minard 1957). It has been adopted by the American Conference of Governmental Industrial Hygienists (2012) and in ISO 7243 (1989). The instrument consists of a black globe thermometer of 150 mm (6″) in diameter (tg), a natural wet bulb thermometer (tnw), and an air temperature sensor (ta), which is shielded from radiation. The instrument is exposed to the environment of interest, and by a weighted summation of WBGT = 0.7tnw + 0.2tg + 0.1ta for conditions in the sun or WBGT = 0.7tnw + 0.3tg for conditions out of the sun, the WBGT value is obtained. For more details and a specification of the sensors, see ISO 7243 (1989). The WBGT values are compared with reference values and if they exceed the reference values, then it is predicted that there may be heat casualties. If the measured values do not exceed the reference values, then no heat casualties are expected. Reference values are shown in Table 2.3.

The reference values are for workers wearing normal clothing (e.g., boiler suit or light shirt and trousers of around 0.6 clo of insulation). A revision of the standard is proposed which may include a wider range of clothing. This is important as heat casualties often involve people wearing protective clothing and equipment.

TABLE 2.3
WBGT Reference Values

Metabolic Rate (W m⁻²)	Acclimatized (°C)		Not Acclimatized (°C)	
Resting M <65		33		32
65 < M < 130		30		29
130 < M < 200		28		26
200 < M < 260	25	26*	22	23*
M > 260	23	25*	18	20*

Source: ISO 7243: Hot Environments—Estimation of the Heat Stress on Working Man, Based on the WBGT-Index (Wet Bulb Globe Temperature), International Organization for Standardization, Geneva, Switzerland, 1989. With permission.

Note: Values assume a maximum rectal temperature of 38°C. Values with asterisks are for sensible air movement and values without asterisks are for insensible air movement.

If the WBGT reference values are exceeded, then a more detailed and more expert assessment is required. This will involve a thermal audit. ISO 7933 (2004) provides a rational method of assessment based upon the calculation and the interpretation of the sweat rate required in the hot environment along with other factors in a method called predicted heat strain (Malchaire et al. 2000). Where environments are extreme and when the thermal strain of individuals is required to be monitored, ISO 9886 (2004) provides guidance on how to use personal monitoring systems.

2.15 COLD STRESS

Cold stress is a combination of the six basic parameters that produce a tendency for the internal body temperature to fall or for local skin temperatures to fall toward unacceptable levels. This can involve significant discomfort, and people respond by behaving to reduce the discomfort and the cold stress if they can and by vasoconstriction and, if severe, shivering. If this is insufficient to preserve or generate enough heat, then the skin and the internal body temperature will fall leading to cold injury and eventually death.

When assessing cold environments, it is important to consider both whole body responses and local responses. Assessment methods are provided in ISO 11079 (2007). For local cooling, a chilling temperature (twc) can be calculated which integrates the effects of air temperature and wind into a single temperature. The temperature values are then interpreted in terms of risk. Whole body cooling is assessed using a rational method and the calculation of the clothing insulation required (IREQ) in any cold environment as defined by the six basic parameters (Holmer 1984). Guidance on personal monitoring of individuals exposed to cold, where environments are extreme, for example, is provided in ISO 9886 (2004). Local skin temperatures as well as internal body temperature will be of particular importance. Working practices for cold environments are described in ISO 15743 (2008).

2.16 THERMAL COMFORT

Thermal comfort is a desirable state that people try to achieve. When they become uncomfortable, they behave in a way that attempts to restore comfort. This complements the physiological system of thermoregulation, which is continuous, automatic, unconscious, and essentially adjusts the temperature and the wetness of the skin to achieve thermal balance and hence maintain internal body temperature. That is, by altering blood flow and distribution, sweating, and in more severe conditions, with increased metabolic heat due to shivering. The conditions for thermal comfort were defined by Fanger (1970). He stated that the conditions for thermal comfort are that the body must be in heat balance and also that skin temperatures and sweat rates must be at comfort levels. This was integrated into the predicted mean vote (PMV), the thermal index, and the associated predicted percentage dissatisfied (PPD) index. The PMV is calculated from the six basic parameters that define the thermal environment and provides a prediction of the mean vote of a large group of people on a 7-point scale (3, hot; 2, warm; 1, slightly warm; 0, neutral; −1, slightly cool; −2, cool; −3, cold) if they had been exposed to the thermal conditions. The PPD is derived from the PMV and provides a prediction of the percentage of people who would be dissatisfied with those conditions. The method has been adopted by ISO 7730 (2005).

Figure 2.1 shows the system of human thermoregulation that includes both physiological responses and behavioral responses. It can be seen that people strive to achieve an optimum state by taking appropriate action through their behavior. This will depend upon whether they are uncomfortably hot or uncomfortably cold. Context and environmental design will determine what adaptive opportunities are available (remove or put on clothing, move around, open or close windows, and so on). A number of methods have been proposed to take account of so-called adaptive methods to achieve thermal comfort (Nicol and Humphreys 1972; Fanger and Toftum 2002; Parsons 2005; Yao et al. 2009; Li and Lim 2013). For more details, see Parsons (2014).

Local discomfort is where particular areas of the body become uncomfortable, usually due to the effects on local areas such as shoulders, feet, and ankles. Drafts, asymmetric radiation, vertical gradients, and contact with hot or cold surfaces are examples. A description and criteria are provided in ISO 7730 (2005). The methods for developing subjective scales and for conducting an environmental survey are provided in ISO 10551 (1995), ISO 28802 (2012), and Parsons (2015a,b).

2.17   HUMAN PERFORMANCE

For practical application, two complementary approaches to predicting the effects of thermal conditions on human performance are useful. The first involves the direct effects of the thermal conditions on human response and function, for example, vasoconstriction in the cold reducing skin temperature with a loss of manual dexterity. These are often considered as the effects on manual performance and the effects on cognitive performance. The second approach is to consider the time off task caused by distraction or off the whole job caused by unacceptable risk and cessation of work (see Parsons [2014]).

The consideration of the direct effects on human performance shows that there are clear effects of cold on manual dexterity and strength. As skin temperatures fall, viscosity of synovial fluid increases, movements become slower, numbness occurs, and muscle strength decreases. These are significant effects with dramatic reductions in human capability. The effects of heat and cold on cognitive performance are less clear and of heat of manual performance, less dramatic, although sweating fingers will reduce grip. As internal body temperatures rise or fall, they will eventually reach limits that are a risk to health. Confusion will clearly have a significant effect on performance as will collapse and death.

Distraction will have a direct effect on performance, as it will lead to time off task. For the periods when people are distracted, performance and productivity will be zero. This is easier to account for than more subtle effects described earlier where the actual degree of performance loss is not clear. It is likely that to a first approximation, time off-task can be related to the degree of dissatisfaction with the thermal environment and hence the PPD index. Performance and productivity however will also depend upon many contextual and individual factors such as motivation and social environment. Whether dissatisfaction causes distraction and time off-task will depend on these factors. For example, a highly motivated person with a vital task that s/he is engaged with will not be as distracted as a person who is not engaged with the task and does not care about the level of performance. A motivation factor along with the PPD value will be required. Further research is needed into this method. It is clear that if environmental conditions exceed acceptable levels and work ceases, then productivity will drop to zero. This has implications for regulations and standards where people should be protected but work should not be unnecessarily stopped. A related point is that where people are very highly motivated, and regulations are not sufficient, then major risk can occur (as sometimes happens in military exercises or sports or when people are paid according to productivity).

2.18   GLOBAL WARMING AND HEALTH

There is increasing acceptance among scientists and policy makers that the earth is becoming warmer. The consequence of this is that weather systems will change and people will be exposed to different and more extreme weather. That is, weather that they are not used to, so their environments will be hotter, colder, or milder with more or less wind, rain, snow, and so on. The principles of assessing human responses to thermal environments will not vary, but the greater prevalence of extreme events has highlighted the risk to vulnerable people and the need for guidelines and working practices in the context of public health systems.

Advice for the avoidance in heat waves is of particular importance as there have been thousands of excess deaths, mainly among elderly and vulnerable people in major cities. Moving away from the heat is a sensible suggestion, and Parsons (2009) has suggested going to the supermarket or other air-conditioned place. He also suggests a technique for cooling in a water bath that avoids thermal shock, and that carefully directed fans should be used, even in hot weather above body temperature, as evaporation of sweat or water sprayed on the face or other part of the body will be greater than heat gain by convection. He notes that it is important to ensure that there are power (for fans, air conditioning, etc.) and water (partial immersion of feet or hands, baths, etc.) available. Plans should be in place in anticipation of a heat wave, and careful monitoring and heat alert systems are required. The World Health Organization (2012) has considered this problem and has provided guidance.

REFERENCES

American Conference of Governmental Industrial Hygienists. 2012. *TLVs and BEIs Based on the Documentation of the Threshold Limit Values for Chemical Substances and Physical Agents and Biological Exposure Indices*. Cincinnati, OH: American Conference of Governmental Industrial Hygienists.

Cabanac, M. 1995. *Human Selective Brain Cooling*. Berlin: Springer Verlag.

Fanger, P. O. 1970. *Thermal Comfort*. Copenhagen: Danish Technical Press.

Fanger, P. O., and J. Toftum. 2002. Extension of the PMV model to non air conditioned buildings in warm climates. *Energ Build*. 34(6):533–6.

Gagge, A. P., Y. Nishi, and R. R. Gonzalez. 1973. Standard effective temperature: A single temperature index of temperature sensation and thermal discomfort. *Proc CIB Comm W45 Human Requirements Symposium: Building Research Station. 5–13 September 1972*. Watford: Her Majesty's Stationery Office.

Goldman, R. F. 1988. Standards for human exposure to heat. In *Environmental Ergonomics*, Mekjavic, I. B., E. W. Banister, and J. B. Morrison (Eds.). London: Taylor & Francis. pp. 99–136.

Holmér, I. 1984. Required clothing insulation (IREQ) as an analytical index of cold stress. *ASHRAE Trans.* 90(1):116–28.

Houghton, F. C., and C. P. Yagloglou. 1923. Determining equal comfort lines. *J ASHVE*. 29:165–76.

ISO 7243. 1989. ED 2. Hot Environments—Estimation of the Heat Stress on Working Man, Based on the WBGT-Index (Wet Bulb Globe Temperature). Geneva: International Organization for Standardization.

ISO 10551. 1995. Ergonomics of the Thermal Environment—Assessment of the Influence of the Thermal Environment Using Subjective Judgement Scales. Geneva: International Organization for Standardization.

ISO 7726. 1998. ED 2. Ergonomics of the Thermal Environment—Instruments for Measuring Physical Quantities. Geneva: International Organization for Standardization.

ISO 7933. 2004. ED 2. Ergonomics of the Thermal Environment—Analytical Determination and Interpretation of Heat Stress Using Calculation of the Predicted Heat Strain. Geneva: International Organization for Standardization.

ISO 8996. 2004. ED 2. Ergonomics of the Thermal Environment—Determination of Metabolic Rate. Geneva: International Organization for Standardization.

ISO 9886. 2004. ED 2. Evaluation of Thermal Strain by Physiological Measurements. Geneva: International Organization for Standardization.

ISO 7730. 2005. ED 3. Ergonomics of the Thermal Environment—Analytical Determination and Interpretation of Thermal Comfort Using Calculation of the PMV and PPD Indices and Local Thermal Comfort Criteria. Geneva: International Organization for Standardization.

ISO 11079. 2007. ED 1. Ergonomics of the Thermal Environment—Determination and Interpretation of Cold Stress When Using Required Clothing Insulation (IREQ) and Local Cooling Effects. Geneva: International Organization for Standardization.

ISO 9920. 2007. ED 2. Estimation of Thermal Insulation and Water Vapour Resistance of a Clothing Ensemble (See also Amended Version 2009). Geneva: International Organization for Standardization.

ISO 15743. 2008. Ergonomics of the Thermal Environment—Cold Workplaces—Risk Assessment and Management. Geneva: International Standards Organization.

ISO 28802. 2012. Ergonomics of the Physical Environment—Assessment of Environments by Means of an Environmental Survey Involving Physical Measurements of the Environment and Subjective Responses of People. Geneva: International Standards Organization.

Kenney, L. W., J. H. Wilmore, and D. L. Costill. 2012. *Physiology in Sport and Exercise*. Fifth ed. Champaign, IL: Human Kinetics.

Leithead, C. S., and A. R. Lind. 1964. *Heat Stress and Heat Disorders*. London: Cassell.

Lewis, T. 1930. Observations upon the reactions of the vessels of the human skin to cold. *Heart*. 15:177–208.

Li, B., and D. Lim. 2013. Occupant behaviour and building performance. In *Design and Management of Sustainable Built Environments*, Yao, R. (Ed.). London: Springer. pp. 279–304.

Malchaire, J., A. Piette, B. Kampmann, G. Havenith, P. Mehnert, I. Holmér, H. Gebhardt, B. Griefahn, G. Alfano, and K. C. Parsons. 2000. Development and validation of the predictive heat strain (PHS) model. In *Environmental Ergonomics IX*, Werner, J., and M. Hexamer (Eds.). Aachen: Shaker Verlag. pp. 133–6.

Nicol, J. F., and M. A. Humphreys. 1972. Thermal comfort as part of a self regulating system. *Thermal Comfort and Moderate Heat Stress: Proc CIB Comm*. Watford, BRE, UK. 263–74.

Olesen, B. W., and F. N. Dukes-Dubos. 1988. International standards for assessing the effect of clothing on heat tolerance and comfort. In *Performance of Protective Clothing*, Mansdorf, S. Z., R. Sager, and A. P. Nielson (Eds.). Philadelphia, PA: American Society for Testing Materials. pp. 17–30.

Parsons, K. C. 1992. The thermal audit. In *Contemporary Ergonomics*, Lovesey, E. J. (Ed.). London: Taylor & Francis. pp. 85–90.

Parsons, K. C. 2005. An adaptive thermal comfort method for people with physical disabilities. *Environmental Ergonomics XI: Proc 11th Intern Conf. Ystad, Sweden. May*. 186–9.

Parsons, K. C. 2009. Maintaining health, comfort and productivity in heat waves. *Global Health Action* 2:39–45.

Parsons, K. C. 2014. *Human Thermal Environments*. Third ed. New York: Taylor & Francis.

Parsons, K. C. 2015a. The environmental ergonomics survey. In *Evaluation of Human Work*, Wilson, J., and S. Sharples (Eds.). Fourth ed. New York: Taylor & Francis.

Parsons, K. C. 2015b. Ergonomics assessment of thermal environments. In *Evaluation of Human Work*, Wilson, J. and S. Sharples (Eds.). Fourth ed. New York: Taylor & Francis.

Vogt, J. J., V. Candas, J. P. Libert, and F. Daull. 1981. Required sweat rate as an index of thermal strain in industry. In *Bioengineering, Thermal Physiology and Comfort*, Cena, K. and J. A. Clark (Eds.). Amsterdam: Elsevier. pp. 99–110.

World Health Organization. 2012. *Public Health Advice on Preventing Health Effects of Heat: New and Updated Information for Different Audiences*. Copenhagen: World Health Organization.

Yaglou, C. P., and D. Minard. 1957. Control of heat casualties at military training centers. *Am Med Assoc Arch Indust Health*. 16:302–16.

Yao, R., B. Li, and J. Lui. 2009. A theoretical adaptive model of thermal comfort: Adaptive predicted mean vote (aPMV). *Built Environ*. 44:2089–96.

3 Ventilation, Indoor Air Quality, Health, and Productivity

Pawel Wargocki

CONTENTS

3.1 INTRODUCTION

Buildings should provide shelter and appropriate conditions for working, learning, leisure, and comfortable living. A built environment should be safe, with no health hazards for its users due to poor design and construction, and/or inadequate operation and maintenance and performance. Negligence and failing to take any actions required to comply with appropriately set quality criteria for the indoor environment can lead to serious indoor environmental problems that result in substantial financial costs and other undesirable consequences for health and comfort. It is the purpose of this chapter to describe how ventilation and indoor air quality (IAQ) in different types of buildings, homes, schools, and offices affects these aspects of human life.

What is IAQ? IAQ depends on the composition of air and the content of airborne contaminants, which are gases, aerosols, particles, and microorganisms. When exposures to these contaminants do not pose risk for adverse health effects and do not result in (olfactory) discomfort, the air quality is rated as high. In contrary, when the risks for health and discomfort are elevated and/or apparent the air quality is judged as poor. There are different definitions of IAQ but the most obvious one relates it to the human responses and the effects that it may have on humans. In other words, air quality can be defined when human requirements are met and these requirements can be differently defined depending on the context and the need. These may refer to the requirements which are needed for relaxing, rest, and recovery (e.g., in homes), for working conditions that promote creativity and high work performance (e.g., in offices), and conditions that are not distracting from proper learning and the teaching process (e.g., in schools). The underlying requirement common for all these different environments is that the air quality should not be detrimental to human health. The earlier human-centric definition of indoor air is somewhat different from the definition taking its roots in the exposure to pollutants that stresses what type of pollutants can be present in indoor environments and at which concentrations. The difference is entailed in the overall message: The human-centric definition motivates and underlines the purpose of the built environment, which should support and safeguard human needs. The definition referring to exposures in this context belongs to the general framework for achieving high IAQ.

IAQ should meet the World Health Organization (WHO)'s mandate (2000) that everyone should have the right to enjoy IAQ that does not endanger his or her health. This mandate should be recognized and treated analogous to the quality of water that we drink. About 11,000 L of air flow through our lungs daily; however, we consume three to four orders of magnitude less food and liquids. For example, an average adult daily consumes no more than 2–5 L of liquids. This is why much effort must be placed on achieving that the air that we breathe does not create danger and hazard to our health and life, similarly as has been done to ensure the quality of drinking water, which has considerably reduced the prevalence of diseases related to contaminated water and the consequent death toll.

Another reason why the air breathed indoors is important is that people spend on average from 80% to 90% and even 95% of their time indoors during their entire lifetime (Klepeis et al. 2001). Employed people spend anywhere from 20 to 60 hours per week in offices or factories of various kinds, and it can be estimated that children spend about one-third of their day in primary or secondary school. Still, the major sites of exposure to indoor air are homes. During different periods of our life, exposures in homes can be from 60% to 95% of our total daily exposure. The latter figure specifically applies to infants, young children, and senior citizens, whereas the former is more or less typical for our total lifetime exposure. It is striking that about 30% of our lifetime-weighted exposures is spent when we sleep; with the current life expectancy in the more developed world being 75 years, sleep constitutes roughly 25 years of our life.

The air breathed indoors may contain dust, fumes, microbes, or aerosolized toxins, and according to different sources, the pollution indoors can often be 2–100 times higher than outdoors. The origin of the pollutants present indoors is both the ambient (outdoor) air and the sources present indoors. Thus, humans staying indoors also inhale outdoor pollutants transported indoors via cracks in construction, openings, or special systems for securing proper ventilation and conditioning of the air indoors. Exposures to outdoor pollutants indoors have significantly longer durations than exposures outdoors; humans staying outdoors on the other hand can be briefly exposed to sparks of pollutants at considerably elevated concentrations.

Outdoor pollutants mainly include particles generated outdoors through traffic and combustion processes, ozone, and organic compounds including polycyclic aromatic hydrocarbons, inorganic compounds, and metals, as well as pollens. Many of these pollutants do not have origins indoors and thus cannot be avoided by simple means of dealing with indoor exposures through, e.g., increased ventilation or control of indoor sources of pollution. Their levels need to be regulated outdoors.

Indoor pollutants originate from humans and their activities, such as cooking or bathing and laundering, as well as cleaning with products containing a variety of chemicals, from tobacco smoking or combustion of fossil fuels, all materials that are used for construction purposes, decoration, finishings and furniture, electronic equipment as well as the equipment and installations that support the environment in buildings, e.g., ventilation and air conditioning systems or systems used to purify and clean air indoors. Sometimes human activities create conditions that can further promote the release and the exacerbation of pollutants. The best example is moisture, which can lead to mold growth or exacerbation of house dust mites, when too high, and spread of infectious diseases, when too low.

Indoor pollutants comprise organic species, inorganic species, particles, fibers, allergens, radioactive gases, or species of microbiological origin such as mold and fungi; they can cause allergies, irritation of mucous membranes, sick building syndrome (SBS) symptoms, toxic reactions, infections, inflammation, cancer, and mutagenic effects. For example, organic pollutants are the major sensory pollutants in indoor air. A typical mixture of organic pollutants can contain up to 6000 compounds, of which at least 500 are emitted by humans and at least 500 are emitted by building materials and furnishing. Thus, it is very difficult to name all the pollutants of concern, and even though they have been identified, the challenge would be on how to measure them. Pollutants known to be particularly hazardous can be controlled via filtration and air cleaning, as well as control of sources emitting these pollutants, but the cocktail of pollutants present indoors is often unknown, and the identification of the air composition is difficult, time consuming, and very expensive. In the case of industrial environments, the hazardous effects are handled by defining the maximum acceptable concentrations or the threshold limit values (TLVs) for individual substances. However, in the case of nonindustrial environments, the situation is more complicated as the pollutants are at concentrations much lower than TLV levels.

No simple index was agreed upon for defining the level of IAQ. Therefore, IAQ is often closely associated with ventilation, which is assumed to reduce exposures to pollutants indoors. It is surmised that when ventilation is high, the air quality should also be high, and when it is low, then the IAQ is presumably poor, assuming at the same time that the air used for ventilation is of a high quality (clean), too. This, however, will not apply in cases when ambient air pollution is high and no efficient methods are used to remove them before this air is used for ventilation. Because ventilation is easier to measure, quantify, and interpret than the hundreds or thousands of pollutants present in indoor air, ventilation is often measured and used as a proxy (surrogate) for IAQ.

Sometimes IAQ is also approximated by defining the percentage of people who find it unacceptable (i.e., percentage dissatisfied). If there are few dissatisfied, then the air quality is high, and if there are many dissatisfied, then the air quality is low. Often the relation between percentage of dissatisfied and ventilation is used to express the underlying effects of improved IAQ, and this is reflected in some ventilation standards that have determined ventilation requirements by defining the percentage of dissatisfied (e.g., EN15251 [2007]; American Society of Heating, Refrigerating, and Air-Conditioning Engineers [ASHRAE] [2013]). It should be recognized that there are large individual differences between the occupants of buildings. Some are easy to satisfy and some are are especially difficult to satisfy or are more sensitive than the general population due to inherited properties and/or chronic diseases or psychological reasons.

Studies by Berglund and Cain (1989), Fang et al. (1998a,b), and Toftum et al. (1998) show that the perception of air quality is also influenced by the humidity and the temperature of the inhaled air, even when the chemical composition of the air is constant and the thermal sensation of the entire body is kept neutral. Keeping the air dry and cool reduces the percentage of dissatisfied persons with air quality and causes the air to be perceived as fresh and pleasant. The effect is probably due to the stimulation of the thermal sense as a result of convective and evaporative coolings of the respiratory tract, if only the temperature is different from the mucosal temperature, which is around 30–32°C. The impact of temperature and the relative humidity on sensory perceptions that cause the estimated percentage of dissatisfied with air quality is not only caused by exposures to pollutants, but also by the thermal properties of air. This metric thus cannot be purely associated with loads of pollutants indoors. This may be considered as a factor limiting to some extent a universal application of this metric for dimensioning ventilation requirements. Impartial observers are used to evaluating the IAQ and the percentage of dissatisfied, as this metric cannot be directly measured with an instrument yet, although there have been attempts to make it happen (Wenger et al. 1993; Müller et al. 2007). There is also no relationship between the percentage of dissatisfied and acute health symptoms, although studies suggest that when the percentage of dissatisfied is reduced, the intensity of symptoms is also reduced (e.g., Wargocki et al. 1999, 2000).

Indoor air pollutants can often cause unspecific effects. Multitudes of biological mechanisms are involved at the same time in response to multiple exposures indoors, and only a few objective measurements are available. The most frequent effects include acute physiological or sensory reactions, psychological reactions, and subacute changes in sensitivity to environmental exposures. The term *SBS* (WHO 1983) is used to describe cases in which building occupants experience acute symptoms and discomfort that are apparently linked to the time they spend in the building, but for which no specific illness or cause can be attributed. Many different symptoms have been associated with SBS, including respiratory complaints, irritation, and fatigue. Sensory perception of odors and mucous irritation lead to perceptions of poor air quality and possible risks thereof and consequently to stress and some behavioral responses (window opening or even leaving the building or changing the workplace). The relevant indoor air pollutants that can cause these effects are those which alone or in combination can stimulate the senses or cause tissue changes.

This chapter will not specifically look at different pollutants present indoors or their origins and the methods for amending the elevated exposures. Instead, it will discuss the consequences for health, comfort, and performance of the exposures indoors to a variety of pollutants. Health will be understood very broadly as mental and social well-being, not merely the absence of disease or infirmity, similarly to how it is defined by WHO (1948). Comfort will express the satisfaction with the environment. In the context of IAQ, it will relate to the sensory response due to the activation of the olfactory sense that is sensitive to around half a million odors and the cranial common chemical sense, which is stimulated by the hundreds of thousands of compounds that cause irritation or pungency response. Performance will be related to the ability of an individual to perform different mentally and physically demanding tasks. Often, in this context, the term *productivity* is used instead of performance, but in this chapter, the former term will be used because *productivity* describes a wider economic term that combines a volume measure of the output in relation to the input and is thus less relevant, as it will also depend on many external factors not directly related to IAQ.

Health, comfort, and performance can be influenced by different factors including physiological, behavioral, and psychological factors, while performance can additionally be influenced by personal and organizational, as well as administrative and managerial variables. These factors are not discussed in the context of the present chapter, which is solely addressing the impact of exposures indoors and IAQ.

IAQ issues received significant interest in the past, especially in the nineteenth century. A renaissance of the interest in IAQ was observed after energy crisis in the 1970s, when the lack of oil caused significant reductions in energy use in buildings and nearly epidemic growth in building-related health symptoms due to exposures to air pollutants indoors beginning with excessive concentrations of formaldehyde and later focusing on a wide variety of air pollutants indoors potentially hazardous to human health. Subsequently, other factors that could contribute to poor air quality were examined as well. It is interesting to notice that similar events are likely to occur in the twenty-first century, when weatherization programs and tightening of buildings, use of excessive insulation, and need to reduce energy, as well as many new chemicals present on the market different than those measured in the past (Weschler 2009), can again create an environmental problem indoors, elevating indoors, exposure to levels at which they may have negative consequences for health and well-being.

3.2 SHORT HISTORICAL PERSPECTIVE ON VENTILATION AND INDOOR AIR QUALITY

Because ventilation is often used as a proxy for IAQ, as indicated earlier, it is instructive to briefly review how the need for ventilation was perceived through history as well as to examine different theories that were put forward to justify the need for ventilation. It should be mentioned at the onset of this short historical review that ventilation has always been treated as a remedy for ill health, i.e., to remove and/or dilute polluted indoor air. Only in the last 100 years has the premise for ventilation been to ascertain comfort, mainly the reduction of foul and/or unpleasant odors (sometimes also avoiding thermal discomfort), assuming that this will also secure health.

Ventilation was recognized as an important factor for providing IAQ by the ancient Egyptians, who required the use of ventilation for stone carvers to avoid exposures to particles and dust generated during this process (Janssen 1999; Addington 2000). Hippocrates (460–377 BC) indicated the adverse effects of polluted air in crowded cities and mines. During Roman times (1 BC), Sergius Orata developed hypocausts, underfloor heating systems, to uniformly distribute heat in a house, which most importantly allowed avoiding combustion indoors and subsequently reduced the risk of harmful exposures. In the case of an open fire indoors, the minimum ratio of window-to-floor area was set and parchment above the window was required during these times to assure infiltration.

During Venetian times, roof windows were developed, while Leonardo da Vinci recognized that no animal could live in an atmosphere where a flame does not burn and that dust can cause damage to health, thus implying a need for ventilation. This was later confirmed by studies of Mayow and Lavoisier. In the seventeenth century, Wargentin expressed the common knowledge of this time that expired air was unfit for breathing until refreshed. In the same century, Gauger, quoting Cardinal Melchior de Polignac, remarked that it is not warmth but inequality of temperature and want for ventilation that cause maladies.

In 1756, Holwell described an accident in the Black Hole of Calcutta, a small dungeon where prisoners and soldiers were kept overnight in poor conditions, and 125 out of 146 died due to suffocation. It was observed during the Crimean War (1853–1855) that there was faster spread of diseases among wounded soldiers in poorly ventilated hospitals. Higher morbidity and mortality were observed in the overcrowded rooms, especially when they were poorly ventilated. The importance of ventilation in small room volumes to avoid the death of people was also informed by Baer in 1882. A few years later, Reid expressed the view that between mental anxiety and defective nutriment, defective ventilation should be considered as one of the evil enemies of the human race. (In this context, it should be noted that even now WHO recognizes poor air quality as one of the most important public health challenges.) A similar view was expressed by Griscom in 1850, who acknowledged that deficient ventilation is fatal, as it leads to the spread of tuberculosis and other diseases. An effective treatment of tuberculosis using country fresh air was then achieved by Trudeau, who opened the first Airdonack Cottage Sanatorium in 1873.

In the early twentieth century, Winslow and Palmer suggested that ill-ventilated rooms do not create large discomfort but result in the loss of appetite. Later Winslow and Herrington observed a similar result of loss of appetite for food when they heated the dust from the vacuum cleaner.

3.2.1 Theories of Ventilation

Different theories have been put up through the last two to three centuries to explain the effects associated with a lack of ventilation (i.e., poor IAQ) (Janssen 1999), starting with the Miasma theory that prevailed for a long time until the eighteenth to the nineteenth century, attributing cholera, chlamydia, and Black Death to a noxious form of "bad air." It was later displaced by the germ theory of disease after the discovery of germs in the nineteenth century. In the early seventeenth century, breathing was attributed to a cool heart. In the same century, Mayow attributed the effects observed from igneo-aerial particles that cause the demise of animals.

One century later, in 1775, Lavoisier discovered two gases in air and attributed the effects of igneo-aerial particles to carbon dioxide (CO_2) and air stuffiness. The theory that CO_2 is a dominant cause of physiological effects of bad air remained dominant for nearly 100 years, although it was acknowledged that other factors could also contribute to the effects observed. The theory prevailed until Pettenkofer, who, in the 1800s, indicated that it is neither the deficiency of oxygen (O_2) nor the excess of CO_2 but the presence or lack of biological pollutants (from humans) that are responsible for the vitiation of air. He said that "the corruption of the air is not caused solely by the carbon dioxide content, we simply use this as a benchmark from which we can then also estimate a higher or lower content of other (pollutant) substances." In 1872, Pettenkofer and Saeltzer suggested CO_2 to be the surrogate for vitiated air, a "stick" for deleterious substances of unknown origin.

Later, in 1887–1889, Brown-Sequard and d'Arsonval attributed anthropotoxin (the toxic effluvia—toxic substances in exhaled air) to be responsible for the negative effects reported through history, when ventilation was lacking. Organic matter from lungs and skins had also been proposed as poisonous by many others prior to the anthropotoxin theory. The theory was rejected by many experiments later performed by Haldane and Smith in 1892–1893, Billings in 1895, and Hill in 1913. They could not confirm that the condensate of expired air could kill the animals, which was claimed by Brown-Sequard. The anthropotoxin theory was then superseded by the idea put up by Billings in 1893 suggesting that the purpose of ventilation is to dilute contagions emitted by humans and thus to reduce the spread of infectious diseases.

The theory of reducing contagions prevailed until the large body of research in the early twentieth century on comfort. Among others, Billings, Flugge, Benedict, and Millner and Hill showed a the lack of ventilation causes discomfort exemplified by unpleasant body odors and temperature, while no negative physiological effects could be observed even at CO_2 levels as high as 1–1.5% (10,000 to 15,000 ppm). A lack of ventilation was consequently associated with thermal effects and discomfort. Since the studies of Lemberg and Yaglou in the 1930s, ventilation was required to merely keep body odors at an acceptable level, defined to be at moderate level.

In the 1980–1990s, it was also acknowledged among others by Fanger and his colleagues that in addition to the body odors emitted by humans, other sources of pollution indoors determine ventilation requirements, but the general principle of providing ventilation to reduce discomfort by achieving acceptable air quality as perceived by humans was not changed. Ventilation has become, as mentioned earlier, merely a question of comfort, not health.

3.2.2 Ventilation Guidelines and Requirements

The actual ventilation requirements have varied throughout history, and also as a result of changes in theories of ventilation.

In 1836, Tredgold suggested the minimum ventilation rate in mines, which should satisfy the needs of a miner. It was set at 1.7 L/s per person, of which 0.2 L/s was for purging the CO_2 from lungs, 1.4 L/s was for removing the moisture produced by the body, and 0.1 L/s was for keeping the candle burning; thus, 1.6 L/s was basically defined to control and remove the body effluents.

In one of the first textbooks on ventilation and heating published in 1893, Billings provided the minimum requirements for ventilation. He was very much concerned about the spread of infectious diseases, as mentioned earlier, particularly tuberculosis, and proposed the minimum rate at 30 cfm/person (~14 L/s per person), while the recommended ventilation rate was set as high as 60 cfm/person (~28.5 L/s per person). He also calculated that 50 cfm/person (~23.5 L/s per person) would keep CO_2 at 0.05% (550 ppm); thus, the exhaled CO_2 would be kept at 0.02% (200 ppm) above outdoor levels (at his time, the average CO_2 levels outdoors were about 300 ppm and less). The ventilation rates proposed by any subsequent ventilation standards or guidelines have never been as high as these recommended by Billings.

One of the very first ventilation guidelines was proposed by the Chicago Commission on Ventilation in 1914; these guidelines were later reconfirmed by the studies and the conclusions of the New York State Commission on Ventilation (1923). Both documents attributed ill health to overheating rather than ventilation by stating: "Had the temperature been controlled well, the ventilation requirements could be reduced." Temperatures of 15–19°C in window-ventilated rooms were observed to cause the lowest prevalence of respiratory illnesses. Consequently, the guidelines recommended 20°C with proper control of relative humidity for living rooms to reduce the spread of infectious diseases. CO_2 was not recognized as a harmful agent, when encountered in working practice; no harmful effects could be designated to the expired air. Relative humidity was recognized as the most important factor regarding ventilation requirements for health. Recirculation was not acceptable, if 100% of air was recirculated. Window-ventilated rooms with natural drafts were the most preferred method for ventilation. These recommendations were made before the widespread use of air conditioning in nonindustrial buildings.

In the time after the recommendation proposed by Billings, the ventilation requirements considerably varied. They were as low as 2.5 L/s per person in the 1981 version of ASHRAE Ventilation Standard, through 4 L/s per person in the Nordic guidelines published the very same year (Sundell 1982), 5 L/s per person in the 1946 American Standard Association Code, reaching finally to 7.5–10 L/s per person, which is approximately the standard of today (Janssen 1999). These rates, to a large extent, reflect studies which determined ventilation requirements for acceptable IAQ to avoid discomfort and odors in the presence of emission from humans (human bioeffluents) as well as in the presence of human bioeffluents and the emissions from building materials and furnishings; for some time, they also acknowledged the effective dilution of odors produced by tobacco smoking. These rates harmonize well with a widely accepted CO_2 concentration of 0.1% (1000 ppm) proposed by Pettenkofer to be an indicator of adequately ventilated rooms. It is worth mentioning that Pettenkofer suggested 0.07% (700 ppm) for bedrooms, which is lower than for other spaces, thus implying that bedrooms require lower exposure levels and higher ventilation with outdoor air.

3.3   AIR QUALITY GUIDELINES AND REQUIREMENTS

It is challenging, and at the same daring, to propose pollutants that need to be regulated for achieving IAQ that secures no negative effects on health, lack of sensory discomfort, and high cognitive performance. The air quality guidelines for avoiding ill health have been proposed by WHO (1987, 2000, and 2005, 2009 and 2010), and they can be considered as a prototypical example for setting the guidelines related to other outcomes. The primary aim of the WHO guidelines is to provide a uniform basis for the protection of public health from the adverse effects of exposure to air pollution, and to eliminate or reduce to a minimum exposure those pollutants that are known or are likely to be hazardous. The guidelines are based on the scientific knowledge available at the time of their development.

The WHO Guidelines for IAQ: Selected Pollutants (WHO 2010) recommends targets for nine air pollutants: carbon monoxide, nitrogen dioxide, benzene, trichloroethylene, tetrachloroethylene, formaldehyde, naphthalene, polycyclic aromatic hydrocarbons (PAHs), and radon. Moreover, the levels of exposures indoors should meet the requirements applicable for pollutants in ambient air, i.e., fine particles (PM2.5), coarse particles (PM10), sulfur dioxide, ozone, styrene, and toluene (WHO 2005). WHO defines the allowable levels of these pollutants, indicating that for radon, benzene, trichloroethylene and PAHs, no safe levels can be established. Table 3.1 shows the guideline values recommended by WHO. Readers are warned that these values might be different from the guidelines recommended by the cognizant authorities relevant to their region and therefore they are requested to consult them. The list presented in Table 3.1 does not include either CO_2 or relative humidity, as neither of them can be considered to be a contaminant. However, both can be and are good indicators of certain processes or circumstances that can potentially create risks for health.

In addition to the guidelines for gaseous pollutants, WHO published guidelines on dampness and mold (WHO 2009). These guidelines concluded that

> (…) persistent dampness and microbial growth on interior surfaces and in building structures should be avoided or minimized, as they may lead to adverse health effects. As the relationships between dampness, microbial exposure and health effects cannot be quantified precisely, no quantitative, health-based guideline values or thresholds can be recommended for acceptable levels of contamination by microorganisms. Instead, it is recommended that dampness and mold-related problems be prevented. When they occur, they should be remediated because they increase the risk of hazardous exposure to microbes and chemicals.

TABLE 3.1

List of Indoor Air Pollutants Defined by WHO to Have Toxic Effects on Humans and Their Recommended Maximum Exposure Levels

	IAQ Guidelines	Ambient Air Guidelines	
Pollutant	IAQ WHO (2010)	AQ WHO (2000)	AQ WHO (2005)
CO (mg/m^3)	100 (15 min) 60 (30 min) 30 (1 h) 10 (8 h) 7 (24 h)	100 (15 min) 60 (30 min) 30 (1 h) 10 (8 h)	
NO$_2$ (µg/m^3)	200 (1 h) 40 (1 year)	200 (1 h) 40 (1 year)	200 (1 h) 40 (1 year)
SO$_2$ (µg/m^3)		500 (10 min) 125 (24 h)	500 (10 min) 20 (24 h)
PM10 (µg/m^3)			50 (24 h) 20 (1 year)
PM2.5 (µg/m^3)			25 (24 h) 10 (1 year)
Ozone (µg/m^3)			100 (8 h)
Benzene (µg/m^3)	No safe level	UR 6 × 10^{-6}	
Trichloroethylene	No safe level	UR 4.3 × 10^{-7}	
Tetrachloroethylene (µg/m^3)	250 (1 year)	250 (1 year) 8000 (30 min)	
Toluene (µg/m^3)		260 (1 week) 1000 (30 min)	
Styrene (µg/m^3)		260 (1 week) 70 (30 min)	
Xylenes (µg/m^3)	Insufficient evidence	Insufficient evidence	Insufficient evidence
Napthalene (µg/m^3)	10 (1 year)		
Formaldehyde (µg/m^3)	100 (30 min)	100 (30 min)	
PAHs	No safe level	8.7 × 10^{-5} per ng/m^3 of B[a]P	

Source: WHO, *The Right to Healthy Indoor Air*, WHO Regional Office for Europe, Copenhagen, 2000; WHO, *WHO Air Quality Guidelines: Global Update 2005*, WHO Regional Office for Europe, Copenhagen, 2005; WHO, *Guidelines for Indoor Air Quality: Selected Pollutants*, WHO Regional Office for Europe, Copenhagen 2010. With permission.

Note: B[a]P, benzo(a)pyrene; UR, unit risk estimated for an air pollutant is defined as "the additional lifetime cancer risk occurring in a hypothetical population in which all individuals are exposed continously from birth throughout their lifetimes to a concentration of 1 µg/m^3 of the agent in the air they breathe."

3.4 VENTILATION, INDOOR AIR QUALITY, PERFORMANCE, AND HEALTH IN OFFICES

3.4.1 HEALTH EFFECTS

The fraction of the incidence/prevalence of reports of discomfort and symptoms that can be related to IAQ in offices is not exactly known. However, in buildings without specific complaints of poor IAQ, the prevalence among occupants is often close to zero and normally below 30%, while in affected buildings, the prevalence often ranges much higher. Several studies and reviews of these studies have been published to date (e.g., Mendell 1993; Seppänen et al. 1999; Seppänen and Fisk

2002; Wargocki et al. 2002b; Li et al. 2007; Sundell et al. 2011; Carrer et al. 2015) to show the relationship between ventilation and air quality and health, and to understand and elucidate which factors could be responsible for the high prevalence of health effects in offices. The interest was on the unspecified weak acute health symptoms (SBS symptoms), being the most common complaint among occupants of the offices, and also on the difference in the prevalence of these symptoms between air-conditioned, mechanically ventilated, and naturally ventilated office buildings, as well as offices with different layouts.

One of the first reports was by Finnegan et al. (1984), who studied nine office buildings in the United Kingdom, and found that the symptoms were more prevalent for those workers in air-conditioned offices. Around the same time, Hedge (1984) also found that eye, nose, and throat symptoms were more prevalent among workers in the air-conditioned offices than in naturally ventilated offices, and additionally that headaches varied by office layout. A comparative study of adjoining air-conditioned and naturally ventilated office buildings (Robertson et al. 1985) showed that SBS symptoms were more prevalent among workers in the air-conditioned building, but the source of this difference could not be pinpointed from the variety of indoor environmental measures that were taken. Hedge et al. (1989a) found higher levels of total volatile organic compounds (TVOCs) in the air-conditioned offices and an association between formaldehyde levels and health reports. A U.K. survey of 47 office buildings and 4473 workers found a significantly higher prevalence of SBS symptoms in the air-conditioned offices compared to those that were simply mechanically ventilated or naturally ventilated (Burge et al. 1987; Wilson and Hedge 1987; Hedge et al. 1989b). Similar findings have been reported by Harrison et al. (1987) for 2587 workers in 27 UK office buildings; Harrison et al. (1990) for 13 U.K. office buildings; for 4369 workers in 14 Danish town halls and 14 nearby office buildings (Skov et al. 1990), and 12 Northern Californian office buildings (Fisk et al. 1993; Mendell et al. 1996).

Fisk et al. (2009) attempted to create a quantitative relationship between SBS and the ventilation rate. They summarized the existing data from studies reporting the ventilation rate and the prevalence and/or intensity of SBS symptoms. It was achieved by integrating slopes depicting the degree of change in symptoms, as a consequence of change in the ventilation rate. The quantitative relationship was established and showed that when the ventilation rate is reduced from 10 to 5 L/s per person, the relative symptom prevalence increases by 12–22% (23% on average), and when it is increased from 10 to 25 L/s per person, the prevalence decreases from 15% to 42% (on average by 29%). At rates higher than 25 L/s per person, the reliability of the established relationship is low.

The relationship of Fisk et al. (2009) confirms to some extent the recommendations of other reviews showing that ventilation rates of up to 25 L/s per person will provide the benefit in reducing symptoms (Wargocki et al. 2002b; Sundell et al. 2011). Their results also matched the observations made in one of the first reviews of the archival literature on SBS by Mendell (1993), who indicated that among the factors contributing to health symptoms among office workers are insufficient ventilation, presence of air-conditioning, presence of carpets, more workers in a space, and video display terminal (VDT) use (VDT is a nonflat computer monitor).

A common reason for the high prevalence of symptoms are improperly designed, operated, and maintained ventilation systems. As already indicated by the first reports of SBS symptoms, which were mentioned earlier, many studies showed that symptom prevalence is low in office buildings where no mechanical ventilation systems are installed compared with mechanically ventilated buildings or with buildings where in addition there is equipment for complete conditioning of the supplied air (air cooling and humidity control) (see, e.g., the review of Seppänen and Fisk [2002]). Poor maintenance of systems supplying and conditioning air, and especially pollution from filters that are dirty and loaded with particulate matter allowing gaseous pollutants to adsorb on their surface and transform into more hazardous pollutants (Bekö et al. 2006, 2007), can be one of the reasons for the observed differences in the prevalence of symptoms in office buildings with and without mechanical ventilation systems as well. In the Base project, which investigated the air quality in 100 office buildings in the United States, the presence of loaded filters was significantly associated with elevated prevalence of SBS symptoms (Buchanan et al. 2008). Other reasons can be the presence of moisture and dirt on other parts of the system (Sieber et al. 1996; Mendel et al. 2003). Yet another explanation could be that office buildings without a mechanical ventilation system had on average higher ventilation rates, which promotes more efficient removal and dilution of pollutants. Many of the buildings in the studies showing the difference between office buildings with and without mechanical ventilation system were actually located in moderate and cold climates, which create a wonderful opportunity to use natural forces due to differences between ambient (outdoor) temperatures and indoor temperatures even during the summer. Very little information is available on whether office buildings without mechanical ventilation would perform better in tropical and subtropical climates. However, there is a good deal of information that air conditioning without sufficient provisions of outdoor air will cause an elevated prevalence of symptoms (Wong and Huang 2004; Sekhar and Goh 2011). In these buildings, reduced outdoor air supply rates are the consequence of an attempt to save energy, considering that cooling and moisture removal are energy-demanding processes and thus consume a lot of energy especially in climates where air conditioning is indispensable.

In the case of buildings with air-conditioning systems, the main reasons for the elevated prevalence of symptoms can be the presence of water in the system, besides the mentioned general poor condition of the system, dirtiness and dust accumulated on ducts and filters, and reduction of outdoor air supply rate to save the energy. The presence of water promotes microbial growth and other potentially unwanted, dangerous, and harmful processes (Sieber et al. 1996). The presence of water in the system does create a risk for SBS symptoms, but even in this case, we have very little information from studies in tropical and subtropical climates about how large this risk actually is. In these climates, air conditioning is practically indispensable to perform modern work indoors. In a study of Sekhar et al. (2003), five air-conditioned office buildings were surveyed in Singapore in the tropical climate. However, this study did not reveal higher dissatisfaction levels with air quality or higher prevalence of SBS symptoms than those observed in the similar study in Europe in air-conditioned and non-air-conditioned buildings (Bluyssen et al. 1996). For example, the building symptom index integrating all symptoms was about two (meaning two symptoms per building) in Singapore and on average two in Europe. Studies in Brazil showed that the age of the ventilation system was a risk factor for upper respiratory symptoms (Graudenz et al. 2002), and that cleaning the ducting system and replacing an air-conditioning system with a new one reduced building-related respiratory symptoms, nasal-ocular symptoms, and persistent coughs (Graudenz et al. 2004, 2005).

There are no relationships that have been agreed upon between the SBS symptoms among office workers and pollutants indoors. The total concentration of volatile organic compounds (TVOCs) was proposed as an index integrating the impact of many volatile organic compounds (VOCs) present indoors at very low concentrations. However, the consensus at present is that it is not a valid proxy for health problems, although in some cases it may provide some indication of the potential problem. The reason is that no clear, systematic, and consistent association between TVOC and its levels has been found in the published literature, as indicated by Andersson et al. (1997), who carefully reviewed and judged the quality of the literature describing TVOC–health relationships and concluded that, although the air including VOCs will affect health, the literature is inconclusive with regard to the relationship between TVOC and health. No specific guidelines regarding TVOC levels and health could be established, suggesting that TVOC may not be an appropriate risk index for health effects associated with exposures in buildings in general and specifically in offices. One of the reasons for the lack of TVOC–health relationships could be the various definitions of TVOC and the different instrumentation used to measure TVOC using different principles and different reference.

Recent ENVIE (Oliveira Fernandes et al. 2009) and IAIAQ (Jantunen et al. 2011) projects do however acknowledge that organic compounds indoors are important contributors to negative health effects of occupants of buildings expressed as disability-adjusted life years. They show that as many as about 2.2 million healthy life years can be lost each year in Europe due to exposure to pollutants in buildings in Europe, of which 0.517 million can be attributed to VOCs causing irritation and odor.

In addition to the fairly consistent evidence that poor air quality in offices results in the elevated prevalence and/or intensity of acute health symptoms (SBS symptoms), studies also show that poor air quality and particularly too low ventilation rates can promote the spread of infectious diseases. Li et al. (2007) reviewed the literature on the role of ventilation in the airborne transmission of infectious agents. They concluded that there is strong evidence that ventilation and air movement in buildings are associated with the spread and the transmission of infectious diseases such as measles, tuberculosis, chickenpox, influenza, smallpox, and severe acute respiratory syndrome. This postulation agrees well with the historical evidence summarized earlier in this chapter. Although the strong link between the ventilation and the risk of infectious diseases is highly likely, the information published in the literature is somewhat weak and inconsistent. Myatt et al. (2004) indirectly showed that such risk exists. They found the probability of detecting airborne rhinovirus in filters and the weekly average CO_2 concentration. When weekly averages of CO_2 concentration above background were higher than 0.01% (100 ppm), the risk significantly increased. This level of CO_2 corresponds to a very high ventilation rate, in the range of 40–50 L/s per person; the level is also similar to the CO_2 of 200 ppm above outdoor level as recommended by Billings and mentioned earlier in this chapter. Infectious diseases can also be associated with increased absence rates. However, in another study, Myatt et al. (2002) could not observe increased absence rates among office workers when the interventions in two office buildings changed the CO_2 levels above outdoor from about 100 ppm to 200 ppm, which was estimated by the authors to correspond to ventilation rates as high as 40–45 L/s per person. The reason for the negative finding could be the too small size of the intervention group considering that Milton et al. (2000) did see the reduced short-term sick leave in many office buildings when outdoor air supply rates were increased from 12 to 24 L/s per person. Laboratory studies with personal ventilation (delivering the air directly to the breathing zone) and with coughing mannequins do provide further evidence of the role of good ventilation in reducing the risks of transmission of infectious diseases (Pantelic et al. 2009). Seppänen and Fisk (2006) showed that short-term sick leave prevalence can be reduced by roughly 10% each by doubling the outdoor air supply rate in offices.

With the current epidemiological data, no clear cutoff point for ventilation rates can be provided, which will certainly reduce the risk for health, infectious diseases, and elevated absence rates in office buildings. It is clear, however, that increased ventilation rates will reduce the risk but this increase may not be uniform for the entire building stock. This is somewhat confirmed in several reviews on ventilation and health (e.g., Godish and Spengler 1996; Seppänen et al. 1999; Wargocki et al. 2002b; Sundell et al. 2011; Carrer et al. 2015), which showed that increased ventilation rates may be effective in mitigating discomfort due to odor and poor perceived air quality, acute health symptoms such as irritation of mucous membranes, and headaches or fatigue associated with exposures in buildings. Some of these reviews also show that ventilation rates below 10 L/s per person in offices can increase the risk for health, but even rates of 15–17 L/s per person would sometimes be needed to reduce the risk, or even ventilation rates higher than 25 L/s per person.

Despite the incapability to select the cutoff point for ventilation in reducing the risk for health, the framework for setting up health-based ventilation requirements was proposed by the HealthVent project (Wargocki et al. 2013), also considering the findings of the review by Carrer et al. (2015) and the ENVIE and IAIAQ projects (Oliveira Fernandes et al. 2009; Jantunen et al. 2011). In this framework, the base ventilation rate of 4 L/s per person is proposed as the basic requirement in any type of indoor non-industrial environment to ensure no health effects related to exposures to poor IAQ, and the ventilation rates cannot be lower than this base rate. This rate was selected based on the epidemiological evidence showing the relationship between ventilation and health, and assuming that the only pollution would be humans. It was selected to create the benchmark or reference point. This ventilation rate is assumed to be sufficient to reduce the risk for chronic health effects if exposures meet the requirements of WHO air quality guidelines (WHO 2005, 2010). If exposures do not meet the guideline requirements at this rate, then the framework stipulates a double sequential approach, in which first, all options of source control resulting in reduced exposures levels are exercised. Only then should the ventilation rates be increased if exposures still do not meet the requirements of the guidelines; the increase should be a multiple of the base rate. Ventilation rates defined by the described approach can be called *health-based*. It has been estimated that if the framework is strictly followed (i.e., indoor sources of pollution are reduced and entrainment of outdoor pollutants into the building is diminished, as well), then the burden of disease due to exposures to pollutants (in all types of buildings, not only offices) can be halved only if the base ventilation rate is used (Asikainen et al. 2012, 2013). The estimations made by Asikainen et al. (2016) showed that the reduced burden of disease can also be obtained by only increasing the ventilation rates; this effect will however be quite small and smaller than according to the framework presented earlier. Ventilation rates up to 14 L/s per person would be needed if only the ventilation is used to mitigate exposures indoors, and the higher rates would be counteractive; they will result in an increase in risk, as more ambient pollutants would be brought indoors.

3.4.2   SBS and Personal and Occupational Factors

Numerous studies of SBS in offices have shown that symptom prevalence is affected by various non-environmental factors including employee's gender, with symptoms being more prevalent among women than men, and this may indicate a greater sensitivity to poor IAQ among women; hours of computer use, which could indicate exposure to VOCs from computer emissions; and higher levels of job stress, which again could change individual susceptibilities to indoor air pollutants (Burge et al. 1987; Wilson and Hedge, 1987; Hedge 1988, 1998; Hedge et al. 1989a, 1995, 1996; Skov et al. 1989; Tamblyn and Menzies 1992; Zweers et al. 1992). An additional complication is that acute health symptoms (SBS symptoms) are very common in the general population and it is likely that no causal relationship can be shown between symptoms and indoor environmental quality, especially when the prevalence is low (Brauer et al. 2006).

3.4.3   Effects on Performance

Employed people spend anywhere from 20 to 60 hours per week in offices or other various kinds of work places. Series of experiments have been carried out in the laboratory and in field to investigate whether air quality and ventilation affect different aspects of office work, and if so, to which extent, i.e., the potential magnitude of the effects is observed.

Laboratory studies were performed under controlled conditions, when IAQ was the only variable/factor that was modified; all other factors defining the quality of an indoor environment such as noise and acoustics, light and temperature, and relative humidity remained unchanged. The observed effects could have thus been mainly attributed to the changes in IAQ. In laboratory experiments, subjects were recruited and exposed to different conditions in exposures lasting up to 5 hours (a bit more than half of a usual working day broken normally by the lunch break lasting from anywhere between 30 minutes to 2 hours, usually though less than an hour); sometimes experiments lasted even 8 hours. Subjects were usually students or young healthy individuals. During exposures, the office work was simulated by engaging the subjects in different tasks typical of office work, from arithmetical calculations to typing and proofreading. In some experiments, subjects performed diagnostic psychological tests, which were presented to them to examine a wide range of cognitive skills and how different motor, cognitive, and other skills were affected by changing the air quality conditions. These tests comprised measurements of reaction time, concentration endurance, and memory, as well as other skills, sometimes requiring higher cognitive demand such as decision-making or creativity. In some studies, the subjects were also asked to evaluate themselves by rating their performance or rating the effort exerted to complete the tasks.

The laboratory experiments generally showed that improving the air quality by either reducing the sources of pollution or increasing the ventilation rates improved the performance of tasks completed by recruited subjects, less so though in the case of some diagnostic psychological tests.

The improvement was usually in terms of increased speed at which the tasks were performed; only rarely was the error rate improved, as well, and it usually remained unchanged. These results suggest that the subjects performed the tasks at such a pace that the error rate could be minimized and they did not want to compromise the quality of the performed work by overexerting the speed at which the work was performed. The studies are described briefly in the following.

Exposures to toluene at 380 mg/m^3 (Bælum et al. 1985) and to a mixture of 22 common indoor air pollutants at concentrations of up to 25 mg/m^3 (Mølhave et al. 1986) have been shown to reduce the performance of diagnostic psychological tests in experiments in which subjects were exposed. However, these studies were performed on selected indoor air pollutants and at concentrations considerably higher than those typically encountered in office buildings (Brown et al. 1994; Wargocki 1998). In a study by Wargocki et al. (1999), the performance of text typing improved as typing speed improved by 6.5%, and the error rate was reduced by 18% when the proportion of dissatisfied with the air quality was reduced from 70% to 25% by removing a 20-year-old carpet. A repetition of this study with the same carpet showed that the performance of text typing improved by 1.5%, and the number of errors in addition reduced by 15%, when the proportion of dissatisfied with the air quality was reduced from 60% to 40% (Wargocki et al. 1999, 2002a; Lagercrantz et al. 2000). In a study by Bakó-Biró et al. (2004), the performance of text processing improved by 9% when the proportion of dissatisfied with the air quality was reduced from 40% to 10% by removing personal computers. In the study by Wargocki et al. (2000), the performance of text typing improved by about 1% for every two-fold increase in the outdoor air supply rate in the range between 3 and 30 L/s per person, causing the proportion of dissatisfied with the air quality to be reduced from 60% to 30%. In another study, in which the ventilation rate was changed, Park et al. (2011) showed improved performance of typing, addition, and memorization by on average of 2.5–5% when the ventilation rates were changed between 5, 10, and 20 L/s per person. In a study of Kaczmarczyk et al. (2004), providing a personalized ventilation providing clean air directly to the breathing rate that increased the amount of unpolluted air supplied to the breathing zone also improved the performance as well; this time, however, the performance was evaluated by the subjects themselves; the authors did not provide information on the size of effect.

There have also been a good deal of field studies examining the effects of improved air quality and ventilation on the performance of office work. In this case, however, it has been more difficult to control all exposures as it has been achieved in laboratory experiments. Many variables can change and eventually affect the performance of office work. The resulting effect on the performance can be simply an integrated effect of changes in many factors. Most of the experiments performed in field experiments were consequently intervention experiments with repetition rather than cross-sectional studies. Interventions with repetitions accounted for, at least to some extent, the many uncontrollable factors. In these intervention experiments, the air quality was modified by changing the ventilation rates, i.e., increasing and/or decreasing the rates at which outdoor air was delivered to office buildings. The field studies were performed in actual buildings with employees in their natural working environment so that they could perform their normal work during the experiments. This is a clear advantage when comparing it with the laboratory studies, especially as the results obtained can be easily transferred into performance/productivity metrics that are relevant for office work. This is much more difficult in the case of results obtained in laboratory studies, as several assumptions need to be made in order to translate the effects on, e.g., typing, to the actual working scenario/context. The reason is that there is a sizable gap between some measures of performance (for instance, the performance of brief diagnostic tests) and actual work performance and productivity that is of an economic significance over longer periods. To this end, the essential step would be to identify critical tasks for office work, which may not be easy due to the variety of tasks performed during office work, and then to make an estimate of the proportion of total work time, for which they are significant (for instance, even if office workers read 30% more slowly because of indoor air pollution, the overall effect on their productivity will be only 3% if the reading speed is only critical for 10% of their working day). An attempt to somewhat deal with that problem was made by Jensen et al. (2009), who used Bayesian models to estimate the contribution of performance on different tasks performed in the laboratory to the average performance relevant for the office working context and economical estimations. But still, the essence of the issue, how the performance measured in laboratory studies translates into actual work performance in buildings, remains unsolved.

In addition to the difficulty of controlling all parameters that may affect the performance in field experiments, it is difficult to identify the exact building and/or office type for experiments, where the work can be reliably quantified. The intention is to use the work performed by the building occupants to estimate the effect on performance, but sometimes it is not possible. Similar tasks can be presented as in the laboratory. This approach can potentially distort the natural aspect of the experiment, and sometimes the employees have to additionally perform the tasks that are neither familiar and customary to them nor often relevant for their work. Field studies performed so far used the former approach, in general. They were carried out in call centers with operators/consultants or nurses, so that their talk time with customers/patients and the wrap-up time to write a brief report after the call could be credibly and reliably measured. None of the studies measured the actual level of air quality, which was approximated by the rate at which ventilation with outdoor air was supplied to office buildings, where the field interventions were performed. The studies are briefly summarized in the following.

In the case of call centers, Wargocki et al. (2004) observed the reduced talk time of operators in the call center when the ventilation rates were increased from 2.5 to 25 L/s per person. Federspiel et al. (2004) observed significant improvement in the average handling time of call center operators only when the measured levels of CO_2 were lower than 100 ppm above outdoor levels; at higher CO_2 levels (up to 300–500 ppm), no effects were observed. The condition of a supply air filter could potentially be an important disturbing factor in the study of Federspiel et al. (2004), because, as shown by Wargocki et al. (2004), the performance of operators did not improve when a used filter was in place in the recirculated airflow after the ventilation rates were increased to 25 L/s per person. The presence of used filters was probably also the reason why the performance of the simulated office work, including addition and typing, could not be shown to be significantly affected by the reduced ventilation rates, resulting in CO_2 concentrations of 3000–4000 ppm in the experiments carried out by the New York State Commission on Ventilation in the 1910s (1923). Tham (2004) showed that the performance of call center operators improved by 9% when the ventilation rate was increased from 10 to 23 L/s per person in an office building with no bag filters (electrostatic filters were used instead).

The results of field experiments thus confirm the results obtained in laboratory experiments and, together with the laboratory experiments, form a very consistent and coherent body of evidence that the performance of office work is expected to be improved when IAQ is improved (outdoor air rates are increased). Based on the results of studies investigating the effects of ventilation on the performance of office work, Seppänen et al. (2006) suggested the quantitative relationship of office work and ventilation rate (Figure 3.1). It shows that work performance will on average increase by approximately 1.5% for each doubling of the outdoor air supply rate.

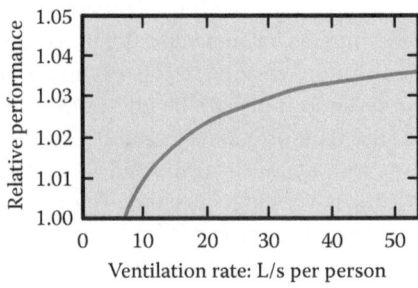

FIGURE 3.1 Ventilation and performance of office work. Quantitative relationship between outdoor air supply rate and performance. (From Seppänen, O. A. and W. Fisk, *HVAC&R Res J.*, 124, 957–73, 2006. With permission.)

The potential mechanisms by which the observed effects occurred are unknown, but it can be argued that those who do not feel very well and experience SBS symptoms such as headaches and difficulty to concentrate and cannot clearly think when the air quality is poor will not work very well. Other possible mechanisms for an effect of poor air quality on performance include distraction by odor, sensory irritation, allergic reactions, or direct toxicological effects. Studies of adult subjects performing simulated office work provide further information on the effects of IAQ on performance. They showed that increased air pollution caused by gaseous emissions from typical building materials, furnishings, and office equipment caused subjects to exhale less CO_2. Some studies suggest that the effect could be due to reduced metabolic rates as the consequence of reduced motivation to perform work in the polluted air, or a consequence of physiological changes leading to inefficient gas exchange in the lungs when polluted air is breathed (Bakó-Biró et al. 2005). The latter mechanism would lead to an increased CO_2 concentration in the blood (a mild acidosis), which is known to cause headaches. It can also be due to elevated CO_2 levels following the results of recent experiments by Satish et al. (2012), who showed that elevated levels of pure CO_2 can reduce the ability of subjects to make decisions at levels of CO_2 as low as 2500 ppm. But few new studies by Zhang et al. (2016a,b) did not observe the effects of CO_2 pn psychological test and office type tasks.

3.4.4 Implications and Costs

By affecting the health and the productivity, indoor air pollution influences the well-being, which, as a result, incurs different costs. By reducing indoor air pollution and improving IAQ, these costs can be reduced and/or avoided.

Reductions in productivity costs due to indoor air pollution were estimated in the 1980s, in the Unites States, to reach ca. $60 billion. This figure is compatible with other estimations carried out in the mid-1990s, also using U.S. data, showing that improving indoor environments can create potential annual savings and productivity gains from $29 to 168 billion by reducing the costs of respiratory illnesses, the costs of asthma and allergy, the costs related to SBS symptoms, and the productivity losses unrelated to health (Fisk and Rosenfeldt 1997).

The total cost of SBS was assumed in the mid-1990s, using U.S. data, to be $50 billion per year and caused by reduced productivity; this corresponds to about $400 per worker annually, which is about 2–3 lost working days of each worker per year. Some of these costs are due to factors not directly related with IAQ and are related to job satisfaction, psychosocial conditions, or personal issues. However, even if half of them are attributed to IAQ, they are still considerable at the level of $7.5–20 billion annually, and 10% reduction in the prevalence of symptoms would result in a significant reduction of costs at $0.75–2 billion annually. For comparison, in Finnish offices, a reduction in the prevalence of SBS symptoms was estimated, in late 1990s, to result in savings of ca. €330 per worker annually, which corresponds to about 1–2 lost working days of each worker per year (Seppännen 1999).

Changing the ventilation rates in U.S. offices would contribute significant benefits due to improved productivity; it could yield benefits of $13 billion from increasing the minimum ventilation rates from 8 to 10 L/s per person and $38 billion from increasing the minimum rates from 8 to 15 L/s per person, which is in significant contrast to savings obtained by reducing the minimum ventilation rates from 8 to 6.5 L/s per person yielding $0.04 billion annually (Fisk et al. 2012). Fisk et al. (2011) postulated that up to 20% reduction in SBS can be obtained by improved ventilation, and it would result in savings of $5 billion annually. The earlier figures are slightly lower than the estimates of Fisk and Rosenfeld (1997) mentioned earlier, but they are still very significant and much higher than the energy savings due to reduced ventilation. Filtration can also bring significant benefits, but they can be offset due to offending pollutants emitted from used filters, which can eventually reduce the performance as indicated earlier (Bekö et al. 2008).

For individuals, the costs of reduced air quality can be associated with reduced wages, time away from work, medical and insurance costs, and generally reduced life quality due to reduced health conditions and potential disability to optimally perform work.

Building owners can enjoy reduced life cycle costs. Wargocki and Djukanovic (2003) compared the life cycle costs of investments that would improve IAQ in an office building with the resulting revenues from increased office productivity that would be predicted from results of experiments investigating the effects of ventilation on performance. Analysis showed that the benefits from improved IAQ can be up to 60 times higher than the investment required to achieve it; the investments can generally be recovered in no more than 2 years (i.e., with payback times similar to the payback of 1.4 years suggested by Dorgan et al. [1998]), and the rate of return can be up to 7 times higher than the minimum acceptable interest rate. In fact, the estimations suggest that the full costs of installing and running the building can be offset by productivity gains of just 10% (Federation of European Heating, Ventilation and Air Conditioning Associations 2006). The benefits estimated by Wargocki and Djukanovic (2003) do not include benefits that result from reduced health costs and reduced absenteeism; lower absenteeism from an increased outdoor air supply rate can result in additional annual savings of $400 per employee, according to a study by Milton et al. (2000). Reduced life cycle costs result in increased property values (Virta et al. 2012). Moreover, improving the air quality can result in extended building and equipment life span, longer tenant occupancy and lease renewals, reduced churn costs, reduced insurance costs, reduced liability risks, and brand value.

Considerable benefits can also be achieved by employers when IAQ is improved. These are due to not only improved productivity and health status of employees, but also a generally satisfactory working environment, lower staff turnover, and more satisfied customers: A 1% increase in productivity corresponds to reduced sick leave of 2 days per year, less breaks from work, improved effective time at work of 5 minutes per day, or 1% increase in the effectiveness of physical and mental works.

In summary, all stakeholders will benefit from improving the air quality, and these benefits outweigh potential energy and investment costs, of course if only the effects of improved performance are taken into account in the calculations. It is, however, surprising to see that despite high economic premiums and rewards, potential health and productivity benefits are not yet integrated in the conventional economic calculations pertaining to building design and operation, which consequently affect indoor air pollution. The potential reason could be the still very low perception of the benefits of improving air quality and low willingness for paying for these improvements. This has been documented by the study of Hamilton et al. (2016) performed among the U.S. building industry. Less than half of the respondents expected that improving ventilation and filtration of buildings would improve productivity, and even less associated these interventions with lower absence rates and risk for health. Large costs were attributed by respondents to the interventions aimed at improving IAQ, larger than the estimated actual costs, and green building owners were less likely to pay for the upgrades.

3.5 VENTILATION, INDOOR AIR QUALITY, LEARNING, AND HEALTH IN SCHOOLS

3.5.1 Effects on Health

Studies show that the environmental conditions in schools are often inadequate, even in developed countries, and that they are frequently much worse than in office buildings. For example, measurements in 39 schools in Sweden showed that 77% of schools did not meet building code regulations (Smedje and Norbäck 2000). The most common defects in schools include insufficient outside air supplied to occupied spaces; water leaks; inadequate exhaust air flows, poor air distribution or balance; and poor maintenance of heating, ventilation and air-conditioning systems, as indicated by the analysis of the National Institute of Occupational Safety and Health Health Hazard Evaluation Reports for educational facilities in the United States where formal complaints had been registered (Angell and Daisey 1997; Daisey et al. 2003). Outdoor air supply rates in schools are considerably lower than in offices, and in many cases even lower than those observed in dwellings (Brelih 2012; Dimitroulopoulou 2012). They are also often much lower than they should be according to current recommendations for classrooms (Daisey et al. 2003; Dijken et al. 2005). For example, ASHRAE Standard 62.1 (2014) recommends for classrooms 5 L/s per person plus 0.6–0.9 L/s per m^2 of floor. Low ventilation rates often lead to carbon dioxide (CO_2) levels being well above the recommended level of 800–1000 ppm (sometimes 1400 ppm) during school hours (Sowa 2002; Dijken et al. 2005; Boxem et al. 2006; Santamouris et al. 2008; Wyon et al. 2010; Gao et al. 2014), implying that the concentration of other pollutants, not only the bioeffluents from children for which CO_2 is a good indicator, will be high, and that classroom air quality is consequently poor. In recent air quality measurements in 320 schools in Denmark, CO_2 concentrations exceeded 1000 ppm in more than 50% of classrooms (Clausen et al. 2014). The air quality in these classrooms did not meet the requirements of the Danish Building Code or the Danish Working Environment Authority, because the outdoor ventilation rates were too low. For comparison, similar measurements in Norway and Sweden showed that only in no more than 20% of classrooms were the CO_2 concentrations above 1000 ppm. Many studies have also reported high concentrations of particles in classrooms (e.g., EFA 2001; Dijken et al. 2005; Simoni et al. 2006).

Higher concentrations of pollutants in classrooms increase the risk of health problems. This has been confirmed by a recent comprehensive review of the measured and reported pollutants in classrooms and the associated health effects (Annessi-Maesano et al. 2013). Among the possible health effects are respiratory problems (both measured and self-estimated) including increased allergic reactions (e.g., Simoni et al. 2010; Zhang et al. 2011), especially for children with asthma, allergy, or any other hypersensitivity, as well as symptoms of fatigue, headache, and poor concentration (e.g., Norbäck et al. 2008a,b). Simoni et al. (2010) found that schoolchildren exposed to CO_2 levels below 1000 ppm had a significantly lower risk of dry cough and rhinitis. Measurements in European schools within the European Sinphonie project confirmed the earlier observations and showed that pupils exposed to an elevated level of indoor pollutants showed higher prevalence of nonspecific respiratory symptoms (Zivkovic et al. 2014). This is particularly worrisome considering that at least every third child suffers from some atopic disease. Studies point toward elevated levels of formaldehyde, ozone, nitrogen oxides, acrolein, and microbiological pollutants due to molds as well particulate matter, having both indoor and outdoor origins. Especially, particular matter is in the focus considering that airborne particles have been shown in many studies to have negative health effects on children. For example, Ward and Ayres (2004), in a meta-analysis of 22 panel studies of the effects of PM10 and PM2.5 values on children's health, found that PM2.5 had a greater effect than PM10 and that nonasthmatic children were more affected than asthmatic children, while Moshammer et al. (2006), in a panel study of 163 healthy children in Austrian schools, reported that their lung function was reduced when the ambient air contained elevated concentrations of particles. In addition, the recent study by Dorizas et al. (2014) confirms the elevated levels of PM10 in school classrooms, especially when chalk to write on blackboards is used.

Pupils, teachers, and other adults working in schools are at an elevated risk when pollutants in schools are higher; however, there are very few data for school personnel. Among them, for example, Wålinder et al. (1998) investigated the influence of the ventilation rates and ventilation system type on the nasal symptoms of school personnel in randomly selected primary schools in Sweden and found that nasal symptoms were worse in mechanically ventilated classrooms (with balanced supply and exhaust) than in naturally ventilated classrooms, even though the former had higher air exchange rates; the only exceptions were mechanically ventilated classrooms with displacement ventilation, in which nasal symptoms were less frequent.

3.5.2 Effects on Performance of Schoolwork and Learning

Poor classroom environmental conditions have been shown to occur frequently. These conditions, in addition to increasing risks for negative health effects, have been shown to reduce the performance of schoolwork. Different methods for measuring the effects on learning performance of students have been used in the reported studies, which all have been completed in schools and not under laboratory conditions. In some cases, psychological and neurobehavioral tests were used; some studies used standard tests for measuring academic achievement and some used absence rates, as the proxy for negative effects on learning. These studies are briefly summarized in the following.

The majority of studies examining the effects of IAQ and ventilation on the performance of schoolwork and learning used psychological and neurobehavioral tests. These tests examine different skills needed for proper learning, such as the ability to concentrate and memorize (Myhrvold et al. 1997; Ribic 2008; Bakó-Biró et al. 2012; Sarbu and Pacurar 2015). They also used shorter tests examining the ability to read, comprehend, and calculate (Wargocki and Wyon 2013). For example, a classroom study by Myhrvold et al. (1996) found a weak association between CO_2 levels and simple reaction time, suggesting a positive effect of increased ventilation on performance. In studies reported by Wargocki and Wyon (2013), pupils performed arithmetic calculations and language-based tasks under different conditions of air quality achieved by changing the ventilation rate between 3 and about 10 L/s per person (Figure 3.2). The speed at which the tasks were solved was improved with increased ventilation, but there were no effects on errors. Similar results were observed by a recent study, which copied the experimental approach used by Wargocki and Wyon (Petersen et al. 2015). In a study by Bakó-Biró et al. (2012), the performance of range of cognitive tasks was improved. Moreover, the time needed to solve simple math tests was reduced when the ventilation rate was increased from about 0.3–0.5 to 13–16 L/s per person. Haverinen-Shaughnessy and Shaughnessy (2015) also confirmed that the results of the tests in math and reading of fifth grade pupils improved with increasing the ventilation rate up to about 7 L/s per person. Ribic (2008) observed improved performance on the d2 test, a standard test for measuring concentration, when the CO_2 concentration was reduced from around 3800 to 870 ppm (absolute). Sarbu and Pacurar (2015) found that the performance of students on two psychological tests requiring concentration and cue utilization (Kraepelin tests and Prague test) linearly improved with reduced CO_2 levels. In contrast to these observations, Mattsson and Hygge (2005) did not observe any positive effect of operation of particle air cleaners on psychological tests despite the measureable effect on reducing the classroom levels of particles.

Although long-term learning outcomes are expected to be affected by the absence of abilities to perform simple psychological tests and ability to read, calculate, and comprehend, the connection between the progress in learning and these abilities is not very well documented. Therefore, some studies measured long-term learning using standardized tests, which are often developed by national education departments. These tests monitor the progress in learning and benchmark both individual pupils and schools, as well as evaluate the effectiveness of teaching methods and curricula over time.

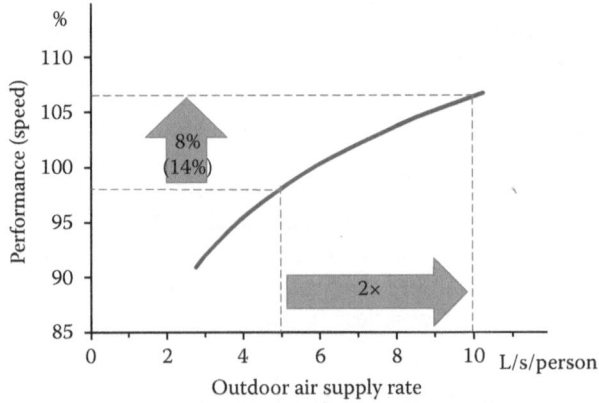

FIGURE 3.2 Ventilation and performance of schoolwork. The effect of increased outdoor air supply rate on the performance on language-based and mathematical tasks. (Reprinted from *Build Environ.*, 59, Wargocki, P. and D. P. Wyon, Providing better thermal and air quality conditions in school classrooms would be cost-effective, 581–9, Copyright (2013), with permission from Elsevier.)

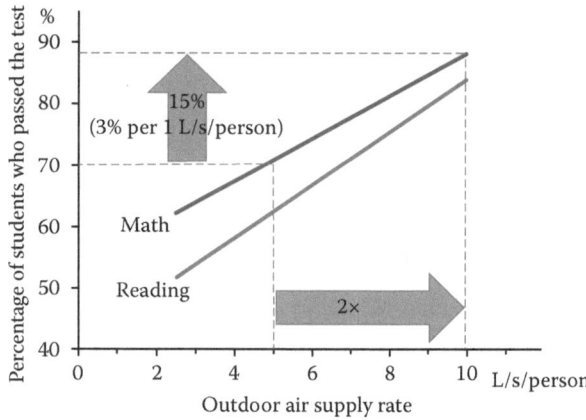

FIGURE 3.3 Ventilation and performance of schoolwork. The effect of increased outdoor air supply rate on the number of students passing the language and math exams. (From Haverinen-Shaughnessy, U., D. J. Moschandreas, and R. J. Shaughnessy, *Indoor Air.*, 21(2), 121–31, 2011. With permission.)

Haverinen-Shaughnessy et al. (2011) showed that poor ventilation in classrooms reduced the number of pupils just passing language and math tests (Figure 3.3). Mendell et al. (2015) showed an increase in mathematics and English scores when ventilation rates were increased, but only in the former case did the effects reach statistical significance. In another study by Toftum et al. (2015), academic achievement was evaluated with the scores from a standardized Danish test scheme, adjusted for a socioeconomic reference score. The lowest national test scores were generally found for pupils in classes with CO_2 concentrations above 2000 ppm, although the association was not significant. Pupils in schools with some means of mechanical ventilation scored on average higher in the national tests than pupils in schools with natural ventilation, probably because the efficacy of ventilation was higher.

Some studies measured illness absence and associated it with poor air quality or poor classroom ventilation to examine indirectly the effects on learning outcomes. Pilotto et al. (1997) showed in a cohort study that air pollutants from gas heaters had a negative effect on attendance at school, which was presumed to be due to a negative effect on children's health. Berner (1993) showed an association between the poor maintenance of schools and the poor academic achievement of the children attending them. Ervasti et al. (2012) found increased short-term sick leave among teachers in schools with poorly perceived air quality. Shendell et al. (2004) found student absence to decrease by 10–20% when the CO_2 concentration was decreased by 1000 ppm in 434 American classrooms. However, a recent study by Gaihre et al. (2014) in Scottish schools showed that an increase of 100 ppm of CO_2 corresponds to only 0.2% increase in absence rates (roughly one order of magnitude lower than the data of Shendell et al. [2004]), corresponding to approximately 0.5 day a year in a 190-day long school year. The study by Gaihre et al. (2014) is in contrast to the studies mentioned earlier, as it did not find any relation between air quality approximated by the levels of CO_2 and educational attainment measured as the percentage of class attaining the average level expected for this group. Simons et al. (2010) found high student absenteeism to be associated with poor ventilation in 2751 New York schools. In a recent work, though performed in day care centers equipped with the balanced mechanical ventilation system, Kolarik et al. (2016) found that increasing the air change rate by 1 h^{-1} would reduce the number of sick days by 12% even though the ventilation rates were quite high in these day care centers, as high as the CO_2 levels were below 1000 ppm, and on average around 640 ppm (absolute). Another recent and very comprehensive study in 162 Californian classrooms observed that illness absence decreased by as much as 1.6% for each additional 1 L/s per person of ventilation rate (Mendell et al. 2013); this is again lower than the data of Shendell et al. (2004) but higher than the data of Gaihre et al. (2014). The earlier findings are quite systematic and suggest that increasing classroom ventilation may substantially decrease illness absence. This may affect the learning experience though there is no clear evidence between the short-term absence of pupils and academic performance (Mendell and Heath 2005). It is also worth mentioning that absence rates can be influenced by many other factors not necessarily related to school environments.

The results from the previous experiments on the effects of classroom air quality on the performance of schoolwork do confirm that these effects are systematic and suggest that improving classroom air quality will have a significant positive effect on some aspects of learning, both on cognitive skills and academic attainment, as well as academic achievements and absence rates. The level of this effect is not the same across different studies as might be expected, but with reasonable confidence, it can be assumed that doubling the ventilation rate would improve the performance of schoolwork by up to 14% and each additional 1 L/s per person would increase the number of students passing the tests by 3% and would reduce the absence rates by at least 1.6% (Figures 3.2 and 3.3).

3.5.3 IMPLICATIONS AND COSTS

It would be interesting to estimate some economic indicators of the expected effects on learning and school performance, as a result of improved classroom air quality. However, this may not be that simple, as the measurable economic effects of quality of educational process cannot be immediately registered, as is the case for offices. The economic effects are first expected to be demonstrated, when pupils begin to work.

Wargocki et al. (2014) tried to estimate future socioeconomic consequences of improved IAQ in Danish primary schools. Assuming that the increased school performance will improve productivity and reduce the duration of primary education (in the Danish system, the pupil can take either 9 or 10 grades in elementary education, depending on the educational attainment) and absenteeism of teachers, the macroeconomic effects were estimated from increasing the ventilation rates from 6 L/s per person required by the Danish Building Code to 8.4 L/s per person required by the Swedish Code. The modeling of benefits showed that increasing the ventilation would yield an average annual increase in the gross domestic product (GDP) of €173 million due to increased productivity of the workforce and more pupils leaving school earlier and an average annual increase in the public budget of €37 million, again through improved productivity and shorter stays in primary school, as well as lower teacher sick leave. These effects correspond to no more than 0.07% of the Danish GDP in 2011. All effects are expected to increase (being higher and higher from year to year over the 20 years from the moment for which the analyses were performed); more students leave schools where the air quality and ventilation are improved.

A different estimation of the effects of reduced absence rates was performed by Mendell et al. (2013). They assumed that the ventilation rates in Californian K–12 schools will be increased from the current levels of 4–7.1 and 9.4 L/s per person, and estimated the benefits from decreased illness absence to school districts (i.e., increased revenue from the state for student attendance, which is the model adopted by the schooling system there) and the benefits to families through decreased costs from lost caregiver wages/time. Such estimated benefits yielded figures from US$33 to US$66 million annually from increased revenue from the state, and from US$80 to US$160 million annually from reduced losses to caregivers.

Both estimations show that the benefits and the potential losses due to reduced learning ability as a consequence of poor air quality are considerable and cannot simply be considered as negligible.

3.6 VENTILATION, INDOOR AIR QUALITY, HEALTH, AND SLEEP QUALITY IN HOMES

3.6.1 EFFECTS ON HEALTH

The data that exist on actual ventilation rates in residential buildings are limited and not representative of the entire residential building stock considering the different typologies of buildings, different building codes, climatic region, and merely cultural and historical merits. Moreover, the data that exist at present were collected during different times, including relatively old and relatively new buildings, and buildings complying with different code and standard requirements. Still, they can provide some information on the levels of ventilation rates that can be expected in residential environments and consequently the levels of exposures to potentially hazardous pollutants in residential building stock. These data indicate that the ventilation rates in dwellings are lower than in office buildings and higher than in schools. This observation is primarily based on the data collected by Brelih and Seppänen (2011) and Brelih (2012), who reviewed studies in dwellings in Europe and showed that the measured mean ventilation rates range from ca. 0.3 to 1 h^{-1}, which is about 5–15 L/s per person, with lower rates being prevalent in naturally ventilated dwellings or dwellings with exhaust ventilation only, and the upper and the middle range being more typically measured in the mechanically ventilated dwellings; the measurements in U.S. homes by Pandian (1988) show very similar levels of air change rates, about 0.5–0.7 h^{-1}. For comparison, the reports of Brelih and Seppänen (2011) and Brelih (2012) showed that the measured ventilation rates in schools in Europe range from ca. 1.5 to 9 L/s per person, and in offices from ca. 9 to 20 L/s per person. Asikainen et al. (2012, 2013) estimated the fraction of dwellings in Europe that do not meet the ventilation rates prescribed by the national standards. Using Bayesian regression, in which the location of the country, the annual mean temperature, and the gross domestic product were treated as explanatory variables, and the existing measurements of ventilation rates and the required ventilation standards by the national standards were the input variables, they showed that about 33% of dwellings in Europe can be expected to have ventilation rates below the national standards. This means that about 20% of European citizens (ca. 110 million) live in housing where there are elevated concentrations of hazardous pollutants.

Ventilation rates in homes that do not meet the standard requirements may not necessarily be reason for concern that the exposures will consequently lead to elevated risk for health problems, since it will all depend on the level of exposures to hazardous pollutants, which will additionally depend on the strength of the pollution sources. Without knowing the actual levels of exposures, many studies performed to date in dwellings suggest that it is beneficial to increase ventilation rates. These studies mainly examined the effects of ventilation on acute health outcomes such as asthma and allergy, building-related symptoms and complaints called SBS symptoms, and respiratory problems. Some were remotely observed by either mortality, cardiovascular and respiratory hospitalizations, obesity, and lead poisoning and can be associated with ventilation levels in homes. The results of these studies show that increasing the ventilation rates will generally reduce health problems (Figure 3.4), though the relationship presented in the figure was developed using data collected in offices; it may be assumed that it would apply also in domestic environments, as well.

Additionally, these studies also show that it is beneficial to retrofit homes with a ventilation system (either mechanical, hybrid, or natural) that secures that the ventilation rates are sufficiently high to deal with the pollutants generated indoors. The installed ventilation systems, however, need proper maintenance, because otherwise they may become the source of pollution and their presence can actually elevate the risk for health or does not bring the expected benefits. This could be one of the possible explanations why, in some studies, increased ventilation rates did not bring the intended positive effect in the form of reduced health risk.

No single ventilation rate can be recommended as protective based on the evidence collected through studies performed in homes, similarly to what has been observed for offices. The effects on health outcomes were significant over a wide range of ventilation rates. At the same time, the results from these studies show that ventilation rates below 0.4 h^{-1} would always increase the risk. This may suggest that ventilation rates in dwellings should not be reduced below 0.4 h^{-1} in case the emissions from sources indoors are similar to the studies, which form the basis for this recommendation. This rate is actually only slightly lower than the rates recommended by many building codes and standards. For example, the Danish Building Regulations Danish Housing and Building Agency (2010) set the requirements at 0.5 h^{-1}, as recommended by some reviews of the effects of ventilation on health (e.g., Wargocki et al. 2002; Sundell et al. 2011).

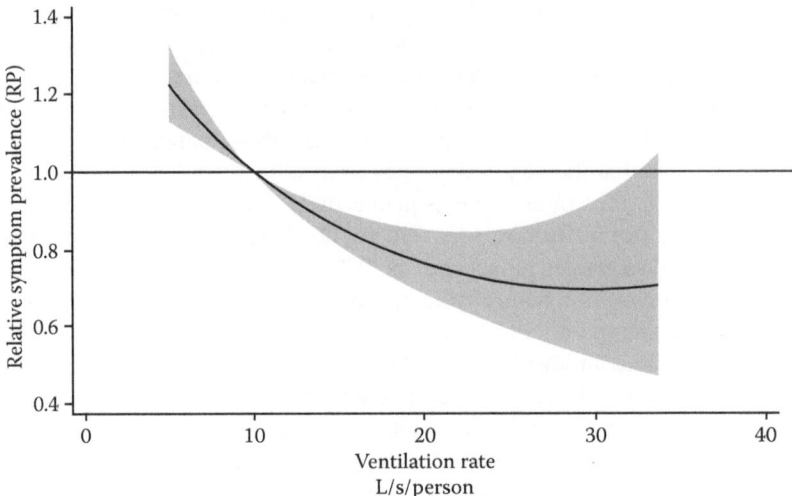

FIGURE 3.4   Ventilation and acute health symptoms in dwellings: The effect of increased outdoor air supply rate on prevalence of acute symptoms. (Fisk, W. J., A. G. Mirer, and M. J. Mendell: Quantitative relationship of sick building syndrome symptoms with ventilation rates. *Indoor Air.* 2009. 19. 159–65. Copyright Wiley-VCH Verlag GmbH & Co. KGaA. Reproduced with permission.)

The strongest evidence of the effects of ventilation can be seen with asthma, allergy, and respiratory symptoms. Any problem with ventilation defined as improper ventilation was associated with the elevated risk of asthma (Ezratty et al. 2003). In studies with children, it was shown that proper ventilation can reduce the risk for wheezing and rhinitis, both approved as reliable indicators of symptoms of asthma and allergy (Bornehag et al. 2005; Hägerhed-Engman et al. 2009; Kovesi et al. 2009). In the case of adults, it has been shown that nocturnal chest tightness, a symptom of problems with the respiratory system as a consequence of asthma, is higher with higher CO_2 levels being an indicator of too low ventilation rates (Norbäck et al. 1995). Retrofitting houses with mechanical ventilation systems was also shown to reduce the risk for asthma symptoms for infants (Kovesi et al. 2009), children (Lajoie et al. 2015), and adolescents (Howieson et al. 2003). The installation of new mechanical ventilation systems with heat recovery in homes without such a system was additionally associated with improved lung functions (Wright et al. 2009; Xu et al. 2010). Emenius et al. (2004) and Clausen et al. (2012) could not, however, confirm the relationship between ventilation rates and asthma and allergy symptoms, and neither could Warner et al. (2000) show that the installation of the mechanical ventilation system would especially improve lung functions. Many studies confirmed that the presence of ventilation in residential buildings does not guarantee a low risk of respiratory and allergic symptoms (e.g., Gustafsson et al. 1996; Jones et al. 1999; Willers et al. 2006). In addition to this Kovesi et al. (2009) showed that, although the levels of relative humidity are reduced when the new ventilation systems are retrofitted, this may not always translate into less health center encounters and hospitalizations due to respiratory problems.

Emenius et al. (2004) showed that humidity and window pane condensation (a marker of elevated humidity) were associated with elevated risks of symptoms. Large amounts of water are produced in homes by their occupants, and their study suggests that all efforts should be made to ensure that the moisture is removed. A typical family of four may produce between 5 and 10 kg water a day or even up to 15 kg can be produced by larger families (BS5250 1989; Ucci et al. 2004). This estimation includes the moisture produced through respiration and during different activities occurring indoors including cooking, washing, and drying clothes. Water is not harmful to health per se but excessive moisture indoors can result in the presence of unwanted contaminants and allergens and can modify the perception of air quality.

High levels of relative humidity (>50–60%) are frequent in residential environments. Relative humidity of >50% will increase survival of the population level of house dust mites (HDMs) and levels >60% will exacerbate their reproduction. HDM's feces cause house dust mite allergy, which is demonstrated in form of allergic reactions, asthma, and rhinitis. Several studies have shown that increased ventilation rates can reduce the concentration of HDMs (Harving et al. 1993, 1994; Sundell et al. 1994); the same conclusions are reached when the houses are retrofitted with the new mechanical ventilation system (Warner et al. 2000), most likely because the ventilation rates are consequently increased. These data constitute additional evidence of benefits of increased ventilation for reducing the risk of asthma and allergy symptoms in homes, but only when the ventilation will reduce the levels of relative humidity to levels that are at least below 50%. The reduction of humidity and moisture to deal with HDMs will additionally bring other benefits by reducing the risk of other health problems related to too high moisture levels that can, e.g., cause mold and related health problems (Bornehag et al. 2001, 2004; WHO 2009).

Relative humidity of >80% in the boundary levels of external walls and partitions increases the risk of condensation and mold growth. The germination of molds will depend on the surface type (which needs to provide sufficient substrate), availability of nutrients, temperature, and moisture. When it does occur, the mold spores can enter the air, and the exposure to spores can cause allergic reactions in the form of bronchial asthma or runny nose. Mold spores and particles containing molds, even when dead, can still emit toxic chemical compounds, so-called mycotoxins. Molds can also emit metabolic VOCs (mVOCs), which are secondary metabolites producing a musty odor typical for houses where molds are suspected. Exposure to mVOCs can cause immune system activation. In homes with verified dampness problems and plasticizer-containing surfaces, low ventilation rates were associated with the risk of bronchial obstruction (Øie et al. 1999).

High levels of moisture can also increase emissions from building materials and furnishings (Fang et al. 2004). Moisture on surfaces and in construction can cause hydrolysis reactions causing decomposition. An example of such a process includes di-2-ethylhexyl phthalate from polyvinyl chloride flooring hydrolyzing on a moist concrete, which produces 2-ethylhexanol, which has a mild odor and may potentially cause strong irritation. This process could be one of the underlying mechanisms explaining the results obtained by Øie et al. (1999) mentioned earlier. Hydrolysis can produce alcohols and monoesters (carboxylic acids), which can contribute to odor problems and irritation. Only limited data are available so far on these processes.

The presence of mechanical ventilation systems in homes and most likely the resulting higher ventilation rates and lower exposure levels were associated with reduced self-estimated health symptoms typical of SBS symptoms among adults compared with homes without mechanical ventilation systems (Ruotsalainen et al. 1991; Engvall et al. 2003; Leech et al. 2004). Ruotsalainen et al. (1991) additionally showed that the important requirement for this to occur is the higher air change rates and not merely the presence of a mechanical ventilation system. Some studies did not, however, find a relation between the building-related (SBS) symptoms and the existence and the operation of mechanical ventilation systems (Kishi et al. 2009). One of the reasons could be the lack of proper maintenance of these systems by the occupants of the dwellings or by simply blocking the outlets and the terminals due to experienced drafts and noise that they were generating (e.g., Coelho et al. 2005; Palonen et al. 2008).

Often the use of air conditioning in homes can result in reducing ventilation rates to minimize the expense for energy. This will also consequently result in an increase in the risk for elevated SBS symptoms (Wong and Huang 2004). The use of central air conditioning in homes may also have benefits by protecting people against ambient sources of pollution and reducing indoor temperatures that can be mortal, especially for the elderly during hot weather (Marmor 1978; Rogot et al. 1992). Bell et al. (2009) showed that the use of air conditioning in homes reduced the exposures to particulate matter mainly having origin in outdoor traffic, and was associated with reduced cardiovascular and respiratory hospitalizations, as well as mortality among the elderly. Deger et al. (2010) showed that children living along streets with highly dense traffic (especially those living on the ground floor) had an increased risk of asthma, so homes need to be sealed against ambient pollution that can be, e.g., transported by ventilation. However, the elimination of ambient pollution by sealing houses should not compromise IAQ and should not elevate indoor exposures. The data from two national longitudinal studies in the United States on house characteristics indirectly suggest this. They show that the increased use of air conditioning resulting in most cases in lowering ventilation rates to achieve energy savings, together with other factors such as lifestyle and nutrition, were associated with obesity and lead poisoning (Jacobs et al. 2009).

The use of air conditioning and/or high ventilation during very cold periods when the absolute humidity is close to zero may reduce the relative humidity in homes below 20–30%. This is occurring very seldom and only for the relatively short periods particularly when houses have very low occupation density. Low relative humidity in clean air has not been shown to cause any negative health effects or sensory discomfort (Wyon et al. 2006), even during half-a-week exposures (Andersen et al. 1974). However, the presence of VOCs with low relative humidity exacerbates the symptoms of dryness and/or irritation (Andersen et al. 1974; Wyon et al. 2006). Low relative humidity lowers the resistance to infections. At low moisture levels, the transmission of infectious particles and diseases is promoted, and short periods with increased humidity have been proposed as an interim solution for handling and inhibiting epidemic outbreaks (ASHRAE 2013).

3.6.2 Effects on Sleep and Next-Day Performance

IAQ in homes can disturb sleep quality and next-day performance, as shown in the studies that observed the correlation between sleep quality and next day concentration (Tynjälä et al. 1999; Meijer et al. 2000). Disturbed sleep is a widespread problem, but essentially, there are marginal data that document whether IAQ in bedrooms plays the important role for these outcomes. Three relatively small experiments provide some support that these can be the case.

In one study, students in dorms slept with windows open and closed so that the resulting ventilation rates were 0.4–1 h^{-1} and 0.2–0.3 h^{-1}. The objective sleep measures showed no differences, while students reported more awakenings during night and that they better remembered dreams with the windows open (Laverge and Janssens 2011). In another experiment, also performed in dorms, students slept with high and low rates of ventilation with outdoor air achieved by a simple fan mounted in the aperture, so that the air change rates were at about 0.3 h^{-1} and 1.4 h^{-1}. This time students reported that the air in their room was perceived to be fresher, that they felt more refreshed, and that their mental state was better when the fan was on. The objectively measured sleep efficiency (time spent asleep) was higher, and their performance on the concentration test taken in the morning was better when the fan was on (Strøm-Tejsen et al. 2014). Yet another study showed that the elderly sleeping with personal ventilation with an air outlet device directly on the head of the sleeping person (thus securing that exposures are reduced had a shorter sleep onset latency (measured objectively) and their heart rate variability was reduced (Zhou et al. 2014). If the results of these studies are confirmed, the changes in the air quality in homes can result in substantial economic benefits.

3.6.3 Implications and Costs

Logue et al. (2012) estimated the burden of disease related to elevated exposures to pollutants in homes based on the American population. They identified major pollutants emitted in houses and their exposure levels and then attributed the health risk related to the inhalation of these pollutants in homes. They showed that 1.3–3.5 million of healthy life years can be expected to be lost due to premature death or disability to perform work each year in the United States due to poor air quality in homes. The pollutants that had the highest attributable risk were PM2.5, second-hand tobacco smoke, formaldehyde, acrolein, radon, and ozone. Similar estimations in Europe (EU-27), as mentioned earlier, though for the entire building stock, i.e., including public buildings (offices, schools, etc.) and residential buildings but excluding second-hand tobacco smoke, resulted in a very similar figure of ca. 2 million healthy life years lost (Jantunen et al. 2011). In addition, in this case, the exposure to PM2.5 had the highest contribution to the burden of disease. These two estimations clearly illustrate that indoor exposures in homes significantly contribute to the burden of disease in the population and create a significant challenge for public health. This burden of disease has also considerable economic consequences. If one healthy life year is valued at €115,000 (Quinet 2013), then even assuming that only 1 million healthy life year is annually lost creates a gigantic loss for the national economy.

Socioeconomical costs due to exposure to indoor air pollution are actually rare, but those that exist confirm that the cost of poor air quality is very high. For example, a recent estimate in France for selected indoor air pollutants (benzene, trichloroethylene, radon, carbon monoxide PM2.5, and environmental tobacco smoke) showed that the loss due to exposures to these pollutants is €20 billion annually, a substantial part of it, around €14 billion, being attributable to exposure to PM2.5 (Kopp et al. 2014).

In addition to costs for health related to exposures to pollutants indoors, there are other costs that can be incurred due to pollution in homes, for example, due to insurance claims or litigation as a result of mold remediation. However, there are no published studies that tried to estimate other economic consequences of poor air quality in homes, e.g., in the form of loss in productivity or impact on academic achievement. These costs can also reach significant figures considering that a large part of education and learning occurs in homes, and more and more work tasks are actually completed at homes as well and not only when not at work (e.g., e-mailing, conference calls, e-meetings, reporting, etc.). As indicated in the previous sections, there are direct links between poor air quality and reduced performance at work and learning. In addition to these effects, IAQ in homes can disturb sleep quality, which is essential for the proper resting and next-day performance, as presented earlier.

3.7 CONCLUDING REMARKS

Despite the ample ventilation rates in office buildings, building occupants suffer from poor air quality, acute health symptoms, and reduced work performance. There is a very consistent body of evidence that indicates that improving air quality will mitigate these problems. Increasing ventilation rates will bring benefits, but what actually should be mitigated are the exposures because ventilation is only the mean to reduce the exposures. The pollution source control and the reduction of sources of pollution both indoors and outdoors should always have the highest priority. There seems to be, however, a low perception about the link between health symptoms and performance and poor air quality, and low perception of the potential significant benefits therein.

A good education system constitutes one of the fundaments of a modern society, because poor learning can have lifelong consequences for a student and for society. A recent Organization for Economic Co-Operation and Development report (Hanushek and Woessmann 2010) shows that countries that perform better in the Program of International Student Assessment (PISA) have higher growth rates. It also shows that

> Foundation skills in mathematics have a major impact on individuals' life chances. The survey shows that poor mathematics skills severely limit people's access to better-paying and more rewarding jobs; at the aggregate level, inequality in the distribution of mathematics skills across populations is closely related to how wealth is shared within nations. Beyond that, the survey shows that people with strong skills in mathematics are also more likely to volunteer, see themselves as actors rather than objects of political processes, and are even more likely to trust others.

The present chapter provided evidence from the archival literature that good classroom air quality is an important prerequisite for learning and educational attainment and, not least, the condition that will secure no health risks for the younger generation and their educators.

Homes should create shelter and conditions fostering restoration and rest. The present chapter showed that the exposures in homes can significantly increase the burden of disease, which may have considerable economic consequences. There are also some data showing that poor air quality conditions in homes may result in reduced next-day performance due to effects on sleep quality, but these data are only preliminary. Increased ventilation could not be shown unequivocally to reduce the exposure so that the risk for health of asthma and allergy symptoms, acute building-related symptoms, and respiratory symptoms could be reduced, but the epidemiological data showed that rates above 0.4 h^{-1} would at least be needed to achieve this goal. Elevated humidity and moisture were seen to consistently increase the risk, but ventilation may not always be effective to reduce their levels. Retrofitting housing with the new mechanical ventilation systems was seen to lower the risk, but malfunctioning and improper maintenance would again increase the risk. Air conditioning in houses may serve a protective role, when the ambient air is polluted, but will otherwise elevate the risk for health due to reduced ventilation rates as a measure taken to conserve energy.

This chapter showed also that the economic benefits of improving IAQ will outweigh the costs needed to implement them and will significantly contribute to improving the quality of life. This provides a strong economic argument and incentive for securing people's lives, work, and rest in air of outstanding quality, the quality that will not be detrimental for health and that will foster our quality of life.

REFERENCES

Addington, D. M. 2000. The history and future of ventillation. *Indoor Air Quality Handbook*. New York: McGraw-Hill Book Co., 1–2.

Andersen, I., G. R. Lundqvist, P. L. Jensen, and D. F. Proctor. 1974. Human response to 78-hour exposure to dry air. *Arch Environ Health*. 29: 319–24.

Andersson, K., J. V. Bakke, O. Björseth, C. G. Bornehag, G. Clausen, J. K. Hongslo et al. 1997. TVOC and health in non-industrial indoor environments. *Indoor Air*. 7: 78–91.

Angell, W. J. and J. Daisey. 1997. Building factors associated with school indoor air quality problems: A perspective. *Proc. of Healthy Buildings/IAQ'97*. Virginia Polytechnic Institute and State University, Washington, DC, 1: 143–8.

Annesi-Maesano, I., N. Baiz, S. Banerjee, P. Rudnai, S. Rive, and SINPHONIE Group. 2013. Indoor air quality and sources in schools and related health effects. *J Toxicol Environ Health. Part B*. 168: 491–550.

American Society of Heating, Refrigerating, and Air-Conditioning Engineers (ASHRAE). 2013. *ASHRAE Position Document on Infectious Diseases*. American Society of Heating, Refrigerating and Air-Conditioning Engineers, Inc., Atlanta, GA.

ASHRAE. 2014. *ASHRAE Standard 62.1-2013, Ventilation for acceptable Indoor Air Quality*. American Society of Heating, Refrigerating and Air-Conditioning Engineers, Inc., Atlanta, GA.

Asikainen, A., O. Hänninen, A. Kuhn, A. Yang, M. Loh, L. Gerharz et al. 2012. Modelling residential ventilation rates in European countries. *Proc Healthy Buildings Conf*. Brisbane, 2012.

Asikainen, A., O. Hänninen, N. Brelih, W. Bischof, T. Hartmann, P. Carrer et al. 2013. The proportion of residences in European countries with ventilation rates below the regulation based limit value. *Int J Vent*. 122: 129–34.

Asikainen, A., P. Carrer, S. Kephalopoulos, E. de Oliveira Fernandes, P. Wargocki, and O. Hänninen. 2016. Reducing burden of disease from residential indoor air exposures in Europe (HEALTHVENT project). *Environ Health*. 15(1): 61.

Bælum, J., I. Andersen, G. R. Lundqvist, L. Mølhave, O. F. Pedersen, M. Væth et al. 1985. Response of solvent exposed printers and unexposed controls to six-hour toluene exposure. *Scand J Work Environ & Health*. 11: 271–80.

Bakó-Biró, Z., D. J. Clements-Croome, N. Kochhar, H. B. Awbi, and M. J. Williams. 2012. Ventilation rates in schools and pupils' performance. *Build Environ*. 48: 215–23.

Bakó-Biró, Z., P. Wargocki, C. Weschler, and P. O. Fanger. 2004. Effects of pollution from personal computers on perceived air quality, SBS symptoms and productivity in offices. *Indoor Air*. 14: 178–87.

Bakó-Biró, Z., P. Wargocki, D. P. Wyon, and P. O. Fanger. 2005. Indoor air quality effects on CO_2 levels in exhaled air during office work. In *Proc Indoor Air 2005, 10th Int Conf Indoor Air Quality and Climate*. Beijing, 1: 76–80.

Bekö, G., G. Clausen, and C. J. Weschler. 2007. Further studies of oxidation processes on filter surfaces: Evidence for oxidation products and the influence of time in service. *Atmos Environ*. 4125: 5202–12.

Bekö, G., G. Clausen, and C. J. Weschler. 2008. Is the use of particle air filtration justified? Costs and benefits of filtration with regard to health effects, building cleaning and occupant productivity. *Build Environ.* 4310: 1647–57.

Bekö, G., O. Halás, G. Clausen, and C. J. Weschler. 2006. Initial studies of oxidation processes on filter surfaces and their impact on perceived air quality. *Indoor Air.* 161: 56–64.

Bell, M. L., K. Ebisu, R. D. Peng, and F. Dominici. 2009. Adverse health effects of particulate air pollution: Modification by air conditioning. *Epidemiol.* 205: 682–6.

Berglund, L. and W. S. Cain. 1989. Perceived air quality and the thermal environment. In *Proc. IAQ '89: The Human Equation: Health and Comfort.* San Diego, CA, 93–9.

Berner, M. 1993. Building conditions, parental involvement, and student achievement in the District of Columbia public school system. *Urban Educ.* 281: 6–29.

Bluyssen, P. M., E. De Oliveira Fernandes, L. Groes, G. Clausen, P. O. Fanger, O. Valbjørn et al. 1996. European Indoor Air Quality Audit Project in 56 Office Buildings. *Indoor Air.* 6: 221–38.

Bornehag, C. G., G. Blomquist, F. Gyntelberg, B. Järvholm, P. Malmberg, A. Nielsen et al. 2001. NORDDAMP: Dampness in buildings and health. *Indoor Air.* 11: 72–86.

Bornehag, C. G., J. Sundell, and C. J. Weschler. 2004. The association between asthma and allergic symptoms in children and phthalates in house dust: A nested case-control study. *Environ Health Perspect.* 112: 1393–97.

Bornehag, C. G., J. Sundell, L. Hägerhed-Engmann, and T. Sigsgaard. 2005. Association between ventilation rates in 390 Swedish homes and allergic symptoms in children. *Indoor Air.* 15: 275–80.

Boxem, G., L. Joosten, M. V. Bruchem, and W. Zeiler. 2006. Ventilation of Dutch schools; an integral approach to improve design. In *Proc 17th Air-conditioning and Ventilation Conf 2006* (Schwarzer, J. and M. Lain (eds.)). Society of Environmental Engineering, Prague, 31–6.

Brauer, C., H. Kolstad, P. Ørbæk, and S. Mikkelsen. 2006. No consistent risk factor pattern for symptoms related to the sick building syndrome: A prospective population based study. *Inter Arch Occ Environ Health.* 79(6): 453–64.

Brelih, N. 2012. Ventilation rates and IAQ in national regulations. *REHVA J.* 1: 24e8.

Brelih, N. and O. Seppänen. 2011. Ventilation rates and IAQ in European standards and national regulations. In *Proc 32nd AIVC Conference and 1st TightVent Conference in Brussels.* Brussels, 12–3.

Brown, S. K., M. R. Sim, M. J. Abramson, and C. Gray. 1994. Concentrations of volatile organic compounds in indoor air. *Indoor Air.* 4: 123–34.

BS 5250. 1989. *Code of practice for control of condensation in buildings.* British Standards Institution, London.

Buchanan, I. S. H., M. J. Mendell, A. G. Mirer, and M. G. Apte. 2008. Air filter materials, outdoor ozone and building-related symptoms in the BASE study. *Indoor Air.* 18: 144–55.

Burge, P. S., A. Hedge, S. Wilson, J. Harris Bass, and A. S. Robertson. 1987. Sick building syndrome: A study of 4373 office workers. *Annals Occup Hyg.* 31: 493–504.

Carrer, P., P. Wargocki, A. Fanetti, A., W. Bischof, E. O. Fernandes, T. Hartmann et al. 2015. What does the scientific literature tell us about the ventilation–health relationship in public and residential buildings? *Build Environ.* 94: 273–86.

Clausen, G., A. Høst, J. Toftum, G. Bekö, C. Weschler, M. Callesen et al. 2012. Children's health and its association with indoor environments in Danish homes and daycare centres—Methods. *Indoor Air.* 22: 467–75.

Clausen, G., J. Toftum, and B. Andersen. 2014. *Indoor environment in classrooms—Results of the mass experiment.* Danish Science Factory, Copenhagen, 19 pp. (in Danish).

Coelho, C., M. Steers, P. Lutzler, and L. Schriver-Mazzuoli. 2005. Indoor air pollution in old people's homes related to some health problems: A survey study. *Indoor Air.* 154: 267–74.

Daisey, J., W. J. Angell, and M. G. Apte. 2003. Indoor air quality, ventilation and health symptoms in schools: An analysis of existing information. *Indoor Air.* 13: 53–64.

Danish Housing and Building Agency. 2010. *Building Regulations.* Danish Ministry of Housing, Copenhagen.

Deger, L., C. Plante, S. Goudreau, A. Smargiassi, S. Perron, R. L. Thivierge et al. 2010. Home environmental factors associated with poor asthma control in Montreal children: A population-based study. *J. Asthma.* 475: 513–20.

Dijken, F. V., J. V. Bronswijk, and J. Sundell. 2005. Indoor environment in Dutch primary schools and health of the pupils. In *Proc Indoor Air 2005* (Yang, X., B. Zhao, and R. Zhao (eds.)). Tsinghua University Press, Beijing, I1: 623–7.

Dimitroulopoulou, C. 2012. Ventilation in European dwellings: A review. *Build Environ.* 47: 109e25.

Dorgan, C. B., C. E. Dorgan, M. S. Kanarek, and A. J. Willman. 1998. Health and productivity benefits of improved indoor air quality. *ASHRAE Trans.* 104 Part 1A: 658–66.

Dorizas, P. V., M. N. Assimakopoulos, C. Helmis, and M. Santamouris. 2015. An integrated evaluation study of the ventilation rate, the exposure and the indoor air quality in naturally ventilated classrooms in the Mediterranean region during spring. *Sci Total Environ.* 502: 557–70.

ECA-IAQ (European Collaborative Action "Urban Air, Indoor Environment and Human Exposure"). 2015. Framework for health-based ventilation guidelines in Europe, Report No 30. EN. Luxembourg: Publications Office of the European Union, in press.

European Federation of Asthma and Allergy Associations (EFA). 2001. *EN15251 2007: Indoor Air Quality in Schools.* EFA, Helsinki.

Emenius, G., M. Svartengren, J. Korsgaard, L. Nordvall, G. Pershagen, and M. Wickman. 2004. Building characteristics, indoor air quality and recurrent wheezing in very young children BAMSE. *Indoor Air.* 14: 34–42.

Engvall, K., C. Norrby, and D. Norbäck. 2003. Ocular, nasal, dermal and respiratory symptoms in relation to heating, ventilation, energy conservation, and reconstruction of older multi-family house. *Indoor Air.* 133: 206–11.

Ervasti, J., M. Kivimäki, I. Kawachi, S. Subramanian, J. Pentti, T. Oksanen et al. 2012. School environment as predictor of teacher sick leave: Data-linked prospective cohort study. *BMC Pub Health.* 12: 770.

Ezratty, V., A. Duburcq, C. Emery, and J. Lambrozo. 2003. Residential thermal comfort, weather-tightness and ventilation: Links with health in a European study LARES. In *Proc. 5th Warwick Healthy Housing Conf.* Warwick, UK. 17–19.

Fang, L., G. Clausen, and P. O. Fanger. 1998a. Impact of temperature and humidity on the perception of indoor air quality. *Indoor Air.* 8: 80–90.

Fang, L., G. Clausen, and P. O. Fanger. 1998b. Impact of temperature and humidity on perception of indoor air quality during immediate and longer whole-body exposures. *Indoor Air.* 8: 276–84.

Fang, L., D. P. Wyon, G. Clausen, and P. O. Fanger. 2004. Impact of indoor air temperature and humidity in an office on perceived air quality, SBS symptoms and performance. *Indoor Air.* 14 Suppl. 7: 74–81.

Federation of European Heating, Ventilation and Air Conditioning Associations. 2006. *REHVA Guidebook 6: Indoor climate and productivity in offices. How to integrate productivity in life cycle costs analysis of building services* (Wargocki, P. and O. Seanen (eds.)). Federation of European Heating and Air-Conditioning Associations, REHVA, Brussels.

Federspiel, C. C., W. J. Fisk, P. N. Price, G. Liu, D. Faulkner, D. L. Dibartolemeo et al. 2004. Worker performance and ventilation in a call-center: Analyses of work performance data for registered nurses. *Indoor Air.* 14 Suppl. 8: 41–50.

Finnegan, M. J., C. A. Pickering, and P. S. Burge. 1984. The sick building syndrome: Prevalence studies. *Brit Med J.* 289: 1573–5.

Fisk, W. J. and A. H. Rosenfeld. 1997. Estimates of improved productivity and health from better indoor environments. *Indoor Air.* 7: 158–72.

Fisk, W. J., D. Black, and G. Brunner. 2011. Benefits and costs of improved IEQ in U.S. offices. *Indoor Air.* 21(5): 357–67.

Fisk, W. J., D. R. Black, and G. Brunner. 2012. Changing ventilation rates in U.S. offices: Implications for health, work performance, energy, and associated economics. *Build Environ.* 47: 368–72.

Fisk, W. J., M. J. Mendell, J. M. Daise, U. Faulkener, A. T. Hodgson, M. Nematollahi et al. 1993. Phase 1 of the California healthy building study: A summary. *Indoor Air.* 3(4): 246–54.

Fisk, W. J., A. G. Mirer, and M. J. Mendell. 2009. Quantitative relationship of sick building syndrome symptoms with ventilation rates. *Indoor Air.* 19: 159–65.

Gaihre, S., S. Semple, J. Miller, S. Fielding, and S. Turner. 2014. Classroom carbon dioxide concentration, school attendance, and educational attainment. *J School Health.* 849: 569–74.

Gao, J., P. Wargocki, and Y. Wang. 2014. Ventilation system type, classroom environmental quality and pupils' perceptions and symptoms. *Build Environ.* 75: 46–57.

Godish, T. and J. D. Spengler. 1996. Relationships between ventilation and indoor air quality: A review. *Indoor Air.* 6: 135–45.

Graudenz, G. S., J. Kalil, P. H. Saldiva, R. Latorre Mdo, and F. F. Morato-Castro. 2002. Upper respiratory symptoms associated with aging of the ventilation system in artificially ventilated offices in São Paulo, Brazil. *Chest.* 122: 729–35.

Graudenz, G. S., J. Kalil, P. H. Saldiva, R. Latorre Mdo, and F. F. Morato-Castro. 2004. Decreased respiratory symptoms after intervention in artificially ventilated offices in São Paulo, Brazil. *Chest.* 125: 326–9.

Graudenz, G. S., C. H. Oliveira, A. Tribess, C. Mendes Jr., R. Latorre Mdo, and J. Kalil. 2005. Association of air-conditioning with respiratory symptoms in office workers in tropical climate. *Indoor Air.* 15: 62–6.

Gustafsson, D., K. Andersson, I. Fagerlund, and N. I. M. Kjellman. 1996. Significance of indoor environment for the development of allergic symptoms in children followed up to 18 months of age. *Allergy*. 51: 789–95.

Hamilton, M., A. Rackes, P. L. Gurian, and M. S. Waring. 2016. Perceptions in the U.S. building industry of the benefits and costs of improving indoor air quality. *Indoor Air*. 26(2): 318–30.

Hanushek, E. A. and L. Woessmann. 2010. The High Cost of Low Educational Performance: The Long-Run Economic Impact of Improving PISA Outcomes. OECD Publishing. 2, rue Andre Pascal, F-75775 Paris Cedex 16, France.

Harrison, J., C. Pickering, M. Finnegan, and P. Austwick. 1987. The sick building syndrome further prevalence studies and investigation of possible causes. In *Proceedings of Indoor Air '87* (Siefert et al. (eds.)). Oraniendruck GmbH, Berlin. 2: 487–91.

Harrison, J. A. C. Pickering, E. B. Faragher, and P. K. C. Austwick. 1990. An investigation of the relationship between microbial and particulate indoor air pollution and the sick building syndrome. In *Indoor Air '90: Proc. 5th Int. Conf. Indoor Air Quality and Climate*. Toronto, Canada. 1: 149–54.

Harving, H., J. Korsgaard, and R. Dahl. 1993. House-dust mites and associated environmental conditions in Danish homes. *Allergy*. 48: 106–9.

Harving, H., J. Korsgaard, and R. Dahl. 1994. Clinical efficacy of reduction in house-dust mite exposure in specially designed, mechanically ventilated "healthy" homes. *Allergy*. 49: 866–70.

Haverinen-Shaughnessy, U. and R. J. Shaughnessy. 2015. Effects of classroom ventilation rate and temperature on students' test scores. *Plos One*. 10(8), e0136165.

Haverinen-Shaughnessy, U., D. J. Moschandreas, and R. J. Shaughnessy. 2011. Association between substandard classroom ventilation rates and students' academic achievement. *Indoor Air*. 21(2): 121–31.

Hägerhed-Engman, L., T. Sigsgaard, I. Samuelson, J. Sundell, S. Janson, and C. G. Bornehag. 2009. Low home ventilation rate in combination with moldy odor from the building structure increase the risk for allergic symptoms in children. *Indoor Air*. 19(3): 184–92.

Hedge, A. 1984. Ill health among office workers: An examination of the relationship between office design and employee well being. In *Ergonomics and Health in Modern Offices* (E. Grandjean (ed.)). Taylor & Francis, London, 46–51.

Hedge, A. 1988. Job stress, job satisfaction, and work related illness in offices. *Proc. 32 HFS Ann Meet*. 32: 777–9.

Hedge, A. 1998. What can we learn about indoor environmental quality concerns from studies. In *Keeping buildings healthy: How to monitor and prevent indoor environmental problems* (O'Reilly, J. T., P. Hagan, R. Gots, and A. Hedge (eds.)). John Wiley & Sons, New York. 119–36.

Hedge, A., P. S. Burge, A. S. Wilson, and J. Harris Bass. 1989a. Work related illness in office workers: A proposed model of the sick building syndrome. *Environ Int*. 15: 143–58.

Hedge, A., T. D. Sterling, E. M. Sterling, C. W. Collett, D. A. Sterling, and V. Nie. 1989b. Indoor air quality and health in two office buildings with different ventilation systems. *Environ Int*. 15: 115–28.

Hedge, A., W. A. Erickson, and G. Rubin. 1995. Psychosocial correlates of sick building syndrome. *Indoor Air*. 5: 10–21.

Hedge, A., W. A. Erickson, and G. Rubin. 1996. Predicting sick building syndrome at the individual and aggregate levels. *Environ Int*. 22(1): 3–19.

Howieson, S., A. Lawson, C. McSharry, G. Morris, E. McKenzie, and J. Jackson. 2003. Domestic ventilation rates, indoor humidity and dust mite allergens: Are our homes causing the asthma pandemic? *Build Serv Eng Res Technol*. 243: 137–47.

Jacobs, D. E., J. Wilson, S. L. Dixon, J. Smith, and A. Evens. 2009. The relationship of housing and population health: A 30-year retrospective analysis. *Environ Health Perspect*. 1174: 597–604.

Janssen, J. E. 1999. The history of ventilation and temperature control. *ASHRAE J*. 9: 47–52.

Jantunen, M., E. De Oliveira Fernandes, P. Carrer, and S. Kephalopoulos. 2011. *Promoting Actions for Healthy Indoor Air IAIAQ*. European Commission Directorate General for Health and Consumers, Luxembourg.

Jensen, K. L., J. Toftum, and P. Friis-Hansen. 2009. A Bayesian Network approach to the evaluation of building design and its consequences for employee performance and operational costs. *Build Environ*. 443: 456–62.

Jones, R. C., C. R. Hughes, D. Wright, and J. H. Baumer. 1999. Early house moves, indoor air, heating methods and asthma. *Res Med*. 9312: 919–22.

Kaczmarczyk, J., A. Melikov, and P. O. Fanger. 2004. Human response to personalized ventilation and mixing ventilation. *Indoor Air*. 14 Suppl. 8: 17–29.

Kishi, R., Y. Saijo, A. Kanazawa, M. Tanaka, T. Yoshimura, H. Chikara et al. 2009. Regional differences in residential environments and the association of dwellings and residential factors with the sick house syndrome: A nationwide cross-sectional questionnaire study in Japan. *Indoor Air*. 19(3): 243–54.

Klepeis, N. E., W. C. Nelson, W. R. Ott, J. P. Robinson, A. M. Tsang, P. Switzer et al. 2001. The National Human Activity Pattern Survey (NHAPS): A resource for assessing exposure to environmental pollutants. *J Expos Anal Environ Epidemiol*. 11(3): 231–52.

Kolarik, B., Z. J. Andersen, T. Ibfelt, E. H. Engelund, E. Møller, and E. V. Bräuner. 2016. Ventilation in day care centers and sick leave among nursery children. *Indoor Air*. 26(2): 157–67.

Kopp, P., G. Boulanger, T. Bayeux, C. Mandin, S. Kirchner, B. Vergriette et al. 2014. Socio-economic costs due to indoor air pollution: A tentative estimation for France. *Proc. Indoor Air 2014*. Hong Kong, HP0955.

Kovesi, T., C. Zaloum, C. Stocco, D. Fugler, R. E. Dales, A. Ni et al. 2009. Heat recovery ventilators prevent respiratory disorders among Inuit children. *Indoor Air*. 19: 489–99.

Lagercrantz, L., M. Wistrand, U. Willén, P. Wargocki, T. Witterseh, and J. Sundell. 2000. Negative impact of air pollution on productivity: Previous Danish findings repeated in new Swedish test room. *Proc Healthy Buildings '2000*. Espoo, 1: 653–8.

Lajoie, P., D. Aubin, V. Gingras, P. Daigneault, F. Ducharme, D. Gauvin et al. 2015. The IVAIRE project—A randomized controlled study of the impact of ventilation on indoor air quality and the respiratory symptoms of asthmatic children in single family homes. *Indoor Air*. 25(6): 582–97.

Laverge, J. and A. Janssens. 2011. Analysis of the influence of ventilation rate on sleep pattern. *Proc Indoor Air 2011*, Austin, TX, a51–3.

Leech, J. A., M. Raizenne, and J. Gusdorf. 2004. Health in occupants of energy efficient new homes. *Indoor Air*. 143: 169–73.

Li, Y., G. M. Leung, J. W. Tang, X. Yang, C. Y. H. Chao, J. Z. Lin et al. 2007. Role of ventilation in airborne transmission of infectious agents in the built environment—A multidisciplinary systematic review. *Indoor Air*. 17: 2–18.

Logue, J. M., P. N. Price, M. H. Sherman, and B. C. Singer. 2012. A method to estimate the chronic health impact of air pollutants in U.S. residences. *Environ Health Perspect*. 1202: 216–22.

Marmor, M. 1978. Heat wave mortality in nursing homes. *Environ Res*. 17: 102–15.

Mattsson, M. and S. Hygge. 2005. Effect of articulate air cleaning on perceived health and cognitive performance in school children during pollen season. In *Proc. Indoor Air 2005* (Yang, X., B. Zhao, and R. Zhao (eds.)). Tsinghua University Press, Beijing, I2: 1111–5.

Meijer, A. M., H. T. Habekothé, and G. L. van den Wittenboer. 2000. Time in bed, quality of sleep and school functioning of children. *J Sleep Res*. 9: 45–153.

Mendell, M. J. 1993. Non-Specific symptoms in office workers: A review and summary of the epidemiologic literature. *Indoor Air*. 34: 227–36.

Mendell, M. J. and G. A. Heath. 2005. Do indoor pollutants and thermal conditions in schools influence student performance? A critical review of the literature. *Indoor Air*. 15: 27–52.

Mendell, M. J., E. A. Eliseeva, M. M. Davies, and A. Lobscheid. 2015. Do classroom ventilation rates in California elementary schools influence standardized test scores? Results from a prospective study. *Indoor Air* (published online).

Mendell, M. J., E. A. Eliseeva, M. M. Davies, M. Spears, A. Lobscheid, W. J. Fisk et al. 2013. Association of classroom ventilation with reduced illness absence: A prospective study in California elementary schools. *Indoor Air*. 236: 515–28.

Mendell, M. J., W. J. Fisk, J. A. Deddens, W. G. Seavey, A. H. Smith, D. F. Smith et al. 1996. Elevated symptom prevalence associated with ventilation type in office buildings. *Epidemiol*. 7: 583–9.

Mendell, M. J., G. M. Naco, T. G. Wilcox, and W. K. Sieber. 2003. Environmental risk factors and work-related lower respiratory symptoms in 80 office buildings: An exploratory analysis of NIOSH data. *Am J Ind Med*. 436: 630–41.

Milton, D., P. Glencross, and M. Walters. 2000. Risk of sick-leave associated with outdoor air supply rate, humidification and occupants' complaints. *Indoor Air*. 10: 212–21.

Moshammer, H., H. P. Hutter, H. Hauck, and M. Neuberger. 2006. Low levels of air pollution induce changes of lung function in a panel of schoolchildren. *Euro Resp J*. 276: 1138–43.

Müller, B., D. Müller, H. N. Knudsen, P. Wargocki, B. Berglund, and O. Ramalho. 2007. A European project SysPAQ. *Proc CLIMA 2007*. Finland. on CD-ROM.

Myatt, T. A., S. L. Johnston, Z. Zuo, M. Wand, T. Kebadze, S. Rudnick et al. 2004. Detection of airborne rhinovirus and its relation to outdoor air supply in office environments. *Am J Resp Crit Care Med*. 169: 1187–90.

Myatt, T. A., J. Staudenmayer, K. Adams, M. Walters, S. N. Rudnick, and D. K. Milton. 2002. A study of indoor carbon dioxide levels and sick leave among office Workers. *Environ Health.* 1: 3.

Myhrvold, A. N. and E. Olsen. 1997. Pupils' health and performance due to renovation of schools. In *Proc Healthy Buildings/IAQ '97* (Woods, J. E., D. T. Grimsrud, and N. Boschi (eds.)). Washington, DC. 1: 81–6.

Myhrvold, A. N., E. Olsen, and Ø. Lauridsen. 1996. Indoor environment in schools—Pupils' health and performance in regard to CO_2 concentration. In *Proc Indoor Air '96* (Yoshizawa, S., K. Kimura, K. Ikeda, S. Tanabe, and T. Iwata (eds.)). Seventh International Conference Indoor Air Quality and Climate, Nagoya, 4: 369–74.

Mølhave, L., B. Bach, and O. F. Pedersen. 1986. Human reactions to low concentrations of volatile organic compounds. *Environ Internat.* 12: 167–75.

New York State Commission on Ventilation. 1923. *Report of the New York State Commission on Ventilation.* Dutton, New York.

Norbäck, D., E. Björnsson, C. Janson, J. Widström, and G. Boman. 1995. Asthmatic symptoms and volatile organic compounds, formaldehyde, and carbon dioxide in dwellings. *Occup Environ Med.* 52: 388–95.

Norbäck, D. and K. Nordstrom. 2008a. Sick building syndrome in relation to air exchange rate, CO_2, room temperature and relative air humidity in university computer classrooms: An experimental study. *Int. Arch. Occup. Environ. Health.* 82: 21–30.

Norbäck, D. and K. Nordstrom. 2008b. An experimental study on effects of increased ventilation flow on students perception of indoor environment in computer classrooms. *Indoor Air.* 18: 293–300.

Øie, L., P. Nafstad, G. Botten, P. Magnus, and J. K. Jaakkola. 1999. Ventilation in homes and bronchial obstruction in young children. *Epidemiol.* 10: 294–9.

Oliveira Fernandes, E., M. Jantunen, P. Carrer, O. Seppänen, P. Harrison, and S. Kephalopoulos. 2009. *Co-ordination Action on Indoor Air Quality and Health Effects.* ENVIE. Final report, 165 pp. http://cordis.europa.eu/documents/documentlibrary/126459681EN6.pdf.

Palonen, J., J. Kurnitski, and L. Eskola. 2008. Thermal comfort and perceived air quality in 102 Finnish single-family houses. *Proc Indoor Air 2008.* Kongens, Lyngby. on CD-ROM.

Pandian, M. D., J. V. Behar, W. R. Ott, L. A. Wallance, A. L. Wilson, S. D. Colome et al. 1998. Correcting errors in the nationwide database of residential air exchange rates. *J. Expos. Anal. Environ. Epidemiol.* 84: 577– 86.

Pantelic, J., G. N. Sze-To, K. W. Tham, C. Y. H. Chao, and Y. C. M. Khoo. 2009. Personalized ventilation as a control measure for airborne transmissible disease spread. *J R Soc Interface.* 6: 715–26.

Park, J. S. and C. H. Yoon. 2011. The effects of outdoor air supply rate on work performance during 8-h work period. *Indoor Air.* 214: 284–90.

Petersen, S., K. L. Jensen, A. L. Pedersen, and H. Smedegaard Rasmussen. 2015. The effect of increased classroom ventilation rate indicated by reduced CO_2-concentration on the performance of schoolwork by children. *Indoor Air* (published online).

Pilotto, L. S., R. M. Douglas, R. G. Attewel, and S. R. Wilson. 1997. Respiratory effects associated with indoor nitrogen dioxide exposure in children. *Int J Epidemiol.* 264: 788–96.

Quinet, E. 2013. *L'évaluation socio-économique en période de transition. Centre d'Analyse Stratégique,* Paris.

Ribic, W. 2008. Nachweis des Zusammenhanges zwischen Leistungsfähigkeit und Luftqualität. Heizung, Lüftung/Klima. *Haustechnik.* 597: 43–6. (in German).

Robertson, A. S., P. S. Burge, A. Hedge, J. Sims, F. S. Gill, M. Finnegan et al. 1985. Comparison of health problems related to work and environment measurements in two office buildings with different ventilation systems. *Brit Med J.* 291: 373–6.

Rogot, E., P. D. Sorlie, and E. Backlund. 1992. Air-conditioning and mortality in hot weather. *Amer J Epidemiol.* 136: 106–16.

Ruotsalainen, R., J. J. K. Jaakkola, R. Rönnberg, A. Majanen, and O. Seppänen. 1991. Symptoms and perceived indoor air quality among occupants of houses and apartments with different ventilation systems. *Indoor Air.* 1: 428–38.

Santamouris, M., A. Synnefa, M. Asssimakopoulos, I. Livada, K. Pavlou, M. Papaglastra et al. 2008. Experimental investigation of the air flow and indoor carbon dioxide concentration in classrooms with intermittent natural ventilation. *Energy Build.* 4010: 1833–43.

Sarbu, I. and C. Pacurar. 2015. Experimental and numerical research to assess indoor environment quality and schoolwork performance in university classrooms. *Build Environ.* 93: 141–54.

Satish, U., M. J. Mendell, K. Shekhar, T. Hotchi, D. Sullivan, S. Streufert et al. 2012. Is CO_2 an indoor pollutant? Direct effects of low-to-moderate CO_2 concentrations on human decision-making performance. *Environ Health Perspect.* 120(12): 1671–7.

Sekhar, S. C. and S. E. Goh. 2011. Thermal comfort and IAQ characteristics of naturally/mechanically venti-lated and air-conditioned bedrooms in a hot and humid climate. *Build Environ.* 4610: 1905–16.

Sekhar, S. C., K. W. Tham, and K. W. Cheong. 2003. Indoor air quality and energy performance of air-conditioned office buildings in Singapore. *Indoor Air.* 134: 315–31.

Seppänen, O. 1999. Estimated cost of indoor climate in Finnish buildings. In *Proc. 8th Int Conf Indoor Air Quality and Climate—Indoor Air '99.* 3: 13–8.

Seppänen, O. and W. Fisk. 2002. Association of ventilation system type with SBS symptoms in office workers. *Indoor Air.* 12: 98–112.

Seppänen, O. A. and W. Fisk. 2006. Some quantitative relations between indoor environmental quality and work performance or health. *HVAC&R Res J.* 124: 957–73.

Seppänen, O., W. Fisk, and Q. H. Lei. 2006. Ventilation and performance in office work. *Indoor Air.* 16: 28–35.

Seppänen, O. A., W. J. Fisk, and M. J. Mendell. 1999. Association of ventilation rates and CO_2-concentrations with health and other responses in commercial and institutional buildings. *Indoor Air.* 9: 226–52.

Shendell, D. G., R. Prill, W. J. Fisk, M. G. Apte, D. Blake, and D. Faulkner. 2004. Associations between class-room CO_2 concentrations and student attendance in Washington and Idaho. *Indoor Air.* 145: 333–41.

Sieber, W. K., L. T. Stayner, R. Malkin, M. R. Petersen, M. J. Mendell, K. M. Wallingford et al. 1996. The National Institute for Occupational Safety and Health indoor environmental evaluation experience: Part three: Associations between environmental factors and self-reported health conditions. *App Occup Environ Hygiene.* 11: 1387–92.

Simoni, M., I. Annesi-Maesano, T. Sigsgaard, D. Norback, G. Wieslander, W. Nysta et al. 2006. Relationships between school indoor environment and respiratory health in children of five European Countries HESE study. In *Proc. 16th ERS Annual Congress. Eur Respir J.* 28Suppl 50: 837.

Simoni, M., I. Annesi-Maesano, T. Sigsgaard, D. Norbäck, G. Wieslander, W. Nystad et al. 2010. School air quality related to dry cough, rhinitis and nasal patency in children. *Eur Respir J.* 35: 742–9.

Skov, P., O. Valbjørn, B. V. Pederson, and Danish Indoor Study Group. 1989. Influence of personal character-istics, job related factors and psychosocial factors on the sick building syndrome. *Scand J Work Environ Health.* 15: 286–96.

Skov, P., O. Valbjørn, B. V. Pederson, and Danish Indoor Study Group. 1990. Influence of indoor climate on the sick building syndrome in an office environment. *Scand J Work Environ Health.* 16: 363–71.

Simons, E., S. A. Hwang, E. F. Fitzgerald, C. Kielb, and S. Lin. 2010. The impact of school building conditions on student absenteeism in Upstate New York. *Am J Public Health.* 100: 1679–86.

Smedje, G. and D. Norbäck. 2000. New ventilation systems at select schools in Sweden—Effects on asthma and exposure. *Arch Environ Health.* 55: 18–25.

Sowa, J. 2002. Air quality and ventilation rates in schools in Poland—Requirements, reality and possible improvements. *Proc Indoor Air 2002.* Monterey, CA, 2: 68–73.

Strøm-Tejsen, P., P. Wargocki, D. P. Wyon, and D. Zukowska. 2014. The effect of CO_2 controlled bedroom ventilation on sleep and next-day performance. *Proc Roomvent 2014.* Sao Paulo. Paper 148.

Sundell, J. 1982. Guidelines for Nordic building regulations regarding indoor air quality. *Environ Internat.* 81: 17–20.

Sundell, J., H. Levin, W. W Nazaroff, W. S. Cain, W. J. Fisk, D. T. Grimsrud et al. 2011. Ventilation rates and health: Multidisciplinary review of the scientific literature. *Indoor Air.* 21: 191–204.

Sundell, J., M. Wickman, G. Pershagen, and S. L. Nordvall. 1994. Ventilation in homes infested by house-dust mites. *Allergy.* 50: 106–12.

Tham, K. W. 2004. Effects of temperature and outdoor air supply rate on the performance of call center opera-tors in the tropics. *Indoor Air.* 14s7: 119–25.

Toftum, J., A. S. Jørgensen, and P. O. Fanger. 1998. Effect of humidity and temperature of inspired air on per-ceived comfort. *Energy Build.* 28(1): 15–23.

Toftum, J., B. U. Kjeldsen, P. Wargocki, H. R. Menå, E. M. N. Hansen, and G. Clausen. 2015. Association between classroom ventilation mode and learning outcome in Danish schools. *Build Environ.* 92: 20–6.

Tynjälä, J., L. Kannas, E. Levälahti, and R.Välimaa. 1999. Perceived sleep quality and its precursors in adoles-cents. *Health Promotion Internat.* 14: 155–66.

Ucci, M., I. Ridley, S. Pretlove, M. Davies, D. Mumovic, T. Oreszczyn et al. 2004. Ventilation rates and moisture-related allergens in UK dwellings. *Proc 2nd WHO Int Housing Health Symp.* WHO, Geneva, 328–34.

Virta, N., F. Hovorka, L. Litiua, and J. Kurniski. 2012. HVAC in sustainable office buildings. *REHVA Design Guide Nro,* 16.

Ward, D. J. and J. G. Ayres. 2004. Particulate air pollution and panel studies in children: A systematic review. *Occup Environ Med.* 61(4): e13.

Wargocki, P. 1998. *Human perception, productivity and symptoms related to indoor air quality*. PhD thesis International Centre for Indoor Environment and Energy, Technical University of Denmark, Kongens Lyngby.

Wargocki, P. and R. Djukanovic. 2003. Estimate of an economic benefit from investment in improved indoor air quality in an office building. In *Proc Healthy Buildings '2003* (Tham, K. W., S. C. Sekhar, and D. Cheong (eds.)). Stallion Press, Singapore, 3: 382–7.

Wargocki, P. and D. P. Wyon. 2013. Providing better thermal and air quality conditions in school classrooms would be cost-effective. *Build Environ*. 59: 581–9.

Wargocki, P., P. Carrer, E. De Oliveira Fernandes, O. O. Hänninen, and S. Kephalopoulos. 2013. Guidelines for health-based ventilation in Europe. *13th Int Conf Indoor Air Quality and Climate: Indoor Air 2013*, Hong Kong.

Wargocki, P., P. Foldbjerg, K. E. Erikse, and L. E.Videbæk. 2014. Socio-economic consequences of improved indoor air quality in Danish primary schools. *Proc Indoor Air 2014*. Paper no. HP0946.

Wargocki, P., L. Lagercrantz, T. Witterseh, J. Sundell, D. P. Wyon, and P. O. Fanger. 2002a. Subjective perceptions, symptom intensity and performance: A comparison of two independent studies, both changing similarly the pollution load in an office. *Indoor Air*. 12: 74–80.

Wargocki, P., J. Sundell, W. Bischo, G. Brundrett, P. O. Fanger, F. Gyntelberg et al. 2002b. Ventilation and health in nonindustrial indoor environments: Report from a European multidisciplinary scientific consensus meeting. *Indoor Air*. 12: 113–28.

Wargocki, P., D. P. Wyon, Y. K. Baik, G. Clausen, and P. O. Fanger. 1999. Perceived air quality, Sick Building Syndrome SBS symptoms and productivity in an office with two different pollution loads. *Indoor Air*. 9: 165–79.

Wargocki, P., D. P. Wyon, and P. O. Fanger. 2004. The performance and subjective responses of call-centre operators with new and used supply air filters at two outdoor air supply rates. *Indoor Air*. 14 (Suppl. 8): 7–16.

Wargocki, P., D. P. Wyon, J. Sundell, G. Clausen, and P. O. Fanger. 2000. The effects of outdoor air supply rate in an office on perceived air quality, Sick Building Syndrome SBS symptoms and productivity. *Indoor Air*. 10: 222–36.

Warner, J. A., J. M. Frederick, T. N. Bryant, C. Wiech, G. J. Raw, C. Hunter et al. 2000. Mechanical ventilation and high-efficiency vacuum cleaning: A combined strategy of mite and mite allergen reduction in the control of mite-sensitive asthma. *J Allergy Clin Immunol*. 105(1): 75–82.

Wålinder, R., D. Norbäck, G. Wieslander, G. Smedje, C. Erwall, and P. Venge. 1998. Nasal patency and biomarkers in nasal lavage—The significance of air exchange rate and type of ventilation in schools. *IntArch Occup Environ Health*. 71: 479–86.

Wenger, J. D., R. C. Miller, and D. Quistgaard. 1993. A gas sensor array for measurement of indoor air pollution—Preliminary results. *Proc Indoor Air '93*. Helsinki, 5: 27–32.

Weschler, C. J. 2009. Changes in indoor pollutants since the 1950s. *Atmos Environ*. 431: 153–69.

Willers, S. M., B. Brunekreef, M. Oldenwening, H. A. Smit, M. Kerkhof, H. De Vries et al. 2006. Gas cooking, kitchen ventilation, and asthma, allergic symptoms and sensitization in young children—The PIAMA study. *Allergy*. 61(5): 563–8.

Wilson, S. and A. Hedge. 1987. *The Office Environment Survey: A Study of Building Sickness*. Building Use Studies Ltd., London.

Wong, N. H. and B. Huang. 2004. Comparative study of the indoor air quality of naturally ventilated and air-conditioned bedrooms of residential buildings in Singapore. *Build Environ*. 39(9): 1115–23.

World Health Organization (WHO). 1948. *The Constitution of World Health Organization*. World Health Organization, WHO Regional Office for Europe, Copenhagen.

World Health Organization (WHO). 2006. WHO Air quality guidelines for particulate matter, ozone, nitrogen dioxide and sulfur dioxide: Global update 2005: Summary risk assessment.

WHO. 1983. *Indoor Air Pollutants: Exposures and Health effects*. EURO Reports and Studies 78, WHO Regional Office for Europe, Copenhagen.

WHO. 1987. *Air Quality Guidelines for Europe*. European Series no. 23. WHO Regional Publications, Copenhagen.

WHO. 2000. *The Right to Healthy Indoor Air*. Report on a WHO meeting, Bilthoven, Netherlands, 15–17 May 2000. WHO Regional Office for Europe, Copenhagen.

WHO. 2005. *WHO Air Quality Guidelines: Global Update 2005*. WHO Regional Office for Europe, Copenhagen.

WHO. 2009. *WHO Guidelines for Indoor Air Quality: Dampness and Mould*. WHO Regional Office for Europe, Copenhagen.

WHO. 2010. *Guidelines for Indoor Air Quality: Selected Pollutants*. WHO Regional Office for Europe, Copenhagen.

Wright, G. R., S. Howieson, C. McSharry, A. D. McMahon, R. Chaudhuri, J. Thompson et al. 2009. Effect of improved home ventilation on asthma control and house dust mite allergen levels. *Allergy*. 6411: 1671–80.

Wyon, D. P., L. Fang, L. Lagercrantz, and P. O. Fanger. 2006. Experimental determination of the limiting criteria for human exposure to low winter humidity indoors. *HVACR & Res J*. 12(2): 201–13.

Wyon, D. P., P. Wargocki, J. Toftum, and G. Clausen. 2010. Classroom ventilation must be improved for better health and learning. *REHVA J*. 6: 35–9.

Xu, Y., S. Raja, A. R. Ferro, P. A. Jaques, P. K. Hopke, C. Gressani et al. 2010. Effectiveness of heating, ventilation and air conditioning system with HEPA filter unit on indoor air quality and asthmatic children's health. *Build Environ*. 452: 330–7.

Zhang, X., Z. Zhao, T. Nordquis, and D. Norbäck. 2011. The prevalence and incidence of sick building syndrome in Chinese pupils in relation to the school environment: A two-year follow-up study. *Indoor Air*. 21: 462–71.

Zhang, X. J., P. Wargocki, and Z. W. Lian. 2016a. Physiological responses during exposure to carbon dioxide and bioeffluents at levels typically occurring indoors. *Indoor Air*. http://dx.doi.org/10.1111/ina.12286.

Zhang, X. J., P. Wargocki, and Z. W. Lian. 2016b. Effects of exposure to carbon dioxide and bioeffluents on perceived air quality, self-assessed acute health symptoms and cognitive performance. *Indoor Air*. http://dx.doi.org/10.1111/ina.12284.

Zhang, X. J., P. Wargocki, and Z. W. Lian. 2016c. Human responses to carbon dioxide, a follow-up study at recommended exposure limits in non-industrial environments. *Build Environ*. 100: 162–71.

Zhou, X., Z. Lian, and L. Lan. 2014. Experimental study on a bedside personalized ventilation system for improving sleep comfort and quality. *Indoor Built Environ*. 232: 313–23.

Zivkovic, Z., S. Cerovic, J. Jocic-Stojanovi, V. Vekovic, and S. Radic. 2014. Respiratory health, indoor air-pollution and asthma burden in school age children. *Euro Resp J*. 44(Suppl 58): P4955.

Zweers, T., L. Preller, B. Brunekreef, and J. S. M. Boleij. 1992. Health and indoor climate complaints of 7,043 office workers in 61 buildings in the Netherlands. *Indoor Air*. 2: 127–36.

4 Psychoacoustics
*Resolving Noise Distractions
in the Workplace*

Nigel Oseland and Paige Hodsman

CONTENTS

4.1 INTRODUCTION

Issues with noise and resolving them go back a long time. Texts written on clay tablets at around the time of the Sumerians (3500–1750 BC) mention how the god Enlil was angered by the noise of an overpopulated city, so he apparently flooded the city to remove the noise problem. Several thousand years later, the Romans passed a law that prohibited chariots driving through the cobblestone streets at night, in order to reduce the noise disturbance. More recently, since the late nineteenth century, much empirical research has been carried out on reducing noise in the workplace.

4.2 NOISE IN THE WORKPLACE

4.2.1 Acoustic Issues in Offices

Acoustician Treasure (2012) reminds us that "despite huge advances in almost every area of architecture and interior design … sound and acoustics, for the most part, have remained secondary concerns. They are possibly the two most pressing issues in architecture today." Similarly, Perham, Banbury, and Jones (2007) commented that "the acoustic design of offices often does not receive the attention that most other architectural systems would. However, unwanted levels of ambient noise, often caused by an excessively reverberant environment, can cause difficulties with communication as well as with concentration at work."

Abbot (2004) reviewed numerous research studies and concluded that noise, in addition to causing nuisance and disturbance in an office environment, is a primary cause of reduction in productivity and can contribute to stress and illness, which in turn can also contribute to absenteeism and turnover of staff. Jensen, Arens, and Zagreus (2005) undertook an extensive postoccupancy evaluation survey of 142 commercial buildings in the United States with 23,450 participants. The primary finding from their study was that dissatisfaction was highest with internal acoustics. Furthermore, they found that half of the respondents reported that poor acoustics interfered with their daily work (Figure 4.1).

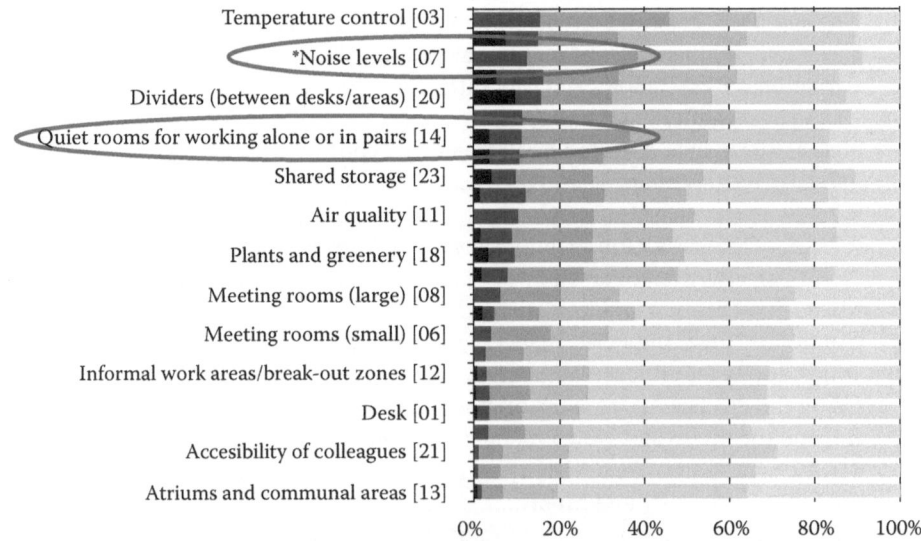

FIGURE 4.1   Satisfaction with office features. (From Oldman, T. and Rothe, P., *Leesman Rev.*, 17, Q2, 2015. With permission.)

The Leesman Index (Oldman and Rothe 2015) is the largest independent measure of workplace effectiveness with ca. 103,000 survey responses from 836 locations. The survey participants were asked, which features do you consider to be an important part of an effective workspace? And then they were asked to rate their satisfaction with their selected important feature. Figure 4.1 shows that noise is considered the second biggest cause of dissatisfaction with almost half (47%) of the occupants dissatisfied with noise levels, and is reported to be the biggest disruptor of performance.

Oseland and Burton (2012) carried out a literature review of the studies showing a quantified impact on productivity from environmental conditions, including temperature, light, and noise (acoustics). They conducted a meta-analysis of 75 studies that they considered credible, including 21 studies exploring the impact of noise. They found that after the noise was reduced, the average increase in productivity was 27.8%. Oseland and Burton went on to weight the results for their relevance to offices, accounting for the environment in which the study was carried out, the type of metrics used, and the relevance of the activity carried out by the participants. The revised impact of noise on productivity is 1.7%. Although this figure appears low, a report published by the British Council for Offices (Richards et al. 2014) suggests that "a 1% improvement in productivity swamps utility costs" and it is estimated that a change in productivity of just 5% may cover annual property costs.

Noise remains a significant problem in office environments, affecting worker satisfaction and productivity, but nevertheless the problem is often ignored. Research on noise in the office environment is often used as part of the on-going debate over the pros and the cons of open-plan versus private office layouts. Our intention is not to enter that debate in this chapter—in most business sectors in the UK and much of Europe, the open-plan office is the norm. Rather, we consider our task here is to help mitigate the noise in these mainstream working environments.

4.2.2 Acoustic Issues in Educational Environments

Two key communication events occur in classrooms: speaking and listening. While primarily seen as a place of learning, classrooms and lecture theaters are also workplaces. Although classroom acoustic problems have been noted since the late 1940s, solving them continues to be widely ignored in design and construction, both in Europe and in the United States.

Heavy workloads and stress in the teaching profession were publicly recognized in Germany in the 1990s. A study by the Institute for Interdisciplinary School Research of the University of Bremen (2003) showed that 80% of teachers said that they experienced stress from pupil noise. Oberdorster and Tielser (2006) took the study further and investigated the effects of acoustics on children and teachers in the classroom. Different working methods were compared to the noise loads and the stress. The results showed the correlations of higher heart rates to higher noise levels, and conversely, lower noise activity levels contributed to reduced heart rates. It was shown that better communication and lower stress levels could be achieved by lowering the sound pressure levels (SPLs). The researchers concluded that although noise in schools is a complex subject, key considerations should include balancing the effect of general noise levels on communication and stress levels.

Another study in the UK (Canning and James 2012) demonstrated how improved acoustic parameters can impact both the learning environment of pupils and the working environment of teachers. The intent of the study was to determine if there was a benefit in designing for the inclusion of hearing-impaired students, as well as all other students and teachers, by changing the physical acoustic conditions of the classroom. The results indicated a strong correlation between the reverberation time (performance of acoustic treatment) and the perceived quality of the teaching environment for both speech and listening. Other key findings of the study are listed in Table 4.1.

The implementation of good acoustics in educational environments is far reaching and twofold. While the pupils' education is enhanced, so too is the health and well-being of the teacher, for whom a classroom is a workplace.

TABLE 4.1

Key Findings from the Essex Study

- Voice levels could be lowered, reducing vocal stress
- Improved signal-to-noise ratio meant the teachers could be understood with a normal speaking voice
- Reported substantial improvements in the behavior and the comprehension of the pupils in classrooms
- Better behavior from the students
- Quieter and calm environments were reported

Source: Canning, D., and A. James, *The Essex Study Optimised Classroom Acoustics for All*, The Association of Noise Consultants, St. Albans, 2012. With permission.

4.2.3 Acoustic Issues for Workers in Hospital Environments

Hospitals are becoming noisier with sound levels doubling since 1972 (Sykes 2009). Noise data from research in hospitals across the United States, from 1960 to 2005, concluded that none of the hospitals in the study actually complied with WHO guidelines on ward noise and that sound levels were found to exceed the typical speech levels for communication between two people (Busch-Vishniac et al. 2005).

The noise in hospitals has many sources including telephones, talking, footstep noise, trolleys, alarms, and machines monitoring the patients. The sterile environment includes many acoustically hard finishes designed to be hard wearing and easy to clean. These reflect sound and increase the background noise level and the reverberation within a space. The medical staff can spend over 40% of their time communicating. This contributes to the high noise levels and emphasizes the importance of a good functional sound environment to avoid misunderstanding that might lead to medical errors.

For full intelligibility in listeners with normal hearing, the speech level should be 15 dB louder than the background noise. So, for example, a nurse working on a hospital ward with a background noise level of 65 dB has to talk at 80 dB. So to be clearly heard, the voice would be raised and this has a long-term effect on voice strain and increased levels of stress.

Communicating in an already noisy environment also has an effect on accuracy. In an observational study in the emergency department of a UK hospital (Woloshynowych et al. 2007), an average of 100.9 communication events per hour, or 1.68 per minute, were registered. Communication multitasking was evident in 14% of these occasions. This is important because the communication load can disrupt the memory and lead to mistakes. Improving the communication between the healthcare staff by reducing the levels of interruptions and minimizing the volume of irrelevant or unnecessary information exchange can have important implications for patient safety.

In a perception study of nurses (Mahmood et al. 2011), almost 60% felt that high noise levels and/ or acoustic problems in the nursing unit led to nursing errors, placing it within the top five environmental factors affecting nursing performance. With further regard to the perceptions of well-being, a study at the Huddinge hospital coronary critical care unit studied the effect of improving the acoustic conditions on the staff and the patients (Blomkvist et al. 2005). From the staff perspective, the study showed that in good acoustic conditions with low background noise levels and low reverberation times, the staff felt that they were working in a better environment. They felt less stressed and had more autonomy over their work and more time to spend with their patients. Their attitude was graded better by patients and the overall quality of care was graded higher. Attention to the acoustic details in hospital environments improves communication, lowers stress levels, enhances staff well-being and job satisfaction, and can have significant implications to the improved quality of care for patients.

There are more than **10 million** people in the UK
with some form of hearing loss, or **one in six of
the population**.

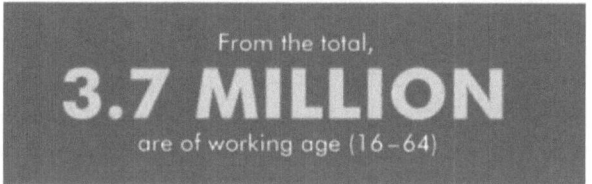

From the total,
3.7 MILLION
are of working age (16–64)

FIGURE 4.2 Current prevalence of hearing loss in the UK population. (Data from Action on Hearing Loss, http://www.actiononhearingloss.org.uk/your-hearing/about-deafness-and-hearing-loss/statistics.aspx, UK; image courtesy of Ecophon. With permission.)

4.3 IMPACT OF NOISE ON PEOPLE

Noise affects people in many ways—it affects our health and well-being, our mental state, and our performance. Sound level can have a physical, a physiological, and a psychological effect.

- Physical effect—Continuous levels of sound above 140 dB* can cause pain and may have physical effects on the body, some of which are immediate. Sound at this level produces mechanical changes in a person, such as heating of the skin, rupture of the eardrum, and vibration of the eyeballs or internal organs. However, the energy created by such sound levels is at least 10 million times more than what is found in the office environment, so physical effects on the human body are unlikely to occur in the office. In workplaces with extreme sound levels, such as factories, airports, or road works, ear defenders are worn to protect the workers rather than alleviate the sound. This strategy does not prevent the high levels of sound affecting the unprotected nonworkers, who may then be physiologically or psychologically affected.
- Physiological effect—Raised sound levels can cause biological changes, such as elevated blood pressure, increased heart rate, hearing loss, and stress. For example, long-term exposure to levels of 85 dB or more during a typical 8-hour workday can damage the eardrums and put people at risk of moderate hearing loss. This level of exposure does not usually occur in offices, so such physiological effects are not a major concern for us. However, Figure 4.2 shows the current prevalence of hearing loss in the UK population and its relevance to the working population. It is estimated by Action on Hearing Loss that by 2035, some 15.6 million people in the UK will suffer some sort of hearing loss (2016).[†] This is a cause for concern if the current workplace design criteria do not often take occupant hearing conditions into account. More people are using personal music systems at work to reduce the distraction. Sound levels below 70 dB pose no known risk of hearing loss, but the extended

---

* The intensity of sound is measured on the decibel (dB) scale; the decibel scale has different weightings, the A weighting dB(A) being most common. The threshold for hearing is 0 dB and the normal conversation is around 40 dB. The noise on a busy street is around 70 dB and a rock band might produce 120 dB.

† Action on Hearing Loss UK https://www.actiononhearingloss.org.uk/your-hearing/about-deafness-and-hearing-loss /statistics.aspx.

intense use of personal stereos in the workplace, or elsewhere, may have a physiological effect. One study reported that the equivalence of 8 hours of continuous noise exposure level for people using personal stereos was 80 dB (Williams 2005). In fact, the European Commission for Electrotechnical Standardisation (Commission for European Communities 2009) has accepted a mandate to control exposure to excessive volume from personal music players to avoid hearing damage. At 80 dB, the exposure is limited to 40 hours per week, where an 89 dB exposure shall be limited to 5 hours per week.

- Psychological effect—This relates to the mental changes in a person due to exposure to sound that they consider unnecessary or disturbing. Psychological effects are mostly manifested as annoyance, heightened stress levels, or reduced performance. Such effects can occur at any sound level. For example, a dripping tap in a home at sound levels of 30 dB may create annoyance, especially at night, whereas sound levels of 120 dB caused by a passing ambulance may be acceptable, depending on the time of day. Attendees at a rock concert generating 120 dB will find the sound level acceptable, whereas neighbors to the venue may not. The response to the sound level is totally subjective and, as such, the psychological effects of sound are the main concern in the office environment.

Recognizing the psychological effects of sound introduces the notion of unnecessary or disturbing sound (noise) having an effect on people, regardless of the actual sound level. The next sections discuss the nuances of noise and sound, and why noise in the workplace is a psychophysical problem. However, we first need to discuss how to actually measure sound and noise.

4.4 MEASURING SOUND

Acoustics is a complex and interdisciplinary science spanning physics, engineering, physiology, and psychology. Even measuring the sound level is not as simple as it may first appear. It is important to keep in mind that simple, controlled sounds are used to describe how sound works, but using them in practice, as noted by Horowitz (2012), "is like asking a physicist to describe the motion of a herd of cows—the behaviour can be modelled as long as the cows are spherical and moving on a frictionless surface in a vacuum." As the purpose of this chapter is to look at the psychological impact of acoustics, we will briefly look at the basic premise behind physical acoustic measurements.

Ideally, one descriptor would be used for solving all room acoustics, but as hearing is multidimensional and room shapes, locations, material content, and activities are so varied, multiple descriptors are necessary to create an optimum acoustic solution. One of the most common descriptors is the SPL. It is defined as the deviation in the pressure from the ambient atmospheric pressure caused by a sound wave. SPL is a logarithmic measure of the effective pressure of a sound wave relative to a reference value, also measured in decibel. It is relatively straightforward to measure SPL using a meter approved by the International Standards. Nevertheless, most acousticians agree that the raw reading from a sound level meter does not correlate with perceived loudness.

For instance, the human ear is less sensitive to low audio frequencies, and so the SPL is adjusted to account for this and provides a measurement that corresponds more closely to the hearing sensation or the loudness. Thus, arithmetic weightings (filters) are applied to the SPL; the A, the B, and the C weightings currently used in sound level meters are aimed at mimicking the perceived loudness over different frequency ranges. The A weighting, expressed as dB(A), is the most commonly used weighting particularly in measuring and specifying sound levels in office environments. The debate continues among acousticians on the appropriate weightings and many recognize that the A weighting was designed (and is possibly only valid) for use at relatively quiet sounds (~40 dB) and for pure tones. It is worth noting that the reported sound level, in dB(A), is only an approximation of the loudness for the average human; it does not account for individual hearing differences due to age or other factors.

Sounds may be ambient (steady) or transient (intermittent). Ambient sound is continuous and long term, such as the background sound of an air-conditioning system. Transient (intermittent) sounds are short term, such as telephone rings, alarms, or even people starting a conversation.

In order to get an approximation of the actual sounds (ambient and transient), an integrating-averaging meter is used to measure time-averaged sound. The time-averaged sound level is usually referred to as the *equivalent continuous sound level* represented by the symbols L_{AT}, L_{eq}, and LA_{eq} (the A-weighted equivalent sound level). The integrating-averaging meter automatically measures the sound levels over set time intervals, divides the sound exposure by the time, and takes the logarithm of the result, presenting a single value in dB(A). Acousticians and standards agencies debate the best methodology for representing sound exposure with peaks and troughs. Although a single dB(A) value can be generated so that it is possible to compare the ambient sound exposures in different environments, the value does not reflect the actual impact of, or the disturbance caused by, unexpected intermittent sounds.

4.4.1 PHYSICAL MEANS OF CONTROLLING SOUND

The physical properties of the materials in a room can have a significant effect on how sound will travel across the space. Utilizing materials to control sound is a crucial part of solving room acoustic problems. *A Guide to Office Acoustics* (Association of Interior Specialists 2011) states that "ceilings have the biggest impact on the acoustic quality of open-plan offices by providing a surface that can be either sound absorbing or sound insulating or a combination of the two." This builds on earlier research by Pirn (1971), who discusses the relative effects of speech effort, speaker orientation, background noise, and speaker-to-listener distance using an articulation index—a measure of speech intelligibility—and supports the need for consistent and efficient absorption. As Pirn states, "flanking surfaces, particularly the ceiling, must be sufficiently absorptive so that transmission by reflection will not seriously impair the barrier's potential shielding qualities."

4.4.2 ACTIVITY-BASED DESIGN FOR IMPROVED ACOUSTICS

Although physical solutions are a necessary part of the acoustic solution, there is more to consider. Hongisto, Keränen, and Larm (2004) note that screens and absorbent surfaces are never so effective that the speech level from the nearest desk could be attenuated below 40 dB(A). Jones and Macken (1993) argue that the main strategy for reducing the effect of irrelevant speech was to reduce it to below the threshold of hearing. This may be possible in some circumstances, but reducing the level of noise by some 40 dB would, in most cases, be technically challenging and financially prohibitive. Although it could be argued the 40 dB threshold is relatively low, perhaps the main point is that the one-size-fits-all approach to office design and acoustics is simply not effective.

The suggestion that activity-based design principles can help to improve room acoustics shows some promise as a solution. It is relatively easy to practically design spaces that provide the freedom to move about, as needed, for speech privacy, concentration, or collaboration. Implementing such principles, however, would require significant changes in organizational culture and design criteria—including the willingness to prioritize acoustic conditions.

There is an approach proposed for office design that may provide some insight into the way forward. Bodin-Danielsson and Bodin (2008) use a comprehensive categorization which allows future office concepts to be more precisely defined and studied, the "people-centered approach to design." The method integrates the complexities of the organization, the people, the processes, and the technology with the construction and architectural aspects of the design by taking a systems view to generate performance and sustainability benefits. The approach includes a flexible framework and a toolkit to support each stage of the design. The authors write that "we believe that theory-based practical methods and toolkits developed through such people-centered multidisciplinary working will ultimately provide a real way forward for improving building design."

4.5   INTRODUCTION TO PSYCHOACOUSTICS

4.5.1   DIFFERENCE BETWEEN SOUND AND NOISE

Noise is often defined as unwanted sound, and occasionally as unwanted or harmful sound. In contrast, desirable and beautiful sound is called euphony. Noise perception starts when sound pressure waves hit the eardrum, and the structures within the ear convert the pressure waves into a stimulus (signal to the brain) and continues as the brain organizes and interprets the signal and applies meaning to it (cognition).

The crux of the matter is that the term *unwanted sound* is totally subjective and based on a range of factors including a person's evaluation of the necessity of the noise, the meaning attached to the noise, whether it can be controlled, and the context (i.e., if the sound is considered normal and expected for the place where the sound is generated).

Benfield et al. (2012) point out that "the rumbling of a thunderstorm can be an exciting and pleasant experience to some but terrifying or depressing to another. Likewise, a parent trying to lull a newborn to sleep or a night shift worker trying to rest during the day perceives bird chirps, garbage trucks, and telephone rings differently from those who are currently less motivated for quiet conditions." Gifford (2007) states that "as the source of the sound becomes more relevant to an employee, as its meaning grows, and as its controllability and predictability decrease, sound is more likely to be perceived as noise and to negatively affect work behavior."

We are always unconsciously listening to sounds and processing information, in the workplace or elsewhere. In a TED talk titled "Why Architects Need to Use Their Ears," Treasure (2010) commented that "your ears are always on," compared with the eyes, which we can shut and thus switch off from visual stimuli. Similarly, Horowitz (2012) stated that "hearing is the only sense which is reliable, even when we sleep."

Psychologists refer to the cocktail party effect as the ability to differentiate important or relevant messages, such as your name, from background noise (Cherry 1953). In the workplace, a natural reflex action means that such unconscious listening to colleagues can be distracting and counterproductive when the information being processed is irrelevant to the performance of the individual (Broadbent 1958). However, the background conversation may not be considered noise if it contains useful information, whereas irrelevant conversation will be perceived as noise and found annoying and distracting, possibly leading to loss of performance.

Jones et al. (2008) commented that the research on the effects of noise on performance can be split into two eras: up to the 1970s, the research was concerned with how loud white noise interfered with cognitive and motor tasks, but from the 1980s, it was recognized that sound need not be loud to be distracting. Jones et al. (2008) noted that our understanding of how mental activities are susceptible to distraction from quieter sounds has appreciably broadened. Researchers are now preoccupied with how the content of the sound together with the nature of the mental activity results in distraction.

Put simply, interpreting sound requires obligatory processing without conscious attention, and this in turn can impair the performance of the concurrent cognitive tasks. This process harps back to how humans evolved—a balancing act is required from the brain's attention system so that we can focus on the task at hand while remaining open to the changes in the environment that might have important consequences for survival.

Summing up his psychoacoustics research, Jones (2014) writes:

Distraction is the price we pay for being able to focus on an event of interest while also gleaning some information from other sources of information. This arrangement has the undoubted advantage of allowing flexibility and adaptability—we can quickly move to new or potentially significant events—but it does mean that extraneous events of no significance can "capture" attention. Distraction from sound is particularly pervasive because we are obliged to process sound—whether we want to or not. Very low levels of sound can be quite damaging to cognitive performance, deficits of 20–30% being commonly found in the laboratory.

In conclusion, noise is clearly a psychophysical matter and it relates as much, if not more, to the interpretation and the meaning attached to the sound and how distracting it becomes as to the sound level per se. Therefore, a well-considered solution to noise in the workplace will facilitate a reduction in the possibility of distraction from the perceived noise rather than simply reducing the sound level, or the perceived loudness.

4.5.2 NONPHYSICAL FACTORS

The reported noise annoyance does correlate with the sound level measurement, but it is generally accepted that the sound level only accounts for 25% of the variance in annoyance. Borsky (1969) suggests that sound level is only a minor factor in noise annoyance, accounting for less than a quarter of the variance in individual noise annoyance reactions. Smith and Jones (1992) propose that noise intensity accounts for only 25% of the variance in annoyance, whereas psychological factors account for 50%, and conclude that the perception and the control of noise is more important than the physical aspects. Job (1988) concurs, writing that "even with the full range of exposure covered and very accurate noise and reaction measurements, noise exposure may only account for 25–40% of the variation in reaction." According to his review of 27 studies, Job found that the sound level accounted for only 18% of the variation in individual annoyance reactions, for those exposed to long-term traffic noise. Marans and Spreckelmeyer (1982) point out that the quantified effects of sound do not necessarily parallel the subjective experience of the same sound.

Tracor Inc. (1971) identified seven nonacoustical variables that are strongly correlated with aircraft noise annoyance: (1) fear of aircraft crashing in the neighborhood, (2) susceptibility to noise or noise sensitivity, (3) distance from the airport, (4) noise adaptability or perceived control, (5) city of residence, (6) belief in misfeasance on the part of those able to do something about the noise problem, and (7) extent to which the airport and the air transportation are seen as important. Sound pressure level measurements explained only 14% of the variance in Tracor's noise annoyance scores. The amount of variance increased to 61% when Tracor included the above-mentioned nonacoustical variables. Although none of these variables is directly relevant to the office environment, the study illustrates the importance of subjective and nonphysical variables.

Similarly, Borsky (1969) observed that the annoyance is heightened when (1) the noise is deemed unnecessary, (2) those making the noise appear unconcerned, (3) those being exposed to the noise dislike other aspects of the environment, or (4) the noise is considered harmful or associated with fear. In their review of population density and noise, Glass and Singer (1972) found that noise affects the behavior depending on the perceived context in which the noise occurs.

In his study of annoyance and sensitivity to noise, Vastfjall (2002) found that people in a bad mood respond more negatively to noise than those who are not. If a person is irritated or annoyed, they will make a more negative evaluation of a perceived annoying noise. So it appears that mood is also an important factor in how a person reacts to noise. For example, Cohen and Spacapan (1984) found that people are less likely to help others under high noise conditions, which may have an impact on the collaboration in the workplace.

Maris (1972) points out that, in general, the models of noise annoyance do not consider the social side of noise annoyance, and the nonacoustic influences may even be treated as an error variance. Maris maintains that sound is usually considered to be an external stimulus, and the evaluation of the perceived sound is studied as if it were an external process taking place in a social vacuum. With this in mind, he proposes that "the social psychological model of noise annoyance (Stallen 1999) considers as external stimuli both the sound ('sounds at source') and a social dimension of the exposure situation ('noise management by source'). The perception of these two stimuli influences an internal evaluation process that can result in noise annoyance. This internal evaluation process includes the appraisal of perceived disturbance and perceived control."

Maris concludes that several attempts have been made to improve our ability to predict the impact of sound levels on noise annoyance, saying that the approach to noise annoyance research remains "purely descriptive and exclusively acoustic."

It is clear that reaction to noise is not simply related to the perceived loudness—psychological factors also play a key role. Based on the research literature, there appear to be four key nonphysical factors relevant to office environments that affect noise perception and performance.

- Task and work activity—Task and work activity mean the nature of the task at hand or work activity, whether it involves cognition or memory; the complexity of the task, whether it involves multitasking; and whether the task requires a quiet environment (e.g., for concentration or sleep).
- Context and attitude—Context and attitude are the feelings toward those creating the noise, the perceived need for the noise, the meaning attached to the noise, and whether the noise source (e.g., conversation) is perceived as being useful.
- Perceived control and predictability—Perceived control and predictability are whether the noise source is intermittent or steady; whether it is predictable; and whether the people who are exposed to the noise can control it.
- Personality and mood—Personality and mood are the differences in those who are more noise sensitive, and in those who seek stimulation versus those that prefer solitude; and the effect of moods such as anger.

Clearly, psychological and social factors affect our response to sound level, and whether we even consider the sound to be noise. A psychophysical, or more specifically a psychoacoustic, approach to workplace noise is required.

4.5.3 NOISE SOURCE AND EFFECT ON PERFORMANCE

Research into the impact of noise on performance has resulted in mixed and often confusing results. As Matthews et al. (2013) point out:

> The study of noise effect on performance is deceptively difficult; noise can affect the efficiency of task performance, usually for the worse but occasionally for the better ... Individuals may not find a particular noise level annoying but their task performance may nevertheless be impaired. Conversely, they may find a particular noise level extremely annoying and yet their task performance may be unaffected.

The reason for the confusing results is the complex interplay between the four factors described above and the difficulty in quantifying the noise source, as previously discussed. In acoustics, much performance research has focused on the impact of ambient versus intermittent sound and relevant versus irrelevant speech.

4.5.3.1 Ambient versus Intermittent Sound

Ambient sound refers to long-term steady background sounds, whereas intermittent sound refers to short-term transient or sporadic sounds. Chanaud (2009) explains that, in general, long-term steady sound becomes normal to the listener and is not noticed. In contrast, transient sounds generally distract a person's attention, and strongly so if the level is high relative to the steady sound (e.g., an increase of 10 dB). Chanaud goes on to say that the distraction is further strengthened if the sound has high information content, such as a meaningful conversation. As Atkinson, Atkinson, and Hilgard (1983) point out in their introduction to basic psychology, predictability is key and "we are much more able to 'tune out' chronic background noise, even if it is quite loud, than to work under circumstances with unexpected intrusions of noise." Their views are backed up by many research studies, but most studies also emphasize the relevance of the task being performed.

Donald Broadbent, an experimental psychologist at Cambridge University, was the leading authority on the impact of noise on performance. After many years of pivotal research, Broadbent (1979) concluded that the performance will not be affected by a continuous loud noise when an employee (1) performs a routine task, (2) merely needs to react to signals at certain times, (3) is informed when to be ready, and (4) is given clear visual signals. However, the performance is affected when the person is multitasking or paying attention to multiple sources. Broadbent found that noise hinders complex tasks but sometimes improves simple tasks. Rabbitt (1968) reported that unpredictable or irregular noise disrupts the performance of mental tasks that require learning or short-term retention of new information.

Another seminal piece of research exploring the impact of noise on performance was conducted by Glass and Singer (1972). They subjected people to soft and loud bursts of sound; for some participants, the sounds were timed 1 min apart, but for others, the sounds were random. Glass and Singer (1972) found that interrupting their participants with unpredictable noise resulted in them making more errors in a proofreading task than the participants who were exposed to regular sound bursts (38.4 errors on average compared with 29.6 errors). In a further experiment, Glass and Singer found that the participants who were given information that allowed them to anticipate loud sound bursts performed better than those who could not predict the intermittent sound. Glass and Singer proposed that uncontrollable noise is a source of stress that results in reduced performance. Their results also have consequences for providing control over noise.

Many studies on the impact of noise on performance take place in a laboratory or simulated office environments. Respected researchers Banbury and Berry (2005) assessed subjective reports of distraction from various office sounds among employees at two different office locations. Their study measured the amount of exposure the workers had to ambient sound in order to determine any evidence of habituation (i.e., workers no longer noticing the background sounds). They found that almost all respondents reported that their concentration was impaired by various components of office noise, particularly unanswered telephones and people talking in the background. Unexpectedly, the study showed that employees are unable to habituate to noise in office environments over time, and the office noise, with or without speech, can disrupt the performance on more complex cognitive tasks, such as memory of prose and mental arithmetic.

So, on the one hand, there is plenty of laboratory-based evidence to indicate that people habituate to background noise, but real-world studies indicate that generalizing this finding is not so straightforward—a fact argued by environmental psychologists for some time (Oseland 2009). For example, the background office sounds used in the earlier laboratory study may have been considered a novel source of sound that the subject knows is not long-term, whereas office workers spend a large amount of their time exposed to the noise in their offices. The participants of the experiments conducted in laboratories will also have different motivations and attitudes from those being studied in the real world. Another important factor, more important than the sound level or time, is whether the background noise (such as that in an actual office) includes relevant speech, as discussed in Section 4.5.3.2.

4.5.3.2   Relevant versus Irrelevant Speech

Relevant speech refers to background speech that is intelligible or possibly has content that has meaning to the listener, whereas irrelevant speech is less intelligible and does not include content that is meaningful for the listener.

Jones et al. (2008) report that memory is particularly sensitive to disruption by background or irrelevant sound, with the negative impact of around 30%. More importantly, the effect on memory underpins many of the other reported effects on performance. For example, short-term memory plays a key role in language skills, particularly when the person is unskilled or stressed, which explains why people who are performing tasks involving memory while being subjected to meaningful speech are more likely to be affected than people who are exposed to irrelevant speech.

Marsh, Hughes, and Jones (2009) conducted four experiments "centered on auditory distraction during tests of memory for visually presented semantic information." Basically, they asked their English-speaking subjects to assign various objects to four different categories and then recall their responses under the states of quiet, pink noise,* meaningful speech (English prose), and irrelevant sound (Welsh prose). They found that irrelevant background sound caused higher distraction and disrupted recall (memory) more than meaningless sound. The effect was exacerbated when the irrelevant speech was semantically related to the material to be remembered. The effects of meaningfulness and semantic relatedness were shown to arise only when the instructions emphasized recall by the category rather than by the serial order. They concluded that their experiments "illustrate the vulnerability of attentional selectivity."

The irrelevant speech effect was first identified by Colle and Welsh (1976), and it has been replicated by a number of researchers using simple serial-recall tasks. Irrelevant speech effects have also been observed using more complex cognitive tasks such as proofreading and text comprehension. The evidence is overwhelming: the irrelevant speech effect that occurs in memory, especially in tasks where the order of information is important, is mostly due to the meaning of the speech and is independent of the intensity of the sound.

The groundbreaking research carried out on speech disruption is that of Banbury and Berry (1997). They examined whether people can become habituated to background noise by testing people's ability to recall a prose under three speech conditions. In experiment 1, they found that background speech can be habituated to after 20 min of exposure and that meaning and repetition had no effect on the degree of habituation seen. Experiment 2 showed that office noise without speech can also be habituated to. Finally, experiment 3 showed that a 5 min period of quiet, but not a change in voice, was sufficient to partially restore the disruptive effects of the background noise previously habituated to. These three experiments showed that irrelevant speech, and office noise that does not contain speech, can be habituated to after a prolonged exposure to the noise stimuli. However, as previously mentioned, a later study by Banbury and Berry (2005), carried out in real offices, actually found that employees are unable to habituate to background noise over time.

Beaman et al. (2012) explored the impact of English and Welsh background speeches on memory of English words among English speakers and bilingual Welsh speakers, and they found that English monolinguals displayed less disruption from the Welsh speech indicating that the meaning of the background speech had an effect on performance. In a second experiment, only English-speaking monolinguals participated and English was used as background speech, but the task complexity was increased. Participants were asked either to simply count the number of vowels in words or to rate them for pleasantness before recalling them. Greater disruption to recall was observed from the meaningful background speech when rating the words for pleasantness compared to counting vowels. These results indicate that background speech is analyzed for meaning, but whether the background speech causes distraction depends on the nature and the complexity of the task.

Jahncke (2012) conducted a series of experiments investigating the impact of speech intelligibility on performance. He actually found decreased word memory performance, increased fatigue, and poorer motivation when the background sound level was increased by 12 dB. More importantly, he showed that the cognitive performance decreased as a function of background speech intelligibility— the higher the intelligibility, the worse the performance. He also demonstrated that the performance is more impaired by background speech if the main task requires "episodic memory and rehearsal" (i.e., word memory and information search tasks).

Veitch et al. (2002) explored the impact of simulated ventilation noise (steady sound) and simulated telephone conversations (intermittent relevant speech) on noise satisfaction. They found that the acoustic satisfaction as subjectively rated speech intelligibility decreases and concluded that office workers require speech privacy.

---

* Pink noise is a variant of white noise (a random signal containing all the frequencies within the human range of hearing) which has been filtered to create sound waves with uniformly distributed energy at each octave.

Many experiments have been conducted to determine whether sound masking (i.e., low background sound using white noise or pink noise) can reduce speech intelligibility and, therefore, reduce noise distraction and increase performance. The research in this area has mixed results. For example, Veitch et al. (2002) found that noise masking, which matches the speech spectrum, is more effective at making speech less intelligible. However, they also noted that simply making the masking noise louder is not a guarantee of improved speech privacy. Indeed, they found that masking sound levels much greater than 45 dB(A) were judged to be too loud.

Haapakangas et al. (2011) examined how the room acoustic design affected cognitive work performance in a full-scale simulated open-plan office (90 m^2; ~970 ft^2). Four acoustic conditions with different speech privacy levels were built, plus a silent condition. The conditions were created by changing the acoustic environment using screens, absorbers, and a speech-masking system. The performance was measured for several cognitive tasks considered essential for office work. As might be expected, they found that the silent condition was the most beneficial acoustic condition for cognitive tasks. In contrast, the condition with the lowest speech privacy (highest intelligibility) was the least beneficial. Unexpectedly, they found that damping (absorption) without sound masking had a similar effect to damping with masking. More importantly, they found that masking, without any acoustic damping, had the worse effect on the cognitive performance. In conclusion, this study places some doubt on the value of sound masking in open-plan offices without adequate absorption.

In the previous section, we presented evidence that showed people are more distracted by intermittent sound than steady background sounds. The evidence also indicated that people could habituate to long-term ambient sound. However, these conclusions were drawn from controlled office simulations, whereas real-world office studies indicate that background sound cannot be habituated to. Further (mostly laboratory) studies suggest that the lack of habituation is more prevalent when the background sound has meaning, such as intelligible and relevant speech. Relevant speech has been found to have a greater impact on disrupting cognitive tasks, in particular those requiring memory (recall) or semantic assessment. So from a practical point of view for offices, the key is to reduce meaningful speech from distracting those carrying out cognitive tasks involving memory, e.g., complex analysis and authoring original prose.

4.5.4 Errors in Predicting Performance

In most scientific studies, the relationship between the dependent variable (Y axis or ordinate) and the independent variable (X axis or abscissa) is explored. In psychophysical studies, the dependent variable, such as noise annoyance, is usually subjective (albeit quantified), and the independent variable is usually more objective, such as the sound level. The objective independent variable may be used to predict the subjective response, and any (error) variation in the prediction is usually associated with the more subjective-dependent variable (Figure 4.3).

As noise (unwanted sound) is both perceived and subjective, it follows that sound level is only a proxy measure of the dependent variable, and as a consequence, it comes with inbuilt (error) variation. If there are measurement errors in both the Y (dependent) and X (independent) axes, then the combined variance due to the product of X and Y will be significantly larger than an error on only one axis and thus results in the unreliability of the predicted relationship of Y and X.

Office worker performance is notoriously difficult to accurately measure. The studies described earlier mostly use laboratory-based performance tasks such as word recall, word searches, basic mathematics, and proofreading. Environmental psychologists question how such tasks relate to the multitasking and complex processes that take place in real office work.

The key measures in performance tasks are speed and accuracy, and there is usually a trade-off between the two. In the real world, although speed and accuracy are important, quality and creativity are also important. Most laboratory studies usually focus on individual performance, but in real offices, the output is often the combined effort of a team. So performance tasks are also a proxy measure and do not completely represent the impact on office worker performance.

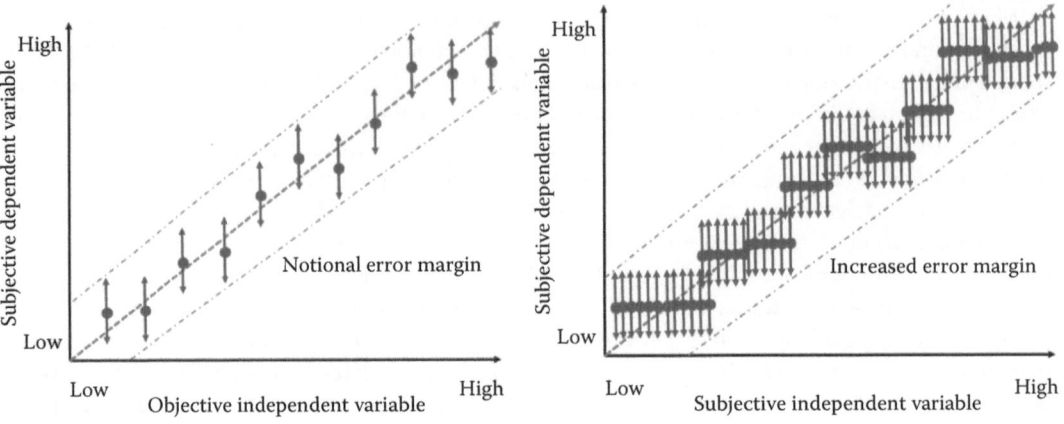

FIGURE 4.3   Subjective independent variables increase error and unreliability.

Performance tasks generally provide objective quantified data. So, in the acoustical studies described earlier, we have the unusual situation that the dependent variable is objective, and the independent variable is a proxy metric for subjective data.

4.6   RELEVANT PSYCHOLOGICAL METATHEORIES

4.6.1   Personality Theory and Arousal Theory

Oseland (2012) explains that

> personality theories date back to ancient Egypt and Mesopotamia but the ancient Greeks are most recognised as developing the first structured theory of personality. At the turn of the century the psychoanalysts, Freud and Jung, developed the psycho-dynamic theory of personality … Eysenck's (1967) two super-traits model is derived directly from Jung's theories and … he proposed two personality dimensions: extraversion (E) and neuroticism (N). Full extroverts and introverts sit on opposing ends of the extraversion dimension: an "extrovert is a friendly person who seeks company, desires excitement, takes risks, and acts on impulse, whereas the introvert is a quiet, reflective person who prefers his or her own company and does not enjoy large social events" (Eysenck and Eysenck 1975). Neuroticism is a dimension of emotional stability that ranges from fairly calm and collected people to ones that experience negative emotional states such as anxiety and nervousness.

Psychologists propose that introverts and extroverts have different innate levels of arousal, which in turn affects how noise affects their performance, as explained by Oseland (2009):

> A key fundamental theory is the Yerkes-Dodson Law (Yerkes and Dodson 1908) which proposes an inverted U-shape relationship between a person's performance and their level of arousal, i.e. excitement or interest. The theory states that people can perform better if they are stimulated or motivated (which increases their level of arousal), but there is a limit, as too much stimulation can lead to stress and thus reduce performance. We might therefore assume that to maximise the performance of office workers we need to design stimulating but not too over-stimulating environments. Unfortunately, one complication is that individuals have a different base level of arousal and therefore need different magnitudes of stimulation for optimal performance. For example, extroverts have a low natural level of arousal and enjoy thrill-rides whereas introverts who have a higher level of arousal might find such rides distressing.

Noise is considered to be a form of stimulation, so it follows that extroverts should perform better than introverts in noisy environments (Figure 4.4).

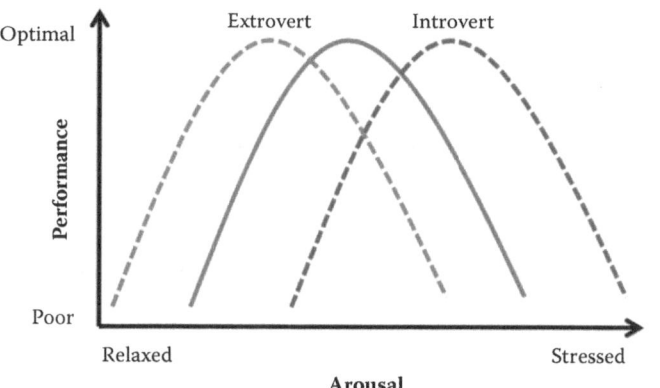

FIGURE 4.4   Performance is affected by arousal, which is affected by personality type.

Oseland (2009) continues:

> A further complication is that difficult and complex tasks (or working under time pressure) are in them-
> selves demanding and therefore increase the level of arousal, thus people need subdued environments to
> maximise performance. In contrast, repetitive or menial tasks require more stimulating environments
> to increase the level of arousal. So, in simplistic terms, stimulating environments with music or noise
> and a buzz of activity may enhance the performance of extroverts or those conducting simple tasks, but
> more calming environments will better suit introverts or those involved in more complex tasks.

Much of the earlier research on the impact of noise on task performance, notably by Broadbent
(1958), generated results in support of the arousal theory. Szalma and Hancock (2011) report that
Broadbent (1978) proposed that noise

> increases arousal which decreases the breadth of attention. At relatively lower levels of arousal, the
> attentional narrowing facilitates performance because it causes the individual to exclude irrelevant
> cues. Beyond an optimal level, however, increases in arousal cause increased narrowing so that task-
> relevant cues are also excluded, and performance is thus impaired.

They also noted that Poulton (1979) supports arousal theory but "argued for a composite model
of noise effects involving arousal and masking of inner speech ... the gains in performance in continuous noise early in the task occur because the increase in arousal compensates for the deleterious
effects of masking."

However, Szalma and Hancock (2011) also point out that the negative effect of intelligible versus
unintelligible speech, explained in the previous section, is inconsistent with Broadbent (1978)'s or
Poulton (1979)'s version of arousal theory. They would expect that noise should increase arousal
regardless of the content of the noise, so they proposed the adoption of the Maximal Adaptability
theory described by Hancock and Warm (1989). This theory relates to the adaptive response of the
individual to noise, as Szalma and Hancock (2011) believe that it is more useful:

> From this perspective, intermittent speech noise of relatively short duration is most disruptive because
> it consumes information-processing resources that the individual cannot effectively replenish through
> compensatory effort because of the limited exposure to the stressor. In contrast, for conditions of con-
> tinuous noise of longer duration, individuals can develop more effective coping strategies.

Regardless of which theory is adopted, it is clear that accounting for introversion/extraversion
and task is of utmost importance when designing environments that minimize the noise distraction.

For example, the theories indicate that an introvert conducting a complex task would thrive in a quiet environment, whereas an extrovert conducting a simple task requires a more stimulating/noisy environment.

Eysenck (1967)'s supertrait model (based on extraversion and neuroticism) underpins the most popular current theories of personality such as the Big Five model (John and Srivastava 1999). The five factors are openness, conscientiousness, extraversion, agreeableness, and neuroticism, and are often referred to as OCEAN.

- Openness—Openness refers to being open to new experiences and is a trait of those who are creative, curious, with broad interests, are imaginative, and artistically sensitive.
- Conscientiousness—Conscientiousness relates to those who are more responsible, hard-working, organized, dependable, self-disciplined, and persistent.
- Extraversion—Extraversion refers to those who are more sociable in nature and prefer keeping company; they are also more impulsive, gregarious, assertive, talkative, and are thrill-seekers.
- Agreeableness—Agreeableness is seen in those who are more cooperative, affectionate, good-natured, helpful, forgiving, caring, and trusting.
- Neuroticism—Neuroticism refers to emotional instability and the tendency to experience negative emotions and anxiety.

The impact of different personality traits on the performance under noisy conditions is explored in Section 4.7.

4.6.2 Environmental Psychology and Behavior

Environmental psychology is a relatively new field of psychology that explores the interrelationship between people and their physical settings. The main focus of this chapter is the research related to perceived noise, and corresponding behavior, in office buildings. Environmental psychology theories provide further insights into psychoacoustics, and why it is important to address psychological as well as physical variables when investigating and resolving noise distraction.

As Oseland (2009) explains:

> Traditional psychology took the view that behaviour is simply a deterministic response to the physical world, but Kurt Lewin (1943) proposed a different perspective, summed up by his equation $B = f(P, E)$, declaring that behaviour is a function of the person (P) as well as the environment (E).

So how we perceive and interact in an environment is dependent upon our individual experiences and different expectations of that environment. Such factors will affect whether we consider a particular sound to be a noise.

Oseland continues:

> Roger Barker (1968) introduced the notion of behavioural settings where the pre-conceived social etiquette associated with a particular setting unconsciously influences the behaviour in that setting, for example consider the behaviour of people in churches and libraries.

In such environments, even quiet sounds are unexpected and considered disturbing. In contrast, the high sound levels at a football match or a rock concert are considered acceptable by participants; they are expected and viewed as the norm.

So our preconception of the working environment will affect our perception of noise in that environment. If workers are going into the office expecting to carry out work requiring concentration, based on previous experience, then a situation where this is not the case will lead to dissatisfaction and likely result in reduced performance. Part of the solution is therefore managing expectations (informing staff what to expect) and agreeing to the acceptable sound levels (from speech and equipment) in different parts of the office.

4.6.3 EVOLUTIONARY PSYCHOLOGY AND BIOPHILIA

Evolutionary psychology is one of the newest fields of psychology, as Oseland (2009) explains:

> Evolutionists believe that over time physiology develops, through natural selection, to ensure the survival of the species. Similarly, evolutionary psychologists argue that innate human behaviour is governed by adaptations of psychological processes which evolved to aid our survival and well-being. Homo sapiens evolved around 400,000 years ago in natural environments, but people have only worked in offices for around 100 years. As a consequence a person's psychological processes are probably more adapted to living on the African savannah than they are to working in offices.

A key theme within evolutionary psychology is *biophilia*, a term coined by Wilson (1984), which explains our affinity to natural environments. Some evolutionary psychologists argue that people feel refreshed after sitting in a natural environment because nature provides a setting for "non-taxing involuntary attention." They also propose that people innately prefer noise to be at a similar level to that found in the natural world (i.e., a slight background buzz of activity).

Alvarsson, Wiens, and Nilsson (2010) observed that sounds from nature, such as birdsong or rippling water, promote faster recovery from stressful tasks compared with traffic noise and ambient building noise, such as that from an air-conditioning equipment. Jahncke (2012) studied the impact of breaks and rest periods on performance. He found that watching a nature movie (with sound) while taking a break increases energy ratings, arithmetic performance, and motivation, compared with just listening to office noise. Continued exposure to office noise gave the lowest ratings of motivation and decreased the task performance. Researchers Fitzgerald and Danner (2012) and Jahncke (2012) suggest that using pleasant sounds from natural environments to mask the background workplace noise could decrease employee stress and increase worker productivity.

The jury is out on whether white (or pink) noise masking helps reduce noise distraction, but an alternative might be more natural sound masking. Soundscaping, or acoustic ecology, is the relationship mediated through sound between living beings and their environment. Organizations such as Sustainable Acoustics and Julian Treasurer's The Sound Agency are exploring the benefits of implementing soundscaping systems in the office in order to strengthen the connection to outside positive sound sources and help mask the unwanted elements of noise.

4.7 ACOUSTICS AND PERSONALITY

4.7.1 TASK PERFORMANCE AFFECTED BY PERSONALITY TYPE

4.7.1.1 Extraversion

Several studies, mostly laboratory-based, have shown that extroverts perform better than introverts at cognitive tasks under noisy conditions. For example, in an early study Morgenstern, Hodgson, and Law (1974) found that extroverts performed better in the presence of distractions than they did in silence, while introverts showed a deficit in performance. The subjects were tasked with remembering words from a list read to them while they were being distracted by German words, versus performing in silence. Similarly, Standing, Lynn, and Moxness (1990) found that introverts and extroverts performed equally well in comprehension tasks in silence, but white noise (at 60 dB) impaired the introverts' performance, whereas the extroverts continued to perform at the same level.

In a mental arithmetic task applied on medical students under quiet laboratory conditions at 42 dB(A) and noisy ones at 88 dB(A), extroverted subjects performed significantly faster in noisy conditions (Belojevic, Slepcevic, and Jakovljevic 2001), and concentration problems and fatigue were more pronounced for the introverts. More recently, Alimohammadi et al. (2013) compared attention and concentration (Cognitrone test by Vienna Test System) in quiet conditions and under road traffic noise of 71 dB(A). Performance, in particular speed, was enhanced in extroverts in the noisy condition, but no significant difference was found in introverts. In a series of experiments, Geen (1984) allowed his subjects to adjust the level of background noise to an optimum during a learning task and discovered that extroverts selected higher sound levels. His studies also showed that the heart rate of introverts increased under noisy conditions, whereas that of extroverts did not, which is in line with arousal theory. The impact of extraversion on performance under noisy conditions has also been observed outside the laboratory. For example, Campbell and Hawley (1982) demonstrated that when studying in a library, introverts were significantly more likely to choose a place to work away from the buzz and the bustle of certain areas, whereas extroverts were more attracted to the latter as their place of work.

4.7.1.2 Neuroticism

More anxious (neurotic) personality types have displayed worse performance in noise compared to emotionally stable individuals, for example, in complex mental tasks such as the retrieval from semantic memory (von Wright and Vauras 1980) or learning prose (Nurmi and von Wright 1983). Eysenck and Graydon (1989) suggest that neurotic introverts are more adversely affected by noise than emotionally stable extroverts when carrying out work–life tasks.

Benfield et al. (2012) propose that "arousal and stress perspectives posit that noxious sound increases negative effect, generalized anxiety, and frustration in the listener, which in turn causes stress and/or arousal with subsequent changes in blood pressure, sleep patterns, immune functioning, and hormone levels." Or as Kryter (1970) explains:

> The general finding that the performance of the more anxious personality types is more affected by noise than that of non-anxious types would attest to the existence of a stimulus-contingency factor. In terms of learning or conditioning, the task becomes disliked and is performed relatively poorly because it is related to or contingent upon the aversive noise. Matthews et al. (2004) propose that, for people with predominantly neurotic and anxious personality types, performance is impaired by noise because of a reduced availability of attentional resources or working memory, due to diversion of attention to processing internal worries.

In conclusion, people with a more neurotic disposition are overly concerned with the source of the unwanted sound and respond to it more negatively, resulting in stress.

4.7.1.3 Openness, Conscientiousness, and Agreeableness

Franklin et al. (2013) examined the relationship between acceptable noise level and personality. The analysis revealed a correlation between acceptable noise and openness or conscientious personality dimensions. They suggest that people who are more open to new experiences may accept more noise, while those who are of the more conscientious personality type, who generally desire fewer distractions when focusing on a task, accept less background noise. No effects on agreeableness were reported.

4.7.2 Noise Sensitivity and Personality Traits

Much research has been done on noise sensitivity, and some researchers have proposed that it is a core personality factor to be considered when reducing noise distraction. In 1978, Weinstein developed his noise sensitivity scale and found that noise sensitivity is not strongly correlated with objective sound level, because some people are simply more sensitive to the same sound levels. However, noise annoyance and noise sensitivity are different. According to Stansfeld (1992), while annoyance is related to sound level, sensitivity is not, and noise sensitive individuals are likely to be more annoyed by noise than nonnoise sensitive individuals at all sound levels.

Although noise sensitivity is an important factor, we do not believe it is necessarily a personality trait per se. Noise sensitivity is determined by Wenstein (1978)'s scale of emotional response to noise, so it is not surprising that those who rate themselves as noise sensitive are then more annoyed than others by what they perceive as noise. Stansfeld (1992) notes that noise sensitivity is considered a self-perceived indicator of vulnerability, and it is linked to the perception of environmental threat and lack of environmental control combined with a tendency to negative affectivity. Stansfield's description is very similar to the personality trait of neuroticism.

Indeed, researchers have found that noise sensitivity is associated with neuroticism (Levy-Leboyer, Vedrenne, and Veyssiere 1976; Goldstein and Dejoy 1980; Job 1988; Stansfeld 1992; Belojevic and Jakovljevic 2001). Luz (2005) reports that "an association between noise sensitivity and neuroticism has been reported from England, Sweden, and Serbia … the statistical connection between noise sensitivity and neuroticism could be used to isolate and ignore the noise sensitive population." In their own research, Belojevic, Jakovljevic, and Slepcevic (2003) found that neuroticism was the best individual predictor of reported noise sensitivity.

Weinstein (1978) previously questioned whether the reported "link between introversion and noise sensitivity is due to direct arousing effect of noise on the central nervous system, or to the fact that noise frequently has interpersonal significance and is seen as an intrusion by those who are ill at ease in social settings and prize privacy (psychosocial effect)." Belojevic, Jakovljevic, and Slepcevic (2003) concurred that age, education level, and introversion were not significantly related to noise sensitivity. They conclude that

> noise sensitivity appears to be a complex, multidimensional personal characteristic but subject to situationally determined (and therefore considerably variable) cognitive and affective factors involving meaning, attitudes, motivation and so on.

As underpinning personality traits tend to be quite stable, noise sensitivity is more likely to be an attribute of a core personality factor, such as neuroticism, than an independent stand-alone personality factor. Noise sensitivity could be related to evolutionary psychology, as previously discussed. Luz (2005) proposes that people who are noise sensitive have a very active orienting response (OR), explaining that

> for our ancestral hunter-gatherers living in natural quiet, an active OR was essential for survival. It helped the hunter keep food on the table and the gatherer to avoid predators. However, in a world filled with roars, buzzes and bangs, an active OR can be a disadvantage, especially for people whose nervous systems have difficulty 'turning off' the OR. The process of 'turning off' the OR is called 'habituation,' and noise sensitive subjects have a harder time habituating to a repeated sound.

4.7.3 Music and Distraction

Listening to music in the workplace is becoming more common; usually people listen through headphones but occasionally music is played in the background. There are increasing concerns that the volume of music played through personal music players could damage hearing.

Music is often considered to be a form of exterior arousal. For example, an electrodermal measurement of arousal showed that playing simple tunes can significantly alter the base rate and the performance of extroverts and introverts (Smith, Wilson, and Davidson 1984). Our concern here is whether music has a beneficial or a detrimental effect on worker performance.

In an early experiment, Daoussis and McKelvie (1986) found that extroverts are reported to be working with music twice as much as introverts, but both groups played background music very softly. Daoussis and McKelvie (1986) administered a reading retention test to the two personality types under quiet conditions and with background music. The performance of the extroverts was not affected under the two conditions, but the performance of the introverts decreased with background music. Furnham and his colleagues at the University College London have carried out many studies on the impact of music on performance for different personality types. Furnham and Bradley (1997) investigated the effect of pop music during a memory test with immediate and delayed recall. They found that the introverts listening to music had a significantly lower recall than the extroverts in the same condition and the introverts who had completed the test in silence. Furnham and Bradley also found that introverts who completed a reading comprehension task when music was being played performed significantly less well than extroverts. Furnham and Bradley concluded that

> there seems little evidence that the presence of background distraction (television, music, talk) actually facilitates performance in complex cognitive tasks, even for extroverts, though it seems clear that it nearly always impairs the performance of introverts.

Furnham, Trew, and Sneade (1999) conducted a similar study examining the effect of vocal and instrumental music on reading comprehension and logic problem tasks. They found that vocal music was more distracting than instrumental music, which is in line with the research on intelligible speech reported earlier. They also found that the performance of the introverts was impaired by the introduction of music whereas that of extroverts was enhanced. In a more recent study, Furnham and Strbac (2002) examined whether background office noise would be as distracting as music. They exposed the participants to silence, background garage music, and office noise while the participants carried out a reading comprehension task, a prose recall task, and a mental arithmetic task. Furnham and Strbac (2002) found that the introverts performed less well on the reading comprehension task than the extroverts in the presence of music and noise. They also confirmed that, in general, the performance was worse in the presence of music and noise than in silence.

Cassidy and MacDonald (2007) reported that introverts were more detrimentally affected by the presence of high-arousal music compared with extroverts. Chamorro-Premuzic et al. (2009) examined the effects of different types of background auditory stimuli on the abilities of introverts and extroverts to perform cognitive and creative tasks. Results showed no significant effects of background auditory stimuli and personality on either cognitive task performance. However, there was a significant effect on the creative performance, with extroverts performing better in the presence of music than introverts. Chamorro-Premuzic et al. (2009) concluded that background music may have a more detrimental effect on the creative task performance of introverts compared with extroverts.

In addition to music, Furnham, Gunter, and Peterson (1994) studied the impact of the presence of an operating television on introverts and extroverts, and found that introverts and extroverts performed equally well at reading comprehension tasks with the television switched off. However, the extroverts performed better than the introverts when the television was on.

Various other studies have examined the distracting effects of television on cognitive processing. Significant performance decrements have been reported for several measures, including spatial problem solving, mental flexibility, and reading comprehension as a function of television (Armstrong and Greenberg 1990; Armstrong, Boiarsky, and Mares 1991). These results are consistent with the idea that background television influences the performance by causing overstimulation when people are performing complex tasks.

4.7.4 Control of Noise and Performance

Quite a few research studies have focused on how the perceived control of noise affects performance. The seminal piece of research in this area is that of Glass and Singer (1972), described earlier, where people in a noisy environment were asked to perform a proofreading task, and they were told either that they could control the noise or that they had no control over the intermittent noise bursts. The subjects performed equally well during the experiment. However, when both groups were tested on a task after being exposed to noise, the performance of the group who had been given control, even if they did not use it, was significantly better. Glass and Singer suggested that this may have been a consequence of the increased tolerance for frustration in conditions in which noise bursts were signaled and could be anticipated. All individuals were exposed to identical circumstances, yet those with the perception that they could alter their circumstances, if they chose to do so, experienced less stress. Another significant finding is that the stress from noise continued to affect the performance of the subjects longer after they were exposed to the noise.

Rather than focus on noise, Carton and Aiello (2009) examined the effects on task performance of the control over social interruptions. They defined an interruption as

> any disruptive event that impedes progress toward accomplishing organizational tasks ... Social interruptions are those that are initiated by human actors.

They showed that participants who were able to anticipate social interruptions performed significantly better than those who could not anticipate them.

Furthermore, participants who had the opportunity to prevent interruptions reported significantly less stress than those who could not. Carton and Aiello (2009) argued that individuals with the knowledge that they may be exposed to interruptions have the ability to use preventive coping tactics to minimize the disruption and the frustration when the interruptions occur. More importantly, from an office perspective, they concluded that individuals need not actually prevent interruptions from happening in order to be beneficial—simply believing that they can prevent interruptions has an effect.

Other researchers have noted that some individuals appear better able than others to cope with the excessive stimulation inherent to the open-plan office environment. Mehrabian (1977) proposed that such individual differences in coping are due to an innate ability to screen. He distinguishes between screeners, who effectively reduce overstimulation by attending to the information on a priority basis, and nonscreeners, who cannot apply this strategy and tend to become overstimulated.

4.8   DESIGN IMPLICATIONS FOR OFFICES

4.8.1   Using the Physical as a Means of Noise Control

From a physical perspective, adequate amounts of absorption are necessary to control sound propagation and aid speech privacy in open-plan environments. The use of sound-absorbing ceilings, screens, carpeting, soft materials, and materials for diffusion (such as plants and trees) are all necessary design criteria (Table 4.2). In addition to the material choices within the space, an activity-based approach allows freedom of the individual to physically go to the appropriate space based on the activity and need for privacy, disturbance reduction, and so on. Adequate acoustic boundaries should also be created between quiet zones for concentration and those where speech may be elevated, such as collaboration spaces. Dedicated quiet booths can be used as an effective solution when space is limited, provided they are not located near vocally intensive areas.

Considerations for the type of space is a first step, and include (1) the size of the room, (2) the location in relation to other spaces, and (3) whether the materials in the space are made of particularly hard surfaces such as concrete (these and other smooth/hard surfaces are like a mirror from the acoustic perspective).

Unless there is a need to project sound, the accommodation should be designed to limit the sound reflections through absorption and aid the sound propagation with barriers capable of absorbing at the speech frequencies. Diffusion (spreading the sound energy) is of importance and will occur due

TABLE 4.2

Examples of Specific Areas and Suggested Physical Design Criteria

Space	Considerations	Physical Solution
Informal meeting areas	Keep sound levels low; reduce sound propagation	Locate away from concentration areas; use partitions or screens, plants, and absorbing walls and ceilings
Open-plan office work area, individual desks	Keep speech and other sounds from spreading and minimize the disturbance of coworkers	Use a sound-absorbing ceiling with good absorption qualities at speech frequencies (400–2000 Hz) and sound-absorbing screens dividing people into groups
Semiopen meeting areas	Prevent sound from spreading, prevent sound levels from escalating, and avoid the need for people to raise their voices (the team should have local speech clarity, so they can talk normally)	Sound-absorbing ceilings with good absorption qualities at speech frequencies; and if people in adjacent areas can be disturbed, use sound-absorbing screens
Project rooms	Hinder wall-to-wall echoes and support communication (speech clarity) (the room should also be properly sound insulated to keep sound from entering or leaving the space)	Use sound insulation, a sound-absorbing ceiling and flooring with good absorption qualities at low frequencies, and wall absorbers
Phone intensive	Speech flows in all directions, resulting in escalating sound levels and impaired speech clarity over the phone, and noise disturbance to other areas	Use a sound-absorbing ceiling with the best absorption qualities at all frequencies, sound-absorbing screens dividing people into groups, and wall absorbers on every possible wall space

to the furniture and other objects in the room; but trees and extensive planting can also be good for the acoustic environment. Consideration should also be given to the type of equipment in the space; fans, projectors, or other mechanical or electrical sound sources should be located away from concentration areas.

4.8.2 Beyond the Physical—New Practical Guidance

Noise is subjective and is as much a psychological as a physical problem. In particular, distraction from noise results in the loss of concentration/focus and memory/recall, which in turn results in the loss of office worker performance.

The interpretation of sound as a noise depends upon a range of factors as previously described. To summarize, noise distraction is related to the following:

- Task and work activity
- Personality and mood
- Perceived control and predictability
- Context and attitude

But as shown earlier in this chapter, different office workers will react differently to the same acoustic conditions in their workplace. Therefore, the actions to resolve the noise distraction need to account for individual differences and not assume that a single physical acoustic solution will work for all office occupants. A psychoacoustic approach to understanding noise distraction indicates that other, people-centered solutions are also required. Such solutions are more behavioral, educational, managerial, and organizational rather than physical.

We will now consider the possible solutions that address the four key factors highlighted earlier.

4.8.2.1   Task and Work Activity

Individuals and teams typically conduct a range of work activities throughout the day. For example, a part of the day may involve meeting colleagues or clients and some of the working day may be spent solo, carrying out processing or analysis tasks. Such activities are better performed in different work environments, which are specifically designed to support the activities, for example, a meeting space is quite different from the space required for focused work. A core principle of activity-based working (or agile working) is that employees are provided with a choice of a range of work settings that support their different work activities.

Activity-based working environments typically include the following:

- Meeting and teleconference rooms that have good acoustic properties to reduce sound transference, offering acoustic privacy and also reducing the noise distraction to and from outside the room.
- Focus rooms or pods for the staff to go to if they wish to carry out work that requires concentration, or a confidential call, and to be free of distraction from their colleagues.

Some activity-based working environments provide phone booths to allow the staff to make personal and confidential calls away from the open-plan desks and their colleagues. Small teleconference rooms allow several staff to join a call without distracting colleagues in the open-plan office.

Rather than offer rooms for focused work, some organizations are now creating larger quiet zones as part of the activity-based working options. Such zones tend to not have desk phones and prohibit impromptu meetings. Part of the agile working approach is to allow occasional remote working, including working from home, where employees can more easily control the level of distraction.

Although activities may vary throughout the day, different teams will usually have core work activities that take up the majority of their day. For example, a sales team is likely to spend more time on the phone than a team of analysts. The working environment for the team can therefore usually be planned around their core work activities, and the teams conducting similar activities could be placed together. Generally, those involved in complex or detailed tasks, task requiring memory and recall, or people who are multitasking are likely to require a quieter environment than those involved in simple single tasks. Obviously, it is preferable to avoid locating more vocally active teams who prefer buzzy environments next to those requiring quiet environments for concentration.

We acknowledge that many organizations are looking to break down team silos and facilitate interaction between teams. Nevertheless, if the primary work activity of the team is heads-down work, then the space should be designed to support that, and additional work settings away from the main open-plan workspace should be provided for interaction and collaboration.

Creativity and innovation are increasingly important attributes of any business. Stimulating spaces are required to promote creativity, but it should also be acknowledged that much of the creative process takes place in solitude away from distraction (Oseland et al. 2011).

4.8.2.2   Personality and Mood

The research literature shows that some personality types are better at coping with noise distraction than others, in particular people who are predominantly extroverted compared to those who are more introverted. Research into collaboration has shown that the most productive teams are those with a rich mix of personality types, but the design of many workplaces is often more suited to extroverts (Oseland 2012).

Psychological profiling is often used to determine whether a person has the relevant personality and attitude for joining an organization. However, they may then be placed in a workspace designed with other personality types in mind. We propose that personality profiling should also be used to cluster people who prefer and function better in similar acoustic environments. Thus, people

TABLE 4.3

Preference of Sound Level Depending on Task and Personality

Personality	Task	Quiet	Loud
Introvert	Simple	😐	🙁
Introvert	Complex	🙂	🙁🙁
Extrovert	Simple	🙁🙁	🙂
Extrovert	Complex	🙁	😐

who are primarily categorized as introvert, neurotic, and conscientious personality types could be accommodated together in spaces that facilitate quiet work. In contrast, those who are primarily extrovert and more open personality types could be allocated to a space in stimulating (loud) environments. Better still, the different personality types could be offered a choice over where they wish to work and select their preferred location.

There is an interaction between personality type and task, as shown in Table 4.3. Introverts generally require quieter spaces to compensate for their innate higher level of arousal, but complex tasks increase arousal, so introverts cope the least well with complex tasks in loud stimulating environments. They perform slightly better at simple tasks in loud environments. In contrast, extroverts prefer loud stimulating environments to counter their innate low level of arousal. Simple tasks in quiet environments do not sufficiently stimulate the extroverts, so they perform least well under these conditions. Complex tasks increase arousal such that the extroverts cope better with these under quieter conditions. Mood affects our willingness to help other people under noisy conditions, and the perception of noise can affect the mood. In organizations seeking to enhance collaboration, it is important that noise annoyance is not increased due to perceived unnecessary noises.

4.8.2.3 Perceived Control and Predictability

The lack of control and the predictability of noise decreases performance, but there is also research evidence to show that the degradation of performance continues after the noise has ceased. The research also indicates that it is perceived control rather than the actual control of noise that has alleviating effects. It is not always practical to give full control over the noise, particularly in open-plan environments, but there are other solutions. Offering a choice of work settings (e.g., by implementing activity-based working) gives employees the option of moving to a quiet zone or room and thus distancing themselves from the noise source. In this solution, it is important that the people affected fully understand that they have options, and they are given a full choice.

Another approach is to introduce some form of office etiquette around noise. Quite often, the people who find loud telephone conversations or nearby discussions distracting believe that they cannot alleviate such problems. These people tend to be the ones who carry out work requiring quiet and also tend to be of the personality types that avoid unnecessary confrontation. Having office protocols, which is a type of charter or policy document, can be particularly helpful to those personality types. The office etiquette should set out the acceptable behavior and acknowledge that the unacceptable behavior can be challenged by all. It can be presented in written format and posted online, similar to office sustainability and other environmental guidelines.

For example, the etiquette document could cover the following:

- What the team member can do when disturbed by unanswered phones, loud teleconference calls, unnecessary chatting, and local meetings
- Guidelines on the acceptable use of mobile phones (e.g., set to voice mail after four rings or put on silent when in the office) and that it is acceptable to switch off unanswered phones
- Protocols that suggest lengthy discussion are continued away from the desks

The agreed protocols would need to be backed up with alternative work settings. The important point is that each team needs to agree on the preferred behavior, and the team members must feel that they have some control over the unnecessary noise.

Finally, it is important to provide methods of controlling the interruption from colleagues. Some organizations use visual cues to indicate when a person is busy, such as small busy flags on the desk. There are mixed views over such techniques, but if a team likes the idea, then it is worth incorporating into the office etiquette. A similar option is to use personal computer presence indicators, such as Microsoft Lync or Cisco Unified Presence, which can be set to "busy" or "available," so that the colleagues refer to the status set by a person before approaching them, or they would ping an instant message to see if they are free. Another visual "do not disturb" cue often used is headphones; they also act as a form of sound masking and can improve the performance of some personality types. Again, there are mixed views on the use of headphones and whether they lead to isolation from the rest of the team. At minimum, we should be cognizant of when a colleague is in midflow before approaching them.

4.8.2.4   Context and Attitude

The perception of noise is affected by the attitudes toward the source of the noise. If people feel that a sound source is justified (e.g., an important announcement) or they are more familiar with those generating the sound (such as close teammates), they will be more tolerant of the distracting noise. So grouping teams such that the background speech may be of value to them rather than a distraction can be helpful. Management should clearly explain to the new members of the team whether it is a noisy or a quiet team and what the norm is. If it is a noisy team, then the manager should justify the business reasons for it and explain the benefits.

It has been demonstrated that relevant speech is more likely to be distracting to a team member than speech that is interpreted as irrelevant, so if a task requires complete focus and concentration, removing noisy team distractions may be preferable (e.g., by working elsewhere).

The facilities management team should announce any unusual planned noises in the workplace (e.g., building works). If they explain the reasoning behind the noise, the resulting benefits, and the timescales, then the occupants are likely to be more tolerant of the noise.

4.8.3   Generic Solutions

The solution to noise distraction is as much to do with the management of the space and the guidance on behavior as it is about the design and acoustic properties. A choice of different types of spaces with different acoustic properties and agreed behaviors is essential for reducing the noise distraction.

People-centered acoustic solutions can be summarized as follows:

- Displace—Displace the noise distraction by providing easy access to informal meeting areas and breakout and brainstorming rooms. Provide quiet areas for the staff to retreat to, including quiet booths, phone-free desk areas, or a library-type space, plus the option to occasionally work from home. Good design and visual cues can be used to indicate how people should behave in a space and the expected noise levels (e.g., consider the layout and the design of a library compared with a café).

- Avoid—Avoid generating noise distraction (e.g., do not provide hands-free speaker phones in open-plan offices or meeting tables in the middle of workstations where people are carrying out work requiring concentration). Locate noisy teams together and away from the quieter teams. Colocate team members, because people are more tolerant of noise from their own team. Consider the personality of the staff and perhaps separate the extroverts who thrive in noisy environments from the introverts who prefer quiet.
- Reduce—Reduce the noise distraction by controlling the desk size and the density (high-density environments with people closer to each other generate more noise distraction). Consider allowing people who are conducting repetitive tasks requiring concentration to use headphones. Use good acoustic design to reduce speech intelligibility across open-plan areas and noise transference between rooms. If sound masking is to be used, consider using more natural soundscapes rather than white or pink noise.
- Educate—Introduce some form of office etiquette which reinforces consideration toward colleagues. The etiquette should cover phone use, loud conversations, music, headphones, managing interruptions, how different work settings are used, and so on. It may also include "do not disturb" signals. Explain to the staff how the office layout works, the facilities available to them, and how they can control noise disruption. If required, explain and justify why there is a noisy environment.

REFERENCES

Abbot, D. 2004. Calming the office cacophony. *Saf Health Pract.* 22(1):34–6.

Action on Hearing Loss. http://www.actiononhearingloss.org.uk/your-hearing/about-deafness-and-hearing-loss /statistics.aspx.

Alimohammadi, I., R. Soltani, S. Sandrock, M. Azkhosh, and M. R. Gohari. 2013. The effects of road traffic noise on mental performance. *Iran J Environmen Health Sci Engin.* 10(18):10–18.

Alvarsson, J. J., S. Wiens, and M. E. Nilsson. 2010. Stress recovery during exposure to nature sound and environmental noise. *Int J Environ Res Pub Health.* 7(3):1036–46.

Armstrong, G. B., G. A. Boiarsky, and M. L. Mares. 1991. Background television and reading performance. *Communic Mono.* 58:235–53.

Armstrong, G. B., and B. S. Greenberg. 1990. Background television as an inhibitor of cognitive processing. *Human Communic Res.* 16(3):355–86.

Association of Interior Specialists. 2011. *A Guide to Office Acoustics.* Solihull: Association of Interior Specialists (AIS).

Atkinson, R. L., R. C. Atkinson, and E. R. Hilgard. 1983. *Introduction to Psychology.* New York: Harcourt Brace Jovanovich.

Banbury, S. P., and D. C. Berry. 1997. Habituation and dishabituation to speech and office noise. *J Exp Psychol App.* 3:181–95.

Banbury, S. P., and D. C. Berry. 2005. Office noise and employees concentration: Identifying causes of disruption and potential improvements. *Ergonomics.* 48(1):25–37.

Barker, R. G. 1968. *Ecological Psychology: Concepts and Methods for Studying the Environment of Human Behavior.* Stamford, CN: Stanford University Press.

Beaman, C. P., J. E. Marsh, and D. M. Jones. 2012. Analyzing the meaning of background speech is obligatory, distraction by meaning is not. *Euronoise 2012.* Prague. 648–53.

Belojevic, G., and B. Jakovljevic. 2001. Factors influencing subjective noise sensitivity in an urban population. *Noise Health.* 4(13):17–24.

Belojevic, G., B. Jakovljevic, and V. Slepcevic. 2003. Noise and mental performance: Personality attributes and noise sensitivity. *Noise Health.* 6(21):77–89.

Belojevic, G., V. Slepcevic, and B. Jakovljevic. 2001. Mental performance in noise: The role of introversion. *J Environ Psychol.* 21(2):209–13.

Benfield, J., G. A. Nurse, R. Jakubowski, A. W. Gibson, B. D. Taff, P. Newman, and P. A. Bell. 2012. Testing noise in the field: A brief measure of individual noise sensitivity. *Environ Behav.* 46(3):353–72.

Blomkvist, V., C. A. Eriksen, T. Theorell, R. Ulrich, and G. Rasmanis. 2005. Acoustics and psychosocial environment in intensive coronary care. *Occup Environ Med.* 62(3):e1.

Bodin-Danielsson, C., and L. Bodin. 2008. Office type in relation to health, well-being, and job satisfaction among employees. *Environ Behav*. 40(5):636–68.

Borsky, P. N. 1969. Effects of noise on community behavior. *Proc Third Int Congress Noise Public Health Hazard*. ASHA Reports 4W. Washington, DC: Speech-Language-Hearing Association.

Broadbent, D. E. 1958. *Perception and Communication*. London: Pergamon Press.

Broadbent, D. E. 1978. The current state of noise research: Reply to Poulton. *Psychol Bull*. 85:1052–67.

Broadbent, D. E. 1979. Human performance and noise. In Harris, C. M. (Ed.), *Handbook of Noise Control*. New York: McGraw Hill, 17.1–17.20.

Busch-Vishniac, I. J., J. E. West, C. Barnhill, T. Hunter, D. Orellana, and R. Chivukula. 2005. Noise levels in Johns Hopkins Hospital. *J Acoustic Soc Amer*. 118(6):3629–45.

Campbell, J. B., and C. W. Hawley. 1982. Study habits and Eysenck's theory of extraversion–introversion. *J Res Person*. 16:139–46.

Canning, D., and A. James. 2012. *The Essex Study Optimised Classroom Acoustics for All*. St. Albans: The Association of Noise Consultants. Accessed August 19, 2015 from http://www.adrianjamesacoustics.co.uk/papers/The%20Essex%20Study.pdf.

Carton, A., and J. R. Aiello. 2009. Control and anticipation of social interruptions: Reduced stress and improved task performance. *J App Soc Psychol*. 39(1):169–85.

Cassidy, G., and R. A. R. MacDonald. 2007. The effect of background music and background noise on the task performance of introverts and extroverts. *Psychol Music*. 35:517–37.

Chamorro-Premuzic, T., and V. Swami. 2009. The effects of background auditory interference and extraversion on creative and cognitive task performance. *Int J Psychol Stud*. 1(2):2–9.

Chanaud, R. C. 2009. *Sound Masking Done Right*. Phoenix, AZ: Mitek Corporation.

Cherry, E. C. 1953. Some experiments on the recognition of speech, with one and with two ears. *J Acoustic Soc Amer*. 25(5):975–9.

Cohen, S., and S. Spacapan. 1984. The social psychology of noise. In Jones, D. M., and A. J. Chapman (Eds.), *Noise and Society*. London: Wiley, 221–45.

Colle, H. A., and A. Welsh. 1976. Acoustic masking in primary memory. *J Verbal Learn Verbal Behav*. 15:17–32.

Commission for European Communities. 2009. Commission decision of 23 June 2009 on the safety requirements to be met by European standards for personal music players pursuant to Directive 2001/95/EC of the European Parliament and of the Council. *Official Journal of the European Union*. L161:38–39.

Daoussis, L., and S. J. McKelvie. 1986. Musical preferences and effects of music on a reading comprehension task for extroverts and introverts. *Percept Motor Skill*. 62:283–9.

Eysenck, H. J. 1967. *The Biological Basis of Personality*. Springfield, IL: Thomas Publishing.

Eysenck, H. J., and S. B. G. Eysenck. 1975. *Manual of the Eysenck Personality Questionnaire*. London: Hodder and Stoughton.

Eysenck, M. W., and J. Graydon. 1989. Susceptibility to distraction as a function of personality. *Person Individ Differ*. 10:681–7.

Fitzgerald, C. J., and K. M. Danner. 2012. Evolution in the office: How evolutionary psychology can increase employee health, happiness, and productivity. *Evol Psychol*. 10(5):770–81.

Franklin, C., L. V. Johnson, L. White, C. Franklin, and L. Smith-Olinde. 2013. The relationship between personality type and acceptable noise levels: A pilot study. *ISRN Otolaryngol*. 2013:902532.

Furnham, A., and A. Bradley. 1997. Music while you work: The differential distraction of background music on the cognitive test performance of introverts and extroverts. *App Cog Psychol*. 11:445–55.

Furnham, A., and L. Strbac. 2002. Music is as distracting as noise: The differential distraction of background music and noise on the cognitive test performance of introverts and extroverts. *Ergonomics*. 45:203–17.

Furnham, A., B. Gunter, and E. Peterson. 1994. Television distraction and the performance of introverts and extroverts. *App Cog Psychol*. 8:705–11.

Furnham, A., S. Trew, and I. Sneade. 1999. The distracting effects of vocal and instrumental music on the cognitive test performance of introverts and extroverts. *Personal Individ Differ*. 27:381–92.

Geen, R. G. 1984. Preferred stimulation levels in introverts and extroverts: Effects on arousal and performance. *J Person Soc Psychol*. 46(6):1303–12.

Gifford, R. 2007. *Environmental Psychology: Principles and Practice*. Fourth ed. Colville, WA: Optimal Books.

Glass, D. C., and J. E. Singer. 1972. Behavioral after effects of unpredictable and uncontrollable aversive events. *Amer Sci*. 60(4):457–65.

Goldstein, J., and D. M. DeJoy. 1980. Behavioral and performance effects of noise: Perspectives for research. In Tobias, J. V., G. Jansen, and W. D. Ward (Eds.), *Proc Int Cong Noise as a Public Health Problem*. Rockville, MD: ASHA Reports, 10:369–74.

Haapakangas, A., V. Hongisto, J. Kokko, D. Oliva, J. Keränen, J. Hakala, and J. Hyönä. 2011. Effects of five speech masking sounds on performance and acoustic satisfaction: Implications for open-plan offices. *Acta Acustica*. 97(4):641–55.

Hancock, P. A., and J. S. Warm. 1989. A dynamic model of stress and sustained attention. *Human Factors*. 31(5):519–37.

Hongisto, V., J. Keränen, and P. Larm. 2004. Simple model for the acoustical design of open-plan offices. *Acta Acustica*. 90:481–95.

Horowitz, S. 2012. *The Universal Sense: How Hearing Shapes the Mind*. London: Bloomsbury.

Jahncke, H. 2012. *Cognitive performance and restoration in open-plan office noise*. PhD Thesis. Luleå: Luleå University of Technology.

Jensen, K. L., E. Arens, and L. Zagreus. 2005. Acoustical quality in office workstations, as assessed by occupant surveys. *Proc Indoor Air 2005*. Beijing. 2401–5.

Job, R. F. S. 1988. Community response to noise: A review of factors influencing the relationship between noise exposure and reaction. *J Acoustic Soc Amer*. 83:991–1001.

John, O. P., and S. Srivastava. 1999. The big five trait taxonomy: History, measurement, and theoretical perspectives. In Pervin, L. A., and O. P. John (Eds.), *Handbook of Personality: Theory and Research*. New York: Guilford Press, 102–38.

Jones, D. 2014. Research Summary. Extracted from Cardiff University website. Accessed October 6, 2014 from http://psych.cf.ac.uk/contactsandpeople/academics/jonesdylan.php.

Jones, D. M., and W. J. Macken. 1993. Irrelevant tones produce an irrelevant speech effect: Implications for phonological coding in working memory. *J Exp Psychol: Learn Memory Cog*. 19:369–31.

Jones, D., R. Hughes, J. Marsh, and W. Macken. 2008. Varieties of auditory distraction. *Proc 9th Int Congress Noise Pub Health Problem (ICBEN)*. Foxwoods. C1T.

Kryter, K. D. 1970. *The Effects of Noise on Man*. New York: Academic Press.

Levy-Leboyer, C., B. Vedrenne, and M. Veyssiere. 1976. Psychologie differentielle des genes dues au bruit. *Annee Psychologique*. 76:245–56 (Abstract).

Lewin, K. 1943. Defining the 'field at a given time.' *Psychol Rev*. 50:292–310.

Luz, G. A. 2005 August. Noise sensitivity rating of individuals. *Sound Vib*. 14–7.

Mahmood, A., H. Chaudhury, and M. Valente. 2011. Nurses' perception of how physical environment affects medication errors in acute care settings. *App Nurs Res*. 24(4):229–37.

Marans, R., and K. Spreckelmeyer. 1982. Evaluating open and conventional office design. *Environ Behav*. 3(14):333–51.

Maris, E. 1972. *The social side of noise annoyance*. Amsterdam: Research Institute for Psychology and Health.

Marsh, J. E., R. W. Hughes, and D. M. Jones. 2009. Interference by process, not content, determines semantic auditory distraction. *Cognition*. 110:23–38.

Matthews, T., A. K. Key, J. Mankoff, S. Carter, and T. Rattenbury. 2004. A toolkit for managing user attention in peripheral displays. *Proc 17th Ann ACM Symp User interface Software and Technology (UIST)*. New York: ACM. pp. 247–56.

Matthews, G., D. R. Davies, S. J. Westerman, and R. B. Stammers. 2013. Noise and irrelevant speech. In *Human Performance: Cognition, Stress, and Individual Differences*. Psychology Press. Hove, UK.

Mehrabian, A. 1977. A questionnaire measure of individual difference in stimulus screening and associated differences in arousability. *Environ Psychol and Nonverb Behav*. 1:89–103.

Morgenstern, S., R. J. Hodgson, and L. Law. 1974. Work efficiency and personality: 2. *Ergonomics*. 17:211–20.

Nurmi, J., and J. von Wright. 1983. Interactive effect of noise, neuroticism and state anxiety in the learning and recall of a textbook passage. *Hum Learn*. 2:119–25.

Oberdörster, M. and G. Tiesler. 2006. Acoustic Ergonomics of School. BAuA Publication Series Research Report Fb 1071, Dortmund.

Oldman, T. and Rothe, P. 2015 June. Noise and its impact on productivity, importance of variety in open plan spaces and Plantronics' 'Soundscape' building. *Leesman Rev*. 17:Q2.

Oseland, N. A. 2009. The impact of psychological needs on office design. *J Corp Real Estate*. 11(4):244–54.

Oseland, N. A. 2012. *The Psychology of Collaboration Space*. London: Herman Miller.

Oseland, N. A., and A. Burton. 2012. Quantifying the impact of environmental conditions on worker performance for inputting to a business case to justify enhanced workplace design features. *J Build Survey, Appraisal Valuat*. 1(2):151–64.

Oseland, N. A., A. Marmot, F. Swaffer, and S. Ceneda. 2011. Environments for successful interaction. *Facilities*. 29(1/2):50–62.

Perham, N., S. Banbury, and D. M. Jones. 2007. Do realistic reverberation levels reduce auditory distraction? *App Cog Psychol*. 21:839–47.

Pirn, R. 1971. Acoustical variables in open planning. *J Acoustic Soc Amer*. 49(5):1339–45.

Poulton, E. C. 1979. Composite model for human performance in continuous noise. *Psychol Rev*. 86:361–75.

Rabbitt, P. 1968. Channel-capacity, intelligibility and immediate memory. *Quart J Exp Psychol: Human Exp Psychol*. 20:241–8.

Richards, D., J. Ward, N. Smith, T. Tweddell, V. Peckett, A. Grudzinski, and M. Lown. 2014. *Improving the Environmental Performance of Offices*. London: British Council for Offices (BCO).

Saint-Gobain Ecophon, Old Brick Kiln, Monk Sherborne Road, Ramsdell, Tadley, RG26 5PP, UK.

Smith, A. P., and D. M. Jones. 1992. Noise and performance. In Jones, D. M., and A. P. Smith (Eds.), *Handbook of Human Performance*. Vol. 1: The Physical Environment. London: Harcourt Brace Jovanovich, 1–28.

Smith, B., R. Wilson, and R. Davidson. 1984. Extrodermal activity and extraversion. *Person Individ Diff*. 5:59–65.

Stallen, P. J. M. 1999. A theoretical framework for environmental noise annoyance. *Noise Health*. 1(3):69–79.

Standing, L., D. Lynn, and K. Moxness. 1990. Effects of noise upon introverts and extroverts. *Bull Psychonom Soc*. 28:138–40.

Stansfeld S. A. 1992. Noise, noise sensitivity and psychiatric disorder: Epidemiological and psychophysiological studies. *Psychol Med Mono Suppl*. 22:1–44.

Sykes, D. M. 2009 May. ANSI S12 Workgroup 44.

Szalma, J. L., and P. A. Hancock. 2011. Noise effects on human performance: A meta-analytic synthesis. *Psychol Bull*. 137:682–707.

Tracor Inc. 1971. *Community Reaction to Airport Noise*. Vol. 1, NASA CR-1761. Washington, DC: National Aeronautics and Space Administration (NASA).

Treasure, J. 2010. Shh! Sound health in 8 steps. *TEDGlobal 2010*. Recorded in July 2010. Oxford.

Treasure, J. 2012. *Building in Sound: Biamp Systems*. White Paper. Beaverton, Beaverton, OR: Biamp Systems.

Vastfjall, D. 2002. *Mood and preference for anticipated emotions*. Doctoral Dissertation. Gothenborg: Chalmers University.

Veitch, J. A., J. S. Bradley, L. M. Legault, S. Norcross, and J. M. Svec. 2002. *Masking Speech in Open-plan Offices with Simulated Ventilation Noise: Noise Level and Spectral Composition Effects on Acoustic Satisfaction*. IRC-IR-846. Ottawa, ON: National Research Council Canada, Institute for Research in Construction.

von Wright, J., and M. Vauras. 1980. Interactive effects of noise and neuroticism on recall from semantic memory. *Scand J Psychol*. 21:97–101.

Weinstein, N. D. 1978. Individual differences in reactions to noise: A longitudinal study in a college dormitory. *J App Psychol*. 63(4):458–66.

Williams, W. 2005. Noise exposure levels from personal stereo use. *Int J Audiol*. 44(4):231–6.

Wilson, E. O. 1984. *Biophilia*. Cambridge, MA: Harvard University Press.

Woloshynowych, M., R. Davis, R. Brown, and C. Vincent. 2007. Communication patterns in a UK emergency department. *Ann Emerg Med*. 50(4):407–13.

Yerkes, R. M., and J. D. Dodson. 1908. The relation of strength of stimulus to rapidity of habit-formation. *J Compar Neurol Psychol*. 18:459–82.

5 Workplace Vibration Effects on Health and Productivity

Robin Burgess-Limerick

CONTENTS

5.1 VIBRATION FUNDAMENTALS

Vibration, or the oscillatory movement of an object about a fixed point, is a characteristic of many occupational settings. Tall buildings vibrate in response to wind, and vibration is also a notable environmental characteristic of moving workplaces such as trucks, trains, cars, buses, and boats. In some contexts, the platforms on which the workers stand vibrate. Hand-held power tools are also a source of occupational exposure to vibration.

Vibrations, which propagate in air, are experienced as sound. This chapter concerns the effects of vibration, which is transferred to the human body via direct contact, typically not only via the seat of a seated worker, but also via the feet of a standing person, or the hands in the case of hand-held power tools.

Vibration is described in terms of the frequency and the amplitude of the oscillatory movements. The vibrations encountered in the workplace typically include multiple frequency components and continuously varying amplitudes. The effects of occupational vibration exposure vary from benign to hazardous depending on the frequency and the amplitude characteristics of the vibration, and the duration of exposure. The exposure to whole-body vibration (WBV) can have short-term consequences such as annoyance, motion sickness, and reduced cognitive performance, or can interfere with manual control. Long-term cumulative health effects of WBV include spinal damage, while irreversible damage to the nerves and the blood vessels of the hand and the arms (or even the feet) may arise as a consequence of the exposure to peripheral vibration.

5.2 MEASUREMENT OF OCCUPATIONAL VIBRATION

As an object vibrates, the rate of change of the displacement of the object (its velocity) is constantly changing direction, and consequently the rate of the change of the velocity of the object (its acceleration) is also constantly changing amplitude and direction. While vibration could be described in terms of any of these (displacement, velocity, or acceleration), it is usually measured by an instrument sensitive to acceleration (an accelerometer) placed at the interface between the body and the vibrating surface (e.g., under the ischial tuberosities of a seated worker or on the handle of a power tool) and described in terms of acceleration.

ISO 2631.1 (ISO 1997, 2010) defines the methods for measuring WBV. The amplitude of the vibration may be described in terms of the maximum acceleration (peak acceleration), or in terms of the root-mean-square (rms) acceleration (m/s^2) which is analogous to an average acceleration. The rms measure may underestimate the likely effects of the vibration if the vibration is intermittent or if the exposure contains instances of transient shock loading. The vibration dose value (VDV) is a measure of the vibration involving the fourth power of acceleration ($m/s^{1.75}$) which puts greater emphasis on the high peak values of acceleration, e.g., jolts and jars. The VDV is a cumulative measure and is typically normalized to an 8-hour exposure duration for reporting, i.e., VDV(8).

ISO 2631-5 (ISO 2004) provides a different measure, the daily equivalent static compression dose (S_{ed}, MPa), which purports to quantify the health risks associated with WBV containing multiple shocks, and predicts the adverse health effects to the lumbar spine and the vertebral endplates based on biodynamic models. The measure was developed by the U.S. military following the observation that civilian standards were insufficiently protective of military personnel subjected to repeated shock loadings during the operation of tactical ground vehicles (Alem 2005).

The effects of the vibration on the human body also depend on the direction in which the vibration occurs. While vibration typically occurs simultaneously in all directions, the measurements are typically made and reported in three orthogonal directions defined with respect to the orientation of the person's body (or hand in the case of hand–arm vibration). For the standing or the seated person, the motion in the fore–aft direction is defined as the movement in the x direction; the lateral movement is defined as the y direction; and the vertical movement as the z direction.

Regardless of the measure employed for describing the amplitude of the vibration, the frequency of the oscillations is expressed in oscillations per second (Hz). All objects, including the body parts, have a natural frequency. If an object is exposed to vibration near the natural frequency, the object resonates, and the amplitude of the vibration is increased with a corresponding increased probability of adverse consequences. If the body is exposed to vibration outside its resonant frequency, the vibration energy will be dampened by the structure, and the vibration wave will be attenuated. A vibration input can be amplified by one part of the body and attenuated in another. For example, vibration frequencies between 0.5 and 80 Hz are important from a WBV perspective (the principal resonance for vertical vibration of a seated person is about 5 Hz); however, frequencies up to 1000 Hz are important from a hand–arm vibration perspective. ISO 2631.1 (ISO 1997) and ISO 5349.1 (ISO 2001) provide frequency weightings for use in measuring vibration (Figure 5.1).

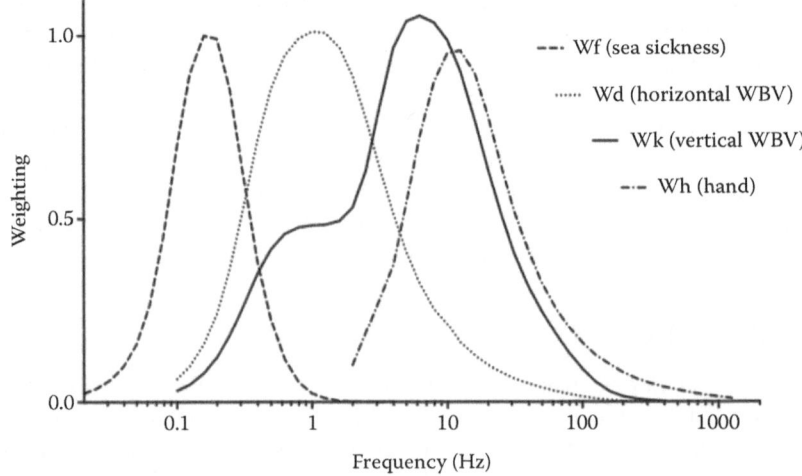

FIGURE 5.1 Frequency weightings provided by ISO standards for use in workplace vibration measurement.

Vibration measurement and analysis devices, which incorporate these weightings and comply with ISO 8041 (ISO 2005), are commercially available from a range of manufacturers (e.g., Brüel and Kjaer, Larson Davis, Svantek). It is also possible to use a consumer electronic device to obtain estimates of the vibration amplitudes suitable for managing the occupational exposures to the WBV (Wolfgang and Burgess-Limerick 2014a; Wolfgang et al. 2014; Burgess-Limerick and Lynas 2016).

5.3 EFFECTS OF LOW-FREQUENCY VIBRATIONS

The oscillations of tall buildings in response to wind can be worrying or unpleasant for the people working within them. These vibrations have relatively low frequency components, typically less than 1 Hz. The threshold of the perception for such vibrations has been put at approximately 0.1 m/s^2 with increasing levels of annoyance associated with amplitudes above 0.15 m/s^2 (Crocker 2007). However, Kwok et al. (2009) note that the human responses to building vibration depend on a mix of psychological and physiological factors including prior experience and habituation and that, while ISO 6897 (ISO 1984) provides guidance, there are no universally accepted design criteria for acceptable building vibration levels.

The adverse effects of exposure to low-frequency vibrations encountered in mobile workplaces include motion sickness. It is believed that the symptoms (nausea, dizziness) arise as a consequence of the conflict between the motion signals arising from the eyes, the vestibular system, and the nonvestibular motion detectors (proprioceptors). Similar symptoms may arise in the absence of any real motion, such as the situation in which visual motion cues are provided by a vehicle simulator in the absence of a corresponding motion. Motions in the range of 0.125–0.25 Hz are most likely to result in symptoms (see Wf in Figure 5.1). The probability of the symptoms increases with increased oscillation magnitude and duration of exposure (Lawther and Griffin 1987), although again, there are considerable individual differences in susceptibility.

5.4 EFFECTS OF VIBRATION ON MOTOR PERFORMANCE

Exposure to vibration can lead to a degraded performance of motor skills such as those required to operate mobile plants and equipment. In particular, vibration can interfere with the acquisition of information via the eyes and the output of information via the hand or foot movements. The effects are dependent on the frequency, the magnitude, and the directions of the vibrations, as well as the complexity of the motor task and the characteristics of the controls involved such as type of control, location, shape, stiffness, gain, and control order (McLeod and Griffin 1989).

The effect of the display vibration on the acquisition of visual information is inversely dependent on the distance of the display from the eyes. The effects of the display vibration are also frequency dependent. Low-frequency oscillations of a display are compensated for by pursuit eye movements. At frequencies of about 1–3 Hz, saccadic eye movements occur instead, and the images are less clear. At higher frequencies, the images may become more blurred. The acquisition of the visual information can also be impaired by the vibration-induced movement of the eyes rather than the display. Again, the effect depends on the frequency of the vibrations. The vestibular–ocular reflex can compensate for the head rotations up to at least 5 Hz but not for the head translation (McLeod and Griffin 1989; Griffin 1990).

Biodynamic feedthrough occurs when vehicle vibrations are involuntarily transmitted through an operator's body to the vehicle controls (e.g., Quaranta et al. 2013; Venrooij 2014). This phenomenon has been particularly noted in situations such as helicopter or backhoe operation. The consequences depend on the sensitivity of the control and the frequency response of the system dynamics at the vibration frequencies. Redesigning the equipment to either reduce the transmission of the vibration to the body part or make the task less susceptible to the vibration by, for example, increasing the control stiffness or reducing the sensitivity, is indicated if these effects become apparent (Griffin 2006). The optimal control sensitivity in conditions that involve the operator being exposed to vibration is likely to be lower than the optimal control sensitivity in static conditions. These effects are likely to be proportional to the vibration amplitude and greatest for vibration frequencies of the order of 2–8 Hz. The dynamic characteristics of most equipment are not sensitive to such relatively high frequencies; however, this is relevant to some sensitive control systems or where precise movements of the hand are required.

Vibration may directly interfere with neuromuscular processes by causing perceptual confusion regarding the forces generated in the controlling limb, and vibration may directly interfere with cognitive processes by increasing arousal. Potential secondary effects of vibration include increased workload, changes in control strategy, or active muscular compensation (McLeod and Griffin 1989).

5.5 HEALTH EFFECTS

People are generally exposed to vibration from a localized source or a source that affects the whole body. The adverse effects on human health associated with the exposure to vibration may be conveniently separated into the effects associated with vibration of the periphery (hand, arms, feet—hence peripheral—or hand–arm vibration) and of the trunk (particularly the back) or WBV.

5.5.1 PERIPHERAL VIBRATION HEALTH EFFECTS

Workers who operate power hand tools are exposed to peripheral vibration. Prolonged and repeated exposure of the hands to vibration is associated with the development of a range of adverse vascular, neural, and connective tissue consequences and symptoms—collectively termed *hand-arm vibration syndrome*. One known potential consequence of peripheral vibration exposure is vibration-induced white finger, which is characterized, as the name suggests, by blanching of the fingers, starting with the fingertips, but potentially affecting all of one or more fingers. The symptoms are also associated with the exposure to cold and often accompanied by numbness and tingling. Reduced dexterity and grip strength have also been reported (Griffin 1990). There is also evidence that workers who are exposed to vibration via the feet could also be at risk for similar health problems. Vibration-induced white feet has been used to describe the tingling, the numbness and the blanching documented in feet associated with working while standing on vibrating platforms (Hedlund 1989; Schweigert 2002; Eger et al. 2014).

Although it is not possible to define the limits of safe exposure, guidelines for the evaluation of hand-transmitted vibration exposure in terms of frequency-weighted acceleration and daily exposure time are provided in ISO 5349 (ISO 2001). The vibration at a wide range of frequencies is implicated (from 8 to 1 KHz) (Griffin 1997). ISO 5349.1 (ISO 2001) provides a frequency weighting (Wh) in which the weighting decreases as the frequency components increase from 16 to 1000 Hz (Figure 5.1), although it has been suggested that this weighting has no empirical basis and may lead to the underestimation of the effects of the exposures to higher frequency vibrations (Griffin et al. 2003).

5.5.2 CONTROLLING PERIPHERAL VIBRATION EXPOSURE

The control measures for reducing the risk of adverse health consequences associated with peripheral vibration include the redesign of tools to reduce vibration amplitude and the limitation of duration of exposure to power tool use. This can be achieved by reducing the magnitude of the forces causing the power tool vibration, making the tool less sensitive to the vibrating forces, isolating the vibration of the tool from the grip surfaces, ensuring proactive maintenance, and improving production processes to reduce the need for the use of power tools (Greenslade and Larsson 1997; Skogsberg 2006; Hewitt et al. 2014).

Whether antivibration gloves are considered likely to effectively attenuate the vibration depends on whether the Wh frequency weighting suggested by ISO 5349.1 (ISO 2001) is employed in the analysis. Gloves typically only attenuate higher frequencies, which are not highly weighted by the ISO standard, and may actually amplify some lower frequencies. Consequently, such protective equipment appears unlikely to be effective if the frequency weighting is utilized during the analysis. However, antivibration gloves may appear to be effective if unweighted accelerations are considered (Griffin 1998; Hewitt et al. 2014). Such gloves are unlikely to be effective for power tools associated with relatively low-frequency components (e.g., pavement tampers) but may provide a modest reduction in the vibration transmitted to the palm of the hand while using power tools with relatively high frequency components (e.g., grinders, sanders). Gloves also increase grip forces required and may reduce dexterity.

5.5.3 WBV HEALTH EFFECTS

A range of adverse consequences has been associated with prolonged exposure to WBV. These include muscular fatigue, gastrointestinal tract problems, autonomic nervous system dysfunction, impaired circulatory function, effects on female reproductive organs, headaches, and nausea (Griffin 1990). However, the most prevalent adverse health consequence of WBV is likely to be the potential for prolonged exposure to WBV to cause or exacerbate the degeneration of spinal structures, resulting in back pain.

A range of epidemiological evidence is available to support the association of WBV with an increased prevalence of intervertebral disk degeneration (Kuisma et al. 2008) and back pain across a wide range of industries (Sandover 1983; Wilder and Pope 1996; Bernard 1997; Bovenzi and Hulshof 1998; Waters et al. 2008). There is also evidence to suggest that poor sitting posture exacerbates the effect of WBV in the causation of back pain (Wikstrom 1993; Hoy et al. 2005) and many mobile equipment operators are known to be exposed to both WBV and poor seated postures (Kittusamy and Buchholz 2004; Eger et al. 2008).

The mechanism/s through which the association between WBV and back pain arises is not completely understood. It is known that the resonant frequency of the spine for vertical vibration in a seated position is in the range of 4–6 Hz (Pope et al. 1998). The bone responds to the impact loading by increasing its thickness. WBV exposures may consequently result in the thickening of the vertebral body endplates. The nutrition of the avascular intervertebral disks is dependent on the diffusion through the endplates (Urban et al. 2004). The increased endplate thickness may result in reduced nutrition and hence reduced rate of repair of these structures, increasing the probability of long-term intervertebral disk damage.

Again, while it is not possible to define the limits for safe exposure, ISO 2631-1 (ISO 1997, 2010) provides guidance regarding the evaluation of health effects, defining a duration-dependent health guidance caution zone. For exposures below the health guidance caution zone, it is suggested that no health effects have been clearly documented. For exposures within the health guidance caution zone, "caution with respect to potential health risks is indicated," and for acceleration amplitudes greater than the health guidance caution zone, it is suggested that "health risks are likely." For an 8-hour daily exposure, the upper and lower bounds of the health guidance caution zone are 0.47 and 0.93 m/s^2 rms, respectively (McPhee et al. 2009). The corresponding values for the VDV measure expressed as an 8-hour equivalent (VDV(8)) are 8.5 and 17 m/s$^{1.75}$.

5.5.4 WBV EXPOSURES

Exposure to elevated levels of WBV for prolonged periods is a hazard for many occupations including bus drivers (Lewis and Johnson 2012), train drivers (Birlik 2009), taxi drivers (Funakoshi 2004), garbage truck drivers (Maeda and Morioka 1998), forklift operators (Blood et al. 2010a), helicopter pilots (Kasin et al. 2011), all-terrain vehicle operators (Rehn et al. 2005), agricultural tractor drivers (Scarlett et al. 2007; Park et al. 2013), and long-haul truck drivers (Blood et al. 2011a), as well as passengers and crew of high-speed watercraft such as rigid hull inflatable boats (Allen et al. 2008).

A range of earth-moving equipment such as dozers, haul trucks, excavators, and graders are used off-road in construction and mining industries and are particularly prone to elevated WBV levels. Twelve-hour shifts are commonplace in industries such as mining and hence considerable potential for harm exists (Burdof and Hulshof 2006). The data collected from such equipment suggest that the vibration amplitudes to which operators are exposed may lie within or above the ISO 2631.1 health guidance caution zone.

For example, Scarlett and Stayner (2005) measured 13 different types of machines used in mining, construction, and quarrying. The vertical rms acceleration values and the VDV(8) values reported included 0.22 m/s^2 and 11.7 m/s$^{1.75}$ for an excavator (both measures below the health guidance caution zone), 0.37 m/s^2 and 14.8 m/s$^{1.75}$ for an 80 t rigid dump truck (below the health guidance caution zone for the rms measure and within the health guidance caution zone for the VDV(8) measure), 0.61 m/s^2 and 15.4 m/s$^{1.75}$ for face shovel loading trucks (within the health guidance caution zone for both measures), and 1.45 m/s^2 and 26 m/s$^{1.75}$ for a bulldozer undertaking civil construction activities (exceeds the health guidance caution zone for both measures).

Eger et al. (2006) collected measurements from 15 types of surface and underground mining equipment. The vertical rms acceleration values reported included 0.37 m/s^2 for a 150 t surface haul truck (below the health guidance caution zone), 0.79 m/s^2 for a grader (within the health guidance caution zone), and 1.64 m/s^2 from a bulldozer (exceeds the health guidance caution zone). Smets et al. (2010) measured eight haul trucks of varying capacities during normal operation at metalliferous surface mines in Canada. Seven of the eight rms measurements were within the ISO 2631.1 health guidance caution zone. All eight VDV(8) measurements were within the health guidance caution zone.

An investigation of the haul trucks in operation at an Australian surface coal mine (Wolfgang and Burgess-Limerick 2014b) concluded that 20 of 32 rms measurements were within the health guidance caution zone while 30 of 32 VDV(8) measurements were within the health guidance caution zone. A strong association was noted between the vibration levels and the roadway roughness (Figures 5.2 and 5.3) indicating that roadway maintenance is likely to be an effective control measure.

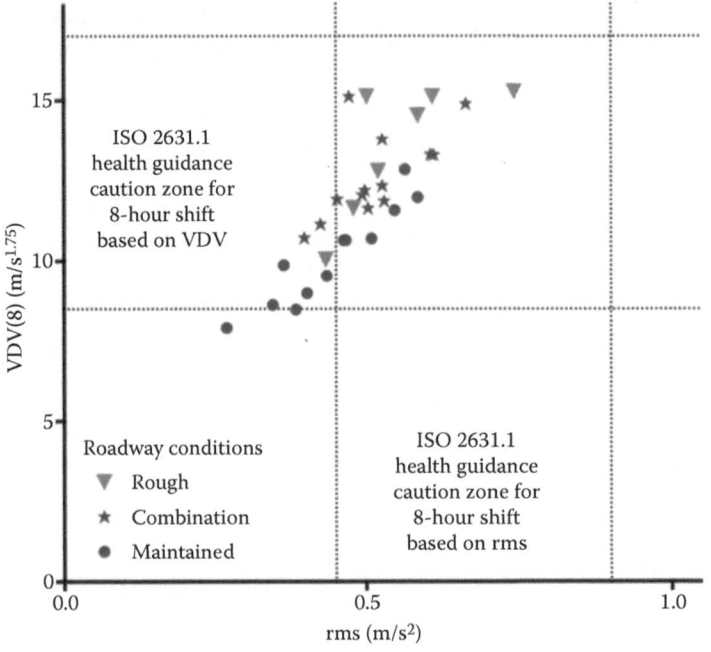

FIGURE 5.2 Vertical WBV measured from 32 haul trucks at a surface coal mine under different roadway conditions. (Redrawn from Wolfgang, R., and R. Burgess-Limerick, *J Occup Environ Hyg.*, 11, 6D77–D81, 2014a. With permission.)

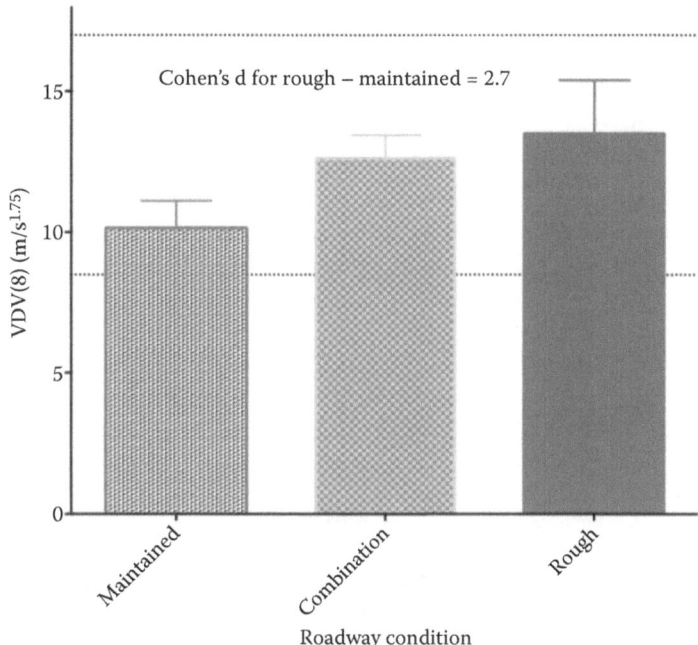

FIGURE 5.3 Mean vertical WBV expressed as VDV(8) measured from 32 haul trucks at a surface coal mine under different roadway conditions. Error bars indicate 95% confidence intervals. (Reprinted from *Appl Ergon.*, 45, Wolfgang, R., and R. Burgess-Limerick, Whole-body vibration exposure of haul truck drivers at a surface coal mine, 1700–4, Copyright 2014b, with permission from Elsevier.)

Burgess-Limerick (2012) reported 26 measurements from dozers performing a range of tasks. Only one of the rms measurements lay within the health guidance caution zone. None of the VDV(8) measurements were below the health guidance caution zone, and one VDV(8) measurement exceeded the health guidance caution zone. A subsequent investigation at a different site revealed an extreme variation in the vibration amplitudes associated with the use of dozers, suggesting that the task performed has a strong influence (Figure 5.4).

5.5.5 Controlling WBV Exposure

The modifiable factors associated with WBV exposure related to occupational contexts include the environment in which the equipment is operated, including the roadway design and maintenance; the activities performed using the equipment; the design of the equipment including suspension design and seat choice; the equipment condition including the maintenance of the suspension and the seating; and the operator behavior, especially the traveling speed.

The design and condition of the roadways play a large role in determining the amplitude of the vibration transmitted to the operator (Blood et al. 2010a, 2011a; Lewis and Johnson 2012), and the importance of roadway maintenance cannot be overestimated. Improving and maintaining roadway standards will have beneficial consequences in reducing the damage to equipment as well as to people. In addition to scheduled roadway maintenance, it is important to ensure that any unexpected deterioration in roadway standards is promptly reported and corrected.

The activities performed have a large influence on the vibration amplitude and the direction (Pinto and Stacchini 2006). For example, ripping using a bulldozer leads to much higher vibration amplitudes than pushing. Village et al. (1989) reported higher vibration amplitudes for load-haul-dump machine operators during loading or dumping, than during traveling. Vibration characteristics may

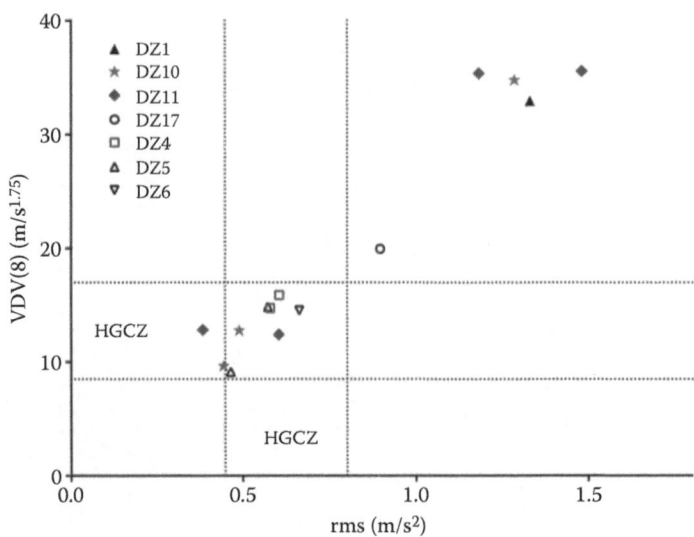

FIGURE 5.4 Vertical WBV expressed as VDV(8) and rms measured from 15 bulldozers in operation during different shifts at a surface coal mine (unpublished data). The trial duration is >100 minutes. The areas marked "HGCZ" on each axis represent the health guidance caution zone defined by ISO 2631.1 for rms and VDV measures of WBV amplitude.

also vary with the load carried by a vehicle. The implication is that the vibration assessment of the equipment must be undertaken for the range of tasks for which equipment is intended to be used, and the duration of the exposure to tasks involving high amplitudes may require limitation.

The design of the equipment is a key area in which WBV exposure can be reduced at the source. Characteristics such as the location of the front axle relative to the cab can have a large influence on the vibration amplitudes. Blood et al. (2011b), for example, reported that the location of the cab over the front axle substantially increased the WBV amplitudes. The location of the cab in front of the front axle may be expected to be associated with even higher vibration levels.

The design of the vehicle suspension is also particularly important (Donati 2002). The suspension is possible between axles and chassis or at the cab. Perversely, in the interests of ensuring equipment reliability, some mining equipment, which operates in the roughest conditions, has been provided with the least effective suspension (McPhee et al. 2009). For example, although shuttle cars were introduced to underground coal mines in 1938, no shuttle cars were equipped with suspension until the last 10 years, and shuttle cars without suspension may still be purchased.

The observations earlier that the vibration characteristics are specific to the tasks undertaken, and the roadway conditions encountered, imply that the design of the suspension should start with an assessment of these vibration characteristics under the whole range of operating tasks, conditions, and vehicle loads likely to be encountered. Design choices such as solid or foam-filled tires also have implications for the vibration characteristics, although even air-filled tires are stiffer than vehicle suspensions (Donati 2002).

Another aspect of the design of the equipment is the choice of seating provided. As with the design of the vehicle suspension, ensuring an appropriate choice of seat requires a detailed understanding of the vibration characteristics, which will be experienced at the seat during use (Donati 2002). Although most seats provide some vibration attenuation (Paddan and Griffin 2002), some seats employed in occupational situations exhibit resonance at low frequencies, resulting in higher vibration amplitudes being experienced on the seat than on the floor. Conventional mechanical seating provides little attenuation below about 6 Hz (Griffin 2007) and may provide reduced protection for lighter drivers (Blood et al. 2010b). Attenuation down to about 2 Hz can be achieved using passive suspension mechanisms below the seat pan, and these seats are commonly found in trucks, coaches, and other work-related vehicles. Active suspension seats are a more recent innovation, which offer a potential for much more effective attenuation of vertical vibrations (Blood et al. 2011b).

A seat suspension mechanism often has limited travel, and adjusting the response of the suspension to suit the mass of the driver is critical. The provision of weight adjustability is likely to be necessary; however, this usually implies providing a manual adjustment mechanism. To be effective, a manual adjustment mechanism must be easy to operate and remain so for the life of the seat under extreme environmental conditions. Even then, operator training in how and why to adjust the seat is necessary (operators frequently mistake the weight adjustment for the height adjustment), and its effectiveness relies on the behavior of operators. Automatic weight sensing and adjustment is more likely to be effective and is available from some seat manufacturers.

The effectiveness of a seat in attenuating the WBV is expressed as the ratio of the frequency-weighted amplitude (either rms or VDV) measured on the seat while occupied by a person expressed as a percentage of the vibration amplitude measured on the floor under the seat, or the seat effective amplitude transmissibility (SEAT). SEAT values less than 100% indicate that the seat is attenuating the vibration amplitudes in that situation, while SEAT values greater than 100% indicate that the vibration levels are being amplified by the seat. The SEAT value measured will vary as a function of operator mass, suspension adjustment, vibration frequency components, and vibration amplitude. Consequently, different SEAT values may be obtained from the same seat in different vehicles, or even the same vehicle being driven over different surfaces. A seat that is suitable in one application may be unsuitable in another context.

Boileau et al. (2006), for example, investigated the vibration characteristics of 16 underground dump trucks. The dominant frequency components of the vibrations during use ranged from 2.6 to 3.4 Hz. A typical suspension seat provided for use on these vehicles was tested using a vibration simulator and found not to attenuate vibration at these frequencies and indeed was more likely to amplify the WBVs experienced under operating conditions. Lewis and Johnson (2012) similarly reported that the seats in metropolitan buses amplified rather than attenuated the vibrations associated with driving in a range of roadway contexts.

Regardless of their suitability when new, the vibration attenuation performance of both vehicle suspension and seating is likely to deteriorate with use. Appropriate maintenance and replacement schedules are required.

A behavioral change program was developed as an intervention to reduce WBV exposure of occupational drivers (Tiemessen et al. 2007), although the program was not found to be effective (Tiemessen et al. 2009).

While a combination of the administrative and design controls discussed earlier has a potential to reduce the exposure of the equipment operators to WBV, the most effective control is to separate the operator from the equipment via the use of remote control (either line of sight, or nonline of sight) or automation. Advances in the areas of teleoperated dozers and underground loaders and automated haul trucks point to this as being an achievable goal in at least some contexts.

REFERENCES

Alem, N. 2005. Application of the new ISO 2631-5 to health hazard assessment of repeated shocks in U.S. army vehicles. *Indust Health*. 43:403–12.

Allen, D. P., D. J. Taunton, and R. Allen. 2008. A study of shock impacts and vibration dose values onboard highspeed marine craft. *Int J Maritime Eng*. 150(A3):1–10.

Bernard, B. P. (Ed.). 1997. *Musculoskeletal Disorders and Workplace Factors: A Critical Review of Epidemiologic Evidence for Work-related Disorders of the Neck, Upper Extremities, and Low Back*. Cincinnatti, OH. US Department of Health and Human Services, National Institute of Occupational Safety and Health. DHHS (NIOSH) Publication No. 97–141.

Birlik, G. 2009. Occupational exposure to whole body vibration—Train drivers. *Indust Health*. 47:5–10.

Blood, R. P., J. Dennerlein, C. Lewis, P. Rynell, and P. W. Johnson. 2011a. Evaluating whole-body vibration reduction by comparison of active and passive suspension seats in semi-trucks. *Hum Fac Erg Soc P*. 55:1750–4.

Blood, R. P., J. D. Ploger, and P. W. Johnson. 2010a. Whole body vibration exposures in forklift operators: Comparison of a mechanical and air suspension seat. *Ergonomics*. 53:1385–94.

Blood, R. P., J. D. Ploger, M. G. Yost et al. 2010b. Whole body vibration exposures in metropolitan bus drivers: A comparison of three seats. *J Sound Vibration*. 329:109–20.

Blood, R. P., P. W. Rynell, and P. W. Johnson. 2011b. Vehicle design influences whole body vibration exposures: Effect of the location of the front axle relative to the cab. *J Occup Environ Hyg*. 8:364–74.

Boileau, P.-E., J. Boutin, T. Eger, and M. Smets. 2006. Vibration spectral class characterization of load-haul-dump mining vehicles and seat performance evaluation. In Dong, R., K. Krajnak, O. Wirth, and J. Wu (Eds.), *Proc First Amer Conf Human Vibration*. 14–15.

Bovenzi, M., and C. T. J. Hulshof. 1998. An updated review of epidemiologic studies of the relationship between exposure to whole body vibration and low back pain. *J Sound Vibration*. 215:595–611.

Burdof, A., and C. T. H. Hulshof. 2006. Modelling the effects of exposure to whole-body vibration on lower-back pain and its long term consequences for sickness absence and associated work disability. *J Sound Vibration*. 298:480–91.

Burgess-Limerick, R. 2012. How on earth moving equipment can ISO2631 be used to evaluate WBV exposure? *J Health Safety Res Pract*. 4(2):13–21.

Burgess-Limerick, R., and D. Lynas. 2016. An iOS application for evaluating whole-body vibration within a workplace risk management process. *J Occup Env Hyg*. doi:10.1080/15459624.2015.1125486.

Crocker, M. J. 2007. Vibration response of structures to fluid flow and wind. In Crocker, M. J. (Ed.), *Handbook of Noise and Vibration Control*. 1375–92. Hoboken, NJ: Wiley.

Donati, P. 2002. Survey of technical preventative measures to reduce whole-body vibration effects when designing mobile machinery. *J Sound Vibration*. 253:169–83.

Eger, T., A. Salmoni, A. Cann, and R. Jack. 2006. Whole-body vibration exposure experienced by mining equipment operators. *Occup Ergon*. 6:121–7.

Eger, T., J. Stevenson, J. P. Callaghan, S. Grenier, and VibRG. 2008. Predictions of health risks associated with the operation of load-haul-dump mining vehicles: Part 2—Evaluation of operator driving postures and associated postural loading. *Int J Indust Ergon*. 38:801–15.

Eger, T., A. Thompson, M. Leduc, Kranjnak, K. Goggins, A. Godwin, and R. House. 2014. Vibration induced white-feet: Overview and field study of vibration exposure and reported symptoms in workers. *Work*. 47:101–10.

Funakoshi, M., K. Taoda, H. Tsujimura, and K. Nishiyama. 2004. Measurement of whole-body vibration in taxi drivers. *J Occup Health*. 46:119–24.

Greenslade, R., and T. J. Larsson. 1997. Reducing vibration exposure from hand-held grinding, sanding and polishing powertools by improvement in equipment and industrial processes. *Safety Sci*. 25:143–52.

Griffin, M. J. 1990. *Handbook of Human Vibration*. London: Academic Press.

Griffin, M. J. 1997. Measurement, evaluation, and assessment of occupational exposures to hand-transmitted vibration. *Occup Environ Med*. 54:73–89.

Griffin, M. J. 1998. Evaluating the effectiveness of gloves in reducing the hazards of hand-transmitted vibration. *Occup Environ Med*. 55:340–8.

Griffin, M. J. 2006. Vibration and motion. In Salvendy, G. (Ed.), *Handbook of Human Factors and Ergonomics*. Third edition. Hoboken, NJ: Wiley.

Griffin, M. J. 2007. Effects of vibration on people. In Crocker, M. J. (Ed.), *Handbook of Noise and Vibration Control*. 343–353. Hoboken, NJ: Wiley.

Griffin, M. J., M. Bovenzi, and C. M. Nelson. 2003. Dose-response patterns for vibration-induced white finger. *Occup Environ Med*. 60:16–26.

Hedlund, U. 1989. Raynaud's phenomenon of fingers and toes of miners exposed to local and whole-body vibration and cold. *Int Arch Occ Env Health*. 61:457–461.

Hewitt, S., R. G. Dong, D. E. Welcome, and T. W. McDowell. 2014. Anti-vibration gloves? *Ann Occup Hygiene*.

Hoy, J., N. Mubarak, S. Nelson, M. Sweerts de Landas, M. Magnusson, O. Okunribido, and M. Pope. 2005. Whole body vibration and posture as risk factors for low back pain among forklift truck drivers. *J Sound Vibration*. 284:933–46.

International Standards Organization (ISO). 1984. ISO 6897: Guidelines for the evaluation of the response of occupants of fixed structures, especially buildings and off-shore structures, to low-frequency horizontal motion (0.063 to 1 Hz). ISO, Geneva.

ISO. 1997. ISO 2631.1: Evaluation of human exposure to whole-body vibration: Part 1: General requirements. ISO, Geneva.

ISO. 2001. ISO 5349-1: Mechanical vibration: Measurement and evaluation of human exposure to hand-transmitted vibration: Part 1: General requirements. ISO, Geneva.

ISO. 2004. ISO 2631-5: Evaluation of human exposure to whole-body vibration: Part 5—Method for evaluation of vibration containing multiple shocks. ISO, Geneva.

ISO. 2005. ISO 8041: Human response to vibration—Measuring instrumentation. ISO, Geneva.

ISO. 2010. ISO 2631.1: Evaluation of human exposure to whole-body vibration: Part 1—General requirements. Amendment 1. ISO, Geneva.

Kasin, J. I., N. Mansfield, and A. Wagstaff. 2011. Whole body vibration in helicopters: Risk assessment in relation to low back pain. *Aviation Space Environ Med.* 82:790–6.

Kittusamy, N., and B. Buchholz. 2004. Whole-body vibration and postural stress among operators of construction equipment: A literature review. *J Safety Res.* 35:255–61.

Kuisma, M., J. Karppinen, M. Haapea, J. Niinimäki, R. Ojala, M. Heliövaara, R. Korpelainen, K. Kaikkonen, S. Taimela, A. Natri, and O. Tervonen. 2008. Are the determinants of vertebral endplate changes and severe disc degeneration in the lumbar spine the same? A magnetic resonance imaging study in middle-aged male workers. *BMC Musculoskel Disorders.* 9:51.

Kwok, K. C. S., P. A. Hitchcock, and M. D. Burton. 2009. Perception of vibration and occupant comfort in wind-excited tall buildings. *J Wind Eng Indust Aerodynam.* 97:368–80.

Lawther, A., and M. J. Griffin. 1987. Prediction of the incidence of motion sickness from the magnitude, frequency, and duration of vertical oscillation. *J Acoust Soc Amer.* 82:957–66.

Lewis, C. A., and P. W. Johnson. 2012. Whole-body vibration exposure in metropolitan bus drivers. *Occup Med.* 62:519–24.

Maeda, S., and M. Morioka. 1998. Measurement of whole-body vibration exposure from garbage trucks. *J Sound Vibration.* 215:959–64.

McLeod, R. W., and M. J. Griffin. 1989. A review of the effects of translational whole-body vibration on continuous manual control performance. *J Sound Vibration.* 133:55–115.

McPhee, B., G. Foster, and A. Long. 2009. *Bad Vibrations*. Second edition. Sydney: Coal Services Health and Safety Trust.

Paddan, G. S., and M. J. Griffin. 2002. Effect of seating on exposures to whole-body vibration in vehicles. *J Sound Vibration.* 253:215–41.

Park, M.-S., T. Fukuda, T.-G. Kim, and S. Maeda. 2013. Health risk evaluation of whole-body vibration by ISO 3631-5 and ISO 2631-1 for operators of agricultural tractors and recreational vehicles. *Indust Health.* 51:364–70.

Pinto, I., and N. Stacchini. 2006. Uncertainty in the evaluation of occupational exposure to whole-body vibration. *J Sound Vibration.* 298:556–62.

Pope, M., D. Wilder, and M. Magnusson. 1998. Possible mechanisms of low back pain due to whole-body vibration. *J Sound Vibration.* 215:687–97.

Quaranta, G., P. Masarati, and J. Venrooij. 2013. Impact of pilots' biodynamic feedthrough on rotorcraft by robust stability. *J Sound Vibration.* 332:4948–62.

Rehn, B., T. Nilsson, B. Olofsson, and R. Lundstrom. 2005. Whole-body vibration exposure and non-neutral neck postures during occupational use of all-terrain vehicles. *Ann Occup Hygiene.* 49:267–275.

Sandover, J. 1983. Dynamic loading as a potential source of low back disorder. *Spine.* 8:652–8.

Scarlett, A. J., and R. M. Stayner. 2005. Whole-body vibration on construction, mining and quarrying machines. Research Report 400. Bootle, UK. Health and Safety Executive.

Scarlett, A. J., J. S. Price, and R. M. Stayner. 2007. Whole-body vibration: Evaluation of emission and exposure levels arising from agricultural tractors. *J Terramech.* 44:65–73.

Schweigert, M. 2002. The relationship between hand-arm vibration and lower extremity clinical manifestations: A review of the literature. *Int Arch Occ Env Health.* 75:79–185.

Skogsberg, L. 2006. Vibration control on hand-held industrial power tools. In Dong, R., K. Krajnak, O.Wirth, and J. Wu. (Eds.). *Proc First Amer Conf Human Vibration.* 108–9. Morgantown, WV. NIOSH.

Smets, M. P. H., T. R. Eger, and S. G. Greiner. 2010. Whole-body vibration experienced by haulage truck operators in surface mining operations: A comparison of various analysis methods utilized in the prediction of health risks. *Appl Ergon.* 41:763–70.

Tiemessen, I. J. H., C. T. J. Hulshof, and H. W. Frings-Dresen. 2007. The development of an intervention programme to reduce whole-body vibration exposure at work induced by a change in behaviour: A study protocol. *BMC Pub Health.* 7:329.

Tiemessen, I. J. H., C. T. J. Hulshof, and H. W. Frings-Dresen. 2009. Effectiveness of an occupational health intervention program to reduce whole-body vibration exposure: An evaluation study with a controlled pretest-posttest design. *Am J Indust Med.* 52:943–52.

Urban, J. P., S. Smith, and J. C. Fairbank. 2004. Nutrition of the intervertebral disc. *Spine.* 29:2700–9.

Venrooij, J. 2014. Measuring, modeling and mitigating biodynamic feedthrough. PhD thesis. Delft: Technical University of Delft.

Village, J., J. Morrison, and D. Leong. 1989. Whole-body vibration in underground load-haul-dump vehicles. *Ergonomics.* 32:1167–83.

Waters, T., A. Genaidy, and H. B. Viruet. 2008. The impact of operating heavy equipment vehicles on lower back disorders. *Ergonomics.* 51:602–36.

Wikstrom, B. 1993. Effects from twisted postures and whole-body vibration during driving. *Int J Indust Ergon.* 12:61–75.

Wilder, D. G., and M. H. Pope. 1996. Epidemiological and aetiological aspects of low back pain in vibration environment—An update. *Clin Biomech.* 11:61–73.

Wolfgang, R., and R. Burgess-Limerick. 2014a. Using consumer electronic devices to estimate whole-body vibration exposure. *J Occup Environ Hyg.* 11:6D77–D81.

Wolfgang, R., and R. Burgess-Limerick. 2014b. Whole-body vibration exposure of haul truck drivers at a surface coal mine. *Appl Ergon.* 45:1700–4.

Wolfgang, R., L. Di Corletto, and R. Burgess-Limerick. 2014. Can an iPod Touch be used to assess whole-body vibration associated with mining equipment? *Ann Occup Hyg.* 58:1200–4.

6 Vision and Lighting

Mariana G. Figueiro and Mark S. Rea

CONTENTS

6.1 INTRODUCTION

Much has been written over the last century on the impact of lighting on productivity. Indeed, it has been the central theme for the lighting industry for most of that time (Luckiesh and Moss 1937; Weston 1945; Blackwell 1972; Smith and Rea 1980; Rea 2013). The research conducted over this period has led to a solid understanding of how lighting affects visual performance (Rea and Ouellette 1991), but the impacts of lighting on nonvisual neural systems affecting circadian rhythms, alertness, and performance are just emerging. Moreover, the impact of these nonvisual effects of lighting on health is at its infancy.

The present chapter begins with a high-level summary of what we know about the impact of light on visual performance and appearance. That summary hardly does justice to the wealth of information on the topic, but there are several good references that provide the reader with much of the background on the topic. In particular, the recent book *Human Factors in Lighting* (third edition) by Dr. Peter Boyce provides an up-to-date, readable, and comprehensive summary of the area. Most of the present chapter is devoted to the rapidly, but still emerging, area of the nonvisual effects of lighting on both productivity and health. Suggestions for future research are also provided.

6.2 LIGHTING, HEALTH, AND PRODUCTIVITY

Health and productivity are broadly defined outcome domains related to physiology and behavior, whereas lighting is a broad stimulus domain covering both technologies and applications. The interrelationships between these domains are many and complicated. Some links are well defined, because they have been alloyed through face-validity studies, controlled psychophysical research, and fundamental neurophysiological investigations. Other links have been forged through logical (or wishful) inference, because the needed data are too expensive and time consuming to gather or, more likely, because the important, interactive variables cannot be easily controlled (Boyce 2004). To elucidate the interrelationships among lighting, health, and productivity as they might affect evidence-based decisions about architectural practice, it is necessary to be as precise as possible with regard to the light stimulus and to the behavioral or the physiological response. It must be stressed, however, that ensuring high precision with regard to the relationship between the light stimulus and the behavioral or the physiological response is no guarantee that the factors that affect health and productivity have been adequately characterized. Again, many factors unrelated to lighting can and do affect productivity and health.

By definition, *light* is the optical radiation incident on the retina that evokes a neural response in humans (Commission Internationale de l'Éclairage 1978). *Lighting* is the application of the optical radiation in a built environment (DiLaura et al. 2012). The decisions about lighting will affect the amount, the spectrum, the distribution, the timing, and the duration of the light stimulus incident on the retina (Figure 6.1). These physical properties of light can differentially affect the various neural channels emanating from the retina to the brain and thus the measured behavioral and physiological outcome measures that might have been affected by the light stimulus. In fact, a multitude of visual and nonvisual pathways and systems simultaneously process the information about the luminous environment, so a given pattern of light on the retina will have many varied consequences. Some of these neural systems evoke a conscious response (e.g., apparent brightness), whereas others do not (e.g., circadian phase change). Obviously, too, the physiological characteristics of an individual's retina and brain can affect the relationship between the light stimulus and the channel response characteristics. For example, the X chromosome codes the color vision capabilities (Boynton 1979). So for males, having just one X chromosome, identical physical properties of light on the retina can appear quite different to different men. Individual differences become particularly important when productivity and health are considered. Lighting may provide adequate support for the visual and nonvisual systems of most people in a building, but because of poor vocational training or a sedentary lifestyle, the productivity and the health of an individual may be poor (Boyce and Rea 2001). Those factors unrelated to lighting, as important as they are for ultimately determining productivity

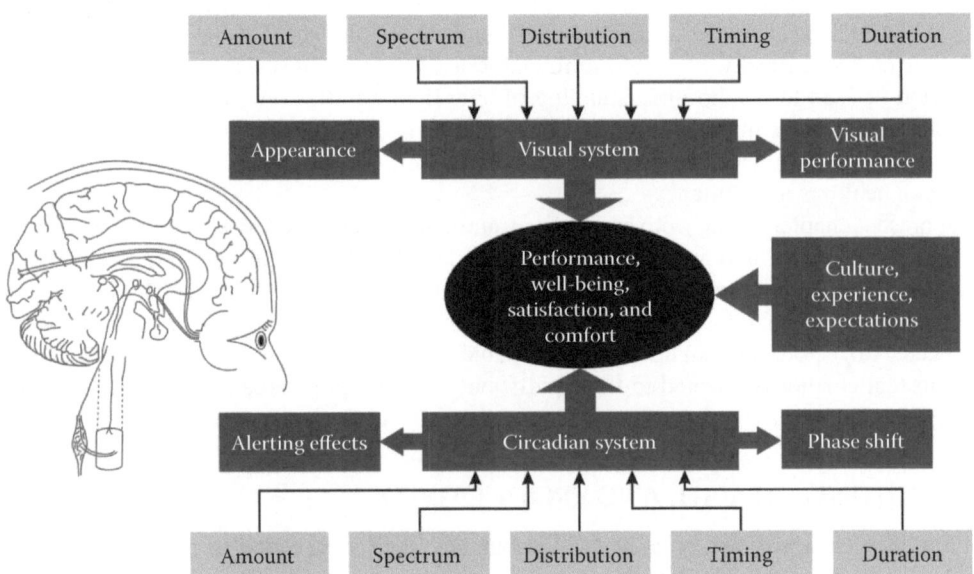

FIGURE 6.1   The physical properties of light (amount, spectrum, distribution, timing, and duration) can differentially affect the various neural channels emanating from the retina to the brain and thus behavioral and physiological outcomes, including performance, well-being, satisfaction, and comfort.

and health, are not addressed in this chapter. Rather, the present discussion is limited to the effects that lighting can have on productivity and health through the neural consequences of light falling on the retina.

6.3 THE VISUAL EFFECTS OF LIGHTING

6.3.1 VISUAL PERFORMANCE

Visual performance is defined as the speed and the accuracy of processing a visual stimulus (Boyce and Rea 2001). Visual performance and productivity, if determined largely by the speed and the accuracy of processing visual information, will be compromised if the visual task background is dim and the targets on the background (e.g., letters on a printed page) are small or of low contrast. These principles apply to both illuminated materials (e.g., printed page) and self-luminous displays (e.g., computer terminals). Figure 6.2 shows one of a family of the functional relationships between the light stimulus, the target size, the target contrast at one background photopic luminance level, and the behavioral response, Relative Visual Performance (RVP) (Rea and Ouellette 1991). This relationship holds for normal people between the ages of 20 and 60 years and is consistent with the electrophysiological recordings from the lateral geniculate nuclei of the rhesus macaque, an old world primate with a visual system very similar to that of humans (Kaplan and Shapley 1982). As the human eye ages, the target contrast, and the background luminance are reduced in predictable

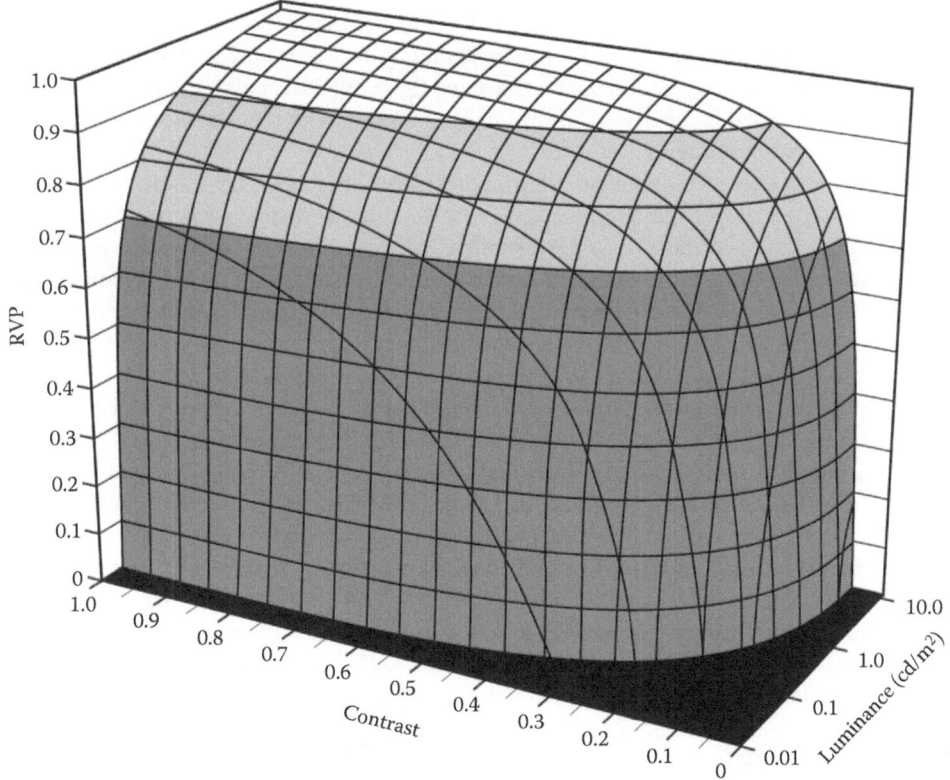

FIGURE 6.2 One of a family of the functional relationships between the light stimulus, target size, target contrast at one background photopic luminance level (Rea and Ouellette 1991), and the behavioral response, Relative Visual Performance (RVP; speed and accuracy). Large differences in light levels are needed to make even modest changes in the speed and the accuracy of processing the visual task.

ways such that, for the same physical properties of the stimulus, the proximal (retinal) stimulus can be computed (Rea and Ouellette 1991). The plateau and escarpment of the visual performance (Boyce and Rea 1987) has been a useful construct for interpreting a wide variety of psychophysical data (Boyce et al. 1990; Bailey, Clear, and Berman 1993; Goodspeed and Rea 1999; Eklund, Boyce, and Simpson 2001; Bullough and Skinner 2009, 2012; Schnell, Yekhshatyan, and Daiker 2009; Bullough et al. 2010; Skinner and Bullough 2011; Bullough, Rea, and Zhang 2012; Bullough, Donnell, and Rea 2013) and for setting the recommended photopic illuminance levels in offices, schools, and factories in recent years (Rea 2000).

Today, for most visual tasks where productivity is important, the size and the contrast of the targets to be processed are sufficiently large and high, respectively, to be on the visual performance plateau. For self-luminous displays, and even for most printed or duplicated materials (Dillon, Pasini, and Rea 1987), the size and the contrast of the targets are well above the escarpment. And for fine surgery or dental examination where low-contrast, small targets must be seen, magnifiers have been part of the working environment for decades. Thus, because contrast and size are usually high, the variations in the visual performance for most practical tasks that require illumination are largely determined by changes in light levels. As can be appreciated from Figure 6.2, however, very large differences in light levels are needed to make even modest changes in the speed and the accuracy of processing the visual task. For example, to make a 1% improvement in the visual performance of a 20-year-old person reading a white page ($\rho = 0.8$) with black, 8-point type (contrast = 0.85), photopic illuminance levels would need to increase from a currently recommended level of 300 to 850 lux (DiLaura et al. 2012), levels close to those recommended in 1972 (Illuminating Engineering Society 1972). This increase in the light level by 283% would require a proportional increase in the electric power needed to energize the same type of ceiling-mounted lighting system, as well as higher capital and maintenance costs. In consideration to Figure 6.2 and modern technologies (aimed at and successfully improving the contrast and the size), the recommended levels of photopic illuminance have dropped by nearly 2/3 over the past 40 years with the aim of reducing the lighting energy demands from buildings.

This reduction in the recommended illuminance levels has been obviously supported by the shift from print-based tasks to tasks presented on self-luminous displays where illumination levels are essentially irrelevant to the visual performance, as it would affect a person's productivity on computer-based tasks. In fact, for most self-luminous displays, less light on the screen will improve the visual performance because the amount of reflected glare is reduced. Computer display screens were highly specular when they were introduced, so the reflected glare played a very large role in the architectural lighting design (Rea 1991). A great many luminaires were specifically marketed as sources of illumination that minimized the reflected glare from early computer screens. In contrast to the screen technologies of 20 or 30 years ago, today's computer monitors used indoors, for the majority of visual tasks performed in offices and factories have effectively eliminated the reflected glare, so the distribution of light from electric light or even daylight from windows is much less important.

As already emphasized, the visual performance is not synonymous with productivity, but it is important to note that there is a distinction between visual performance and task performance (Boyce and Rea 2001). Task performance is affected by cognitive and motor functions as well as by vision. Figure 6.2 represents the relationship between visual performance and the characteristics of the physical stimulus. However, the physical stimulus characteristics can change considerably with changes in the eye–task–lighting geometry. Under dim illumination, people (without presbyopia) will lean forward to enlarge the visual angle of the target, thereby bringing it onto the plateau of visual performance (Rea 1983; Rea, Ouellette, and Kennedy 1985). People will also tilt the page, adjust the electronic display or the luminaire (Rea 1983), or else lean to one side to eliminate the reflected glare on the visual task. These behavioral changes aimed at improving visual performance are helpful in the short run, but there may be untold negative consequences on posture and fatigue, thereby decreasing productivity in the long run. Modern improvements in the visual task characteristics, such as those previously noted, may have been much more important for supporting and improving task performance and productivity than has been previously and systematically considered.

Finally, the impact of the other dimensions of light, temporal, and spectral characteristics should be noted. For achromatic foveal tasks, such as reading, the light spectrum of the illuminance plays a very minor role, as long as the (photopic) background luminance and the contrast are unchanged (Boyce et al. 2003). Obviously, however, if the visual task requires discrimination among subtle shades of hue or lightness, the spectral composition of the light source plays an essential role (Section 6.3.2). The visual system responds very quickly (<500 ms), so the temporal characteristics of the lighting are relatively unimportant as long as the light source does not perceptibly flicker or does not create a stroboscopic effect with objects moving in the environment (Rea and Ouellette 1988; Bullough et al. 2012). The current standards limit flicker from light sources (Rea 2000; DiLaura et al. 2012; Eighth Census Conference on Newborn ICU Design 2012).

6.3.2 Appearance

Productivity sometimes depends less on visual performance than on the appearance of the task. How fresh fish or meat appears can affect the profitability of a grocery store. Assessing the quality of a reproduced image by a printer or the color of a patient's skin by a physician can be essential for their respective measures of productivity. In these cases, the lighting must be able to reveal the subtle variations in contrast or hue, no matter how long it takes to make the assessment. As already noted, there are different neural channels for visual performance and for appearance (Kaplan and Shapley 1982).

Figure 6.3 shows the results of one study where the same achromatic visual stimuli were presented to subjects, but they were asked to perform different tasks (Rea 1990). One curve is essentially a slice through the plateau and escarpment of visual performance in Figure 6.2, where the light level and the size of the printed materials were held constant, while the contrast of the print on the paper was systematically varied. The other curve represents the subjective ratings of the appearance of contrast for those very same stimulus conditions (Rea 1990). As can be seen from Figure 6.3,

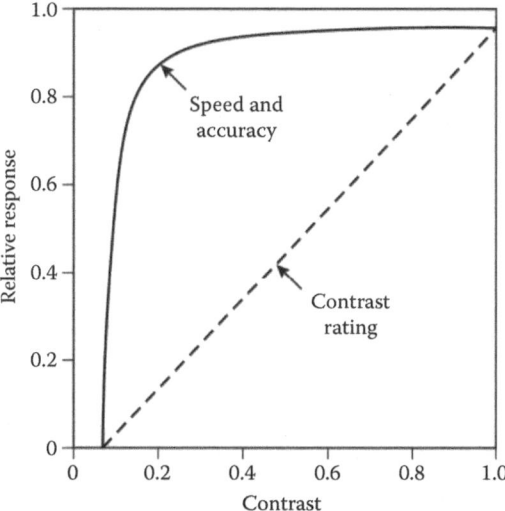

FIGURE 6.3 Results of one study where the same achromatic visual stimuli were presented to subjects, but they were asked to perform different tasks (Rea 1990). Perceived contrast (quality of print) is linearly related to the target contrast, but speed and accuracy of reading are much like a step function.

the functional relationship between the stimulus and the response varies with the neural channel responsible for performing the task. Our ability to make fine discriminations of contrast (quality of print) is linearly related to the contrast of the target, but speed and accuracy are much like a step function—you see it (on the plateau) or you do not (beneath the escarpment).

Whereas the speed and the accuracy of processing achromatic tasks is largely unaffected by the spectral characteristics of the background photopic luminance (Boyce et al. 2003), the spectral composition of the light source affects the apparent brightness of the illumination. At the same measured photopic illuminance level, cool (blue-white) light sources will make a room appear brighter than warm (yellow-white) light sources (Fotios 2001; Fotios, Cheal, and Boyce 2005; Bullough, Radetsky, and Rea 2011; Fotios and Cheal 2011; Rea, Radetsky, and Bullough 2011; Bullough et al. 2014). Thus, even though the visual performance is largely unaffected by the spectral composition of the source, the apparent brightness of the task or the room will be strongly related to the correlated color temperature (CCT) of the illumination. Again, this is because different neural channels in the visual system process optical radiation on the retina very differently.

The spectral composition of the light source also affects the hue, the lightness, and the saturation of the objects being illuminated. In the past, color rendering index (CRI) has been the standard metric for quantifying the color rendering properties of light sources (Kinney 1962; DiLaura et al. 2012), but recent technological advances in light-emitting diode (LED) sources have shown that CRI, while still useful, does not completely predict the appearance and the acceptability of colored objects like fruit, vegetables, fabrics, and skin (Narendran and Deng 2002). Where color plays an essential role in sales, and thus productivity, color rendering can be more important than any other benefit provided by the lighting system.

Current research has shown that the appearance of colored objects is affected by not only how natural they appear, but also how vivid they appear. Different cultures, and different individuals within those cultures, vary with respect to how vivid objects should appear (Rea and Freyssinier 2013). For retail applications where, for example, fresh fruits and vegetables are sold, color saturation is a very important consideration, particularly for some people. A new metric quantifying the ability of a light source to increase the color saturation, called gamut area index (GAI), has been developed as an adjunct to the traditional color rendering metric, CRI (Rea and Freyssinier 2008, 2010). This two-metric system has been shown to be very useful for characterizing the color rendering properties of a light source; CRI reflects how natural objects appear, and GAI reflects how saturated the colors will appear. The GAI should be thought of as a measure of the color spiciness provided by the illumination. For example, people in India tend to prefer more color spice (saturation) than those in Sweden (Rea and Freyssinier 2008, 2010, 2013).

6.4 THE NONVISUAL EFFECTS OF LIGHTING

In addition to affecting productivity through visual performance and appearance, retinal light exposures affect our brain via nonvisual retinal pathways. As illustrated in Figure 6.1, there are two ways that lighting affects the behavior and the physiology through nonvisual systems. Perhaps the more notable and better understood is the impact of light on circadian system regulation. The light on the retina is converted to neural signals that travel to the brain via the retinohypothalamic tract (RHT). The RHT directly innervates our biological clock in the suprachiasmatic nuclei (SCN). The SCN, in fact, govern the timing of every circadian rhythm exhibited by our behavior and physiology, such as the sleep–wake cycle. The RHT is comprised of long axons from intrinsically photosensitive retinal ganglion cells (ipRGCs). Although the ipRGCs are directly coupled with the SCN, all retinal photoreceptors contribute to the neural signals sent to the SCN. The SCN can regulate our circadian rhythms without the light stimulation of the retina; however, the regular, 24-hour pattern of light and dark most of us experience every day coordinates our circadian-varying behavior and physiology to our local position on Earth. The symptoms of jet lag reflect the asynchrony between the intrinsic timing of the SCN and the new light/dark pattern on the retina. Under normal conditions, however, the close coupling of the internally oscillating SCN to the externally oscillating light/dark pattern on the retina ensures our survival on Earth (Klein 1993).

Light also impacts a second, more poorly understood alerting system. Recent research suggests that the pathway from the retina to the brain is not always synonymous with acute melatonin suppression, a marker of circadian system activation. This does not mean, however, that light, which activates the circadian clock, does not affect alertness but, rather, that other pathways appear to also drive mental activity and physiology. Obviously then, a system that elevates alertness can affect performance at certain tasks. Almost nothing is known, however, about whether this system has any direct bearing on productivity and health.

6.4.1   LIGHT AS A STIMULUS FOR THE CIRCADIAN SYSTEM

Currently, lighting technologies, standards, measurement devices, and applications have been largely based on how our foveae respond to a light stimulus (i.e., photopic light levels). However, there are key differences between the light's effect on the foveae and the light's effect on the circadian system. In 1980, Lewy et al. demonstrated that a 2-hour exposure to high levels of light at night (2500 lux at the cornea of an incandescent light source) significantly suppressed melatonin production in a healthy subjects. After this seminal study, a series of studies were performed to better understand the absolute and the spectral sensitivity of acute melatonin suppression by light at night (LAN) exposure. It has now been demonstrated in laboratory studies that light levels lower than those used by Lewy et al. can suppress melatonin and phase shift the circadian clock as measured by the onset of evening melatonin (Zeitzer et al. 2000). Nevertheless, the absolute sensitivity of the human circadian system, as measured by acute melatonin suppression, is still lower than the absolute sensitivity of the human visual system. For example, we can use our eyes for navigation under starlight, but the circadian system is blind to such low levels of light. Even at home, more light is needed to affect the melatonin levels at night than to read black fonts on white paper (Rea, Figueiro, and Bullough 2002).

In terms of the spectral sensitivity of the circadian system, a major discovery in the past decade was made in 2002 by Berson, Dunn, and Takao, who identified a new class of photoreceptor in the retina: the ipRGCs. The discovery of the ipRGCs confirmed earlier studies by Foster and Hankins (2002) and Lucas and Foster (1999), who showed that genetically manipulated rodents lacking functional rods and cones acutely suppressed the melatonin and phase shifted the onset of the wheel-running activity in the dark after a pulse of light (Freedman et al. 1999; Lucas, Freedman, and Munoz 1999). Melanopsin, an opsin identified by Provencio et al. (1998), is believed to be the photopigment that provides the intrinsic photosensitivity to the ipRGCs. Subsequent work showed that rods and cones as well as ipRGCs are involved in the circadian phototransduction (Panda et al. 2002; Hattar et al. 2003). Based on these studies, it is now well accepted that the circadian system is a blue sky detector, with a peak spectral sensitivity close to 460 nm, while the fovea exhibits a photopic luminous efficiency based on the spectral sensitivities of long (L)- and middle (M)-wavelength sensitive cones that, together, have a peak sensitivity at 555 nm.

Unlike the visual system, which will respond to a light stimulus in milliseconds, it takes a few minutes before an acute response by the circadian system can be measured (McIntyre et al. 1989; Figueiro, Lesniak, and Rea 2011). It may take a few hours or even days before a more sustained, phase-shifting response can be measured (Zeitzer et al. 2000). Moreover, depending on the timing of light exposure, light can phase advance or phase delay the timing of the biological clock. Light that is applied before the minimum core body temperature (CBT_{min}), which is reached approximately 1.5–2.5 hours before we naturally awaken, will delay the clock (e.g., one will wake up later the following day), and light applied after CBT_{min} is reached will advance the clock (e.g., one will wake up earlier the following day). Because the SCN in humans have an intrinsic period, on average, of 24.2 hours, light in the morning will advance the clock, thereby entraining the SCN to our local position on Earth every day. Light at night can delay the clock, so morning light becomes particularly important for promoting circadian entrainment and, presumably, our health and well-being. Although light is the main synchronizer of the biological clock to the solar day, it is not the only one. Exercise, social activities, timing of the sleep/wake cycle, and scheduled meals have also been shown to shift and synchronize the clock, although their impact on the circadian rhythms seems to be weaker than the impact of light/dark cycles.

Recently, however, new research showed that a 30-second flash of light applied through closed eyelids every 30 seconds or every 60 seconds over a period of 1 hour was also effective at suppressing the hormone melatonin and phase shifting the dim light melatonin onset, a marker of the circadian phase. This recent research suggests that the circadian system does not need to have sustained light to be activated and that brief flashes of light continuously delivered over the course of 60 minutes can acutely suppress the melatonin and phase shift the circadian clock (Figueiro and Rea 2012; Figueiro, Bierman, and Rea 2013; Figueiro, Plitnick, and Rea 2014).

Finally, prior light history seems to reduce the absolute sensitivity of the circadian system in terms of both acute and phase shifting responses (Hebert et al. 2002; Smith, Schoen, and Czeisler 2004; Chang, Scheer, and Czeisler 2011). For example, people continuously exposed to bright light during daytime are less sensitive to light exposure at night that might suppress the melatonin production. In summary, while research is ongoing to better understand the lighting characteristics affecting the circadian system, it is safe to state that the 24-hour patterns of light and dark affect the circadian regulation and, thus, our health and well-being. New instruments and standards based upon circadian light exposures need to be applied and integrated into lighting standards in the future.

6.4.2 Light as a Stimulus for the Alerting System

Previous studies have shown that, in addition to resetting the timing of the circadian pacemaker, light can also have an acute effect on subjective and objective measures of alertness. Most studies performed to date examined the impact of light on alertness at night where the detrimental effects of fatigue can be easily observed. Compared to dim light or darkness, bright LAN (>2500 lux of white light at the cornea) increased high-frequency (beta range) and reduced low-frequency (theta and alpha ranges) electroencephalographic (EEG) activity, as well as reduced slow-eye movements (SEMs) and subjective sleepiness—all measures that have been associated with decreasing sleepiness and increasing alertness at night. Until recently, the observed alerting effects of light were associated with the light's ability to suppress nocturnal melatonin. However, as detailed in this section, recent studies showed that melatonin suppression is not needed to elicit an alerting response in humans, suggesting that the lighting characteristics affecting the melatonin suppression are not necessarily the same as those affecting alertness. While, as discussed in this section, the acute alerting effects of light are well documented, the temporal, spectral and absolute sensitivities of this alerting system are not yet well understood.

One of the first studies to demonstrate the acute alerting effects of LAN was performed by Badia et al. (1991), who investigated the effects of bright white light (5000–10,000 lux at the cornea) on CBT_{min} and alertness, during the day and the night. The subjects were exposed to 90-minute blocks of alternating bright and dim (<50 lux at the cornea) light during daytime and nighttime hours. The CBT_{min} and the beta power (power in the 18–21 Hz range) from the EEG recordings were higher after the exposure to bright light than after the exposure to dim LAN, but not during the day. The relatively high power in the beta range is associated with higher cognition.

Cajochen et al. (2000) examined the dose–response relationship for light level and alertness. The study indicated that a 6.5-hour exposure to corneal illuminance levels ranging from 3 to 9100 lux of a 4100 K light source systematically increased alertness as measured by a reduction in self-reported sleepiness, the incidence of SEMs, and a reduction of EEG activity in the alpha–theta frequency (5–9 Hz); the relatively lower power in the alpha–theta range is associated with less sleepiness. These alerting responses were positively correlated with the degree of melatonin suppression by light, suggesting that melatonin suppression might be mediating the observed effect. Rüger et al. (2006) presented 24 male subjects with either bright light (5000 lux at the cornea) or dim light (<10 lux at the cornea) between either noon to 4:00 p.m. or midnight to 4:00 a.m. The researchers found that bright light exposure at night (midnight to 4:00 a.m.) increased heart rate and reduced sleepiness as measured by the Karolinska sleepiness scale (KSS).

These laboratory findings have been confirmed in more realistic environments. Figueiro et al. (2001) showed that the subjective feelings of wakefulness, alertness, and overall well-being of nurses working in a newborn intensive care unit were improved after 15-minute exposures to bright white light (>2500 lux at the cornea) at the start, middle, and end of their shift by comparison to exposure to dim light (<50 lux at the cornea). In a similar study the following year, Yoon et al. (2002) presented 12 female nurses with either dim light (100–500 lux) or bright light (4000–6000 lux) for 4 hours during the night shift. The subjective alertness measured by the visual analog scale (VAS) was significantly increased after the exposure to bright light in comparison to dim light.

In a recently published study, Lowden and Åkerstedt (2012) tested the effects of a dynamic lighting regime on alertness, sleep quality, and adaptation to rotating shifts in those working in a nuclear power control room during a Nordic winter season. A new light source was designed to deliver three light regimes, including one with high light levels and high CCT (745 lux at the cornea, 6000 K). The subjects experienced the control lighting condition (200 lux at the cornea, 3000 K) and the experimental lighting condition during a sequence of three night shifts, two free days, two morning shifts, and one afternoon shift. The results showed that the dynamic lighting regime increased alertness and delayed wake times. This study seems to offer promising results for the development and the implementation of dynamic indoor lighting to promote alertness, sleep, and adaptation to shift work.

The impact of nocturnal exposure to low levels (5, 10, 20, and 40 lux at the cornea) of short-wavelength (470 nm, blue) light on subjective and objective alertness, performance, and biomarkers (melatonin and cortisol) has also been investigated. Figueiro et al. (2007) demonstrated that subjective (Norris scale) and objective (reduction in the alpha power between 8–12 Hz) measures of alertness were highly correlated, and that both measures monotonically increase with four light levels of blue light. In addition, Cajochen et al. (2005) found a significant decrease in the subjective sleepiness per KSS and significantly greater melatonin suppression after exposure to 5 lux of 460 nm (blue) light by comparison to after exposure to a 550 nm (green) light or a dark condition.

While the alerting effects of LAN are linked to melatonin suppression, interesting new data demonstrated that lights that do not suppress melatonin can also exert an alerting effect in humans. Figueiro et al. (2009) demonstrated that exposures to 40 lux of both short-wavelength (470 nm, blue) and long-wavelength (640 nm, red) lights in the middle of the night increased the beta power (13–30 Hz) and reduced the alpha power (8–12 Hz) relative to preceding dark conditions, although only the blue light significantly suppressed the melatonin relative to darkness. The exposures to both the red and blue lights also significantly increased the heart rate relative to darkness. These findings suggest that the melatonin pathway does not seem to be the only light-sensitive pathway that can affect the alertness at night.

These results are consistent with studies showing the alerting effects of light during the daytime, when the melatonin levels are low. Phipps-Nelson et al. (2003) examined the effects of bright light (1000 lux) on daytime subjective (KSS) and objective sleepiness (SEM) in 16 subjects who were sleep deprived for 2 days. They concluded that the bright light exposure reduced the subjective sleepiness and decreased the variance of the self-reports (SEMs). Vandewalle et al. (2006) exposed individuals to high levels (>7000 lux) of polychromatic white light for 20 minutes during the daytime. They found an enhancement in the cortical activity during an oddball task, and the subjective alertness improved in a dynamic manner, where the effect declined within minutes after the end of the light stimulus. In the previously mentioned study, Rüger et al. (2006) found that bright light exposure, as seen at night, also significantly reduced sleepiness and fatigue during daytime hours. The findings of these studies suggested that the alerting effects of light during daytime appear to be mediated by that mechanisms that are separate from melatonin suppression.

The effects of light on daytime alertness were also investigated in field studies. In a 14-week field study, Mills, Tomkins, and Schlangen (2007) exposed shift workers to two different types of light luminaires, one of which was at 17,000 K (170 lux at the cornea) while the other was at 2900 K (128 lux at the cornea). The 2900 K lighting condition served as the baseline condition. Despite the factors such as the seasonal effects, the uneven distribution of subjects in the two experimental groups, and the possibility of biased responses to the questionnaires due to the visually observable light level differences between the two floors, the light at 17,000 K indicated better subjective alertness and well-being, as measured by self-assessed alertness and a modification of Columbia Jet Lag scale compared to the light at 2900 K (Mills, Tomkins, and Schlangen 2007). In another similar 8-week field experiment by Viola et al. (2008), 104 office workers in two similar floors went through the same lighting conditions for 4 weeks each. The 17,000 K lighting conditions (310 lux on the work plane) reduced the daytime sleepiness and improved the subjective measures of alertness compared to 4000 K (421 lux on the work plane) in the white light condition (Viola et al. 2008).

More recently, Smolders, de Kort, and Cluitmans (2012) investigated the alerting and the vitalizing effect of illuminance at 200 and 1000 lux (at the eye level) on subjective and physiological measures both in the morning and in the afternoon for 60 minutes, divided into four 15-minute blocks. The results showed an effect of lighting condition on subjective sleepiness (KSS) and feelings of vitality. The results showed that the exposure to light at 1000 lux reduced the sleepiness and increased the heart rate compared to light at 200 lux.

Sahin and Figueiro (2013) investigated the effect of blue and red lights on alertness and sleepiness during the day, close to the postlunch dip. In a lab study, 13 participants were presented with one of three conditions: (1) no active LEDs (dark), (2) 630 nm LEDs (40 lux), or (3) 470 nm LEDs (40 lux) for 1 hour in the early afternoon. EEG readings were taken for each condition, and sleepiness was measured by KSS. The results showed that the power in the alpha, alpha–theta, and theta ranges (all measures of sleepiness) were significantly lower after the exposure to the 630 nm light compared to the dark condition; even though the power in the alpha and alpha–theta ranges was lower than in the dark condition, this difference did not reach statistical significance. No significance was shown in the KSS scores, but the scores indicated that the red light showed lower sleepiness ratings, compared to blue light and dark conditions.

These recent results suggest that long-wavelength, red light positively affects alertness measures during the day, and it can be used to promote alertness during the postlunch dip. While the mechanisms associated with this alerting effect of long-wavelength light are still unknown, it seems reasonable to infer that long-wavelength light activates the sympathetic system, separate from the acute melatonin suppression response, which is maximally sensitive to short-wavelength light. This arousal system would support our fight-or-flight response. One way to test this hypothesis is to measure heart rate variability (HRV). Accordingly, the visual information about ambient illumination is detected by the retina and neural signals are then directed to various areas of the brain to help regulate the sympathetic nervous systems.

In one study, Schäfer and Kratky (2006) investigated whether colored fluorescent light can affect the sympathetic nervous system by measuring HRV. In a lab study, 12 participants underwent a 15-minute period of darkness, followed by a 10-minute period of illumination with one of three lighting conditions, and another 15-minute period of darkness. The HRV was recorded during the last 2–3 minutes of each period. Lighting conditions consisted of a light box containing fluorescent light tubes, providing either (1) a single phosphor red light (700 lux, no distance was specified), (2) a single phosphor green light (700 lux at 2.7 m distance from the eyes), or (3) a single phosphor blue light (700 lux at 1.2 m distance from the eyes). The results showed that the colored light affected the HRV within minutes. There was a strong positive correlation between the detrended fluctuation analysis and the very low frequency (VLF) components, indicating that the red and green lights increased the sympathetic modulation with the increase of VLF, while blue light increased the parasympathetic modulation when VLF components decreased. The participants perceived the red light and green light as stimulating or activating, while blue light was perceived as inducing an awake but more relaxed state.

These laboratory and field studies suggest that light has an acute alerting effect on humans during both the daytime and the nighttime. These effects may be mediated through neural connections to the sympathetic nervous system. One challenge is to demonstrate, as discussed in Section 6.4.3, whether this alerting effect of light can lead to better task performance and productivity.

6.4.3 Light and Measures of Performance

As with alertness, performance exhibits circadian patterns that are influenced by the amount of prior sleep. Performance is also dependent on the level of mental load, sleep restriction, and on the time of day that the task is being performed. In general, performance, particularly when high mental load and prolonged attention are required, is at its lowest in the middle of the night (Belenky et al. 2003; Tucker and Fishbein 2009).

Studies have shown that light can impact cognitive performance via two pathways: through promoting circadian entrainment or through an acute alerting effect. Campbell and Dawson (1990) showed that bright light (>1000 lux) at night improved cognitive performance, while Daurat et al. (2000) showed that bright light improved the speed of performance without affecting the accuracy. Chellappa et al. (2011) showed that exposure to a 6500 K light significantly decreased reaction times in tasks associated with sustained attention (psychomotor vigilance and go/no-go tasks), but not in tasks associated with executive function (paced visual serial addition task). Interestingly, Figueiro, Nonaka, and Rea (2014) showed that exposure to daylight during the daytime hours has a carryover effect on performance during the second half of the night when subjects were sleep deprived for 27 hours.

More recently, Figueiro et al. (2016) investigated whether exposure to 200 lux of a long-wavelength (630 nm, red) light and 360 lux of a warm (2568 K) white light would increase measures of nocturnal alertness and improve performance in sleep-restricted subjects. The authors hypothesized that exposure to saturated red or white light would improve performance, and that only white light would significantly affect melatonin levels. Seventeen individuals participated in a 3-week, within-subjects, nighttime laboratory study. Compared to remaining in dim light, the participants had significantly faster reaction times in the go/no-go test after the exposure to both red and white light. Compared to dim light, the power in the alpha and alpha–theta regions was significantly decreased after exposure to red light, suggesting a decrease in sleepiness. Melatonin levels were significantly suppressed by the white light only. These results showed that not only can red light improve measures of alertness, but can also improve certain types of performance at night without affecting melatonin levels.

While it is reasonable to expect that an increase in daytime and nighttime alertness should also result in an increase in performance, not all of the studies performed to date suggest a tight relationship between measures of alertness and performance. Yoon et al. (2002) found a significant increase in measures of alertness (VAS) after a bright light exposure (4000–6000 lux at the cornea), but they did not find a significant improvement in performance as measured by short-term performance tests. Similarly, Figueiro et al. (2009) found an alerting effect of both red light (630 nm) and blue light (470 nm) as measured by EEG, but not significant improvement in performance under either condition.

6.4.4 Light and Health

The long-term effects of the lighted environment on the health of daytime workers are not well documented. Most of the effects of light on health have been more extensively studied in shift workers, because shift work obviously disrupts the regular light/dark cycle. Deleterious biological effects related to shift work and its associated sleep deprivation range from digestive problems to an increase in the risk of breast and colon cancers (Haus and Smolensky 2006). In a review of the literature, Keller (2009) reported that extended duration shifts could make workers more prone to chronic diseases; studies showed a relationship between long work hours and a depressed immune system.

Since much of the published work on the health effects of shift work has been focused on better understanding the relationships between shift work, light, and cancer risks, a review of the literature in this topic is summarized in the following sections.

6.4.5 LIGHT AT NIGHT, CIRCADIAN DISRUPTION, AND CANCER RISKS

In 1987, Stevens put forward the melatonin hypothesis, which postulated that melatonin suppression by LAN increased estrogen production, which led to faster breast cancer cell turnover. A series of research projects were initiated after this hypothesis was postulated, and based on the results of these studies, there are currently two parallel theories concerning the relationship between shift work, LAN exposure, and hormone-related diseases such as breast cancer. One theory focuses on LAN as a disruptor to the normal melatonin hormone production at night, which was the original hypothesis postulated by Stevens (1987). A second theory focuses on the negative health effects of light-induced circadian disruption, which evolved from the studies published to date.

In 2010, the WHO International Agency for Research on Cancer (IARC) classified shift work that involved circadian disruption as "probably carcinogenic to humans" (WHO IARC 2010). This conclusion was based on data from animal studies, which showed that light exposure during the dark period that leads to acute melatonin suppression and/or circadian disruption, is associated with the development and growth of tumors. Although the IARC decided that the data from animal models were sufficient to conclude that there was a causal link between shift work and circadian disruption resulting from exposure to LAN, they felt that the epidemiological evidence for the carcinogenicity of light during the daily dark period was limited.

6.4.6 EVIDENCE FROM BASIC STUDIES

Hill and Blask (1988) investigated the effects of melatonin on the growth characteristics of breast cancer cells (MCF-7) in vitro. The researchers found that the melatonin-treated cell numbers were significantly lower and that the effects continued for days after incubation. Hill and Blask's study showed that the melatonin directly acts on breast cancer cells, and suggested that the melatonin concentrations normally present in human blood offer a direct oncostatic protection; conversely, a deviation in the blood levels of melatonin from the physiological range might release once-dormant estrogen-responsive breast tumor stem cells from this inhibitory state.

Cos et al. (2002) conducted an in vitro study on melatonin-regulated MCF-7 human breast cancer cell proliferation as well as apoptosis, the death of cells. The results showed that melatonin and vitamin D, used as a positive control for the study, both act in similar ways on MCF-7 cells, as both decreased cell proliferation and increased the expression of various proteins in MCF-7 cells. No significant increase in the apoptotic cells was observed after the melatonin treatment, but the proportion of apoptotic cells was significantly higher in cells cultured in vitamin D than in the melatonin or in the control conditions. The data supported the hypothesis that melatonin limits MCF-7 cell proliferation by modulating the cell cycle length through control of the p53-p21 pathway, but without clearly inducing apoptosis; while apoptosis increased in relation to the controls, there was no significant increase (1%). The cell cycle phase distribution after the melatonin treatment was not observed, possibly because of the short incubation time. The authors concluded that under the conditions of this experiment, melatonin does not induce MCF-7 cells to fall into apoptosis.

The two earlier experiments were completed in vitro, using MCF-7 human breast cancer cells. Another method of human cancer research involves using live rodents, typically mice, as test subjects, as well as using cancer cells from mice in vitro. Farriol et al. (2000) studied the inhibitory effect of melatonin on a mouse carcinoma cell line (CT-26). The cells were exposed to either low or high doses of melatonin. The authors found an inhibitory effect of melatonin on the cell proliferation for the high concentrations of melatonin. A statistically significant inverse relationship between DNA synthesis and added melatonin was found in the cell lines. High levels of melatonin did not induce cell membrane damage; no statistical differences were found between the lactic acid dehydrogenase levels of cells with melatonin or controls. The authors noted that the inhibitory effect of melatonin seen in this study was contradictory to other research looking at melatonin effects on the cell lines of other origins (e.g., MCF-7).

Filipski et al. (2002) investigated whether the circadian disruption accelerated the tumor growth in mice. The researchers found that the body temperature rhythm was suppressed in 60 of 75 mice with verified SCN destruction, suggesting circadian disruption in these animals. The SCN regulate the daily rhythms of many important biological functions, including sleep, alertness, cell division, hormone production, and DNA repair. The light/dark pattern incident on the retina sets the timing of the SCN. Mice with an SCN lesion had statistically significant shorter life spans and accelerated tumor growth rates. The authors called for the further investigation of the mechanisms through which the SCN lesion contributes to accelerated tumor growth, specifically the endocrine and immune effects of SCN lesions.

Following up on the earlier study, Filipski, Li, and Levi (2006) investigated whether tumor growth in mice could be increased by disrupted circadian function resulting from either complete destruction of the SCN via bilateral electrolytic lesion, or chronic jet lagging of the mice via advancing light onset by 8 hours every 2 days. The team compared the tumor growth when the mice experienced jet lag (advance of the light onset by 8 hours every 2 days for 10 days) or when the mice were exposed to a regular 12-hour-on/12-hour-off light schedule. The authors hypothesized that the tumor progression would accelerate faster under the jet-lagged condition compared to the control condition.

Blask et al. (2005) demonstrated that rats bearing rat hepatoma or breast cancer xenografts that were exposed to various levels of light had a dose-dependent suppression of nocturnal melatonin (higher light levels suppressed more melatonin) and a stimulation of tumor growth and linoleic acid uptake/metabolism (which feeds the tumor). In addition, blood samples were collected from healthy, premenopausal women during the daytime, at night when subjects were in darkness, and at night when subjects were exposed to 2800 lux of white light for 90 minutes. The tumors that were perfused with melatonin-deficient blood had high tumor activity.

6.4.7 EVIDENCE FROM EPIDEMIOLOGICAL STUDIES

Schernhammer et al. (2001) examined the relationship between breast cancer and working rotating night shifts. They showed a moderate increase in breast cancer risk among nurses who worked 1–14 years and 15–29 years on rotating shifts. The risk increased even further if they had worked 30 years or more. In 2003, Schernhammer et al. examined the relationship between working rotating night shifts and colorectal cancer. Their data suggest that women who worked at least three nights per month for 15 years or more on rotating night shifts had a higher risk for colorectal cancer. Since their studies were reported, several other investigators have found similar effects. Recently, Pesch et al. (2010) examined the risk of breast cancer among German women and found a larger risk of breast cancer among women working the night shift for more than 20 years. The study included 857 women diagnosed with breast cancer and 892 control participants without cancer.

Grundy et al. (2011) investigated the effects of nighttime light exposure on melatonin levels in fast-rotating shift nurses. The nurses worked a rotating shift pattern consisting of two 12-hour day shifts, two 12-hour night shifts, and 5 days off. The data were collected over four 48-hour periods in summer and winter. The participants wore light loggers to measure the ambient light in their workplace and also provided personal information about their health and work habits and completed the morningness–eveningness questionnaire. The results showed that melatonin production was not significantly different between the day and night shifts in the summer or winter, and melatonin did not show a significant relation to light exposure, years of shift work, or chronotype. The authors noted that the light levels in the hospital used for the study may also have been too low to show a strong effect on the melatonin. The study's findings could mean that the rapidly rotating shift schedule is not very disruptive to the melatonin production, or simply that there are other biological pathways contributing to the relationship between shift work and cancer.

Langley et al. (2012) investigated the connection between melatonin levels and those of other hormones for workers with different shift work history to determine if a biomarker could be found for breast cancer. In a field study that took place in the summer and the winter, 82 nurses provided data about their menstruation, caffeine intake, activity levels, and any oral contraceptive use. The levels of melatonin, creatine, oestradiol, oestrone, progesterone, and prolactin were measured; no correlation was found between melatonin and any of these hormones. The participants who had been on shift work for 15 years or more had higher levels of oestradiol, but this proved to be not significant. No data were collected on the participants' light exposure. The authors noted that the study was limited by the short shift work time not disrupting the circadian rhythms enough to show a change in levels, and by the small number of participants.

Lie et al. (2011) analyzed the relationship between breast cancer risk and shift work in nurses. Women who had been in nursing school between 1914 and 1985 were sent questionnaires to provide basic information about age, diagnoses, hormone treatment, alcohol consumption, exposure to x-rays, and so on. Of these, 699 had experienced breast cancer, and 895 were cancer-free control participants. Two-thirds of the nurses in both groups were postmenopausal. The results showed that the risk of cancer increased in nurses who had worked a minimum of 5 years and had worked night shifts; the risk was slight and nonsignificant.

6.4.8 POSSIBLE LIGHTING SOLUTIONS TO DECREASE THE CIRCADIAN DISRUPTION OR THE ACUTE MELATONIN SUPPRESSION IN SHIFT WORKERS

While more research is needed to firmly establish the direct link between LAN, melatonin suppression, circadian disruption, and breast cancer risks in rotating shift workers, it is prudent to adopt lighting schemes that will minimize both acute melatonin suppression and circadian disruption, while still maintaining alertness and performance.

Smith, Fogg, and Eastman (2009) proposed a compromise solution to increase alertness without disrupting circadian rhythms in rotating shift workers. According to their proposition, rotating shift workers would receive high levels of light in the early evening to delay their circadian clock and remove light in the morning on their way home. This would make rotating-shift workers night owls and would delay the dim light melatonin onset to the second half of their shift. In entrained individuals, the onset of melatonin occurs about 2 hours prior to bedtimes; therefore, if the evening bright light exposure delays the onset of melatonin, it will also delay their bedtimes and allow shift workers to perform better during their shift.

Van de Werken et al. (2013) studied the effect of short-wavelength attenuated polychromatic white light at night on the melatonin levels of 33 subjects. In the study, the short-wavelength light was selectively removed by covering a Philips TL-D 36W/830 lamp with short-wavelength attenuating foil (heat-shrinkable tubing). The other two study conditions were dim light and bright full-spectrum light. The research team found that short-wavelength attenuated polychromatic white light at night only marginally suppressed nighttime melatonin levels (6% melatonin suppression). The performance levels were similar to those found with bright light exposure; however, the alertness was moderately reduced. The authors noted that in certain safety-sensitive professions and situations where nighttime alertness is key, it would not be advisable to drastically attenuate the short-wavelength light at night due to the reduction in alertness. Otherwise, short-wavelength-attenuated light appears to benefit night shift workers by allowing them to maintain high levels of melatonin at night while gaining a level of performance on par with that found under bright full-spectrum lighting.

Finally, as discussed earlier, it has been shown that long-wavelength (red) light, which does not change the circadian phase or acutely suppress melatonin, can increase objective and subjective measures of alertness at night (Figueiro et al. 2009, 2014); therefore, the use of long-wavelength light may be an ideal solution for maintaining alertness and performance in those working at night without disrupting the circadian system. Future research should investigate the effectiveness of this solution in the field.

6.5 CONCLUSIONS

The light incident on the retina can affect both visual and nonvisual systems. Although much more can be learned through research, the impact of light on productivity through visual systems has been extensively studied over the past century. These studies have led to current lighting practice that, coupled with modern visual display technologies, provides an excellent visual environment in offices, factories, and schools.

Only recently has the impact of retinal light exposures on the nonvisual systems been extensively studied. In particular, our understanding of the importance of the 24-hour light/dark cycle on productivity and health has considerably grown over the past 15 years. This rapidly developing science has only started to impact the lighting practice, but it seems inevitable that this area of research will directly affect how future buildings are constructed and operated.

The successful application of light for productivity and health via nonvisual systems will fundamentally depend upon controlling the light stimulus, which is quite different from controlling the light stimulus for visual systems. New tools for quantifying (measurement and calculation) the nonvisual light stimulus will need to be developed. These tools will enable architects and designers to precisely specify the light to support both visual and nonvisual systems. It must be recognized, however, that precisely specifying and controlling the light stimulus will not necessarily guarantee high productivity and good health. Both of these broadly defined outcome domains are influenced by factors completely unrelated to lighting, such as training and motivation. Nevertheless, the precise characterization of the retinal light stimulus for visual and nonvisual systems will significantly increase the likelihood that productivity and health outcomes can be reliably realized. Moreover, new research will undoubtedly provide new insights into how light can be used to affect human health and productivity.

REFERENCES

Badia, P., B. Myers, M. Boecker et al. 1991. Bright light effects on body temperature, alertness, EEG and behavior. *Physiol Behav.* 50(3):583–8.

Bailey, I., R. Clear, and S. Berman. 1993. Size as a determinant of reading speed. *J. Illum Eng.* 22(2):102–17.

Belenky, G., N. J. Wesensten, D. R. Thorne et al. 2003. Patterns of performance degradation and restoration during sleep restriction and subsequent recovery: A sleep dose-response study. *J Sleep Res.* 12(1):1–12.

Berson, D. M., F. A. Dunn, and M. Takao. 2002. Phototransduction by retinal ganglion cells that set the circadian clock. *Science.* 295(5557):1070–3.

Blackwell, H. R. 1972. A human factors approach to lighting recommendations and standards. In *Technology for Man 72; Proceedings of the Sixteenth Annual Meeting.* Los Angeles, CA.

Blask, D., G. Brainard, R. Dauchy et al. 2005. Melatonin-depleted blood from premenopausal women exposed to light at night stimulates growth of human breast cancer xenografts in nude rats. *Cancer Res.* 65(23):11174–84.

Boyce, P., and M. Rea. 1987. Plateau and escarpment: The shape of visual performance. In *Proc CIE 21st Session.* Venice.

Boyce, P. R. 2004. *Reviews of Technical Reports on Daylight and Productivity.* Lighting Research Center, Rensselaer Polytechnic Institute.

Boyce, P. R. 2014. *Human Factors in Lighting.* 3rd ed. Boca Raton, FL: CRC Press.

Boyce, P. R., and M. S. Rea. 2001. *Lighting and Human Performance. II: Beyond Visibility Models toward a Unified Human Factors Approach to Performance*: Electric Power Research Institute, Palo Alto, CA, National Electrical Manufacturers Association, VA, Environmental Protection Agency Office of Air and Radiation, Washington, DC.

Boyce, P. R., Y. Akashi, C. M. Hunter et al. 2003. The impact of spectral power distribution on the performance of an achromatic visual task. *Light Res Technol.* 35(2):141–56.

Boyce, P. R., S. M. Berman, B. L. Collins et al. 1990. *Lighting and Human Performance: A Review.* Washington, DC: National Electrical Manufacturers Association (NEMA).

Boynton, R. M. 1979. *Human Color Vision.* New York: Holt, Rinehart and Winston.

Bullough, J. D., and N. P. Skinner. 2009. Predicting stopping distances under different types of headlamp illumination. In *8th Internat Symp Automotive Lighting.* München, Germany: Herbert Utz Verlag.

Bullough, J., K. S. Hickcox, T. Klein et al. 2012. Detection and acceptability of stroboscopic effects from flicker. *Light Res Technol.* 44(4):477–83.

Bullough, J. D., E. T. Donnell, and M. S. Rea. 2013. To illuminate or not to illuminate: Roadway lighting as it affects traffic safety at intersections. *Accid Anal Prev.* 53:65–77.

Bullough, J. D., L. C. Radetsky, U. C. Besenecker et al. 2014. Influence of spectral power distribution on scene brightness at different light levels. *LEUKOS* 10(1):3–9.

Bullough, J. D., L. C. Radetsky, and M. S. Rea. 2011. Testing a model of scene brightness with and without objects of different colours. *Light Res Technol.* 43(2):173–84.

Bullough, J. D., M. S. Rea, and X. Zhang. 2012. Evaluation of visual performance from pedestrian crosswalk lighting. In *Transportation Research Board 91st Annual Meeting.* Washington, DC.

Bullough, J. D., and N. P. Skinner. 2009. Predicting stopping distances under different types of headlamp illumination. In *8th International Symposium on Automotive Lighting.* München, Germany: Herbert Utz Verlag.

Bullough, J. D., and N. P. Skinner. 2012. Vehicle lighting and modern roundabouts: Implications for pedestrian safety. *SAE Int J Passenger Cars Mech Syst.* 5:195–8.

Bullough, J. D., X. Zhang, N. P. Skinner et al. 2010. Design and demonstration of pedestrian crosswalk lighting. In *Transport Res Board 89th Ann Meet.* Washington, DC: Transportation Research Board.

Cajochen, C., M. Munch, S. Kobialka et al. 2005. High sensitivity of human melatonin, alertness, thermoregulation and heart rate to short wavelength light. *J Clin Endocrinol Metab.* 90:1311–6.

Cajochen, C., J. M. Zeitzer, C. A. Czeisler et al. 2000. Dose-response relationship for light intensity and ocular and electroencephalographic correlates of human alertness. *Behav Brain Res.* 115(1):75–83.

Campbell, S. S., and D. Dawson. 1990. Enhancement of nighttime alertness and performance with bright ambient light. *Physiol Behav.* 48(2):317–20.

Chang, A. M., F. A. Scheer, and C. A. Czeisler. 2011. The human circadian system adapts to prior photic history. *J Physiol.* 589(Pt 5):1095–102.

Chellappa, S. L., R. Steiner, P. Blattner et al. 2011. Non-visual effects of light on melatonin, alertness and cognitive performance: Can blue-enriched light keep us alert? *PLoS One.* 6(1):e16429.

Commission Internationale de l'Éclairage. 1978. *Light as a True Visual Quantity: Principles of Measurement.* Paris: Commission Internationale de l'Éclairage.

Cos, S., M. D. Mediavilla, R. Fernandez et al. 2002. Does melatonin induce apoptosis in MCF-7 human breast cancer cells in vitro? *J Pineal Res.* 32(2):90–6.

Daurat, A., J. Foret, O. Benoit et al. 2000. Bright light during nighttime: Effects on the circadian regulation of alertness and performance. *Biol Signals Recept.* 9(6):309–18.

DiLaura, D., K. Houser, R. Mistrick et al. (Eds.). 2012. *IES Lighting Handbook: Reference and Application.* 10th ed. New York: Illuminating Engineering Society of North America.

Dillon, R. F., I. C. Pasini, and M. S. Rea. 1987. Survey of visual contrast in office forms. Paper read at *21st Session of the Commission Internationale de l'Éclairage,* Venice, Italy, June 17–25.

Eighth Census Conference on Newborn ICU Design. 2012. *Recommended Standards for Newborn ICU Design.* Clearwater Beach, FL: Committee to Establish Recommended Standards for Newborn ICU Design.

Eklund, N., P. Boyce, and S. Simpson. 2001. Lighting and sustained performance: Modeling data-entry task performance. *J Indoeur Stud.* 30(2):126–41.

Farriol, M., Y. Venereo, X. Orta et al. 2000. In vitro effects of melatonin on cell proliferation in a colon adenocarcinoma line. *J App Toxicol.* 20(1):21–4.

Figueiro, M. G., and M. S. Rea. 2012. Preliminary evidence that light through the eyelids can suppress melatonin and phase shift dim light melatonin onset. *BMC Research Notes.* 5(1):221.

Figueiro, M. G., A. Bierman, B. Plitnick et al. 2009. Preliminary evidence that both blue and red light can induce alertness at night. *BMC Neuroscience.* 10:105.

Figueiro, M. G., A. Bierman, and M. S. Rea. 2013. A train of blue light pulses delivered through closed eyelids suppresses melatonin and phase shifts the human circadian system. *Nature and Science of Sleep.* 5:133–41.

Figueiro, M. G., J. D. Bullough, A. Bierman et al. 2007. On light as an alerting stimulus at night. *Acta Neurobiol Experiment.* 67(2):171–8.

Figueiro, M. G., N. Z. Lesniak, and M. S. Rea. 2011. Implications of controlled short-wavelength light exposure for sleep in older adults. *BMC Research Notes*. 4(1):334.

Figueiro, M. G., S. Nonaka, and M. S. Rea. 2014. Daylight exposure has a positive carryover effect on nighttime performance and subjective sleepiness. *Light Res Technol*. 46(5):506–19.

Figueiro, M. G., B. Plitnick, and M. S. Rea. 2014. Pulsing blue light through closed eyelids: Effects on phase shifting of dim light melatonin onset in older adults living in a home setting. *Nature Sci Sleep*. 6:149–56.

Figueiro, M. G., M. S. Rea, P. R. Boyce et al. 2001. The effects of bright light on day and night shift nurses' performance and well-being in the NICU. *Neonatal Intensive Care*. 14(1):29–32.

Figueiro, M. G., L. Sahin, B. Wood, and B. Plitnick. 2016. Light at night and measures of alertness and performance: Implications for shift workers. *Biol Res Nursing*. 18(1):90–100.

Filipski, E., V. M. King, X. Li et al. 2002. Host circadian clock as a control point in tumor progression. *J Nat Cancer Instit*. 94(9):690–7.

Filipski, E., X. M. Li, and F. Levi. 2006. Disruption of circadian coordination and malignant growth. *Cancer Causes Control*. 17(4):509–14.

Foster, R. G., and M. W. Hankins. 2002. Non-rod, non-cone photoreception in the vertebrates. *Prog Retinal Eye Res*. 21(6):507–27.

Fotios, S. A. 2001. Lamp colour properties and apparent brightness: A review. *Light Res Technol*. 33(3):163–78.

Fotios, S., and C. Cheal. 2011. Predicting lamp spectrum effects at mesopic levels: Part 1: Spatial brightness. *Light Res Technol*. 43(2):143–57.

Fotios, S., C. Cheal, and P. Boyce. 2005. Light source spectrum, brightness perception and visual performance in pedestrian environments: A review. *Light Res Technol*. 37(4):271–91.

Freedman, M., R. Lucas, B. Soni et al. 1999. Regulation of mammalian circadian behavior by non-rod, non-cone, ocular photoreceptors. *Science*. 284:502–4.

Goodspeed, C. H., and M. S. Rea. 1999. The significance of surround conditions for roadway signs. *J Indoeur Stud*. 28(1):164–73.

Grundy, A., J. Tranmer, H. Richardson et al. 2011. The influence of light at night exposure on melatonin levels among Canadian rotating shift nurses. *Cancer Epidemiol Biomarkers Prevent*. 20(11):2404–12.

Hattar, S., R. J. Lucas, N. Mrosovsky et al. 2003. Melanopsin and rod-cone photoreceptive systems account for all major accessory visual functions in mice. *Nature*. 424:75–81.

Haus, E., and M. Smolensky. 2006. Biological clocks and shift work: Circadian dysregulation and potential long-term effects. *Cancer Causes Control*. 17(4):489–500.

Hebert, M., S. K. Martin, C. Lee et al. 2002. The effects of prior light history on the suppression of melatonin by light in humans. *J Pineal Res*. 33(4):198–203.

Hill, S. M., and D. E. Blask. 1988. Effects of the pineal hormone melatonin on the proliferation and morphological characteristics of human breast cancer cells (MCF-7) in culture. *Cancer Res*. 48(21):6121–6.

Illuminating Engineering Society. 1972. *IES Lighting Handbook*. New York: Illuminating Engineering Society.

Kaplan, E., and R. M. Shapley. 1982. X and Y cells in the lateral geniculate nucleus of macaque monkeys. *J Physiol*. 330(1):125–43.

Keller, S. M. 2009. Effects of extended work shifts and shift work on patient safety, productivity, and employee health. *J Amer Assoc Occupat Health Nurses*. 57(12):497–502, quiz 3–4.

Kinney, J. S. 1962. Factors affecting induced color. *Vision Res*. 2(12):503–25.

Klein, D. 1993. The mammalian melatonin rhythm-generating system. In *Light and Biological Rhythms in Man*, L. Wetterberg (Ed.). Oxford: Pergamon Press.

Langley, A. R., C. H. Graham, A. L. Grundy et al. 2012. A cross-sectional study of breast cancer biomarkers among shift working nurses. *BMJ Open*. 2(1):e000532.

Lewy, A., T. Wehr, T. Goodwin et al. 1980. Light suppresses melatonin secretion in humans. *Science*. 210(4475):1267–9.

Lie, J. A., H. Kjuus, S. Zienolddiny et al. 2011. Night work and breast cancer risk among Norwegian nurses: Assessment by different exposure metrics. *Amer J Epidemiol*. 173(11):1272–9.

Lowden, A., and T. Åkerstedt. 2012. Assessment of a new dynamic light regimen in a nuclear power control room without windows on quickly rotating shiftworkers—Effects on health, wakefulness, and circadian alignment: A pilot study. *Chronobiol Internat*. 29(5):641–9.

Lucas, R. J., and R. G. Foster. 1999. Neither functional rod photoreceptors nor rod or cone outer segments are required for the photic inhibition of pineal melatonin. *Endocrinol*. 140(4):1520–4.

Lucas, R., M. Freedman, and M. Munoz. 1999. Regulation of the mammalian pineal by non-rod, non-cone, ocular photoreceptors. *Science*. 284:505–7.

Luckiesh, M., and F. Moss. 1937. *The Science of Seeing*. New York: Van Nostrand.

McIntyre, I., T. Norman, G. Burrows et al. 1989. Quantal melatonin suppression by exposure to low intensity light in man. *Life Sci.* 45(4):327–32.

Mills, P. R., S. C. Tomkins, and L. J. Schlangen. 2007. The effect of high correlated colour temperature office lighting on employee wellbeing and work performance. *J Circad Rhythms.* 5(1):2.

Narendran, N., and L. Deng. 2002. Color rendering properties of LED light sources. In *Solid State Lighting II: Proc SPIE.* 4776:61–67.

Panda, S., T. K. Sato, A. M. Castrucci et al. 2002. Melanopsin (Opn4) requirement for normal light-induced circadian phase shifting. *Science.* 298(5601):2213–6.

Pesch, B., V. Harth, S. Rabstein et al. 2010. Night work and breast cancer—Results from the German GENICA study. *Scand J Work Env Health.* 36(2):134–41.

Phipps-Nelson, J., J. R. Redman, D. J. Dijk et al. 2003. Daytime exposure to bright light, as compared to dim light, decreases sleepiness and improves psychomotor vigilance performance. *Sleep.* 26(6):695–700.

Provencio, I., G. Jiang, W. De Grip et al. 1998. Melanopsin: An opsin in melanophores, brain, and eye. *Proc Nat Acad Sci of the USA.* 95:340–5.

Rea, M. S. 1983. Behavioral responses to a flexible desk luminaire. *J Illumin Eng Soc.* 13(1):174–90.

Rea, M. S. 1990. Some basic concepts and field applications for lighting, color, and vision. In *Glare and Contrast Sensitivity for Clinicians,* Nadler, M., D. Miller, and D. Nadler (Eds.). New York: Springer-Verlag.

Rea, M. S. 1991. Solving the problem of VDT reflections. *Progress Arch.* 72(10):35–40.

Rea, M. S. (Ed.). 2000. *IESNA Lighting Handbook: Reference and Application.* Ninth ed. New York: Illuminating Engineering Society of North America.

Rea, M. S. 2013. *Value Metrics for Better Lighting.* Bellingham, WA: SPIE—The International Society for Optical Engineering.

Rea, M. S., and J. P. Freyssinier. 2008. Color rendering: A tale of two metrics. *Color Res Applic.* 33(3):192–202.

Rea, M. S., and J. P. Freyssinier. 2010. Color rendering: Beyond pride and prejudice. *Color Res Applic.* 35(6):401–9.

Rea, M. S., and J. P. Freyssinier. 2013. White lighting: A theoretical and empirical framework. In *Proc 12th Internat AIC Colour Congress.* 2:651–654. Newcastle upon Tyne, UK: Association Internationale de la Couleur (AIC).

Rea, M. S., and M. Ouellette. 1988. Table-tennis under high intensity discharge (HID) lighting. *J Indoeur Stud.* 17(1):29–35.

Rea, M. S., and M. J. Ouellette. 1991. Relative visual performance: A basis for application. *Light Res Technol.* 23(3):135–44.

Rea, M. S., M. G. Figueiro, and J. D. Bullough. 2002. Circadian photobiology: An emerging framework for lighting practice and research. *Light Res Technol.* 34(3):177–87.

Rea, M. S., M. J. Ouellette, and M. E. Kennedy. 1985. Lighting and task parameters affecting posture, performance and subjective ratings. *J Indoeur Stud.* 15(1):231–8.

Rea, M. S., L. C. Radetsky, and J. D. Bullough. 2011. Toward a model of outdoor lighting scene brightness. *Light Res Technol.* 43(1):7–30.

Rüger, M., M. C. Gordijn, D. G. Beersma et al. 2006. Time-of-day-dependent effects of bright light exposure on human psychophysiology: Comparison of daytime and nighttime exposure. *Amer J Physiol.* 290(5):R1413–20.

Sahin, L., and M. G. Figueiro. 2013. Alerting effects of short-wavelength (blue) and long-wavelength (red) lights in the afternoon. *Physiol Behav.* 116–117:1–7.

Schäfer, A., and K. W. Kratky. 2006. The effect of colored illumination on heart rate variability. *Forsch Komplementmed.* 13(3):167–73.

Schernhammer, E. S., F. Laden, F. E. Speizer et al. 2001. Rotating night shifts and risk of breast cancer in women participating in the Nurses' Health Study. *J Nat Cancer Instit.* 93(20):1563–8.

Schernhammer, E. S., F. Laden, F. E. Speizer et al. 2003. Night-shift work and risk of colorectal cancer in the Nurses' Health Study. *J Nat Cancer Instit.* 95:825–88.

Schnell, T., L. Yekhshatyan, and R. Daiker. 2009. Effect of luminance and text size on information acquisition time from traffic signs. *Transport Res Rec.* 2122:52–62.

Skinner, N. P., and J. D. Bullough. 2011. Influence of intelligent vehicle headlamps on pedestrian visibility in roundabouts. In *9th Internat Sympos Automotive Lighting.* München: Herbert Utz Verlag.

Smith, K. A., M. W. Schoen, and C. A. Czeisler. 2004. Adaptation of human pineal melatonin suppression by recent photic history. *J Clin Endocrinol Metab.* 89(7):3610–4.

Smith, M. R., L. F. Fogg, and C. I. Eastman. 2009. A compromise circadian phase position for permanent night work improves mood, fatigue, and performance. *Sleep.* 32(11):1481–9.

Smith, S. W., and M. S. Rea. 1980. Relationships between office task performance and ratings of feelings and task evaluations under different light sources and levels. *Proc 19th Sess Commission Internationale de l'Éclairage.* Kyoto: Commission Internationale de l'Éclairage.

Smolders, K. C., Y. A. de Kort, and P. J. Cluitmans. 2012. A higher illuminance induces alertness even during office hours: Findings on subjective measures, task performance and heart rate measures. *Physiol Behav.* 107(1):7–16.

Stevens, R. G. 1987. Electric power use and breast cancer: A hypothesis. *Amer J Epidemiol.* 125:556–61.

Tucker, M. A., and W. Fishbein. 2009. The impact of sleep duration and subject intelligence on declarative and motor memory performance: How much is enough? *J Sleep Res.* 18(3):304–12.

Van de Werken, M., M. C. Gimenez, B. de Vries et al. 2013. Short-wavelength attenuated polychromatic white light during work at night: Limited melatonin suppression without substantial decline of alertness. *Chronobiol Internat.* 30(7):843–54.

Vandewalle, G., E. Balteau, C. Phillips et al. 2006. Daytime light exposure dynamically enhances brain responses. *Current Biol.* 16(16):1616–21.

Viola, A. U., L. M. James, L. J. Schlangen et al. 2008. Blue-enriched white light in the workplace improves self-reported alertness, performance and sleep quality. *Scand J Work, Environ Health.* 34(4):297–306.

Weston, H. C. 1945. *The Relation between Illumination and Industrial Efficiency: The Effect of Brightness Contrast.* Industrial Health Research Board of the Medical Research Council Report 87. London: His Majesty's Stationary Office.

WHO International Agency for Research on Cancer (IARC). 2010. *IARC Monographs on the Evaluation of Carcinogenic Risks to Humans: Painting, Firefighting, and Shiftwork.* Lyon: WHO.

Yoon, I. Y., D. U. Jeong, K. B. Kwon et al. 2002. Bright light exposure at night and light attenuation in the morning improve adaptation of night shift workers. *Sleep.* 25(3):351–6.

Zeitzer, J. M., D-J. Dijk, R. Kronauer et al. 2000. Sensitivity of the human circadian pacemaker to nocturnal light: Melatonin phase resetting and suppression. *J Physiol.* 526(3):695–702.

7 Shift Work Effects on Health and Productivity

Sampsa Puttonen

CONTENTS

7.1 INTRODUCTION

As a result of the evolving 24-hour society, about 15–30% of the workforce works outside the normal business hours. About 15% of the workers in the United States are employed on shift work and of the European Union (EU) workforce, 19% are engaged in shift work that includes night work, with rather more men (23%) than women (14%) involved. Shift work is related to a wide range of public health problems ranging from poor sleep and accidents to cardiovascular disease (CVD) and cancer. Although the risk estimates for the outcomes are low (odds ratio [OR] ≤2), the extensive exposure of workers (about 20%) combined with the high prevalence of the diseases means that shift work is definitely among one of the most serious occupational health problems of our time. This chapter reviews the major health risk factors of shift work, its effects on productivity, and risk reduction strategies to alleviate the risks of shift work.

7.2 WORKING TIMES IN MODERN SOCIETY

Shift work has been a public health issue for over 100 years. Still, even today, the association of shift work with major chronic diseases is under debate. In the 1920s, night work was banned for women in several European countries based on an early International Labour Organization (ILO) convention (ILO 1919). Society has changed to 24/7 work hours during a few decades: Working hours have become more irregular, demanding, and fast paced, resulting in increased occupational stress and sleeping difficulties (Harma and Kecklund 2010). As a result of the evolving 24-hour society, increasing proportion of the workforce works outside the normal business hours. The number of individuals working in shift work has steadily increased during the last decades; about 15% of the workers in the United States are employed on shift work and of the EU workforce, 19% are engaged in shift work that includes night work, with rather more men (23%) than women (14%) involved, the corresponding number being, e.g., in China as high as 36% (European Communities 2008).

The changes in the working time reflect societal reasons, economic demands, as well as individual preferences. On the one hand, companies look for a prompt adaptation of production and service systems to the increasing market demands and technological innovations; while on the other hand, employees ask for a more balanced pattern between working and leisure times to improve their working and social life. Individual risk of shift work can be considered low. However, shift work increases the risk of major chronic diseases and affects a large proportion of the workforce. This makes shift work as one of the most serious occupational health problems of the society.

7.3 DEFINITION OF SHIFT WORK

Shift work can be defined as either work at changing hours of the day (e.g., morning shift, afternoon shift, and night shift) or work at constant but unusual hours of the day (e.g., permanent afternoon shift or permanent night shift) (Knauth 1996). Shift systems vary with respect to several characteristics such as the duration of the duty period (e.g., from 8 to 12 hours), the interruption of weekend/Sunday (continuous/discontinuous), the speed (fast/slow) and the direction of the shift rotation (clockwise/counterclockwise), and the start and finish times of the shifts (Costa 2003). The definition of a night shift varies in the literature. A recommended definition has been given by the IARC working group, which defines *night shift* as a work period that includes at least 3 hours of work between midnight and 5:00 a.m. (Stevens et al. 2011).

The way shift work is organized affects its potential health and productivity effects. Night shift work causes sleep loss, disruption of biological rhythms, and perturbation of the social and family life, which can negatively influence mental and physiological health and productivity.

In addition to night work, other unusual or irregular working hours may also cause circadian misalignment and significant sleep loss (e.g., the early morning shifts).

7.4 HOW SHIFT WORK INFLUENCES THE DEVELOPMENT OF POOR HEALTH

7.4.1 Shift Work and Sleep

Several internal and external factors influence our alertness as well as our mental and physiological performances. For circadian rhythms to benefit an organism, they must be synchronized to the daily changes in the environment. Light–dark rhythm is a strong natural zeitgeber that synchronizes our biological rhythms in accord with the 24-hour day length of the earth. However, various nonphotic signals are also able to act as circadian synchronizers. Of these, the physical activity and the timing of feeding can have profound effects on the circadian rhythms. In the modern world, the impact of social factors to our rhythms, such as working times, has vastly increased.

Circadian rhythms are endogenously controlled events that occur on a daily basis and are driven by intracellular clocks. These rhythms regulate many aspects of behavior and physiology, from sleep–wake cycles to metabolism. Our circadian system contains a master clock in the SCN located in the anterior hypothalamus. In humans, the circadian clocks have been identified in cells throughout the body.

Any shift worker knows the acute effects of night shift work on compromised alertness, cognitive capacity, and physiological need for sleep and recovery. The disturbances in sleep–wakefulness are probably the best-established health problems of shift work. Disturbed sleep, which itself is a major health risk of shift work, mediates the long-term health effects of night shift work. Previous studies have shown that insomnia symptoms predict early retirement and development of depression (Tsuno, Besset, and Ritchie 2005; Sivertsen et al. 2006). Prospective epidemiological studies indicate that disturbed or shortened sleep predicts not only sickness absence, work-related stress, and burnout but also many diseases too: obesity and type 2 diabetes (T2D), coronary heart disease and brain infarctions, musculoskeletal diseases, and minor psychiatric problems (Nilsson, Roost, Engstrom, Hedblad, and Berglund 2004; Åkerstedt 2007; Patel, Malhotra, White, Gottlieb, and Hu 2006).

Individuals whose work schedules overlap with the normal sleep period experience misalignment between the endogenous circadian clock and the time at which the worker is able to rest. Restriction of sleep, desynchrony of bodily circadian systems, and decreased alertness do not go without harmful consequences. Research has shown a decrease in (work) performance, deficits in memory function, and an increase in suffer from occasional sleeping difficulties restricted sleep (for review, see Banks and Dinges [2007]). The variation of sleep-wakefulness in shift work is influenced by several factors. The major factors predicting sleep (sleep length, sleep latency, sleep structure) and sleepiness (sleep propensity and sleepiness) at work are related to the circadian variation of body functions, time spent awake, jet lag, and sleep recovery (Folkard and Åkerstedt 1992). In addition, aging, chronotype (morningness–eveningness), living habits, and social and personality factors may modify the circadian variation of body functions and tolerance to shift work (Saksvik, Bjorvatn, Hetland, Sandal, and Pallesen 2011). For example, it is more difficult for morning types, with an earlier circadian phase and sleep–wake rhythm, to phase delay their circadian rhythms during consecutive night shifts than it is for evening types (Harma 1995).

Insomnia is prevalent among shift workers, especially in relation to morning and night shifts, and most shift workers suffer from occasional sleeping difficulties at some point in their career. Additionally, the typical prevalence of excessive sleepiness during work shifts (showing sleep intrusions) is around 25%, and those with reduced alertness constitute an additional 50%. So a conservative estimate of the proportion of shift workers with severely reduced sleep or alertness lies somewhere above 50% (Åkerstedt 2005). In addition, shift work–related sleep deprivation may increase appetite and food intake and consequently predispose to the development of obesity (Arble, Bass, Laposky, Vitaterna, and Turek 2009).

It is estimated that approximately 10% of shift workers suffer from recurrent insomnia and/ or excessive fatigue known as the shift work disorder (SWD) (Drake, Roehrs, Richardson, Walsh, and Roth 2004). The *International Classification of Sleep Disorders* (third edition) defines SWD as shortened sleep with complaint of insomnia or excessive sleepiness temporally associated with a recurring work schedule that overlaps the usual time for sleep. The symptoms must be associated with the shift work schedule over the course of at least 3 months, and the circadian and sleep time misalignment has to be demonstrated by a sleep log or an actigraphic monitoring for at least 1 week. Finally, sleep disturbance may not be explained by another sleep disorder, a medical or neurological disorder, mental disorder, medication use, or substance misuse.

Lifestyle and health behavioral factors, individual susceptibility, morbidity, and genetic factors are all potential contributors to circadian rhythm pathology in SWD. For example, cognitive deficits from sleep debt show large and stable interindividual variation (Van Dongen, Baynard, Maislin, and Dinges 2004), and this vulnerability to sleep debt greatly varies even in a highly selective population of shift workers and does not match well with self-reported increases in sleepiness (Van Dongen 2006).

Significant daytime sleepiness measured by the Epworth Sleepiness scale, melatonin use or bright light therapy, and depressive symptoms have been identified as predictors of SWD (Waage et al. 2014). Of the work schedule-related factors, the number of nights worked the last year predicted the risk of future SWD in the study. SWD is underrecognized in clinical settings (Sack et al. 2007) partly due to the insufficient research on its clinical definition, prevalence, and diagnosis.

7.4.2 SHIFT WORK AND PHYSICAL HEALTH

In today's society, shift work is a major work-related disease risk factor. Shift work is prevalent; about 20% of the employees are engaged in shift work and shift work is a risk factor for several noncommunicable diseases (NCDs). Shift work that includes night work has been especially linked to the risk of obesity, CVDs, T2D, metabolic syndrome, breast cancer, and gastric ulcers, and shift work may also be linked to some other cancers and articular diseases.

Shift work disturbs our normal sleep–wake rhythms that result in increased risk of traffic and work injuries, poorer work performance, and productivity via, e.g., decreased alertness.

7.4.3 SHIFT WORK AND CVDS

Globally, NCDs are the leading cause of mortality, with CVDs comprising up to 46% of all NCD deaths. Shift work and CVDs have been studied in dozens of studies and even after three decades of reviews, there exists no consistent evidence that shift work increases the risk of CVD (Kristensen 1989; Boggild and Knutsson 1999; Frost, Kolstad, and Bonde 2009; Vyas et al. 2012). In their systematic review of 16 studies, Frost, Kolstad, and Bonde (2009) concluded that there is limited evidence for a causal relationship between night shift work and ischemic heart disease. The data showed a stronger association between shift work and morbidity than mortality, suggesting that the selection out of shift work and the use of occupational health screening could weaken the association with mortality. The most recent review and meta-analysis was published by Vyas et al. (2012) and this analysis of 34 studies with over 2 million participants showed that shift work significantly predicted the risk of myocardial infarction (risk ratio [RR]: 1.23; 95% confidence interval [CI]: 1.15–1.31) and coronary events (RR: 1.24; 95% CI: 1.10–1.39), but not cardiovascular mortality. Taken together, the cumulating evidence quite clearly suggests that night shift work is related to CVD morbidity, but the evidence for mortality risk is nonconclusive.

Shift work can increase the risk of CVD by its known influence on several interrelated psychosocial, behavioral, and physiological mechanisms and risk factors of CVD (Puttonen, Härmä, and Hublin 2010). The suggested psychosocial mechanisms relate to the difficulties in controlling the working hours, poor work–life balance, and insufficient recovery from work. The most probable behavioral changes are weight gain, smoking, and possibly eating habits (Hemio et al. 2015). The suggested physiological and biological mechanisms are, e.g., activation of the autonomic nervous system, inflammation, changed lipid, and glucose metabolism (Puttonen, Härmä, and Hublin 2010).

7.4.4 SHIFT WORK AND CANCER

Exposure to light at night disturbs the circadian system with alterations of the normal sleep–wake patterns, suppression of melatonin hormone production, and regulation of genes involved in cancer-related pathways (Stevens et al. 2011; Aho et al. 2013). In addition, sleep deprivation common in night shift workers suppresses the natural killer cell activity reducing cellular immune defense (Wilder-Smith, Mustafa, Earnest, Gen, and Macary 2013). In 2007, on the basis of sufficient evidence in animals and limited evidence in humans, the IARC classified night shift work as a probable human carcinogen (group 2A) (Straif et al. 2007). Epidemiological evidence has linked night shift work to increased breast cancer risk, and in some studies, the association between shift work and prostate cancer (Sigurdardottir et al. 2012) has been reported. Ijaz et al. (2013) synthesized the evidence on the relationship between night shift work and breast cancer and focused on the cumulative exposure to night work in years and in that of number of night shifts. They identified 16 studies (12 case-control and 4 cohort studies) and found a 9% risk increase per 5 years of night shift work exposure in 9 case-control studies, but not in the 3 included cohort studies (RR: 1.01; 95% CI: 0.97–1.05). They concluded that the findings indicate insufficient evidence for a link between night shift work and breast cancer. Another systematic review and meta-analysis of 15 studies by Kamdar et al. (2013) ended in a quite similar conclusion that there is only weak evidence that night shift work increases the risk of breast cancer (pooled RR for every night work: 1.21; 95% CI: 1.00–1.27). Both of the reviews highlighted the apparent methodological weaknesses of the studies and called for better exposure assessment and the need for objective prospective exposure measurement in future studies.

7.4.5   SHIFT WORK AND METABOLIC SYNDROME

Metabolic syndrome (MetS) is a clustering of the risk factors for CVDs such as insulin resistance, hypertension, blood lipid and glucose abnormalities, and central obesity (Galassi, Reynolds, and He 2006). The possible role of shift work on the development of MetS has been explored in a number of studies over the last two decades. The observations in the cross-sectional data have suggested that the occurrence is higher in rotating night shift workers compared to their day-working counterparts. In a study of healthy male blue-collar workers (877 day workers, 474 rotating shift workers in 12-hour shifts), the odds ratio for MetS in shift workers compared to that of day workers was 1.5 when the effect of age and physical activity were taken into account (Sookoian et al. 2007; Esquirol et al. 2009). The study by Esquirol et al. (2009) focused on male chemical plant employees who had worked 10 or more years in their current shift system (8-hour rotating shift or day shift) and reported a positive association between shift work and MetS, but the association depended on the criteria used for MetS. Puttonen, Viitasalo, and Harma (2012) found a marginally significant correlation between night shift work and MetS in a sample of male employees of an airline (OR: 1.5; 95% CI: 0.95–2.34), and the association was further attenuated with additional adjustments. In the latter study, compared to day workers, MetS was more prevalent in former male workers and two-shift workers (OR: 2.0; 95% CI: 1.26–3.19). The association with two-shift workers was not significant after adjusting for several potential confounders (OR: 1.5; 95% CI: 0.93–2.24). In women, no differences in the prevalence of MetS between day workers and former or current shift workers were observed. The follow-up studies have showed that the rotating shift workers may have an elevated risk for developing MetS (De Bacquer et al. 2009; Lin, Hsiao, and Chen 2009; Pietroiusti et al. 2009). Two published reviews that focused on the association between shift work and MetS ended up with different conclusions. The first systematic review in 2013 with 10 studies concluded that there is insufficient evidence for an association between shift work exposure and MetS (Canuto, Garcez, and Olinto 2013). A meta-analysis on night shift work and MetS found that shift work including night work is significantly associated with MetS (pooled RR: 1.6; 95% CI: 1.24–1.98) (Wang et al. 2014). The review included a total of 13 studies, of which 4 were longitudinal. The risk of MetS was also higher in shift workers with longer exposures to night work. A number of factors may have affected the differential results of the studies. Most of the selected studies in the reviews were cross-sectional, not enabling the investigating direction of association; the definitions of MetS varied between studies; and the shift schedules and strenuousness of shifts systems also varied between studies.

7.4.6   SHIFT WORK AND T2D

T2D has an increasing trend all over the world, and it is also an emerging health problem among workforce. The WHO forecasts that diabetes will be the seventh leading cause of death in 2030. T2D usually manifests at the ages above 40 and, for example, in the Finnish population, its estimated prevalence according to the WHO criteria is 10% in men and 7% in women (Yliharsila et al. 2005). Obesity, smoking, sedentary lifestyle, and poor dietary habits are well-known risk factors for T2D. Recently the role of shift work as a risk factor for the development of T2D has also gained interest in the scientific community.

The study by Pan, Schernhammer, Sun, and Hu (2011) demonstrated a gradual increase in T2D risk with increased exposure to night shift work (working ≥3 night shifts per month) in a sample with over 177,000 women from two large prospective cohorts. Compared with day workers, the risk for having worked 1–2, 3–9, 10–19, and ≥20 years of shift work were 1.05, 1.20, 1.40, and 1.58, respectively. The exposure to shift work was also related to weight gain in the sample. The findings of a meta-analytic study of 12 studies with 226,652 participants by Gan et al. (2015) supported the earlier findings and confirmed a positive association between shift work and T2D. In the total sample, the risk of T2D was increased by 9%, and in the subsample of rotating night shift workers, the risk was significantly higher (42%) than in other working hour groups.

7.4.7 Shift Work and Mental Health

Shift workers are known to show higher levels of health complaints including sleep disturbances. Insomnia is a typical symptom of depression, but insomnia can be an independent predictor of depression. Indeed, data have demonstrated that sleep disturbance is one of the most important predictors of a depressive episode, and individuals with sleeping problems have a two-fold risk of developing depression compared to individuals without sleeping problems (Baglioni et al. 2011). Based on the effects of sleeping difficulties on mental health, researchers may be grounded to hypothesize that shift work and depression are interrelated.

One of the early studies by Skipper, Jung, and Coffey (1990) reported no association between shift work and mental health in a sample of 463 nurses, and depression, according to the Beck depression scale, did not differ between day and shift workers in a sample public workers (Luca, Bellia, Bellia, Luca, and Calandra 2014). Instead, Scott, Monk, and Brink (1997) found an increased prevalence of a major depressive disorder in shift workers in their pilot study that included 89 current and former shift workers. In the longitudinal data of the British Household Panel Survey, Bara and Arber (2009) studied the results based on the subjectively rated symptoms of anxiety/depression and found that men working nights and women working varied shifts were adversely affected. Another prospective study did not find a major relationship between the shift work and the development of a depressed mood in a 10-year follow-up (Driesen, Jansen, van Amelsvoort, and Kant 2011). The significant finding between shift work and depression was limited to older male workers (≥45 years), but the association disappeared when demographic and work-related factors were adjusted. There is an indication that shift work sleep disorder is associated with mental symptoms and depression, but most results come from cross-sectional studies, and the direction of relation is unclear (Flo et al. 2012). Taken together, the current findings do not consistently point to poorer mental health of shift workers, nor is there sufficient evidence to indicate that working shifts increases the future risk of poor mental health. It also possible that selection processes into and out of shift work have affected the results. For example, individuals who have mental health symptoms and low tolerance for irregular working times tend to change to day work. In all, compared with studies of physical health, studies exploring the potential effects of shift work on mental health are limited in numbers, and the studies have mainly been correlative and mostly based on the subjective questionnaire data of mental health. These methodological shortcomings have contributed to the inconsistencies in the overall picture.

7.5 INFLAMMATION AND INFLAMMATORY DISEASES

7.5.1 Shift Work and Atherosclerosis and Inflammation

Atherosclerotic vascular disease is the greatest cause of death in developed countries (Mathers and Loncar 2006). Atherosclerosis is an inflammatory disease in which a plaque builds up in the arterial walls, making them gradually narrower and less elastic (Hansson 2005). The manifested coronary heart disease (CHD) develops over a long time span with physical changes beginning decades before the disease manifests. Some of the health hazards of shift work may remain undetected when CHD is used as an endpoint. Employees may change to day work when the first symptoms of CHD occur, or they may leave the workforce after getting the disease, highlighting the importance to focus on the subclinical phase. Puttonen et al. (2009) studied the relationship between shift work and subclinical atherosclerosis in a population-based sample of 1543 (712 men and 831 women) young adults (24–39 years) free of CVD. Atherosclerosis was assessed by measuring the thickness of the common carotid artery intima–media complex with ultrasound. Shift work was associated with a thicker mean intima–media, and 2.2-fold odds of carotid plaque only in men. The prevalence of plaque was 21% in the shift-working and 14% in the day-working men. In agreement, the study by Haupt et al. (2008) using a similar ultrasonic methodology to assess atherosclerotic process reported a higher risk in former shift workers (*n* = 698) who were 45 years of age or older. These findings provide support for the hypothesis that shift work contributes to increased CHD risk by accelerating the atherosclerotic process, making preventive actions against CVDs especially important in shift workers.

Our diurnal sleep–wake rhythm plays an important role in the homeostatic regulation of the activity and the expression of immune cells. However, only a small number of studies have tested whether the mechanism is active in the observed shift work–CVD association. (For a review see Faraut, Bayon, and Leger [2013].) High sensitive C-reactive protein (CRP), leukocyte count, and NK-cell activity have received some interest in shift work studies. Sookoian et al. (2007) and Nishitani and Sakakibara (2007) reported increased leukocyte counts in rotating shift workers compared to daytime workers and Puttonen, Viitasalo, and Harma (2011) found increased CRP levels both in two- and three-shift workers. In addition, three-shift work that includes night work has been associated with reduced T-lymphocyte function (Curti, Radice, Cesana, Zanettini, and Grieco 1982; Nakano et al. 1982), and NK-cell activity has been found to be lower in shift workers compared to day workers and lower during a night shift compared to day shift (Kobayashi, Furui, Akamatsu, Watanabe, and Horibe 1997; Okamoto et al. 2008).

7.5.2 Shift Work and Rheumatoid Arthritis

Rheumatoid arthritis (RA) is a chronic inflammatory disease with both genetic and environmental determinants (Lee and Weinblatt 2001). The onset of the disease is often the result of a long-term process, and the inflammatory biomarker levels have been reported to be increased years before the RA becomes clinically apparent. The genetic influence in RA is estimated to be as high as 50–60% (Silman and Pearson 2002), and the concordance for RA is only about 12–15% in monozygotic twins (Aho, Koskenvuo, Tuominen, and Kaprio 1986). Although environmental factors play an important role in the disease process, they remain poorly established. Several risk factors including smoking, socioeconomic status, diet, infectious agents, and hormonal factors have been suggested as risk factors of RA.

The possible role of shift work in the development of RA has been studied very little. Previous studies on the physiological changes in shift work suggest that shift work may induce changes in the biological markers of systemic inflammation (see the previous paragraph), such as leukocyte count and CRP. T-cell responses, which are known to play a key role in initiating the cascade of immunological events potentially leading to the development of manifested RA (Monaco, Andreakos, Kiriakidis, Feldmann, and Paleolog 2004), have been found to differ between shift workers and day workers (Curti, Radice, Cesana, Zanettini, and Grieco 1982; Nakano et al. 1982). Shift work has also been associated with a higher risk for common infections (Mohren et al. 2002). Puttonen et al. (2010) explored whether shift work is associated with RA in samples of 67,402 Finnish local government employees. The results indicated that female shift workers may have a higher risk of developing RA than day workers (RR: 1.4; 95% CI: 1.02–1.82). In order to verify the inflammatory pathway between shift work and RA, further studies linking the adverse changes in the immune system with both shift working and disease risk are needed.

7.6 SHIFT WORK, COGNITION, AND PRODUCTIVITY

Laboratory and questionnaire studies suggest that sleep loss and sleepiness are associated with a decreased threshold for taking a risk in a variety of cognitive tasks (Killgore, Balkin, and Wesensten 2006). With regard to the ability to self-monitor performance, a simulation study of night shift work suggested that individuals can monitor their performance only to a moderate degree when working night shifts work (Baranski, Pigeau, and Angus 1994; Baranski and Pigeau 1997; Lamond et al. 2003). Folkard and Tucker (2003) reported in their review of five studies that injury risk is increased by 8% in an afternoon shift and by 30% in a night shift compared to a morning shift. Work injury risk was also estimated to be 6% higher on the second night, 17% higher on the third night, and 30% higher on the fourth consecutive night shift. Permanent night work may bring some protection against this effect due to resynchronization (Wagstaff and Sigstad Lie 2011). The night shift seems to have a substantial effect of accidents, but the shifts starting very early in the morning also increase the risk of injury (Wagstaff and Sigstad Lie 2011). The acute effects of circadian disruption and sleep loss on cognitive performance are relatively well described in the literature. However, studies assessing other than the acute effects of shift work on cognition are emerging little by little. Özdemir et al. (2013) studied the effect of shift work on the participants' memory, attention, and learning functions. Night shift workers were tested after working 2 weeks of night shifts, and particularly, the working memory function was poorer in the group of shift workers compared to that in day workers. A large-scale follow-up study with 3232 participants showed that the association between shift work and cognitive decline, assessed by the memory and the speed of processing tasks, was stronger in the participants who had been exposed to rotating shift work for more than 10 years (Marquie, Tucker, Folkard, Gentil, and Ansiau 2015). The recovery of cognitive functioning after leaving any form of shift work took at least 5 years.

Figure 7.1 gives an overview of the factors that play a role in the association between shift work, productivity, and health. Other work and non-work-related factors and mechanism that influence a

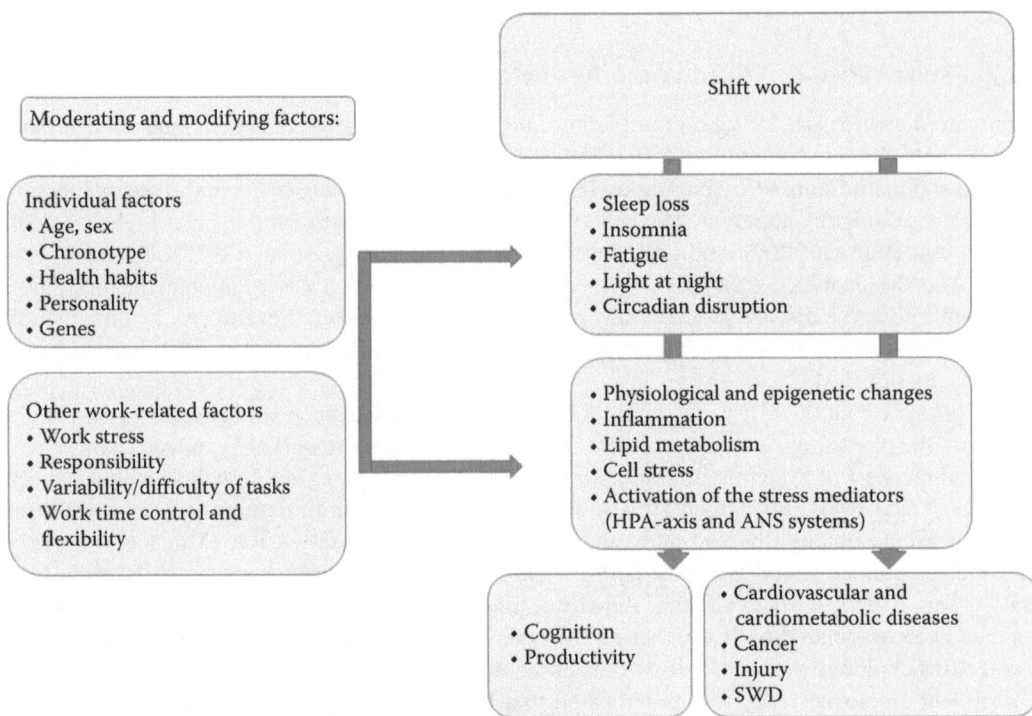

FIGURE 7.1 Schematic representation on how shift work affects health and productivity. ANS, autonomic nervous system; HPA, hypothalamic–pituitary–adrenal.

person's resources and tolerance to shift work can also influence the health and productivity outcomes in shift work. These include, e.g., a second job, work–life balance, social support, life stress, and physical and mental health.

7.7 STRATEGIES TO ALLEVIATE THE HEALTH RISKS OF SHIFT WORK

Several strategies can reduce the health risks among shift workers. These strategies can be divided into two groups according to whether they are targeted at work circumstances (e.g., ergonomic shift scheduling) or to the individual (e.g., health habits). Effective managing requires both types of actions. The first step for the organization is to acknowledge that shift work is a potential health risk factor; then it is important to identify the specific risks of the work tasks and actively develop an operations model to tackle the hazards and increase the productivity and the well-being at work. Health screening for individuals who may be vulnerable to shift work or its health effects are needed at different phases of a work career, starting as early as during the recruitment of the employees. Regular health checkups with health screenings are needed in all age groups including young shift workers. Even a simple screening questionnaire can efficiently detect the risk for, e.g., metabolic disorders (Viitasalo et al. 2012). Shift workers would also benefit from education to better cope with shift work (eating, sleeping, and exercising), and the use of countermeasures should be widened (e.g., taking naps before the night shift or adopting scheduled napping at night).

Developing better-tolerated shift systems together with individual-based flexibility is a suggested way to decrease the risks of night shift work, and the use of ergonomic working schedules and individual solutions should be more widely adopted (Sallinen and Kecklund 2010). Shift work schedules markedly vary with respect to timing and duration of shifts, occurrence of circadian disruption, disturbed sleep, and fatigue. A number of interventions have evaluated the effects of the changes in shift schedule, and the results quite clearly indicate that health risk and fatigue can be alleviated via adopting better-tolerated shift schedules (Sallinen and Kecklund 2010). The findings generally support the benefits of fast-forward rotating shift systems, in which the number of consecutive morning, evening, and night shifts is three or less. The positive effects of fast-forward shift systems are related to better sleep quality and increased sleep length (Czeisler, Moore-Ede, and Coleman 1982; Hakola and Harma 2001; Härmä et al. 2006), as well as lower levels in chronic disease risk factors (e.g., blood lipids, blood pressure) (Orth-Gomer 1983; Viitasalo, Kuosma, Laitinen, and Härmä 2008). In another study, the change to a more regular and predictable shift schedule predicted the decrease in triglyceride and high-density lipoprotein cholesterol levels (Boggild and Jeppesen 2001).

7.8 INDIVIDUAL TARGETED INTERVENTIONS

People with traditional CVD risk factors, such as smoking, overweight, and high cholesterol, are more likely to be affected by shift work (Tenkanen, Sjöblom, Kalimo, Alikoski, and Härmä 1997). On the other hand, irregular working hours or shift work can make it more difficult to maintain a healthy life style. Shift workers can have an increased risk of several conditions and diseases such as gastrointestinal disorders and cardiometabolic diseases that are known to be influenced by health habits. It is not clear based on scientific research how much working shifts may change the employees' health habits and what is the role of circadian disruption in the association between shift work health habits and health.

Health protection programs to help employees cope with health risk factors and stressors of shift work have proven to be relatively effective. A program that included comprehensive medical examinations and health promotion activities targeted at shift workers was able to slightly reduce the mortality risks among shift working employees (Oberlinner et al. 2009). A study by Viitasalo et al. (2015) evaluated the feasibility and the effectiveness of lifestyle counseling in an occupational setting on decreasing the risk for diabetes and CVD. The health checkup including low-intensity lifestyle intervention was feasible in an occupational setting and resulted in some beneficial effects for men with elevated T2D risk.

Based on a meta-analytic study, both acute and long-term physical activities have beneficial effects on sleep (Kubitz, Landers, Petruzzello, and Han 1996), and physical fitness also decreases shift work-related sleep problems (Neil-Sztramko, Pahwa, Demers, and Gotay 2014). However, heavy nocturnal exercise can have undesired effects, as it can induce phase delays in the circadian rhythm (Atkinson, Edwards, Reilly, and Waterhouse 2007). A 4-month intervention study involving a physical training specifically designed for shift workers resulted in improved sleep and alertness, especially during night shifts among nurses in an irregular shift system (Harma, Ilmarinen, Knauth, Rutenfranz, and Hanninen 1988a,b). Physical fitness seems to decrease sleepiness, because improved sleep and physical exercise at night hasten the circadian adaptation to night work.

7.9 DIETARY HABITS OF SHIFT WORKERS

Health behaviors such as healthy diet have an important influence on long-term health outcomes, but they can also influence the workers' alertness and performance on duty. The prevalence of gastrointestinal complaints has been suggested to vary between 20% and 75% in shift workers, compared to 10–25% of day workers, and shift workers may have a higher prevalence of irritable bowel syndrome (Lennernas, Hambraeus, and Åkerstedt 1995; Nojkov, Rubenstein, Chey, and Hoogerwerf 2010). There are very few randomized controlled studies on the efficacy of dietary interventions during shift work. Some favorable effects of such interventions on fatigue levels at work have been reported, but the effects on long-term health and energy balance have not been adequately studied (Lowden, Moreno, Holmback, Lennernas, and Tucker 2010). A study to explore food and nutrient intake differences in 1478 employees (55% men) reported that shift-working men used less vegetables and fruits daily than male day and in-flight workers. In women, the energy intake from saturated fat was higher among shift workers compared with day workers. In older female participants, the energy intake from fat and saturated fat was higher in the shift work and in-flight work groups than in the day work group. It is suggested that lifestyle counseling including advice on nutrition should be incorporated in the routine occupational healthcare of a shift worker.

7.10 DISCUSSION

Shift work that includes night work is related to a wide range of public health problems ranging from poor sleep to CVD and breast cancer. Shift work disturbs our normal sleep–activity rhythms and increases fatigue-related traffic and work injuries. It influences work performance and productivity via fatigue and excessive sleepiness, and irregular working times make sticking to a regular and healthy life style difficult. Although the risk estimates for the outcomes are low, the extensive exposure of workers combined with the high prevalence of the diseases means that shift work is definitely among one of the most serious occupational health problems of our time.

The observed inconsistencies in the epidemiologic evidence on the association between shift work and health reflect the common problems in the exposure assessment and the selection of a proper control group. Inconsistencies also arise from primary, secondary, and tertiary selection (Boggild 2009). A major step toward more systematic shift work data collecting and reporting was the 2011 consensus report on what aspects of shift work should epidemiological studies capture (Stevens et al. 2011). Recently, Härmä et al. (2015) utilized detailed working hour data of 12,391 nurses and physicians from employers' registers containing a total 14.5 million separate work shifts to develop a method and algorithms that allowed a detailed characterization of working time patterns potentially relevant for health. All together, 29 algorithms characterizing four potentially health-relevant working time patterns were designed: (1) length of the working hours, (2) time of the day, (3) shift intensity, and (4) social aspects of the working hours. This method that enables objective shift work exposure measurement can be used in future large-scale epidemiological studies.

Remedies to mitigate the effects of shift work

Individual

- Individual-based working time flexibility
- Napping (off and on duty) and sleep schedules
- Health behavior (stimulants, physical activity, healthy diet)
- Light therapy
- Sleep hygiene

Organization

- Health/fatigue risk management system
- Shift scheduling
- Individual-based working time flexibility
- Education and training programs
- Health promotion
- Health screening

FIGURE 7.2 Measures to support health and productivity in shift work.

Whether shift work predisposes to mental health problems remains an open issue. This is partly due to the shortage of high-quality studies. The indirect evidence from sleep research showing that insomnia increases the risk of depression would suggest that shift work also carries an increased risk of depression. It is evident that shift work causes circadian disruption, sleep loss, and sleeping difficulties. However, data show that shift workers do not necessarily report more general insomnia compared to day workers (Åkerstedt, Ingre, Broman, and Kecklund 2008). What they suffer mostly is shortage of sleep and difficulties in staying awake. Typical sleeping difficulties are shift specific such as problems falling asleep before early morning shifts and after night shifts. It may be that shift workers' risk of depression is limited to a subgroup of employees with SWD.

Shift work, especially night shift work, inevitably induces fatigue and burdens both the body and the brain. However, several recommendable and effective strategies and measures exist to alleviate the acute and long-term effects of shift work (Figure 7.2). A structured multidimensional approach with an appropriate follow-up is likely to yield the best results. For example, timed bright light, light blocking glasses, health promotion, healthy diet, and napping have been found to be effective in reducing fatigue and health risks in shift workers. In addition to targeted individual support, work organization-level programs and support are needed. Occupational healthcare should take a vital role in the preventive actions. The first step to get a grip on this is identifying the risks and developing a fatigue/health risk management system for the organization. Research data show that shift scheduling, individual-based working time flexibility, health promotion, and education and training programs are all effective to decrease the known risks of shift work.

REFERENCES

Aho, K., Koskenvuo, M., Tuominen, J., and Kaprio, J. (1986). Occurrence of rheumatoid arthritis in a nationwide series of twins. *J Rheumatol*, *13*(5), 899–902.

Aho, V., Ollila, H. M., Rantanen, V., Kronholm, E., Surakka, I., van Leeuwen, W. M. et al. (2013). Partial sleep restriction activates immune response-related gene expression pathways: Experimental and epidemiological studies in humans. *PloS One*, *8*(10), e77184.

Åkerstedt, T. (2005). Shift work and sleep disorders. *Sleep*, *28*(1), 9–11.

Åkerstedt, T. (2007). Altered sleep/wake patterns and mental performance. *Physiol Behav*, *90*(2–3), 209–218.

Åkerstedt, T., Ingre, M., Broman, J. E., and Kecklund, G. (2008). Disturbed sleep in shift workers, day workers, and insomniacs. *Chronobiol Int*, *25*(2), 333–348.

Arble, D. M., Bass, J., Laposky, A. D., Vitaterna, M. H., and Turek, F. W. (2009). Circadian timing of food intake contributes to weight gain. *Obesity (Silver Spring)*, *17*(11), 2100–2102.

Atkinson, G., Edwards, B., Reilly, T., and Waterhouse, J. (2007). Exercise as a synchroniser of human circadian rhythms: An update and discussion of the methodological problems. *Eur J Appl Physiol*, *99*(4), 331–341.

Baglioni, C., Battagliese, G., Feige, B., Spiegelhalder, K., Nissen, C., Voderholzer, U. et al. (2011). Insomnia as a predictor of depression: A meta-analytic evaluation of longitudinal epidemiological studies. *J Affect Disord*, *135*(1–3), 10–19.

Banks, S., and Dinges, D. F. (2007). Behavioral and physiological consequences of sleep restriction. *J Clin Sleep Med*, *3*(5), 519–528.

Bara, A. C., and Arber, S. (2009). Working shifts and mental health—Findings from the British Household Panel Survey (1995–2005). *Scand J Work Environ Health*, *35*(5), 361–367.

Baranski, J. V., Pigeau, R. A., and Angus, R. G. (1994). On the ability to self-monitor cognitive performance during sleep deprivation: A calibration study. *J Sleep Res*, *3*(1), 36–44.

Baranski, J. V., and Pigeau, R. A. (1997). Self-monitoring cognitive performance during sleep deprivation: Effects of modafinil, d-amphetamine and placebo. *J Sleep Res*, *6*(2), 84–91.

Boggild, H. (2009). Settling the question—The next review on shift work and heart disease in 2019. *Scand J Work Environ Health*, *35*(3), 157–161.

Boggild, H., and Jeppesen, H. J. (2001). Intervention in shift scheduling and changes in biomarkers of heart disease in hospital wards. *Scand J Work Environ Health*, *27*(2), 87–96.

Boggild, H., and Knutsson, A. (1999). Shift work, risk factors and cardiovascular disease. *Scand J Work Environ Health*, *25*(2), 85–99.

Canuto, R., Garcez, A. S., and Olinto, M. T. (2013). Metabolic syndrome and shift work: A systematic review. *Sleep Med Rev*, *17*(6), 425–431.

Costa, G. (2003). Shift work and occupational medicine: An overview. *Occup Med (Lond)*, *53*(2), 83–88.

Curti, R., Radice, L., Cesana, G. C., Zanettini, R., and Grieco, A. (1982). Work stress and immune system: Lymphocyte reactions during rotating shift work. Preliminary results. *Med Lav*, *73*(6), 564–569.

Czeisler, C. A., Moore-Ede, M. C., and Coleman, R. H. (1982). Rotating shift work schedules that disrupt sleep are improved by applying circadian principles. *Science*, *217*(4558), 460–463.

De Bacquer, D., Van Risseghem, M., Clays, E., Kittel, F., De Backer, G., and Braeckman, L. (2009). Rotating shift work and the metabolic syndrome: A prospective study. *Int J Epidemiol*, *38*(3), 848–854.

Drake, C. L., Roehrs, T., Richardson, G., Walsh, J. K., and Roth, T. (2004). Shift work sleep disorder: Prevalence and consequences beyond that of symptomatic day workers. *Sleep*, *27*(8), 1453–1462.

Driesen, K., Jansen, N. W., van Amelsvoort, L. G., and Kant, I. (2011). The mutual relationship between shift work and depressive complaints—A prospective cohort study. *Scand J Work Environ Health*, *37*(5), 402–410.

Esquirol, Y., Bongard, V., Mabile, L., Jonnier, B., Soulat, J. M., and Perret, B. (2009). Shift work and metabolic syndrome: Respective impacts of job strain, physical activity, and dietary rhythms. *Chronobiol Int*, *26*(3), 544–559.

European Communities (2008). Working time in the EU and other global economies. Luxembourg: Office for Official Publications of the European Communities.

Faraut, B., Bayon, V., and Leger, D. (2013). Neuroendocrine, immune and oxidative stress in shift workers. *Sleep Med Rev*, *17*(6), 433–444.

Flo, E., Pallesen, S., Mageroy, N., Moen, B. E., Gronli, J., Hilde Nordhus, I. et al. (2012). Shift work disorder in nurses—Assessment, prevalence and related health problems. *PLoS One*, *7*(4), e33981.

Folkard, S., and Åkerstedt, T. (1992). A three-process model of the regulation of alertness-sleepiness. In Broughton R. J., and Ogilvie, R. D. (Eds.), *Sleep, Arousal, and Performance* (pp. 11–26). Boston: Birkhauser.

Folkard, S., and Tucker, P. (2003). Shift work, safety and productivity. *Occup Med (Lond)*, *53*(2), 95–101.

Frost, P., Kolstad, H. A., and Bonde, J. P. (2009). Shift work and the risk of ischemic heart disease—A systematic review of the epidemiologic evidence. *Scand J Work Environ Health*, *35*(3), 163–179.

Galassi, A., Reynolds, K., and He, J. (2006). Metabolic syndrome and risk of cardiovascular disease: A meta-analysis. *Am J Med*, *119*(10), 812–819.

Gan, Y., Yang, C., Tong, X., Sun, H., Cong, Y., Yin, X. et al. (2015). Shift work and diabetes mellitus: A meta-analysis of observational studies. *Occup Environ Med*, *72*(1), 72–78.

Hakola, T., and Harma, M. (2001). Evaluation of a fast forward rotating shift schedule in the steel industry with a special focus on ageing and sleep. *J Hum Ergol (Tokyo)*, *30*(1–2), 315–319.

Hansson, G. K. (2005). Inflammation, atherosclerosis, and coronary artery disease. *N Engl J Med*, *352*(16), 1685–1695.

Harma, M. (1995). Sleepiness and shiftwork: Individual differences. *J Sleep Res*, *4*(S2), 57–61.

Harma, M., and Kecklund, G. (2010). Shift work and health—How to proceed? *Scand J Work Environ Health*, *36*(2), 81–84.

Härmä, M., Hakola, T., Kandolin, I., Sallinen, M., Virkkala, J., Bonnefond, A. et al. (2006). A controlled intervention study on the effects of a very rapidly forward rotating shift system on sleep-wakefulness and well-being among young and elderly shift workers. *Int J Psychophysiology: Off J Int Organ Psychophysiology*, *59*(1), 70–79.

Harma, M. I., Ilmarinen, J., Knauth, P., Rutenfranz, J., and Hanninen, O. (1988a). Physical training intervention in female shift workers: I: The effects of intervention on fitness, fatigue, sleep, and psychosomatic symptoms. *Ergonomics*, *31*(1), 39–50.

Harma, M. I., Ilmarinen, J., Knauth, P., Rutenfranz, J., and Hanninen, O. (1988b). Physical training intervention in female shift workers: II: The effects of intervention on the circadian rhythms of alertness, short-term memory, and body temperature. *Ergonomics*, *31*(1), 51–63.

Härmä, M., Ropponen, A., Hakola, T., Koskinen, A., Vanttola, P., Puttonen, S. et al. (2015). Developing register-based measures for assessment of working time patterns for epidemiologic studies. *Scand J Work Environ Health*, *41*(3), 268–279.

Haupt, C. M., Alte, D., Dorr, M., Robinson, D. M., Felix, S. B., John, U. et al. (2008). The relation of exposure to shift work with atherosclerosis and myocardial infarction in a general population. *Atherosclerosis*, *201*(1), 205–211.

Hemio, K., Puttonen, S., Viitasalo, K., Harma, M., Peltonen, M., and Lindstrom, J. (2015). Food and nutrient intake among workers with different shift systems. *Occup Environ Med*, *72*(7), 513–520.

Ijaz, S., Verbeek, J., Seidler, A., Lindbohm, M. L., Ojajarvi, A., Orsini, N. et al. (2013). Night-shift work and breast cancer—A systematic review and meta-analysis. *Scand J Work Environ Health*, *39*(5), 431–447.

International Labour Organization (ILO). (1919). Night work (women) convention concerning employment of women during the night. ILO, Washington, D.C.

Kamdar, B. B., Tergas, A. I., Mateen, F. J., Bhayani, N. H., and Oh, J. (2013). Night-shift work and risk of breast cancer: A systematic review and meta-analysis. *Breast Cancer Res Treat*, *138*(1), 291–301.

Killgore, W. D., Balkin, T. J., and Wesensten, N. J. (2006). Impaired decision making following 49 h of sleep deprivation. *J Sleep Res*, *15*(1), 7–13.

Knauth, P. (1996). Categories and parameters of shiftwork systems. In Colquhoun, P. W., Costa, G., Folkard, S., and Knauth, P. (Eds.), *Shiftwork: Problems and Solutions*. (Vol. 7, pp. 17–28). Frankfurt.

Kobayashi, F., Furui, H., Akamatsu, Y., Watanabe, T., and Horibe, H. (1997). Changes in psychophysiological functions during night shift in nurses: Influence of changing from a full-day to a half-day work shift before night duty. *Int Arch Occup Environ Health*, *69*(2), 83–90.

Kristensen, T. S. (1989). Cardiovascular diseases and the work environment: A critical review of the epidemiologic literature on nonchemical factors. *Scand J Work Environ Health*, *15*(3), 165–179.

Kubitz, K. A., Landers, D. M., Petruzzello, S. J., and Han, M. (1996). The effects of acute and chronic exercise on sleep. A meta-analytic review. *Sports Med*, *21*(4), 277–291.

Lamond, N., Dorrian, J., Roach, G. D., McCulloch, K., Holmes, A. L., Burgess, H. J. et al. (2003). The impact of a week of simulated night work on sleep, circadian phase, and performance. *Occup Environ Med*, *60*(11), e13.

Lee, D. M., and Weinblatt, M. E. (2001). Rheumatoid arthritis. *Lancet*, *358*(9285), 903–911.

Lennernas, M., Hambraeus, L., and Åkerstedt, T. (1995). Shift related dietary intake in day and shift workers. *Appetite*, *25*(3), 253–265.

Lin, Y. C., Hsiao, T. J., and Chen, P. C. (2009). Persistent rotating shift-work exposure accelerates development of metabolic syndrome among middle-aged female employees: A five-year follow-up. *Chronobiol Int*, *26*(4), 740–755.

Lowden, A., Moreno, C., Holmback, U., Lennernas, M., and Tucker, P. (2010). Eating and shift work—Effects on habits, metabolism and performance. *Scand J Work Environ Health*, *36*(2), 150–162.

Luca, M., Bellia, S., Bellia, M., Luca, A., and Calandra, C. (2014). Prevalence of depression and its relationship with work characteristics in a sample of public workers. *Neuropsychiatr Dis Treat*, *10*, 519–525.

Marquie, J. C., Tucker, P., Folkard, S., Gentil, C., and Ansiau, D. (2015). Chronic effects of shift work on cognition: Findings from the VISAT longitudinal study. *Occup Environ Med*, *72*(4), 258–264.

Mathers, C. D., and Loncar, D. (2006). Projections of global mortality and burden of disease from 2002 to 2030. *PLoS Medicine*, *3*(11), e442.

Mohren, D. C., Jansen, N. W., Kant, I. J., Galama, J., van den Brandt, P. A., and Swaen, G. M. (2002). Prevalence of common infections among employees in different work schedules. *J Occup Environ Med*, *44*(11), 1003–1011.

Monaco, C., Andreakos, E., Kiriakidis, S., Feldmann, M., and Paleolog, E. (2004). T-cell-mediated signalling in immune, inflammatory and angiogenic processes: The cascade of events leading to inflammatory diseases. *Curr Drug Targets Inflamm Allergy*, *3*(1), 35–42.

Nakano, Y., Miura, T., Hara, I., Aono, H., Miyano, N., Miyajima, K. et al. (1982). The effect of shift work on cellular immune function. *J Hum Ergol (Tokyo)*, *11 Suppl*, 131–137.

Neil-Sztramko, S. E., Pahwa, M., Demers, P. A., and Gotay, C. C. (2014). Health-related interventions among night shift workers: A critical review of the literature. *Scand J Work Environ Health*, *40*(6), 543–556.

Nilsson, P. M., Roost, M., Engstrom, G., Hedblad, B., and Berglund, G. (2004). Incidence of diabetes in middle-aged men is related to sleep disturbances. *Diabetes Care*, *27*(10), 2464–2469.

Nishitani, N., and Sakakibara, H. (2007). Subjective poor sleep and white blood cell count in male Japanese workers. *Ind Health*, *45*(2), 296–300.

Nojkov, B., Rubenstein, J. H., Chey, W. D., and Hoogerwerf, W. A. (2010). The impact of rotating shift work on the prevalence of irritable bowel syndrome in nurses. *Am J Gastroenterol*, *105*(4), 842–847.

Oberlinner, C., Ott, M. G., Nasterlack, M., Yong, M., Messerer, P., Zober, A. et al. (2009). Medical program for shift workers—Impacts on chronic disease and mortality outcomes. *Scand J Work Environ Health*, *35*(4), 309–318.

Okamoto, H., Tsunoda, T., Teruya, K., Takeda, N., Uemura, T., Matsui, T. et al. (2008). An occupational health study of emergency physicians in Japan: Health assessment by immune variables (CD4, CD8, CD56, and NK cell activity) at the beginning of work. *J Occup Health*, *50*(2), 136–146.

Orth-Gomer, K. (1983). Intervention on coronary risk factors by adapting a shift work schedule to biologic rhythmicity. *Psychosom Med*, *45*(5), 407–415.

Özdemir, P. G., Selvi, Y., Ozkol, H., Aydin, A., Tuluce, Y., Boysan, M. et al. (2013). The influence of shift work on cognitive functions and oxidative stress. *Psychiatry Res*, *210*(3), 1219–1225.

Pan, A., Schernhammer, E. S., Sun, Q., and Hu, F. B. (2011). Rotating night shift work and risk of type 2 diabetes: Two prospective cohort studies in women. *PLoS Medicine*, *8*(12), e1001141.

Patel, S. R., Malhotra, A., White, D. P., Gottlieb, D. J., and Hu, F. B. (2006). Association between reduced sleep and weight gain in women. *Am J Epidemiol*, *164*(10), 947–954.

Pietroiusti, A., Neri, A., Somma, G., Coppeta, L., Iavicoli, I., Bergamaschi, A. et al. (2009). Incidence of metabolic syndrome among night shift health care workers. *Occup Environ Med*, *67*(1), 54–57.

Puttonen, S., Härmä, M., and Hublin, C. (2010). Shift work and cardiovascular disease—Pathways from circadian stress to morbidity. *Scand J Work Environ Health*, *26*(2), 96–108.

Puttonen, S., Kivimaki, M., Elovainio, M., Pulkki-Raback, L., Hintsanen, M., Vahtera, J. et al. (2009). Shift work in young adults and carotid artery intima-media thickness: The cardiovascular risk in young Finns study. *Atherosclerosis*, *205*(2), 608–613.

Puttonen, S., Oksanen, T., Vahtera, J., Pentti, J., Virtanen, M., Salo, P. et al. (2010). Is shift work a risk factor for rheumatoid arthritis? The Finnish Public Sector study. *Ann Rheum Dis*, *69*(4), 779–780.

Puttonen, S., Viitasalo, K., and Harma, M. (2011). Effect of shiftwork on systemic markers of inflammation. *Chronobiol Int*, *28*(6), 528–535.

Puttonen, S., Viitasalo, K., and Harma, M. (2012). The relationship between current and former shift work and the metabolic syndrome. *Scand J Work Environ Health*, *38*(4), 343–348.

Sack, R. L., Auckley, D., Auger, R. R., Carskadon, M. A., Wright, K. P., Jr., Vitiello, M. V. et al. (2007). Circadian rhythm sleep disorders: Part I: Basic principles, shift work and jet lag disorders: An American Academy of Sleep Medicine review. *Sleep*, *30*(11), 1460–1483.

Saksvik, I. B., Bjorvatn, B., Hetland, H., Sandal, G. M., and Pallesen, S. (2011). Individual differences in tolerance to shift work—A systematic review. *Sleep Med Rev*, *15*(4), 221–235.

Sallinen, M., and Kecklund, G. (2010). Shift work, sleep and sleepiness—Differences between shift schedules and systems. *Scand J Work Environ Health*, *36*(2), 121–133.

Scott, A. J., Monk, T. H., and Brink, L. L. (1997). Shiftwork as a Risk Factor for Depression: A pilot study. *Int J Occup Environ Health*, *3*(Supplement 2), S2–S9.

Sigurdardottir, L. G., Valdimarsdottir, U. A., Fall, K., Rider, J. R., Lockley, S. W., Schernhammer, E. et al. (2012). Circadian disruption, sleep loss, and prostate cancer risk: A systematic review of epidemiologic studies. *Cancer Epidemiol Biomarkers Prev*, *21*(7), 1002–1011.

Silman, A. J., and Pearson, J. E. (2002). Epidemiology and genetics of rheumatoid arthritis. *Arthritis Res*, *4 Suppl 3*, S265–S272.

Sivertsen, B., Overland, S., Neckelmann, D., Glozier, N., Krokstad, S., Pallesen, S. et al. (2006). The long-term effect of insomnia on work disability: The HUNT-2 historical cohort study. *Am J Epidemiol*, *163*(11), 1018–1024.

Skipper, J. K., Jr., Jung, F. D., and Coffey, L. C. (1990). Nurses and shiftwork: Effects on physical health and mental depression. *J Adv Nurs*, *15*(7), 835–842.

Sookoian, S., Gemma, C., Fernandez Gianotti, T., Burgueno, A., Alvarez, A., Gonzalez, C. D. et al. (2007). Effects of rotating shift work on biomarkers of metabolic syndrome and inflammation. *J Intern Med*, *261*(3), 285–292.

Stevens, R. G., Hansen, J., Costa, G., Haus, E., Kauppinen, T., Aronson, K. J. et al. (2011). Considerations of circadian impact for defining "shift work" in cancer studies: IARC working group report. *Occup Environ Med*, *68*(2), 154–162.

Straif, K., Baan, R., Grosse, Y., Secretan, B., El Ghissassi, F., Bouvard, V. et al. (2007). Carcinogenicity of shift-work, painting, and fire-fighting. *Lancet Oncol*, *8*(12), 1065–1066.

Tenkanen, L., Sjöblom, T., Kalimo, R., Alikoski, T., and Härmä, M. (1997). Shift work, occupation and coronary heart disease over 6 years of follow-up in the Helsinki Heart Study. *Scand J Work Environ Health*, *23*(4), 257–265.

Tsuno, N., Besset, A., and Ritchie, K. (2005). Sleep and depression. *J Clin Psychiatry*, *66*(10), 1254–1269.

Van Dongen, H. P. (2006). Shift work and inter-individual differences in sleep and sleepiness. *Chronobiol Int*, *23*(6), 1139–1147.

Van Dongen, H. P., Baynard, M. D., Maislin, G., and Dinges, D. F. (2004). Systematic interindividual differences in neurobehavioral impairment from sleep loss: Evidence of trait-like differential vulnerability. *Sleep*, *27*(3), 423–433.

Viitasalo, K., Hemio, K., Puttonen, S., Hyvarinen, H. K., Leiviska, J., Harma, M. et al. (2015). Prevention of diabetes and cardiovascular diseases in occupational health care: Feasibility and effectiveness. *Prim Care Diabetes*, *9*(2), 96–104.

Viitasalo, K., Kuosma, E., Laitinen, J., and Härmä, M. (2008). Effects of shift rotation and the flexibility of a shift system on daytime alertness and cardiovascular risk factors. *Scand J Work Environ Health*, *34*(3), 198–205.

Viitasalo, K., Lindstrom, J., Hemio, K., Puttonen, S., Koho, A., Harma, M. et al. (2012). Occupational health care identifies risk for type 2 diabetes and cardiovascular disease. *Prim Care Diabetes*, *6*(2), 95–102.

Vyas, M. V., Garg, A. X., Iansavichus, A. V., Costella, J., Donner, A., Laugsand, L. E. et al. (2012). Shift work and vascular events: Systematic review and meta-analysis. *Brit Med J*, *345*, e4800.

Waage, S., Pallesen, S., Moen, B. E., Mageroy, N., Flo, E., Di Milia, L. et al. (2014). Predictors of shift work disorder among nurses: A longitudinal study. *Sleep Med*, *15*(12), 1449–1455.

Wagstaff, A. S., and Sigstad Lie, J. A. (2011). Shift and night work and long working hours—A systematic review of safety implications. *Scand J Work Environ Health*, *37*(3), 173–185.

Wang, F., Zhang, L., Zhang, Y., Zhang, B., He, Y., Xie, S. et al. (2014). Meta-analysis on night shift work and risk of metabolic syndrome. *Obes Rev*, *15*(9), 709–720.

Wilder-Smith, A., Mustafa, F. B., Earnest, A., Gen, L., and Macary, P. A. (2013). Impact of partial sleep deprivation on immune markers. *Sleep Med*, *14*(10), 1031–1034.

Yliharsila, H., Lindstrom, J., Eriksson, J. G., Jousilahti, P., Valle, T. T., Sundvall, J. et al. (2005). Prevalence of diabetes and impaired glucose regulation in 45-to-64-year-old individuals in three areas of Finland. *Diabet Med*, *22*(1), 88–91.

Section II

Ergonomic Design Issues in Workplaces

8 Office Workplaces

Peter Vink, Iris Bakker, and Liesbeth Groenesteijn

CONTENTS

8.1 MUCH KNOWLEDGE WORK AND OFFICE WORK

In 2012, 230 million knowledge workers were employed worldwide (Manyika et al. 2013). Kastelein (2014), Bakker (2014), and Groenesteijn (2015) all report in their PhD theses that new attention is needed for the ergonomic design of office interiors because of the increase in knowledge workers. Knowledge workers work in new office environments like the combi-office (see Figure 8.1) and use computer workstations and new information and communication technologies. These technologies provide alternatives for where, when, and how to do the work, and they reduce the necessity of coming to the office. Sometimes knowledge workers are displaced from their leaders working in various departments and time zones or from remote sites, such as home offices, hotels, and airport lounges, etc. (Groenesteijn 2015). Interestingly, Kastelein (2014) reports that while teleworking is becoming more widely accepted, a number of leading companies like Apple, Yahoo!, and Google are holding on to (or have started embracing) the belief that having workers in the same place is crucial to their success. This appears to be based on the view that physical proximity can lead to casual exchanges (Hagstrom 1965; Kraut et al. 1988; Appel-Meulenbroek 2014), which in turn can lead to breakthroughs for products and services.

What all these workers have in common is that, despite where or when they work, they are performing their work seated for many hours and often use (mobile) computers or other mobile devices. Therefore, it is important to focus on the functional ergonomic design of seating to support the growing number of users and their various ways of working (Groenesteijn 2015). Also, the design of office areas, like meeting rooms, lounge areas, and desks, is relevant in the context of knowledge work.

This chapter presents research outcomes on health and productivity relevant to knowledge workers and people working in offices. Attention is paid to the environment around the office (see Section 8.3), followed by the layout and the environment of the interior (see Sections 8.4 and 8.5), but most attention will be given to the direct environment of the worker: the work place (see Sections 8.6 and 8.7) and the meeting rooms (see Section 8.8). First attention is given to the sustainable productivity at the office (see Section 8.2).

(a)   (b)

FIGURE 8.1 An example of a Dutch combi-office. (a) The open area where communicative tasks can be performed and (b) an example of a room for work where concentration is possible.

8.2 SUSTAINABLE PRODUCTIVITY

The only reason to build an office or facilitate workers with information and communication technologies is to help them to be sustainably productive. Of course working in a barn may be sufficient for work that is productive for a while, but in the long run, essential facilities and high-quality communication with direct eye contact with colleagues (Hatfield et al. 1993; Miles et al. 2009) are missing, which impairs the quality of communication, and the health of the workers could be threatened as well. In this chapter, *sustainable productivity* is defined as productivity that can be performed by workers that can also maintain their health. *Health* is defined as a state of complete physical, mental, and social well-being (WHO 2003). For productivity, a common definition is the one from Sink and De Vries (1984): productivity is strictly a relationship between the resources that come into an organizational system over a given period and the output generated with those resources over the same period. A lot has been written on productivity in offices (Voordt 2003). A study by Van Rhijn et al. (2005) on assembly work showed that the effects on productivity are strongly dependent on the way productivity is measured (see Table 8.1). In this example, the productivity increase can be defined as 31%, but also as 96%. The reduction of 31% was in time per person per product, while 96% was the increase in product per person per day per square meter. For knowledge work, this is more complex, but there is some evidence for the following statement: adapt the environment to the task. This could mean creating various areas in the office where workers can do specific tasks, like a space for concentrated work and a space for more communicative tasks. In a study of 706 office workers in Germany, the office where workers could select the environment fitting to the task was significantly more productive (Rieck and Kelter 2005).

TABLE 8.1
Effects in Various Productivity Measures of a New Way of Working, a New Flow Concept, and a New Design of the Interior in an Assembly Case

	Traditional	New	Change (%)
Order lead time (minutes)	155	83	−46
Lead time per person/product (minutes)	5.2	3.6	−31
Productivity in the number of products/person/day	93.3	134.7	44
Required work space (m²)	80.5	45	−45
Number of products per person/day/m²	4.6	9	96

Source: Van Rhijn, J. W., M. P. de Looze, G. H. Tuinzaad, L. Groenesteijn, M. D. de Groot, and P. Vink., *Int J Prod Res*, 43, 3687–701, 2005. With permission.

Transferring assembly productivity measuring methods to the service industry should be done with care. Grönroos and Ojasalo (2004) developed a productivity model for the service industry showing that more factors than internal efficiency play a role. According to their model, service productivity is a function of (1) how effectively input resources into the service (production) process are transformed to outputs in the form of services (internal efficiency), (2) how well the quality of the service process and its outcome are perceived (external efficiency or effectiveness), and (3) how effectively the capacity of the service process is utilized (capacity efficiency). Measuring knowledge productivity has an extra complicating dimension, as it can be conceived as the result of individual efforts and close cooperation between people and groups over a longer time (Bakker 2014). Kastelein (2013) showed in evaluating the Google office interior in Zurich that people that walked more and visited more coffee areas meeting other colleagues came up with more relevant ideas, which could increase the income of Google. In assembly, more walking usually means a loss of working time, as it is not adding value to a product. Therefore, in this chapter, productivity is not only seen as making services or products fast, but also coming up with new ideas, which is a part of innovation and creativity. In contrast to the productivity of the assembly employee, Drucker (1999) characterizes the productivity of the knowledge worker in terms of autonomy, responsibility for innovation, continuous learning, and quality next to quantity, which in itself could also contribute to health and well-being.

In the next sections, the effects of office interiors on health and productivity will be described. Some of these effects focus on changes at the departmental level, some at the local level or even individual workstations, but having the abovementioned productivity issues in mind, the measurements of the effects are always limited and therefore generalization should be done with care. Also, productivity and health are defined in a broad sense in these cases. For instance, productivity also includes creativity, and health includes comfort and well-being.

8.3   ENVIRONMENT AROUND THE OFFICE

The environment of the office building also has an effect on productivity and health. An office surrounded by nature or parks can relieve worker stress. Walking outside during a lunch break can reduce stress levels. Park et al. (2010) showed the difference in effects between a forest and a city in an experiment that had 280 subjects walk in a forest and a city. The salivary cortisol was significantly lower (15.8% decrease), the average systolic blood pressure was significantly lower (1.9% decrease), and the heart rate was lower for people after walking in the forest as compared to those walking in the city. These physical parameters are related to stress, and they indicate a reduction in stress. Additionally, the experienced mood states were improved, e.g., less depression and lower tension. Another interesting observation by Park et al. (2010) showed that after 14 minutes of viewing a forest, the same effects were shown as mentioned above. This means that the view from the window also influences stress. Bazley and Vink (2008) found that workers that had a view of the mountains were more satisfied than those who had no windows. The health effects of views of nature are well established. Ulrich (1984) showed 30 years ago that patients recovered significantly faster from gallbladder with views of nature as compared to those viewing a wall. A more recent study among 785 office workers showed that productivity was higher for those subjects having a window with a view of nature (Nicol 2006). Some countries even had regulations for windows. Dutch regulations stipulate that offices that are being used for 2 hours or more per day have to have daylight access (Ministry of Social Affairs and Labour 1998).

8.4   ENVIRONMENT IN THE OFFICE

The indoor climate has an effect on productivity. Based on measuring the opinion of workers, Kosonen and Tan (2004) state that an indoor temperature of 27°C has a 30% lower productivity compared with that of 21°C, but Hedge and Gaygen (2010) found that productivity was higher at

25°C than at lower temperatures, which suggests an optimum range. Indoor climate studies have shown that there is no singular comfortable indoor temperature. For example, the comfortable indoor temperature is dependent on the outside temperature (De Dear and Brager 2002). In the northern hemisphere, higher indoor temperatures are preferred in the summer than in the winter. The tolerances of thermal perceptions are not fixed within one season. People who live or work in naturally ventilated buildings, where they are able to open the windows, become used to this thermal diversity. Their thermal perceptions extend over a wider range of temperatures, and the preferred comfort range is broader (De Dear and Brager 2002). To say it simply, people get used to a wider range of temperatures and thereby broaden their comfort zone. Roelofsen et al. (2013) showed that thermal comfort is age dependent. Older people have a smaller range of comfortable indoor temperatures than younger people. Perhaps the most important advice is that humans should be in control of changing the indoor temperature (Roelofsen 2016), which means, for instance, the possibility of opening the windows or locally adapting the temperature by thermostats. Temperature is not the only factor of importance in the office environment. Satish et al. (2012) showed that high CO_2 concentrations significantly impair the decision-making performance, but noise and light are also of great importance. Hongisto (2005) states that productivity for a task where concentration is needed is reduced by 7% when there is much noise in the open space around them. Aries and Zonneveldt (2009) state that light is needed to perform the task and also to create a good atmosphere, and glare should be avoided as it reduces productivity. Also, working with a view on green plants has positive effects on creativity (Shibata and Suzuki 2002; Klein Hesselink et al. 2006), which is a form of productivity needed in knowledge work, although for other repetitive office work, plants have been shown to impair productivity by 12% (Larsen et al. 1998).

8.5 LAYOUT OF THE OFFICE

The layout of the office can be arranged in many ways, but it should always facilitate the activities and the tasks of the employees. In fact, the only reason to build an office is to support workers. There are smaller companies that have no office, work from home, or specially dedicated shared offices like "seats2meets" in the Netherlands and have meetings in restaurants, but the majority of companies and organizations have an office. Usually first, a vision of what the activities and the tasks will be in the office is needed sometimes based on the measurements of the current activities performed (Bakker 2014). It will always be a mix of concentration and communication tasks, which have different demands for the layout (Vink 2009). So ideally, there are dedicated areas for concentrated working and for collaboration and communication. For concentration, silent areas without distraction are needed, and for collaboration and communication open space office areas, and two-, three-, or more-person meeting rooms are needed. Rieck and Kelter (2005) showed in a study of 706 employees that comfort and productivity are highest in a combi-office (see Figure 8.2), which is an office where employees can choose their own work space that best fits their task. A study by Blok et al. (2009) showed that productivity and well-being significantly improved in a group of 1125 office workers by changing from six-person rooms to a combi-office (see Figure 8.3).

8.6 THE INDIVIDUAL WORKSTATION

The choice of a workstation is strongly dependent on the task. For a computer-aided design task, larger screens are needed, and for an information and communication technology worker, more powerful computers or laptops are often needed. If work is done on a tablet computer, it is probably better to use an additional keyboard and to have an adjustable support to have the tablet positioned at a comfortable viewing angle. Albin (2014) demonstrated that, for a tablet computer, employees preferred an angle of 33° with respect to the horizontal. Additionally, a height-adjustable table is needed. A study among 45 employees (Vink and Kompier 1997) showed that there is no one ideal table height. The height is dependent on the anthropometrics of the employee, the devices used, and

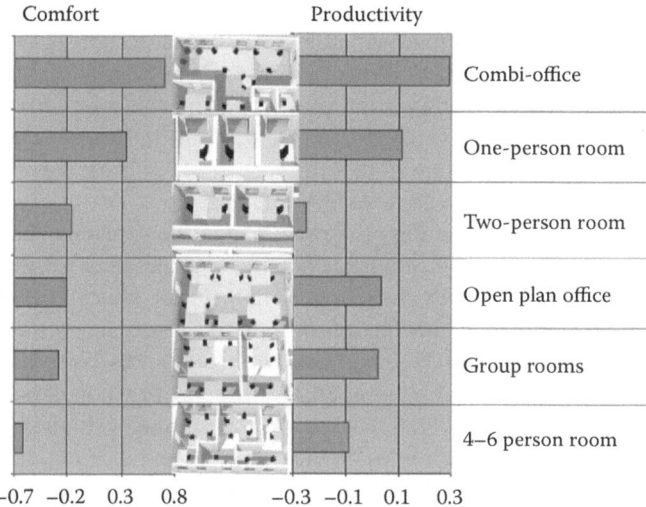

FIGURE 8.2   Comfort and productivity according to 706 employees working in different offices in the study by Rieck and Kelter (2005); the higher the number, the higher the comfort or the productivity.

(a)   (b)

FIGURE 8.3   A six-person room in the old situation (a) and a part of the office garden in the new situation (b) of the experiment by Blok et al. (2009).

the type of task. Training is also needed to have employees use this equipment correctly (Vink and Kompier 1997; Robertson and O'Neill 2003; Robertston et al. 2009), but training pays off; as in both studies, beneficial effects on health and productivity were found.

In many textbooks and Internet sites (e.g., http://ergo.human.cornell.edu/ergoguide.html), the ideal workstation is seen as one where the employee can work in a neutral posture which is one where the feet rest firmly on the ground; the upper legs are approximately horizontal; the feet are in front of the knees, so that the knee angle is greater than 90°; the trunk is slightly reclined from vertical by up to 20°; the head and the neck are balanced and vertical; the upper arms are vertical and the lower arms are horizontal; and the hands and wrist are straight and flat. The seat and the desk must be adjustable to support this posture.

While the above posture is a good-seated posture for productivity, it is only a healthy posture if the duration of the sitting is managed, and it is much more important to vary the employee's posture throughout the day (e.g., Ellegast et al. 2012; Groenesteijn 2015). For reading work, a more reclined posture, where the back and head/neck are supported, is preferable. Office workers prefer a 123°

reclined backrest for reading (Groenesteijn 2015). Wilke et al. (1999) showed that the pressure in the lumbar disk is lower in this position compared with that in upright sitting. Hosea et al. (1986) showed that in this position, the back muscle activity is low, and Gscheidle and Reed (2004) found that a back-to-seat recline angle of 114° is the preferred office-sitting position. This means that a good office seat should be adjustable in height, and the seat pan and the backrest should rotate in the sagittal plane. Ellegast et al. (2012) showed that the type of office chair is not of major importance, but rather the fact that it moves and is adjustable is important.

Regarding the armrest and the position of the seat pan, there is much debate. However, it seems that there is literature supporting the importance of having the arms supported for some of the time, and armrests definitely help with chair ingress and egress, which is especially important for older workers. Zhu et al. (2011) show that chair arm support reduces upper trapezius muscle activity and discomfort, and Hedge et al. (2011) saw a reduction in shoulder complaints from 24% to 16% comparing workers working with and without an arm support in a population of 1504 employees. Goossens and Snijders (1995) showed that in a seat pan where the position is horizontal or with the front of the seat lower, there are shear forces on the upper leg and the buttock, which increases discomfort. By having a seat slightly tilted backward, this can be prevented. In the United States, there is an ergonomics standard that presents important chair design requirements (ANSI-HFES 100 2007). There is also an international ergonomics standard that applies to the design of an ergonomic workstation arrangement (ISO 9241-5 1998).

The adjustability of seats is an important issue. Vink et al. (2007) showed that 63% of the office workers in the Netherlands do not adjust their chairs, and part of this group were workers who do not have their own desk. These so-called flexworkers share workstations. Groenesteijn (2015) studied two seat designs: a complex adjustable one with more controls and a simple one with fewer controls, and asked the flexworkers and the workers with their own desks to use the chairs. Most of the workers adjusted the office chairs the first time for seat height, armrest height, and backrest inclination. The adjustment times for seat height and armrest height were shorter for the simple chair. She also showed that the backrest pressure adjustment takes considerable time, and it is difficult to adjust this without instruction. The flexworkers adjusted their chair more often and were faster in their adjustment of the backrest pressure compared with workers with their own desks. The quality of the adjustments of seat height, armrest, and backrest pressure was improved for 32% of the subjects by the instruction given by an expert while the subject was seated and experienced the difference in adjustment. Also, this study shows the important of training.

8.7   WORKING PARTLY STANDING AND VARYING POSTURE

A risk of office work is physical inactivity. According to Commissaris et al. (2014), physical inactivity is associated with CVDs, T2D, depression, obesity, and some forms of cancer, and 3.2 million people die prematurely because of an inactive work style. Long periods of uninterrupted seated work is one of the risk factors (Buckley et al. 2015). Hu et al. (2003) state that each 2-hour sitting time increases the risk of obesity by 5% and the risk of diabetes by 7% in female workers. This is one of the reasons to encourage variation in the working posture throughout the day. One possibility is to do the work while sitting for part of the time and then standing for part of the time. In some situations, a standing posture is even preferable over a seating-working position. Sengupta and Das (2000) show that the reach envelopes of workers are larger while standing, and the capability to exert forces is often greater while standing (Yates and Karwowksi 1992). On the other hand, standing for the largest part of the working day is not advised. A review of 17 studies (McCulloch 2002) shows that working while standing during the largest part of the day increases the risk of vein problems in the legs, lower back and feet problems, and premature birth, and this is confirmed by Krause et al. (2000) and Tüchsen et al. (2000). Postural variation seems to have positive effects. Some health effects of sit–stand tables have been reported. Aaras et al. (2001) showed, for instance, that standing-VDU (working with visual display units) work significantly reduced the activity of the trapezius muscle compared with sitting.

Vellinga (2001) showed that in 84% of the subjects, the well-being improved after introducing sit–stand tables. The beneficial results for sit–stand workstations were reported by several studies (Hedge and Ray 2004; Pronk et al. 2012; Robertson et al. 2013; Karakolis and Callaghan 2014).

However, there is still skepticism on sit–stand tables in the field. The study of Vellinga (2001) showed, for instance, that half of the user group only stood for 15 minutes a day, and 10% never did use the standing option. Wilks et al. (2006) studied 90 workers in four companies and found that after 1 month, over 60% of the workers did not adjust their sit–stand workstations. The question is if this will have any effect on health. In a study by Vink et al. (2005), workers could stand, sit, and have the bar stool position (see Figure 8.4). It appears that working in this position resulted in more variations in the posture, and 8% of the time, work was done in the standing position, and discomfort in the neck and the back was significantly reduced. The main conclusion is that it is worthwhile trying office interiors with additional possibilities for standing or part-sitting, as this could contribute to the desirable variation in working postures, and although sit–stand workstation are often relatively little used, the beneficial effects still can be shown on discomfort and well-being.

Recently, TNO, the second largest applied scientific research institute in Europe, also studied the effect of a bicycle desk: the oxidesk (see Figure 8.5). The first results of the tests (Oxidesk 2016) show that especially reading can be done quicker on the oxidesk compared with that at a normal desk. Other tasks were possible as well. The results showed that 36% reported that the concentration was better at the oxidesk. Moreover, 68% would want to use the oxidesk on a daily basis. However, Straker et al. (2009) tested 30 participants (16 men, 14 women) aged 22–64, 15 of whom were touch typists, in six conditions: standing, sitting, walking (1.6 km/h), walking (3.2 km/h), cycling (1.6 km/h), and cycling (3.2 km/h); and they found that for both typing and mousing, the speed and errors were significantly reduced when on a treadmill workstation and on a bicycle workstation, although the detrimental effects of the bicycle workstation were 50% less than those for the treadmill. Commissaris et al. (2014) tested 15 participants in six conditions: sit, stand, treadmill, elliptical, bicycle (25% effort), and bicycle (40% effort), with seven tasks: type (5 minutes), edit (5 minutes), use of mouse (5 minutes); use of phone (3 minutes); attention perceptual performance is one task, flanker test, working memory (6–8 minutes). They found minimal differences in cognitive task performance; the quality of phone conversations was rated lower in dynamic working; the typing speed was significantly slower when walking; and the mousing deteriorated in all dynamic working conditions. It seems that certain tasks (typing, mousing) are performed faster and more accurately when sitting in a stable posture; others like creative thinking benefit from dynamic working, and yet others show no difference between static and dynamic working patterns.

Using a treadmill, a bicycle, an elliptical, or other dynamic technology at the workstation is one option to stimulate movement during work. Of course other options are possible as well as walking during bilateral meetings and having standing meetings. Also, the design of the interior plays a role.

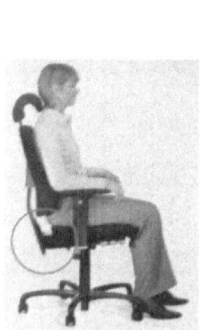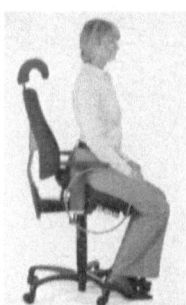

FIGURE 8.4   A workstation with three possible working positions, which results in more postural variation and movement throughout the day.

FIGURE 8.5   The oxidesk combining bicycling and working.

FIGURE 8.6   The only way to reach the first floor area is by stairs, which stimulates human movement.

If the stairs are hidden, it is not to be expected that these will be used. Well-designed stairs or if the stairs are the only option for reaching an area in the office (see Figure 8.6) this is also a way to stimulate movement.

8.8   THE MEETING ROOM INTERIORS

Thanks to new mobile technology, working remotely is easy, and this reduces the need to go to the office. Indeed, in the United States, 2.6% of the U.S. employee workforce (3.3 million people, not including the self-employed or unpaid volunteers) considered home as their primary place of work (http://globalworkplaceanalytics.com/telecommuting-statistics). To bind employees to the company and to exchange information, the office is more often considered as a meeting place (Finch 2012). This means paying more attention to spaces for meetings. However, meetings are not always a pleasure. Almost half of the employees in the Netherlands see meetings as a waste of time, and 44% of the workers in the Netherlands define meetings as not useful (Duijvestijn et al. 2007). Therefore, to increase productivity, meetings should be effective, and in terms of health, meetings should be a pleasure in which workers experience feedback (Csikszentmihalyi 2003; Gaillard 2003; Giancola 2011) and the respect of colleagues.

Generally, professionals in the Netherlands spend between 20% and 40% of their working time in meetings. For top managers, this percentage is higher; they spend an average of 70% of their working time in meetings (Nobis 2001). In the United States, the situation is somewhat similar, and middle managers spend ~35% of their week in meetings, and upper management spends ~50% of their week in meetings (The Muse 2016). The fact that meetings are necessary to compensate for the lack of contact due to distant work and the fact that there is a need for having more effective meetings led to the following experiment. In this experiment, the goal was to define meeting room interiors that better facilitate the meetings. A unique environment was found to perform the experiment: an enterprise with four exactly the same meeting rooms. The windows were on the same side of the building (southeast); the rooms had the same width, height, and length; the doors were at the same position; and the rooms had the same floor and light and temperature. After discussing with various architects and facility managers working in the field, it was decided to make four interiors. The first one was a traditional interior with square tables put in a square with 14 seats around it (Square), 3 seats on each side opposite to each other and 4 seats on the other two sides also opposite to each other. The second one was a large oval meeting table (Oval), where 14 seats were positioned along the table. The third one was a meeting room with couches, plants, and a low table, which created a more homely atmosphere (Home). And the last one was a meeting room with 14 stools and a high table, with the height that is often seen in bars (Stools). The furniture for these rooms was delivered free of costs by Ahrend, a Dutch office furniture company. The employees of the enterprise were willing to have their meetings in these rooms and complete questionnaires after each meeting. The first research question was whether significant differences could be found between the four meeting rooms regarding satisfaction with the meeting, meeting atmosphere, group cohesion, and involvement in the meeting. It was also interesting to check the opinions on which meeting room was the most stimulating for sharing information, selecting/reducing information, promoting creativity (generating ideas), and evaluating processes, as these were elements of meetings often seen in the enterprise at hand. Briggs et al. (2001) also distinguished these types of meetings. One hundred and twelve participants were divided over various groups (3–12 in size) and held their meetings over a 12-week period in the four meetings rooms as described earlier (Square, Oval, Home, and Stools; see Figure 8.7). The order in which they had their meetings varied over the groups, and care was taken that each participant did have their meeting in all rooms. The meeting times varied between 30 and 120 minutes. Figures 8.8 and 8.9 show that the participants were significantly more satisfied with the Stool interior, and the atmosphere was significantly less in the Square interior. Figures 8.8 and 8.9 show that the Square interior had a significantly lower group cohesion and the Stool interior, a significantly higher group cohesion. The involvement in the meeting was significantly lower in the Square and Home interiors and higher in the Oval interior.

Figure 8.10 shows that the Oval interior is seen as appropriate for sharing and reducing information and for evaluating. The study showed that 34% of the participants commented that for very delicate and sensitive evaluations, the Home interior would be better. For generating ideas, the Home interior was seen as more suitable. The study also showed that 50% of the participants commented that for generating ideas during a short time, the Stool interior would be preferable.

Of course this study was only done in one enterprise with its specific characteristics, which means that generalization should be done with care. On the other hand, other studies show likewise

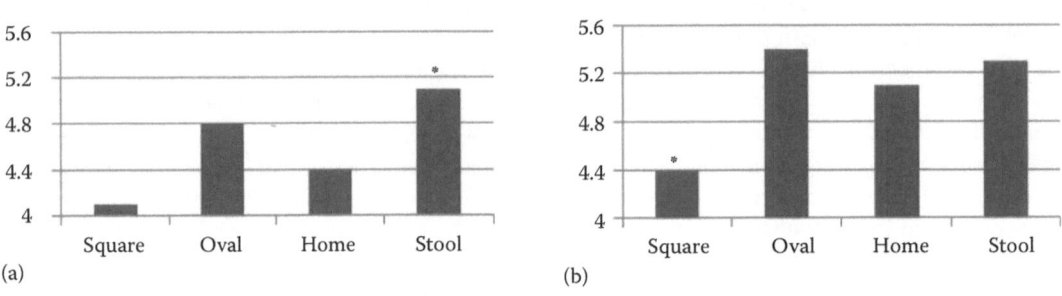

FIGURE 8.7 Four types of meeting rooms tested: Square (a), Oval (b), Home (c), and Stools (d).

FIGURE 8.8 Average score for (a) meeting satisfaction and (b) meeting atmosphere. An asterisk means that it is significantly different from the average.

results. The fact that the Square meeting room is in fact the least suitable for several meetings is described, for instance, by Kooij-de Bode et al. (2009). They showed pictures of the comparable four meeting room interiors to 114 participants, and 54% preferred the Oval interior for sharing information and 50% preferred the Oval for reducing information. For generating ideas, the Oval, Home, and Stool interiors scored best in the study by Kooij-de Bode et al., which is a bit different, as the Oval did not score so high in our study. Kooij-de Bode et al. (2009) found a preference for the Oval and Home interiors for evaluating, which is comparable to ours.

The results of this study indicate that the Square interior should be avoided, because it results in lower atmosphere, lower involvement of the participants, and lower group cohesion compared with

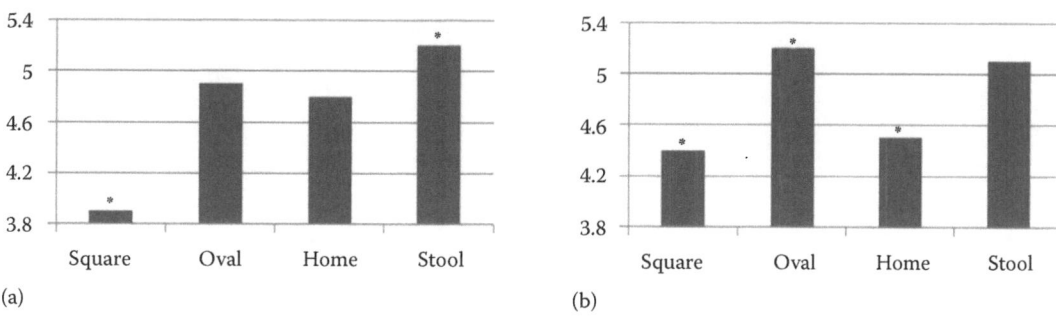

FIGURE 8.9 Average score for (a) group cohesion and (b) involvement in the meeting. An asterisk means that it is significantly different from the average.

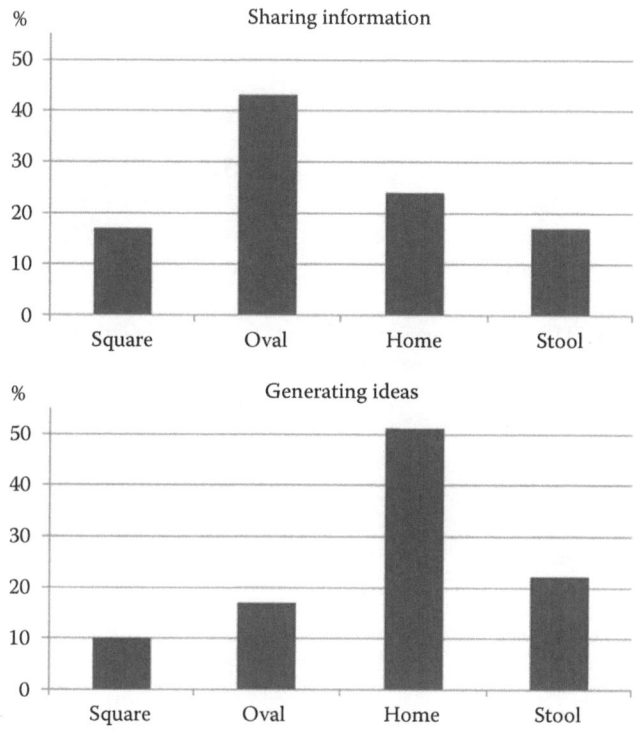

FIGURE 8.10 Percentage of participants that have the opinion that a certain meeting interior is better occupied for a specific type of meeting. (*Continued*)

the other meeting room interiors tested in the experiment. And for many types of meetings, the Oval interior is preferable. The Oval also created more involvement of the participants. To stimulate group cohesion and satisfaction with the meeting, a Stool interior could be used; for long creativity sessions, a Home interior could be preferable; and for short creativity sessions, the Stool interior could be used.

Another type of meeting room is promising as well: the standing meeting room (see Figure 8.11). Bluedorn et al. (2003) showed that a standing meeting is 34% shorter with the same outcome and quality. In 2008, a similar experiment in the Netherlands showed likewise benefits (Managers Online 2016).

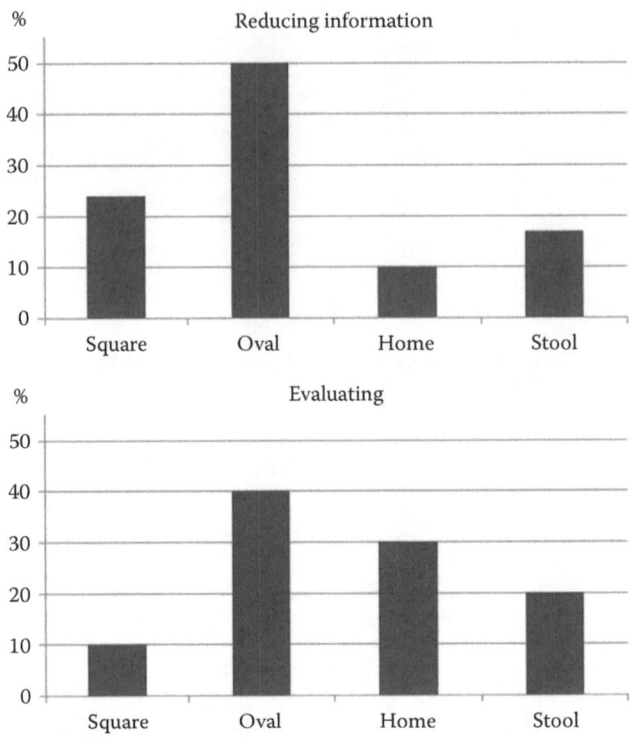

FIGURE 8.10 (CONTINUED)   Percentage of participants that have the opinion that a certain meeting interior is better occupied for a specific type of meeting.

FIGURE 8.11   An example of a standing meeting.

8.9 CONCLUSION

The vision in this chapter is that an office interior is primarily there to support the employees in performing their task for many years. This means that it is focused on sustainable productivity. Work tasks and activities should be supported by the office environment, and health should be stimulated as well to create productivity on the long run. Office work is changing from routine clerical work to more knowledge work, which means that communication and creativity will play a large role, but concentrated work will always be there as well. New devices will be introduced which will have consequences as well. The tablet computer is now already seen in the workplace and has consequences for the work environment, but other new devices and new activities will certainly be introduced as well. This chapter focused on the current state of the art. Based on literature, suggestions for interiors to improve health and productivity can be made.

In summary, health and productivity in offices can be improved by choosing the following:

- The location of the office, which is preferably surrounded by nature or a park
- The indoor environment, which is preferably adapted to the tasks and controllable by the employees
- The layout, which is adapted to the tasks (a combi-office is, for many tasks, a good solution)
- The work station, which is ideally not only adapted to the task and is adjustable, but also stimulates to vary the posture
- An environment that stimulates movement with bicycle desks or stairs that tempt people to use them
- Various meeting rooms, which fit to the type of meeting: oval or round for most meetings, but stools or standing meeting tables for more active participation

REFERENCES

Aaras, A., G. Horgen, H. Bjorset, O. Ro, and H. Walsoe. 2001. Musculoskeletal, visual and psychosocial stress in VDU operators before and after multidisciplinary ergonomic interventions: A 6-year prospective study—Part II. *Appl Ergon.* 32:559–71.

Albin, T. 2014. *Quantitative approaches to physical ergonomic issues encountered while assessing workplace designs.* PhD thesis. Delft: Delft University of Technology.

ANSI/HFES 100-2007 Human Factors Engineering of Computer Workstations. From http://www.hfes.org/publications/ProductDetail.aspx?ProductID=69.

Appel-Meulenbroek, R. 2014. *How to measure added value of corporate real estate and building design.* Dissertation. Bouwstenen 191. Eindhoven: Technical University Eindhoven.

Aries, M. B. C., and L. Zonneveldt. 2009. Effects of light (in Dutch: Effecten van verlichting) In *Aangetoonde effecten van het kantoor-interieur: Naar comfortabele, innovatieve, productieve en duurzame kantoren.* P. Vink (ed.). 13–21. Alphen aan den Rijn: Kluwer.

Bakker, I. C. 2014. *Uncovering the secrets of a productive work environment, a journey through the impact of plants and colour.* PhD dissertation, Delft: Delft University of Technology.

Bazley, C., and P. Vink. 2008. Changing the View of Workspace from Inside to Outside. In: *Advances in Social and Organizational Factors.* P. Vink. (ed.). 154–62. Boca Raton, FL: CRC Press Holistic.

Blok, M., E. de Korte, L. Groenesteijn, M. Formanoy, and P. Vink. 2009. The effects of a task facilitating working environment on office space use, communication, concentration, collaboration, privacy and distraction. *17th World Congress on Ergonomics. (IEA 2009).* CD rom 2OF0003. Beijing.

Bluedorn, A. C., D. B. Turban, and M. S. Love. 2003. The effects of stand-up and sit-down meeting formats on meeting outcomes. *J App Psychol.* 84:277–85.

Briggs, R. O., G. J. De Vreede, J. F. Nunamaker, and D. Tobey. 2001. ThinkLets: Achieving predictable, repeatable patterns of group interaction with group support systems (GSS). *Conference on System Sciences 2001: Proc 34th Ann Hawaii Int Conf.* Maui, HI.

Buckley, J. P., A. Hedge, T. Yates, R. J. Copeland, M. Loosemore, M. Hamer, G. Bradley, and D. W. Dunstan. 2015. The sedentary office: A growing case for change towards better health and productivity. *Brit J Sports Med.* 49(21):1357–62.

Commissaris, D. A. C. M., R. Könemann, S. Hiemstra-van Mastrigt, E. Burford, J. Botter, M. Douwes, and R. P. Ellegast. 2014. Effects of a standing and three dynamic workstations on computer task performance and cognitive function tests. *Appl Ergon*. 45:1570–78.

Csikszentmihalyi, M. 2003. *Flow in zaken, over leiderschap en betekenisgeving*, Amsterdam: Uitgeverij Boom.

De Dear, R. J., and G. S. Brager. 2002. Thermal comfort in naturally ventilated buildings: Revisions to ASHRAE Standard 55. *Energy Build*. 34:549–61.

Drucker, P. F. 1999. Knowledge-worker productivity: The biggest challenge. *Calif Manage Rev*. 41:79–94.

Duijvestijn, J., F. Vennik, and C. Gabreels. 2007. *NS Vergaderbarometer*. Utrecht: Nederlandse Spoorwegen.

Ellegast, R. P., K. Kraft, L. Groenesteijn, F. Krause, H. Berger, and P. Vink. 2012. Comparison of four specific dynamic office chairs with a conventional office chair: Impact upon muscle activation, physical activity and posture. *Appl Ergon*. 43:296–307.

Finch, E. 2012. *Facilities Change Management in Context*. Oxford: Blackwell Publishing Ltd.

Gaillard, A. W. K. 2003. *Stress, Productiviteit en gezondheid*. Amsterdam: Uitgeverij Nieuwezijds.

Giancola, F. L. 2011. Examining the job itself as a source of employee motivation. *Compens Benefits Rev*. 43:23–29.

Goossens, R. H. M., and C. J. Snijders. 1995. Design criteria for the reduction of shear forces in beds and seats. *J Biomech*. 28:225–30.

Groenesteijn, L. 2015. *Seat design in the context of knowledge work*. PhD thesis. Delft: Delft University of Technology.

Grönroos, C., and K. Ojasalo. 2004. Service productivity—Towards a conceptualization of the transformation of inputs into economic results in services. *J Bus Res*. 57:414–23.

Gscheidle, G. M., and P. M. Reed. 2004. Sitter-selected postures in an office chair with minimal task constraints. *Hum Fac Erg Soc P*. 48(8):1086–90.

Hagstrom, W. O. 1965. *The scientific community*. In *Patterns of Contact and Communication in Scientific Research Collaboration*. Kraut, R., C. Egido, and J. Galegher (eds.). 1988. *Proc 1988 ACM Conf Computer-Supported Cooperative Work*. 1–12. Carbondale, IL: Southern Illinois University Press.

Hatfield, E., J. L. Cacioppo, and R. L. Rapson.1993. Emotional contagion. *Curr Dir Psychol Sci*. 2:96–9.

Hedge, A., and D. Gaygen. 2010. Indoor environment conditions and computer work in an office. *HVAC & R Res*. 16:123–38.

Hedge, A., and E. J. Ray. 2004. Effects of an electronic height-adjustable worksurface on self-assessed musculoskeletal discomfort and productivity among computer workers. *Hum Fac Erg Soc P*. 48(8):1091–5.

Hedge, A., J. Puleio, and V. Wang. 2011. Evaluating the impact of an office ergonomics program. *Hum Fac Erg Soc P*. 55(1):594–8.

Hongisto, V. 2005. A model predicting the effect of speech of varying intelligibility on work performance. *Indoor Air*. 15:458–68.

Hosea, T. M., S. R. Simon, J. Delatizky, M. A. Wong, and C. C. Hsieh. 1986. Myoelectric analysis of the paraspinal musculature in relation to automobile driving. *Spine*. 11:928–36.

Hu, F. B., T. Y. Li, G. A. Colditz, W. C. Willett, and J. E. Manson. 2003. Television watching and other sedentary behaviors in relation to risk of obesity and type II diabetes mellitus in woman. *J Am Med Assoc*. 289:1785–91.

International Standards Organization (ISO). 1998. ISO 9241-5. *Ergonomic requirements for office work with visual display terminals (VDTs)—Part 5: Workstation layout and postural requirements*. Geneva: ISO.

Karakolis, T., and J. P. Callaghan. 2014. The impact of sit-stand office workstations on worker discomfort and productivity: A review. *Appl Ergon*. 45:799–806.

Kastelein, J. P. 2014. *Space Meets Knowledge: The Impact of Workplace Design on Knowledge Sharing*. Dissertation, Breukelen: Nyenrode Business University.

Klein Hesselink, J., E. de Groot, M. Loomans, and A. Kremer, A. 2006. *Fysiologische en psychische en gezondheidseffecten van planten in de werksituatie op gezondheid en welbevinden van mensen* (Physiological and Mental and Health Consequences of Plants in the Work Situation on Health and Well-being of People). TNO rapport 21573/018.10311. Hoofddorp: TNO Kwaliteit van Leven.

Kooij-de Bode, H., M. Blok, L. Groenesteijn, M. Formanoy, J. Zonneveld, and P. Vink. 2009. More satisfaction in meeting by choosing the right interior (in Dutch). In *Aangetoonde effecten van het kantoor-interieur: Naar comfortabele, innovatieve, productieve en duurzame kantoren*. P. Vink (ed.). 139–158. Alphen aan den Rijn: Kluwer.

Kosonen, R., and F. Tan. 2004. Assessment of productivity loss in air-conditioned buildings using PMV index. *Energy Build*. 36:987–93.

Krause, N., J. W. Lynch, G. A. Kaplan, R. D. Cohen, R. Salonen, and J. T. Salonen. 2000. Standing at work and progression of carotid atherosclerosis. *Scand J Work Environ Health*. 26:227–36.

Kraut, R., C. Egido, and J. Galegher. 1988. Patterns of contact and communication in scientific research collaboration. *Proc 1988 ACM Conf Computer-Supported Cooperative Work*. 1–12.

Larsen, L., J. Adams, B. Deal, B. S. Kweon, and E. Tyler. 1998. Plants in the workplace: The effects of plant density on productivity, attitudes, and perceptions. *Environ Behav*. 30:261–81.

Managers Online. 2016. http://www.managersonline.nl/nieuws/7762/staand-vergaderen-scheelt-miljarden.html.

Manyika, J., M. Chui, J. Bughin, R. Dobbs, P. Bisson, and A. Marrs. 2013 May. *Disruptive Technologies: Advances That Will Transform Life, Business, and the Global Cconomy*. McKinsey Global Institute.

McCulloch, J. 2002. Health risks associated with prolonged standing. *Work*. 19:201–5.

Miles, L. K., N. K. Nind, and C. N. Macrae. 2009. The rhythm of rapport: Interpersonal synchrony and social perception. *J Exp Soc Psychol*. 45:585–9.

Ministry of Social affairs and Labour. 1998. *Arbo informatieblad kantoren*. Den Haag: SDU uitgevers.

Nicol, F., M. Wilson, and C. Chiancarella. 2006. Using field measurements of desktop illuminance in European offices to investigate its dependence on outdoor conditions and its effect on occupant satisfaction, and the use of lights and blinds. *Energy Build*. 39:802–13.

Nobis, E. 2001. *Ambtenaar vergadert veel en vrijblijvend*. The Hague: Het Financiele Dagblad.

Oxidesk. 2016. http://www.markantoffice.com/oxidesk/nl/nieuws/26-newsflash/75-oxidesk-pilot-test-bij-ach -mea.

Park, B. J., Y. Tsunetsugu, T. Kasetani, T. Kagawa, and Y. Miyazaki. 2010. The physiological effects of Shinrin-yoku (taking in the forest atmosphere or forest bathing): Evidence from field experiments in 24 forests across Japan. *Environ Health Prevent Med*. 15:18–26.

Pronk, N. P., A. S. Katz, M. Lowry, and J. R. Payfer. 2012. Reducing occupational sitting time and improving worker health: The take-a-stand project. *Prevent Chron Dis*. 9:E154.

Rieck, A., and J. Kelter. 2005. The empirical OFFICE 21® study "soft success f." In *HCI International 2005: 1—Engineering Psychology, Health and Computer System Design*. Stuttgart.

Robertson, M. M., B. C. Amick, K. DeRango, T. Rooney, L. Bazzanid, L. Harriste, and A. Moore. 2009. The effects of an office ergonomics training and chair intervention on worker knowledge, behavior and musculoskeletal risk. *Appl Ergon*. 40:124–35.

Robertson, M. M., and M. J. O'Neill. 2003. Reducing musculoskeletal discomfort: Effects of an office ergonomics workplace and training intervention. *Int J Occup Saf Ergon*. 9:491–502.

Robertson, M. M., V. M. Ciriello, and A. Garabet. 2013. Office ergonomics training and a sit-stand workstation: Effects on musculoskeletal and visual symptoms and performance of office workers. *Appl Ergon*. 44:73–85.

Roelofsen, P. 2013. Healthy ageing-design criteria for the indoor environment for vital elderly. *Intell Build Int*. From http://www.researchgate.net/profile/C_Roelofsen/publications.

Roelofsen, P. 2016. *A healthy indoor environment as a strategy for productivity enhancement*. PhD thesis Delft: Delft University of Technology.

Satish, U., M. J. Mendell, K. Shekhar, T. Hotchi, D. Sullivan, S. Streufert, and W. J. Fisk. 2012. Is CO_2 an indoor pollutant? Direct effects of low-to-moderate CO2 concentrations on human decision-making performance. *Environ Health Perspect*. 120(12):1671–7.

Sengupta, A. K., and B. Das. 2000. Maximum reach envelope for the seated and standing male and female for industrial workstation design. *Ergonomics*. 43:1390–404.

Shibata, S., and N. Suzuki. 2002. Effects of the foliage plant on task performance and mood. *J Environ Psychol*. 22:265–72.

Sink, D. S., and S. J. de Vries. 1984. An in-depth study and review of state-of-the-art and practice productivity measurement techniques. *Proc Inst Indust Engin*. 1:335–447.

Straker, L., J. Levine, and A. Campbell. 2009. The effects of walking and cycling computer workstations on keyboard and mouse performance. *Human Factors*. 51:831–44.

The Muse. 2016. https://www.themuse.com/advice/how-much-time-do-we-spend-in-meetings-hint-its-scary.

The State of Telework in the U.S.—How Individuals, Business, and Government Benefit (redacted). http:// globalworkplaceanalytics.com/telecommuting-statistics.

Tüchsen, F., N. Krause, H. Hannerz, H. Burr, and T. S. Kristensen. 2000. Standing at work and varicose veins. *Scand J Work Environ Health*. 26:414–20.

Ulrich, R. S. 1984. View through a window may influence recovery from surgery. *Science*. 224:420–1.

Van der Voordt, D. J. M. 2003. *Kosten en baten van werkplek innovatie*. Delft: Center for People and Buildings.

Van Rhijn, J. W., M. P. de Looze, G. H. Tuinzaad, L. Groenesteijn, M. D. de Groot, and P. Vink. 2005. Changing from batch to flow assembly in the production of emergency lighting devices. *Int J Prod Res*. 43:3687–701.

Vellinga, R. 2001. *Research on Sit/Stand Tables* (in Dutch: *Onderzoek Zit/Sta-tafels*). Ouderkerk: Witteveen, Project-Inrichtingen.

Vink, P. 2009. The taylored office interior (In Dutch: De kantoorwerkplek op maat: een voorbeeld). In *Aangetoonde effecten van het kantoor-interieur: Naar comfortabele, innovatieve, productieve en duurzame kantoren*. P. Vink (ed.). 59–65. Alphen aan den Rijn: Kluwer.

Vink, P., and M. A. J. Kompier. 1997. Improving office work: A participatory ergonomic experiment in a naturalistic setting. *Ergonomics*. 40:435–49.

Vink, P., L. Groenesteijn, M. M. Blok, and M. den Hengst. 2008. Effects of a meeting table and chairs making half standing possible. In *Proc 2nd Int Conf AHFE*. Karwowski, W., and G. Salvendy (eds.). Louisville, KY: AHFE Intern. CD Rom.

Vink, P., I. Konijn, B. Jongejan, and M. Berger. 2005. Varying the office work posture between standing, half-standing and sitting results in less discomfort. In *Ergonomics and health aspects of working with computers*. Karsh, B., (ed.). 115–20. Berlin/Heidelberg: Springer Verlag; *Proc HCII 2009 congress*. San Diego, CA.

Vink, P., R. Porcar-Seder, A. Page de Poso, and F. Krause. 2007. Office chairs are often not adjusted by end-users. *Hum Fac Erg Soc P*. 51(17):1015–19.

Yates, J. W., and W. Karwowski. 1992. An electromyographic analysis of seated and standing lifting tasks. *Ergonomics*. 35:889–98.

Wilke, H. J., P. Neef, M. Caimi, T. Hoogland, and L. E. Claes. 1999. New in vivo measurements of pressures in the intervertebral disc in daily life. *Spine*. 24:755–62.

Wilks, S., M. Mortimer, and P. Nylen. 2006. The introduction of sit-stand worktables: Aspects of attitudes, compliance and satisfaction. *Appl Ergon*. 37:359–65.

World Health Organization (WHO). 2003. WHO definition of health. From http://www.who.int/about/definition/en/print.html. Accessed October 15, 2014.

Zhu, C., S. Shimazu, M. Yoshioka, and T. Nishikawa. 2011. Power assistance for human elbow motion support using minimal EMG signals with admittance control. *2011 Int Conf Mechatronics Automation (ICMA)*. 276–81. Beijing.

9 Healthcare Settings, Health, and Productivity

Tim Springer

CONTENTS

9.1 INTRODUCTION

Simply stated, healthcare involves the provision of patient care. However, underlying that simple premise is the reality of interacting with, diagnosing, treating, healing, rehabilitating, and caring for the fullest range of the human population representing every imaginable size, shape, age, and level of wellness, health, injury, illness, or infirmity from before birth to after death. Consequently, healthcare work settings encompass a range of activities and locations wider and more varied than any other industry sector. From back streets to battlefields, hospitals to homes, the highest tech to the highest touch, healthcare workers occupy and labor in some of the most challenging work settings imaginable.

Healthcare work and work settings present several unique challenges:

- The cost of errors—Simply put, mistakes can cost lives, affect health or, at a minimum, substantially increase the duration, complexity, and cost of treatment and care.
- Nonstop operation—Many healthcare settings operate 24 hours a day, 7 days a week, 365 days a year. Consequently, shift work and shared workplaces are prevalent and present challenges for worker health, safety, and performance.

- Predominantly female workforce—According to the Bureau of Labor Statistics 2013 data, women comprise 78.4% of the healthcare workforce (U.S. Department of Labor 2013a).
- Regulations and information security requirements—Regulations and information security requirements impose unique and potentially conflicting demands on healthcare workplaces. Information sharing, collaboration, and teamwork must be balanced against patient privacy and restrictions on the access to patient information.
- Patient care—Patient care involves moving patients who may be unable to assist or even resistant. Associated physical actions are often performed from less than ideal positions and postures, thereby increasing the risk of injury to healthcare workers.
- Multimodality of healthcare jobs—Many healthcare jobs are multimodal involving physical activity, highly skilled knowledge work, and technology use.

This set of circumstances presents many opportunities and challenges for applying ergonomics.

9.2 HEALTHCARE CONTEXT

9.2.1 HEALTHCARE IS BIG

Healthcare is the largest sector in the U.S. economy. Healthcare spending accounts for 17% of the gross domestic product (GDP) and will grow to one-fifth (20% of the GDP) of the U.S. economy, by 2021. In terms of dollars spent, total healthcare spending in the United States is expected to reach $4.8 trillion in 2021, up from $2.6 trillion in 2010 and $75 billion in 1970 (Centers for Medicare and Medicaid Services [CMS] 2011). Historically, the care delivered in a hospital setting has taken the lion's share of total healthcare spending, at around one-third, or $814 billion (The Henry J. Kaiser Family Foundation 2010). The healthcare industry employs 16 million people or one in every eight workers. This number does not include jobs in the pharmaceutical, insurance, or device-manufacturing industries.

9.2.2 HEALTHCARE IS COMPLEX

The advances in science and technology have improved the ability to treat an expanding scope of human health concerns. Accompanying the advance of medicine and healthcare has been an increase in the complexity of professions, work, and work settings. Healthcare services can be broadly classified into six categories:

1. Preventative care—Preventive care involves disease prevention, health and wellness, and behavioral and lifestyle interventions (e.g., disease screening, genetic testing, immunization, nutrition, exercise, weight control, smoking cessation).
2. Curative care—Curative care provides diagnosis and treatment (e.g., annual physicals, diagnostic testing, prescription treatments, surgery).
3. Rehabilitative care—Rehabilitative care involves mental and physical therapies to restore health and functionality.
4. Palliative care—Palliative care focuses on providing comfort and relief from symptoms, pain, and associated stress from injury or illness.
5. Hospice—A hospice provides holistic care for terminally ill patients and family.
6. Public health—Public health protects the general public from a variety of health-related threats (e.g., epidemic and pandemic protocols).

An illustration of the complexity involved in providing patient care is the number of specialties, certificates, and licenses of healthcare professionals.

The Association of American Medical Colleges identifies 36 broad categories and 63 specialties (Center for Workforce Studies 2012). The American Board of Medical Specialties lists 37 board certificates and 125 subspecialties including such diverse practices as undersea and hyperbaric medicine, surgery of the hand, and pediatric transplant hepatology (American Board of Medical Specialties 2015). Similarly, nursing identifies 104 areas of specialization, 92 of which are patient facing, and 34 of which involve work settings outside of a hospital (Johnson & Johnson 2015).

Further complicating any discussion of healthcare work, workers, and work settings is the question of, who pays? The central role of employment benefits and insurance coverage often defines the access to and the demand for healthcare. Services and activities related to overall health, such as nutrition, dentistry, or vision care, are not always part of employee healthcare benefits. Consequently, these elements of healthcare may be excluded in certain discussions simply because of the limits imposed by health insurance. For the purposes of this chapter, the question, who pays for healthcare? will not define the scope or the scale of the discussion.

That said, the limits of space and time make it impossible for a detailed examination of all types of healthcare work and work settings. Challenges and opportunities affecting the health, safety, and performance of healthcare workers will be illustrated using representative examples.

9.2.3 Healthcare Is Changing

The healthcare industry is undergoing rapid, fundamental change. The magnitude and the speed of change are driven by a number of forces discussed in the following sections.

9.2.3.1 Government Forces: Regulations, Legislation, Policies, and Practice

Healthcare is among the most regulated industries:

> The array of regulations that govern health care can seem overwhelming to people who work in the industry. Almost every aspect of the field is overseen by one regulatory body or another, and sometimes by several. Health care professionals may feel that they spend more time complying with rules that direct their work than actually doing the work itself. (Field 2008)

Most of the power governing day-to-day activities resides with the states against a backdrop of broader federal mandates. Each state imposes its own regulations on practitioners, facilities, and organizations. At the federal level, rules and regulations are issued and monitored by an operating division within the U.S. Department of Health and Human Services (USDHHS 2015a). These range from the Administration for Children and Families to the Centers for Disease Control, and from the Food and Drug Administration to the CMS. Two federal acts have had significant impact on how and where healthcare is delivered:

1. The Health Insurance Portability and Accountability Act (HIPAA) of 1996 provides federal protections for individually identifiable health information and specifies a series of administrative, physical, and technical safeguards for use by healthcare providers and their business associates to assure the confidentiality, the integrity, and the availability of electronic health information (USDHHS 2015b). HIPAA poses challenges to healthcare providers to meet the information security requirements in an environment that is adopting mobile information technology and increasingly open, and the effectiveness of which relies on good communication and informed collaboration.

2. The Patient Protection and Affordable Care Act of 2010 (Public Law 111-148 2010), called the Affordable Care Act (ACA) or "Obamacare," places the emphasis on accountability and evidence-based practice accompanied by a greater role for the federal government in monitoring and spending. Under ACA rules, an estimated 32 million newly insured citizens will have increased access to healthcare services. The CMS will shift its emphasis from quantity of services to quality of services. Accompanying this emphasis on quality and accountability is a focus on patient-centered care and adoption of shared decision-making.

9.2.3.2 Economic Forces: Jobs, Growth, and Competition

Occupations and industries related to healthcare are projected to add 5 million new jobs between 2012 and 2022. The total healthcare employment is projected to increase by 10.8%, or 15.6 million, during the decade. This accounts for nearly one-third of the overall projected job growth.

Growth is driven, in part, by the demand for healthcare workers to address the needs of an aging population and, in part, by the increased demand from the 32 million newly insured citizens under the ACA. Three major healthcare-related occupational groups are projected to grow to more than 20%, nearly double the overall growth from 2012 to 2022: a 28.1% growth in healthcare support occupations, a 21.5% growth in healthcare practitioners and technical occupations, and a 20.9% growth in personal care and service occupations (U.S. Department of Labor 2013b).

The U.S. Federal Trade Commission (2004) predicted that the ACA would benefit patients as consumers following from the increased competition among providers. Following public hearings on competition in healthcare, they issued a report recommending a number of ways to increase the competition in the delivery of healthcare that would result in increased accountability, more transparency, lower costs, and higher quality of care (US Federal Trade Commission 2004).

9.2.3.3 The Obesity Epidemic

Obesity is a serious problem affecting the health of a growing number of people in developed countries around the world.

- The worldwide prevalence of obesity has nearly doubled between 1980 and 2008 (WHO 2015).
- More than one-third (34.9% or 78.6 million) of U.S. adults are obese.
- Childhood obesity has more than doubled in children and quadrupled in adolescents in the past 30 years.
- In 2012, more than one-third of children and adolescents were overweight or obese (Ogden et al. 2014).
- The percentage of children aged 6–11 years in the United States who were obese increased from 7% in 1980 to nearly 18% in 2012. Similarly, the percentage of adolescents aged 12–19 years who were obese increased from 5% to nearly 21% over the same period (Health Statistics 2012; Trust for America's Health 2015).
- Rising obesity rates have significant health consequences, contributing to increased rates of more than 30 serious diseases. These conditions create a major strain on the healthcare system. More than one-quarter of healthcare costs are now related to obesity (Trust for America's Health 2015).

The impact on healthcare is felt both in terms of the maladies associated with morbid obesity and the necessity for accommodating bariatric patients in healthcare settings. Providing care to bariatric patients increases the risk of injury to healthcare workers. The excessive size of bariatric patients necessitates specialized equipment for all forms of patient care from physical transfer, transport, and movement in beds, to diagnostic and monitoring instrumentation, to basic hygiene.

9.2.3.4 The Aging Population

People 65 years and older represented 12.4% of the population in the year 2000. That number grew to 12.9% or about 1 in every 8 Americans (39.6 million) in 2009—the latest year for which data are available. By 2030, there will be about 72.1 million older persons (19% of the total population), more than twice their number in 2000 (USDHHS 2015c).

1. Aging workforce

    The number of individuals in the labor force who are 65 years or older is expected to grow by 75%, while the number of individuals in the workforce who are 25–54 is only expected to grow by 2% (Administration on Aging 2012). By 2016, one-third of the total U.S. workforce will be 50 years or older—a group that may reach 115 million by 2020 (Heidkamp, Mabe, and DeGraaf 2012). The baby boom generation will be driving this graying of the worker population. As they have throughout their lives, this generation will place increased and novel demands on society and healthcare specifically. Unlike those of their forebears, baby boomers expect to lead more active lives with sustained levels of health into their senior years. As this population ages, they will redefine long-term care and how healthcare can provide assistance either with equipment, personnel, or both (USDHHS 2015d).

2. Aging healthcare workforce

The demographic trend of the aging population is magnified when considering healthcare workers. According to the Institute of Medicine, the average age of a nurse in the United States is 50 years old. By 2020, nearly half of all registered nurses will reach the traditional retirement age. Nearly one-quarter of physicians in a 2007 nationwide survey were 60 years or older. In 2001, more than 80% of all dentists in the United States were older than 45; the number of dentists expected to enter the field by 2020 will not be sufficient to replace the number of dentists likely to retire, and older workers are at higher risk for injury and disability posing problems for both healthcare workers and their employers (Institute of Medicine of the National Academies 2008).

9.2.3.5 Technology Forces

From the evolution of the litters of Ancient Egypt and Rome to the early hospital beds of the 1800s with mechanical adjustments to today's high-tech patient care environments, healthcare has a long history of integrating technology into its practice (Hospital Beds 2015). Today healthcare is driving and being driven by rapid advances in technologies. From digital imaging to nanotechnologies, robotics to three-dimensional printing to nuclear medicine, state-of-the-art technologies are at the forefront of patient care. As technologies and techniques have progressed, the complexity of the sociotechnical systems of healthcare has grown. Two technology trends promise to pose challenges and opportunities when considered in the context of healthcare. The mobility provided by smart, powerful personal devices must be juxtaposed with regulations covering the use of technology and patient information security. Similarly, the advances in informatics and its application to digital records promise an improvement in both efficiencies and accuracy; however, those benefits must be balanced with concerns over information security and regulations regarding access and sharing. The increased use of healthcare information technologies, including electronic medical records, computerized provided-order entry systems, e-prescription systems, picture archiving and communications systems (PACS), etc., is placing a greater burden on the healthcare staff, and the frequent failure to address ergonomic design issues is resulting in musculoskeletal injuries to these professionals (Ruess et al. 2003; Hedge, James, and Pavlovic 2011) and also impacting cognitive performance and medical errors (Lawler, Hedge, and Pavlovic 2011). Attention needs to be paid to the ergonomic design of both the medical devices and the places where they are to be used. Many healthcare facilities use mobile computer carts to record patient information (Figure 9.1), and ergonomic design guidelines have been developed to help with the selection of these devices (Hedge 2008a, 2014).

Computer wall stations also are becoming more commonplace in patient rooms, but these need to provide the appropriate adjustability to fit the users (Figure 9.2), and Hedge (2008b) has developed an ergonomic design checklist for computer wall stations.

9.2.3.6 Industry Trends

Four key industry initiatives are beginning to change the way healthcare organizations operate and especially how they view the physical environment:

1. Evidence-based practice—Evidence-based practice is the initiative of the USDHHS, Agency for Healthcare Research and Quality (AHRQ). The AHRQ began in 1997. The AHRQ launched a program to promote evidence-based practice in everyday care by establishing

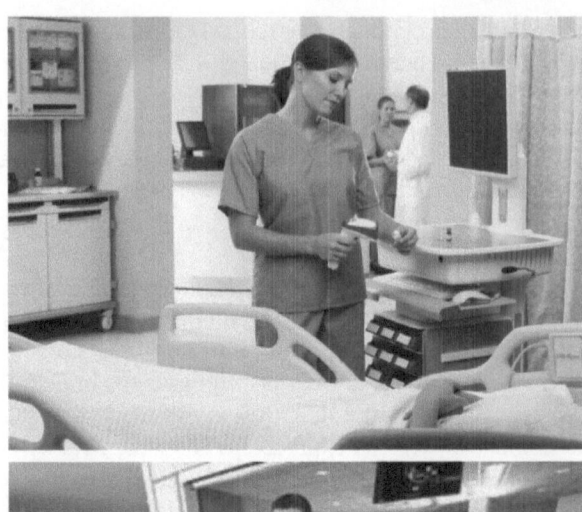

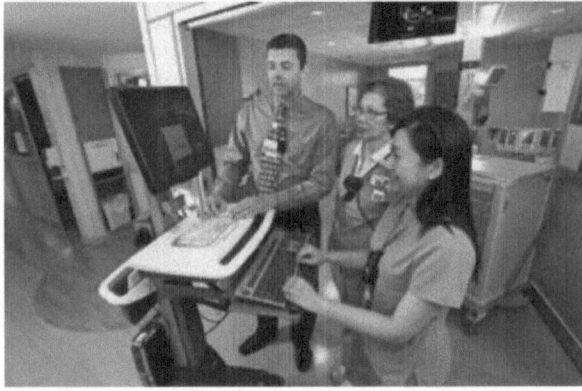

FIGURE 9.1 Computers on wheels (COWS). (Courtesy of ASR Healthcare, https://asr4u.wordpress.com/2012 /05/05/point-of-care-and-microban-cows-or-wows-need-a-microban-keyboard-mouse/; Courtesy of Advance Healthcare Network, http://health-information.advanceweb.com/Web-Extras/Online-Extras/No-Strings-or-Cables -Attached.aspx.)

evidence-based practice centers (EPCs). The EPCs develop evidence on Medicare and Medicaid populations. Under the EPC program, 5-year contracts are awarded to institutions in the United States and Canada to review all relevant scientific literature on a wide spectrum of clinical and health services topics to produce various types of evidence reports. These reports may be used for informing and developing coverage decisions, quality measures, educational materials and tools, clinical practice guidelines, and research agendas. The EPCs also conduct research on the methodology of evidence synthesis (AHRQ 2015).

2. Patient-centered care focusing on the patient experience—The accountability provisions of the ACA have prompted a renewed interest in a seemingly simple idea—patient-centered care. This idea shifts the focus away from the diseases and back to the patient and the family (Gerteis et al. 1993). This approach is central to efforts that improve the quality of healthcare. A 2001 report by the American Institute of Medicine listed patient-centered care as one of the six key factors necessary to close the quality gap in healthcare (Committee on Quality of Health Care in America 2001). An essential element of patient-centered care is the notion of shared decision making, where patients and families are enlisted as allies in designing, implementing, and evaluating care plans and systems (Gerteis et al. 1993). This shifts the focus from the quantity of services driven by technology and the skills of medical professionals to the quality of care emphasizing the involvement of the patient as a team member and the commitment to the best outcomes in terms of patient experience.

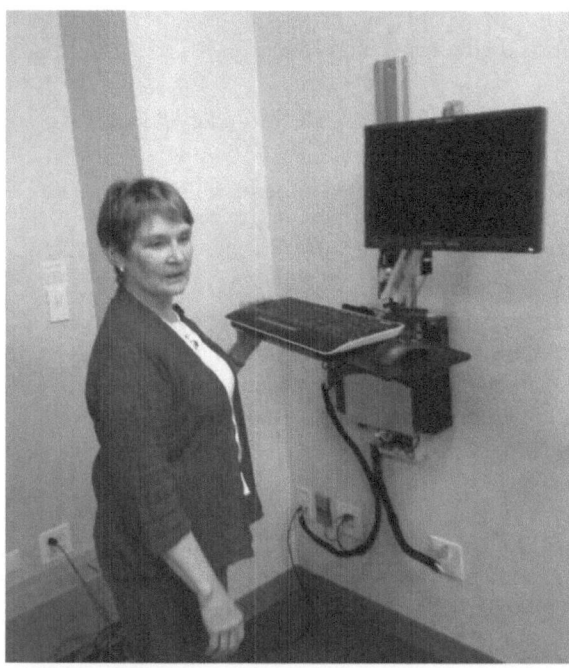

FIGURE 9.2 In-room computer wall station. Note: Although an adjustable platform has been provided, installation yields an unworkable height for many (10th percentile stature of female shown). (From Springer, T., Ergonomics for healthcare environments, Knoll, http://www.knoll.com/media/760/617/healthcare_ergonomics .pdf, 2014. With permission.)

3. Metrics and measures—To support the drive toward improved quality and accountability, the CMS developed a hospital consumer assessment of healthcare providers and systems survey (HCAHPS). While many hospitals have collected information on patient satisfaction for their own internal use until HCAHPS, there was no national standard for collecting and publicly reporting information about patient experience of care that allowed for valid comparisons to be made across hospitals locally, regionally, and nationally. HCAHPS is the first national, standardized, publicly reported survey of the patients' perspectives of hospital care. HCAHPS was developed to meet three broad goals. First, the survey is designed to produce data about the patients' perspectives of care that allow for objective and meaningful comparisons of hospitals on topics that are important to consumers. Second, the public reporting of the survey results creates new incentives for hospitals to improve the quality of care. Third, the public reporting serves to enhance the accountability in health care by increasing the transparency of the quality of hospital care provided in return for the public investment.

4. Evidence-based environmental design—Evidence-based practice and patient-centered care converge when investigating the impact of the physical environment on patients and staff. The Center for Healthcare Design (CHD) and Planetree are two organizations dedicated to fostering evidence-based environmental design and a more comprehensive approach to healthcare environments. The CHD, which started in 1993, states that its goal is to use design to improve patient outcomes in healthcare. CHD sponsors projects that measure the effects of the built environment on healthcare outcomes. They aim to leverage the effect by sharing documented examples of healthcare facilities in which the design has improved the quality of care and the financial performance of the institution. The CDH reports well-designed healthcare environments that can
   - Improve the quality of care
   - Attract more patients in an increasingly competitive market

- Attract and retain skilled staff
- Enhance operational efficiency and productivity

The CHD maintains a knowledge base of articles and evidence illustrating how their approach yields these benefits (CHD 2015).

Planetree membership includes more than 500 organizations from eight countries. Their philosophy is based on a simple premise: care should be organized first and foremost around the needs of the patients. The Planetree model (Planetree 2014) includes ten points:

1. Human interactions/independence, dignity, and choice
2. Importance of family, friends, and social support
3. Patient/resident education and community access to information
4. Healing environment: Architecture and interior design
5. Nutritional and nurturing aspects of food
6. Arts program/meaningful activities and entertainment
7. Spirituality and diversity
8. Importance of human touch
9. Integrative therapies/paths to well-being
10. Healthy communities/enhancement of life's journey

Of particular relevance to this chapter is the recognition of the importance of the physical environment to patient experience and healing in the Planetree model (Planetree 2014):

Healing Environment: Architecture and Interior Design—The physical environment is vital to healing and well-being. Each hospital and continuing care community is designed to incorporate the comforts of home, clearly valuing humans, not just technology. By removing architectural barriers, the design encourages patient and family involvement. An awareness of the symbolic messages communicated by the design is an essential part of planning. Spaces are provided for both solitude and social activities, including libraries, kitchens, lounges, activity rooms, chapels, gardens and overnight accommodation for families.

As the healthcare systems gain experience with HCAHPS patient satisfaction scores, a strong case can be made to apply evidence-based changes to healthcare environments to improve quality and reduce errors, noise, anxiety, confusion, and stress. All of which, both families and patients understand, feel, and contribute to a positive patient experience. As stated by Sadler et al. (2009):

Leaders need to understand the clear connection between constructing well-designed healing environments and improved health care safety and quality for patients, families, and staff, as well as the compelling business case for doing so. The physical environment in which people work and patients receive their care is one of the essential elements in reducing a number of preventable hospital acquired conditions.

Ulrich and Zimring (2004) reviewed the research evidence regarding the design of healthcare environments and its effect on patients and staff and found rigorous studies that link the physical environment to patient and staff outcomes in four areas:

1. Increase in staff effectiveness, reduction of errors, and increase in staff satisfaction by designing better workplaces
2. Improvement of patient (and staff) safety
3. Reduction of stress and improvement of outcomes
4. Improvement of overall healthcare quality

Four years later, Ulrich et al. (2008) conducted a second review of over 1000 studies:

Compared to 2004, the body of evidence has grown rapidly and substantially.... It is now widely recognized that well designed physical settings play an important role in making hospitals less risky and stressful, promoting more healing for patients, and providing better places for staff to work.

The Joint Commission, an independent, not-for-profit organization that accredits and certifies more than 20,500 healthcare organizations and programs in the United States, released a 2008 report containing a chapter on the design of the physical environment that relies on much of the same published research (The Joint Commission 2008). Evidence shows a link between the design of healthcare settings and the improved outcomes.

1. Infection control
   a. Installing hand sanitizer dispensers at each point of patient contact (e.g., in each room) improves hand hygiene—a major factor in improved infection control.
   b. Staff training on effective hand hygiene protocols helps control the infection.
2. Acoustics
   a. Reducing the ambient noise through environmental treatments such as ceiling tiles reduced stress among both patients and staff and allowed patients to sleep better and increased patient satisfaction.
   b. Improved acoustics minimizes the distractions and facilitates effective communication in patient handoff.
3. Lighting
   a. Effective task lighting reduces errors—especially those associated with medication in both the pharmacy and on hospital wards.
   b. Daylighting reduces patient and staff stress and length of stay and improves patient recovery.
4. Room design and configuration
   a. Single bedrooms versus multibed patient rooms reduce infection, increase actual usable beds, and raise patient satisfaction.
   b. The room design can reduce the number and the frequency of patient transfers, freeing the staff from the times associated with transfer and lowering the risk of injury to the staff.
5. Effective wayfinding reduces the time spent by the staff in directing visitors (Sadler et al. 2009)

Research directly focusing on the impact of work setting on nurses' performance reported similar findings. Based on a review of literature regarding nurses' performance and work settings, researchers identified four key physical environmental variables that have the most real or potential effect on workplace errors:

- Noise
- Lighting
- Ergonomics/furniture/equipment
- Design/layout (Chaudhury, Mahmood, and Valente 2009)

A follow-up study (Mahmood, Chaudhury, and Valente 2011) specifically looked at medication errors in the larger context of the quality of care. The workplace elements contributing to errors included

- Inadequate space in the charting and documentation areas
- Lengthy walking distances to the patient rooms
- Insufficient patient surveillance opportunity/lack of visibility to all parts of the nursing unit
- Small size of the medication room

- Inappropriate organization of the medical supplies
- High noise levels in the nursing unit
- Poor lighting
- Lack of privacy in the nursing stations

Anjali (2006) conducted an extensive review of peer-reviewed journal articles and research reports published in medicine, nursing, psychology, ergonomics, and architecture periodicals, and books. Anjali found that staff effectiveness is undermined by poorly designed work systems resulting in

- Multiple patient transfers within the hospital
- Frequent communication breakdowns
- Medical errors
- Increased costs
- Reduced quality of care
- Reduced patient and staff satisfaction
- Wastage of staff time
- Reduced staff productivity

9.3   ERGONOMICS IN HEALTHCARE ENVIRONMENTS

Table 9.1 illustrates some of the activities and work settings in which ergonomics has been and can be applied.

The following sections illustrate both successful ergonomic interventions and opportunities to apply ergonomics to current issues affecting healthcare quality, staff safety, and performance.

TABLE 9.1
Activities and Work Settings

Patient Handling	Manual Material Handling	Workplace Design	Equipment/Tool Design
Ambulance transport	Facilities maintenance	Nurses' stations	Patient rooms
Emergency rooms	Housekeeping	Computer workstations	Surgery
Patient rooms	Food service	Mobile computers (COWS)	Imaging (X-ray, magnetic resonance imaging, computerized axial tomography)
Admission + discharge/transport	Laundry	Laboratory	Rehab/therapies
Rehab	Sterilization services	Pharmacy	Pharmacy
Radiology	Deliveries	Dental exam room	Facilities maintenance
Surgery		Office	Dentist
Morgue			
Clinic			
Nursing home			
Extended care			
Physician office/exam room			
Dentist			

Source: Adapted from Enos, L., *Overcoming Barriers to Implementing Ergonomics Programs in Healthcare: Case Studies from the Field*, Oregon Nurses Association, Tualatin, OH, 2004; Hedge, A., Best practices for site-wide hospital ergonomics, National Ergonomics Conference & ErgoExpo, Las Vegas, NV, 2005. With permission.

9.3.1 SAFE PATIENT HANDLING AND TRANSFER

Moving patients poses the biggest risk for injury among healthcare workers. Musculoskeletal disorders are prevalent among staff who are engaged in patient handling, most often affecting the lower back and the shoulders. These injuries are due in large part to overexertion related to repeated manual patient handling activities, often involving heavy manual lifting associated with transferring and repositioning patients and working in extremely awkward postures. Some examples of patient handling tasks that may be identified as high risk include transferring from toilet to chair, transferring from chair to bed, transferring from bathtub to chair, repositioning from side to side in bed, lifting a patient in bed, repositioning a patient in a chair, or making a bed with a patient in it (Occupational Safety and Health Administration [OSHA] 2015a). Increasing numbers of bariatric patients, staffing shortages, and an aging healthcare workforce have increased the concern regarding accidents and injuries associated with patient movement and transfers. The staff in hospitals and nursing and extended care facilities engage in more frequent patient movement and transport and are at the highest risk of injury. However, the effects were found across all healthcare settings. Many organizations employ lifting teams and put in place policies to minimize lifting while moving patients.

Another approach adopted by hospitals and nursing and extended care facilities are zero-lift policies and practices integrating work process evaluation, aiding devices, and training. Recognizing the potential risks associated with patient transport, a national movement is underway combining state regulations, National Safety standards, and Joint Commission guidelines, aimed at achieving "safe patient handling—no manual lift." The goal is to require mechanical lifting equipment and friction-reducing devices for all healthcare workers, patients, and residents across all healthcare settings (Figure 9.3).

The Puget Sound Chapter of the Human Factors and Ergonomics Society (PSHFES) in cooperation with the Washington State Department of Labor and Industries published a table listing 39 projects by type of workplace, interventions, costs, measurements, and savings (PSHFES 2009). Of the 39 projects listed, 33 involved applying ergonomics via lift-assist equipment, process redesign, and/or training. In all cases, work-related injury was reduced, and in some cases, nearly eliminated, resulting in substantial savings leveraged by the modest cost invested. Hamrick et al. (2005) examined OSHA records for 111 facilities implementing ergonomic programs to reduce musculoskeletal disorders (MSDs). Sixty four percent (64%) experienced a significant decrease in MSDs following the implementation of ergonomic interventions. The median reduction in MSDs was 37%. Interventions included reduced bending, reduced carrying/lifting, and zero lifting. Multimodal interventions yielded the greatest effect. The decreases in back-related injuries (44%) were much higher than that in the national data (17% reduction) for the same time. The reduction increased (i.e., additional reduction in MSDs) during the second year of the programs. The researchers concluded that ergonomic interventions can be very effective in reducing MSDs.

The Joint Commission (2012) published a monograph highlighting evidence-based practice to improve patient and worker safety. Using a holistic, systems-based approach, the monograph includes specific examples of activities and interventions to improve safety. The primary example is reducing MSDs through safe patient handling.

9.3.2 DESIGN OF MEDICAL TECHNOLOGIES

Historically, ergonomics has addressed the issues of human–machine systems with the goal of improving safety and performance. The integration of technology in patient care offers many examples of ways in which ergonomics can improve the design of medical devices. The following are two recent examples.

1. Infusion pump—Device design is a logical area in which ergonomic improvement lead to reduced errors and improved performance. Wiseman, Cox, and Brumby (2013) looked at how infusion pumps, a common device in hospitals, were actually used. By evaluating

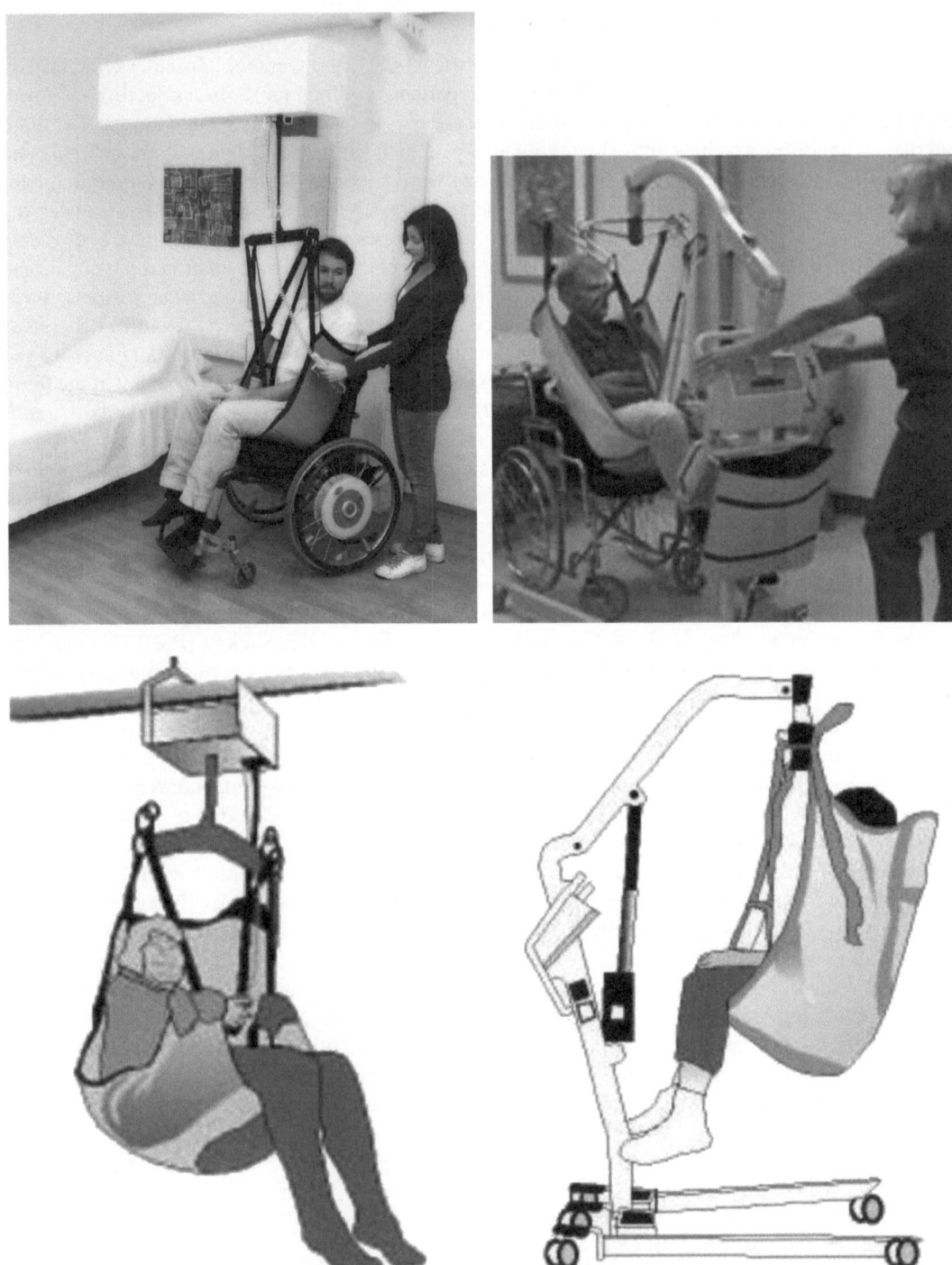

FIGURE 9.3 Patient lift devices. (Courtesy of CDC, Preventing back injuries in healthcare settings, National Institute for Occupational Safety and Health Science Blog, http://blogs.cdc.gov/niosh-science-blog /2008/09/22/lifting/, Atlanta, GA; Courtesy of OSHA, Guidelines for nursing homes: OSHA 3182-3R 2009, https://www.osha.gov/ergonomics/guidelines/nursinghome/final_nh_guidelines.pdf, Washington, DC.)

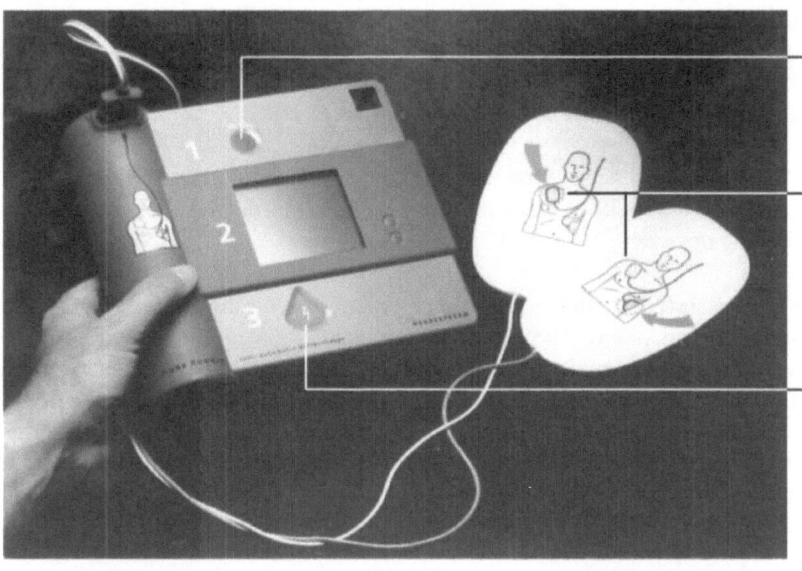

The "on" button is prominent and colored green to invite interaction.

Each electrode indicates independently where it should be located.

The "shock" button is colored and shaped to draw attention to its significance but is clearly not a standard stop button.

- The face of the unit is organized and leveled to show clearly three sequential steps.

- The "book" form suggests the appropriate orientation for carrying and use.

- To minimize distraction, other controls are downplayed and located elsewhere.

FIGURE 9.4 Portable defibrillator. (From Fulton Suri, J., *Ergonomics in Design*, 8, 4–12, 2000. With permission.)

traditional infusion pump keypad design, the researchers showed that the layout was not optimized for and did not accurately reflect the way the devices are most commonly used. "The digit 0 is used far more frequently than any other digit. The numbers 1000, 100, and 50 are used in nearly half of all infusions." The researchers then developed four heuristics based on digit and number distribution analysis that should be followed to reduce the key presses needed when programming the infusion pumps. These guidelines reduce the number of chances for error.

2. Portable defibrillator—A commonplace device now present in many public facilities, the portable defibrillator, is a good example of how human factors and ergonomics applied to the design of devices can yield simple, smart consumer/patient devices that not long ago required extensive training and high skill to operate. The advances in technology and Human Factors and Ergonomics (HFE)-focused design approaches allow novices to follow voice and symbol guidance to use the device to save lives. Fulton Suri (2000) offers a fascinating account of how HFE techniques evaluated and improved such a design (Figure 9.4).

9.3.3 Opportunities

The size and the complexity of healthcare work settings continue to present opportunities to apply ergonomics to the benefit of the patients, the staff, and the provider organizations. Four challenges and opportunities are highlighted in the following sections.

9.3.3.1 Alarm Fatigue

Alarm fatigue occurs when the staff become desensitized to the constant noise of alarms, are overwhelmed by the sounds, and turn alarms down or off. The Joint Commission (2013) issued a warning, in April 8, 2013, that healthcare workers can become numb to the incessant beeping of medical devices, creating life-threatening situations for patients. In 2014, alarm management was identified as a national safety goal. The ECRI Institute Listed alarm fatigue as one of the top two technology hazards for 2013 and 2014 (Wong, Mabuyi, and Gonzalez 2013).

- More than 19 in 20 hospitals (95.1%) say that they are concerned about alarm fatigue.
- Almost 1 in 10 hospitals (87.8%) believe that a reduction of false alarms would increase the use of patient-monitoring devices.

A review of literature by Cvach (2012) includes evidence practice recommendations incorporating smarter technology, staff training, practice protocols, and organizational culture and policy. This problem seems ripe for the application of human factors and ergonomic principles. A recent exchange of views by physicians in the Journal of the American Medical Association reinforced the need:

> A human factors approach [to alarm management] based around the hospital culture should be used that engages architects, designers, acoustical engineers, facility engineering, staff and clinicians to address alarm fatigue and its implication on the physical built environment. (Barach and Arora 2014)

9.4 PATIENT HANDOVER/HANDOFF—COMMUNICATION AND TEAMWORK

Transitions are characteristically fraught with potential for errors. Nowhere is that more apparent than when considering transition in care and care coordination. When transitioning from one level of care to another (e.g., from surgery to recovery; from hospital to home) or from one shift to another (Joseph et al. 2008), effective communication and teamwork are essential to insure accurate transmission of critical information, responsibility, and authority necessary for the effective handoff. Inadequate patient handover consistently appears as a factor contributing to adverse events including delays in diagnosis and treatment, redundant or unnecessary procedures and tests, longer hospital stays, and decreased patient and provider satisfactions. Studies of patient handoffs reveal a wide variation in the procedures and the quality that exists in handoffs from one level of care or team to another. There seems to be a lack of consensus about the primary purpose of patient handoffs resulting in a lack of consistent measures of effectiveness and a broad spectrum of interventions to improve the process (Manser and Howard 2010; Hesselink et al. 2012; Laugaland, Aase, and Barach 2012). The identified barriers to effective handoffs include medium of communication, training and education, social setting, time constraints, and physical setting. Balancing the concerns for patient confidentiality concerns (see HIPPAA) with the need for collaboration and teamwork can be a challenge. The physical settings best suited for such communication are private and quiet minimizing the impact of distraction as a barrier to the transfer of information. In short, the complexity of the cases and the attention needed to ensure a smooth transfer require a physical location that reduces potential interruptions and background noise. Appropriate lighting should be available along with ample writing space to take notes (Solet et al. 2005).

9.5 CONSTRAINED POSTURES

Dentistry, often overlooked as one part of the larger healthcare industry, offers significant opportunities to apply ergonomics to improve safety, health, and performance. Dentists and dental hygienists commonly assume awkward and constrained postures to obtain better views of the intraoral cavity (Figure 9.5).

The instruments used in dentistry often introduce vibration and require forceful hand exertions and repetitive movements. All of which raise the risk of musculoskeletal disorders and repetitive stress injuries (Hedge 1998; American Dental Association 2004; Oregon Coalition for Healthcare Ergonomics 2015).

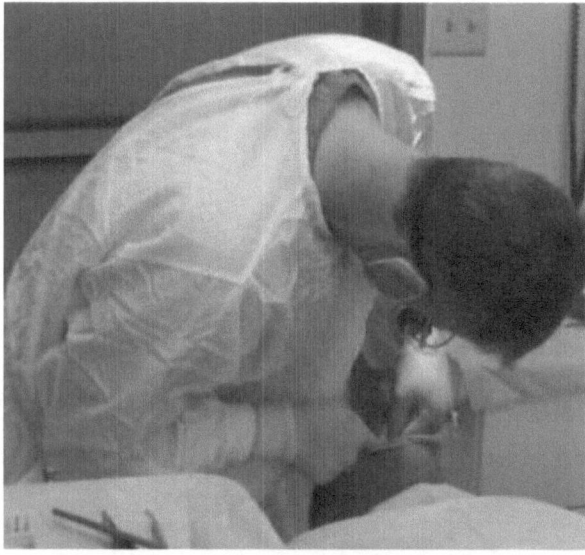

FIGURE 9.5 Dentist posture in need of improvement. (Courtesy of U.S. Air Force Dental Evaluation and Consultation Services, http://www.osap.org/resource/resmgr/DOCS/ergonomics_1.ppt, U.S. Air Force, Great Lakes, IL.)

9.6 WORKPLACE DESIGN

One example of workplace design presenting the opportunity for improvement through the application of ergonomics is diagnostic imaging, where advances in technology have radically changed the nature of the tasks. With the transition from a film-based to a soft copy, filmless environment presents an opportunity to redesign not only the radiologists' workflow, but also the work environments. A typical radiologist's workroom features poorly designed overhead fluorescent lights, cramped space, and inadequate chair and table. The reliance on multiple high-resolution displays has led to cave-like workplaces where technicians and physicians scan through layered images (Figure 9.6).

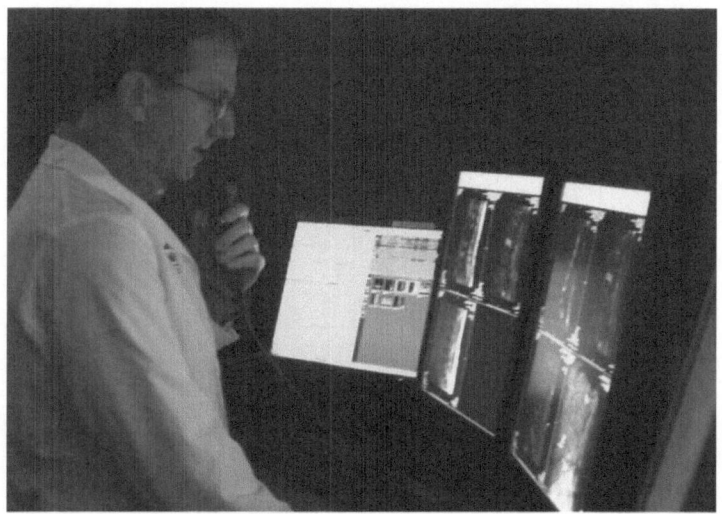

FIGURE 9.6 A radiologist interprets medical images on a PACS workstation. (Courtesy of Zackstarr, http://en.wikipedia.org/wiki/Radiology#mediaviewer/File:Radiologist_in_San_Diego_CA_2010.jpg, 2010.)

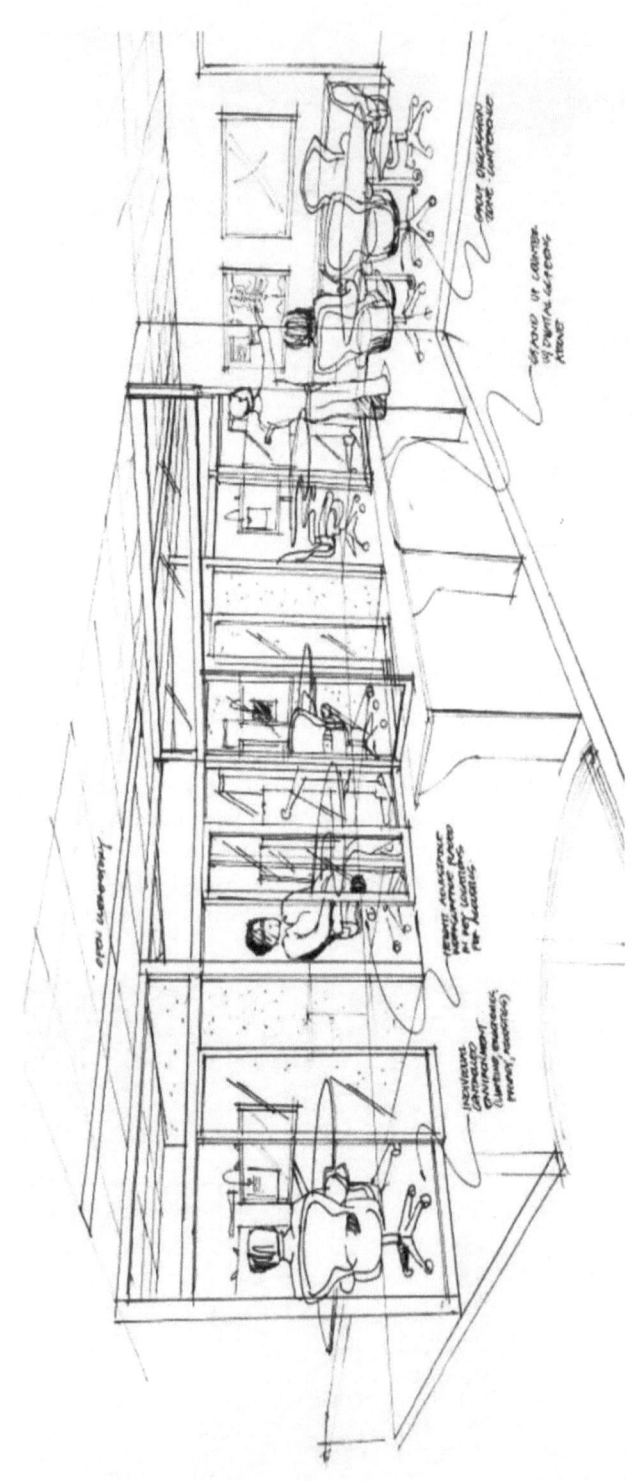

FIGURE 9.7 Redesign for a radiology reading room. (From Radiology Society of North America, Redesigning reading room helps combat ergonomic injuries, http://www.rsna.org/NewsDetail.aspx?id=8947, Oakbrook, IL.)

Effective redesign is likely to result in substantial improvements in radiologist performance resulting in reduction of fatigue, increased productivity, increased diagnostic accuracy, and possibly increased job satisfaction (Figure 9.7). Several studies have documented the deficiencies in the ergonomic design of radiology reading rooms (Siegel and Reiner 2002; Ruess et al. 2003; Hedge 2013). Brynjarsdóttir and Hedge (2006) developed a digital reading room ergonomics checklist to help to improve the reading room design. Hugine, Guerlin, and Hedge (2012) evaluated an innovative reading room design that incorporated a variety of ergonomic design features and found that it enhanced the comfort and the performance of radiologists.

9.7   GUIDELINES

The vast majority of evidence regarding healthcare settings focuses on patient-facing settings.

Most patient-facing work settings are unique to healthcare—e.g., nurses' stations, surgical suites, digital imaging, examination rooms, and emergency rooms. Some are less so, e.g., admissions, doctor's offices, and waiting rooms. Because these patient-facing settings have direct links to the patient experience as expressed by patient satisfaction measures and revenue, most of the investment in technology and facilities is made in these settings. Beyond the examples and the evidence presented in this chapter, a number of organizations have published guidelines for the design of healthcare workplaces:

- Hospital eTool (OSHA 2015b)
- Guidelines for Nursing Homes: Ergonomics for the Prevention of Musculoskeletal Disorder (OSHA 2009)
- The CHD (Joseph 2006; Joseph and Ulrich 2007; Sadler et al. 2009)
- The National Institute of Buildings—Whole Building Design Guide: Health Care Facilities (Carr 2014)
- Accountable Design for Accountable Care (Watkins et al. 2013)

Like most industries, healthcare organizations also include support or off-stage settings. Many of these workplaces are typical office work settings, e.g., medical records, insurance and billing, and human resources. They include functional space types to support individual, shared, and collaborative work. Support workplaces can benefit from good design and attention to ergonomics. Returns include improved performance, improved worker satisfaction, positive effect on recruiting and retention, and improved quality of patient care. The author has written a white paper reviewing the evidence and providing guidelines (Springer 2007).

9.8   CONCLUSION

Healthcare as an industry is large and complex. A growing body of evidence substantiates the impact of the physical environment on improved quality of patient care and worker health, safety, satisfaction, and performance. Healthcare environments possess significant untapped potential for benefits to organizations willing to apply ergonomics to their environments.

REFERENCES

Administration on Aging. 2012. Older Americans 2012: Key indicators of well-being. Washington, DC: Administration on Aging.
AHRQ (Agency for Healthcare Research and Quality). 2005 October. EPC evidence-based reports. Rockville, MD: Agency for Healthcare Research and Quality. From http://www.ahrq.gov/research/findings/evidence -based-reports/index.html.
American Board of Medical Specialties. 2015. Specialty and subspecialty certificates. From http://www.abms .org/member-boards/specialty-subspecialty-certificates/.

American Dental Association. 2004. An introduction to ergonomics: Risk factors, MSDs, approaches and inter-ventions. Report of the Ergonomics and Disability Support Advisory Committee (EDSAC) to Council on Dental Practice (CDP). From http://rgpdental.com/pdfs/topics_ergonomics_paper(2).pdf.

Anjali, J. 2006. The role of the physical and social environment in promoting health, safety, and effectiveness in the healthcare workplace. *The Center for Health Design.* Issue Paper #3.

Barach, P., and V. Arora. 2014. Hospital alarms and patient safety. *J Am Med Soc.* 312(6):651.

Brynjarsdóttir, H., and A. Hedge. 2006. Cornell digital reading room ergonomics checklist. From http://ergo .human.cornell.edu/AHProjects/Hronn06/CDREC.pdf.

Carr, R. 2014. The National Building Institute: Whole building design guide: Health care facilities. From http://www.wbdg.org/design/health_care.php.

Center for Workforce Studies. 2012 November. *2012 Physician Specialty Data Book.* Washington DC: American Association of Medical Colleges.

Chaudhury, H., A. Mahmood, and M. Valente. 2009. The effect of environmental design on reducing nursing errors and increasing efficiency in acute care settings: A review and analysis of the literature. *Environ Behav.* 41(6):755–86.

CHD (Center for Health Design). 2015. Knowledge repository. From https://www.healthdesign.org/search /articles.

CMS (Centers for Medicare and Medicaid Services). 2011. National health expenditures projections 2011–2021. From http://www.cms.gov/Research-Statistics-data-and-Systems/Statistics-Trends-and-Reports /NationalHealthExpendData/Downloads/Proj2011PDF.pdf.

Committee on Quality of Health Care in America. 2001. *Crossing the Quality Chasm: A New Health System for the 21st Century.* Institute of Medicine. Washington, DC: National Academy Press. From https:// iom.nationalacademies.org/Reports/2001/Crossing-the-Quality-Chasm-A-New-Health-System-for-the -21st-Century.aspx.

Cvach, M. 2012. Monitor alarm fatigue an integrative review. *Biomed Instrum Technol.* 46:268–277.

Enos, L. 2004. Overcoming barriers to implementing ergonomics programs in healthcare: Case studies from the field. From http://www.iienet2.org/uploadedfiles/ergo_community/case_studies/264pres.pdf.

Field, R. 2008. Why is health care regulation so complex? *Pharm and Ther.* 33(10):607–8.

Fulton Suri, J. 2000. Saving lives through design ergonomics in design. *Ergonomics in Design.* 8(3):4–12.

Gerteis M., S. Edgman-Levitan, J. Daley, and T. Delbanco. 1993. *Through the Patient's Eyes.* San Francisco: Jossey-Bass.

Hamrick, C. A., K. Fujishiro, J. Weaver, W. S. Marras, and C. A. Heaney. 2005. Implementation of ergonomic interventions in healthcare: Results from 111 facilities. *Hum Fac Erg Soc P.* 49(14):1365–9.

Health Statistics. 2012. *Health, United States 2011: With Special Features on Socioeconomic Status and Health.* Hyattsville, MD; U.S. Department of Health and Human Services. From http://www.cdc.gov/nchs/data /hus/hus11.pdf.

Hedge, A. 1998. Introduction to ergonomics. In D. Murphy (ed.) *Ergonomics and the Dental Health Profes-sional.* New York: American Public Health Association. 1–23.

Hedge, A. 2005. Best practices for site-wide hospital ergonomics. National Ergonomics Conference & ErgoExpo Las Vegas Dec. 1. From http://ergo.human.cornell.edu/Conferences/NECE05/AH-Best%20Practices%20 for%20Site-Wide%20Hospital%20Ergonomics.pdf.

Hedge, A. 2008a. Cornell healthcare computer cart ergonomic checklist. From http://ergo.human.cornell.edu /Pub/AHquest/CUCompCartEval.pdf.

Hedge, A. 2008b. Cornell healthcare computer wall-station ergonomic checklist. From http://ergo.human .cornell.edu/Pub/AHquest/CUCompWallStationEval.pdf.

Hedge, A. 2013. Evaluating ergonomics risks for digital radiologists. In V. G. Duffy (ed.) *Digital Human Modeling and Applications in Health, Safety, Ergonomics, and Risk Management: Human Body Modeling and Ergonomic.* Fourth International Conference, Las Vegas, NV, July 21–26, 2013. Proceedings Part II:50–8.

Hedge, A. 2014. Cornell healthcare workstation on wheels (wows) ergonomic checklist. From http://ergo .human.cornell.edu/Pub/AHquest/CUWOWEval.pdf.

Hedge, A., T. James, and S. Pavlovic. 2011. Ergonomics concerns and the impact of healthcare IT. *Int J Indust Ergon.* 41(4):345–51.

Heidkamp, M., W. Mabe, and B. DeGraaf. 2012. *The Public Workforce System: Serving Older Job Seekers and the Disability Implications of an Aging Workforce.* New Brunswick, NJ: NTAR Leadership Center, Rutgers University.

Hesselink, G., L. Schoonhoven, P. Barach, A. Spijker, P. Gademan, C. Kalkman, J. Liefers, M. Vernooij-Dassen, and H. Wollersheim. 2012. Improving patient handovers from hospital to primary care: A systematic review. *Ann Intern Med.* 157:417–28.

Hospital Beds. 2013. History of hospital beds. From http://www.hospitalbeds.org.uk/hospital-beds-information /history-of-hospital-beds.html (accessed November 13, 2015).

Hugine, A., S. Guerlain, and A. Hedge. 2012. User evaluation of an innovative digital reading room. *J Dig Imag*. 25(3):337–46.

Institute of Medicine of the National Academies. 2008. *Retooling for an Aging America: Building the Health Care Workforce*. Washington, DC: National Academies Press.

Johnson & Johnson. 2015. The campaign for nursing's future. From http://www.discovernursing.com/explore -specialties#no-filters.

Joseph, A. 2006. The role of the physical and social environment in promoting health, safety, and effectiveness in the healthcare workplace. *The Center for Health Design*. Issue Paper #3.

Joseph, A., and R. Ulrich. 2007. Sound control for improved outcomes in healthcare settings. *The Center for Health Design*. Issue Paper #4.

Joseph S., L. McCane, D. M. Thevenin, and P. Barach. 2008. Examining links between sign-out reporting during shift changeovers and patient management risks. *Risk Anal*. 28(4):1–13.

Laugaland, K., K. Aase, and P. Barach. 2012. Interventions to improve patient safety in transitional care— A review of the evidence. *Work*. 41:2915–24.

Lawler, E. K., A. Hedge, and S. Pavlovic. 2011. Cognitive ergonomics, socio-technical systems, and the impact of healthcare IT. *Int J Indust Ergon*. 41(4):336–44.

Mahmood, A., H. Chaudhury, and M. Valente. 2011. Nurses' perceptions of how physical environment affects medication errors in acute care settings. *Appl Nurs Res*. 24(4):229–37.

Manser, T., and S. Howard. 2010. Healthcare handover and patient safety. *Hum Fac Erg Soc P*. 54(12):947–8.

Ogden C. L., M. D. Carroll, B. K. Kit, and K. M. Flegal. 2014. Prevalence of childhood and adult obesity in the United States: 2011–2012. *J Am Med Assoc*. 311(8):806–14.

Oregon Coalition for Healthcare Ergonomics. 2015. Ergonomics for clinical support & diagnostic services: Dental. From http://hcergo.org/dental.htm.

OSHA (Occupational Safety and Health Administration). 2009. OSHA 3182: Guidelines for nursing homes: Ergonomics for the prevention of musculoskeletal disorders. Revised. From https://www.osha.gov/ergo nomics/guidelines/nursinghome/final_nh_guidelines.html.

OSHA. 2015a. Safe patient handling. From https://www.osha.gov/SLTC/healthcarefacilities/safepatient handling.html.

OSHA. 2015b. Hospital eTool. From https://www.osha.gov/SLTC/etools/hospital/index.html.

Planetree. 2014. Components of the planetree model. http://planetree.org/about-planetree/.

Public Law 111-148. 2010. Patient protection and affordable care act. From http://www.gpo.gov/fdsys/pkg /PLAW-111publ148/content-detail.html.

Puget Sound Chapter of the Human Factors and Ergonomics Society (PSHFES). 2009 Washington State Department of Labor and Industries: Examples of costs and benefits of ergonomics. From http://www .pshfes.org/resources/documents/ergonomics_cost_benefit_case_study_collection.pdf.

Ruess, L., S. C. O'Connor, K. H. Cho, R. Slaughter, F. H. Husain, and A. Hedge. 2003. Carpal tunnel syndrome and cubital tunnel syndrome: Musculoskeletal disorders in four symptomatic radiologist. *Am J Roentgen*. 181(1):37–42.

Sadler B. L., A. Joseph, A. Keller, and B. Rostenberg. 2009. Using evidence-based environmental design to enhance safety and quality. *IHI Innovation Series*. White paper. Cambridge, MA: Institute for Healthcare Improvement. From http://www.IHI.org.

Siegel, E., and B. Reiner. 2002. Radiology reading room design: The next generation. *Appl Radiol*. 31(4):11–16.

Solet, D. J., M. Norvell, G. H. Rutan, and R. M. Frankel. 2005. Lost in translation: Challenges and opportunities in physician-to-physician communication during patient handoffs. *Acad Med*. 80(12):1094–9.

Springer, T. 2007. Ergonomics for healthcare environments. Knoll. From http://www.knoll.com/media/760/617 /healthcare_ergonomics.pdf.

The Henry J. Kaiser Family Foundation. 2010 May 1. Health care costs: A primer. From http://kff.org/health -costs/issue-brief/health-care-costs-a-primer/.

The Joint Commission. 2008. Guiding principles for the development of the hospital of the future. From http:// www.jointcommission.org/guiding_principles_for_the_development_of_the_hospital_of_the_future_/.

The Joint Commission. 2012 November. Improving patient and worker safety: Opportunities for synergy, collaboration and innovation. Oakbrook, Terrace IL. From http://www.jointcommission.org/.

The Joint Commission. 2013 April 8. Sentinel event alert. From https://www.ecri.org/2014hazards.

Trust for America's Health. 2015. Obesity. From http://healthyamericans.org/obesity/.

Ulrich, R., and C. Zimring. 2004. The role of the physical environment in the hospital of the 21st century. *The Center for Health Design*.

Ulrich R. S., C. Zimring, X. Zhu, J. DuBose, H. B. Seo, Y. S. Choi, X. Quan, and A. Joseph. 2008. A review of the research literature on evidence-based healthcare design. *Health Environ Res Design*. 1(3):61–125.

U.S. Federal Trade Commission. 2004 July 23. Improving health care: A dose of competition: A report by the Federal Trade Commission and the Department of Justice. From http://www.ftc.gov/reports/improving -health-care-dose-competition-report-federal-trade-commission-department-justice.

U.S. Department of Labor. 2013a February. *Women in the Labor Force: A Data Book*. Washington, DC: Bureau of Labor Statistics.

U.S. Department of Labor. 2013b December 19. Bureau of Labor Statistics: Employment projections 2012–2022 summary. From http://www.bls.gov/news.release/ecopro.nr0.htm.

USDHHS (U.S. Department of Health and Human Services). 2015a. Laws & regulations. From http://www .hhs.gov/regulations/.

USDHHS. 2015b. Health information privacy. From http://www.hhs.gov/ocr/privacy/hipaa/understanding/index .html.

USDHHS. 2015c. Administration on aging: Aging statistics. From http://www.aoa.gov/Aging_Statistics/.

USDHHS. 2015d. Administration for community living. From http://www.acl.gov/.

Watkins, N., J. Zook, W. Gray, R. Saravay, E. Peavey, T. Gorton, and D. Clarke. 2013. *Accountable Design for Accountable Care*. McGraw Hill Financial Research Foundation + HOK, New York. From http:// mcgraw-hillresearchfoundation.org/2013/03/04/accountable-design-for-accountable-care/.

WHO. 2015 January. Obesity and overweight: Fact sheet No. 311. From http://www.who.int/mediacentre /factsheets/fs311/en/.

Wiseman, S., A. L. Cox, and D. P. Brumby. 2013. Designing devices with the task in mind: Which numbers are really used in hospitals? *Human Factors* 55:61–74.

Wong, M., A. Mabuyi, and B. Gonzalez. 2013 October 16. First national survey of patient-controlled analgesia practices. *The Physician-Patient Alliance for Health & Safety*. From http://ppahs.files.wordpress .com/2013/10/ppahs-sasm-handout.pdf.

10 Ergonomics Design in Control Rooms

Matko Papic

CONTENTS

10.1   INTRODUCTION

This chapter will discuss the application and the impact of ergonomics in control room environments globally and in various industry applications. Today, ergonomics plays a prominent role in the design of mission critical facilities, and its role continues to grow as companies and agencies start to see the impact that it has on the efficiency and the safety of operations. Control Rooms are both operationally complex and technologically sophisticated environments. As such, they are often seen as early adopters of either breakthrough technology or innovative approaches to resolving complex monitoring applications. The human–machine interface (HMI) is continuously evolving, and ergonomics, no doubt, is taking a more prominent role. On the other hand, control rooms often have to deal with legacy equipment and processes that are often decades behind the current technology trends. This is largely due to the complexity of the processes and the ongoing challenge of replacing infrastructure that has been in place for a long time. These and many others operating realities and challenges make control rooms a very interesting environment when viewed from the perspective of ergonomics.

This chapter will provide some background on control room environments and what exactly 24/7 mission critical applications are. It will discuss some of the history and how the technology and environment have evolved over the years to what they are today. Finally, the various aspects of ergonomics and their application in control rooms will be highlighted, including where technology and operational models are heading in the future.

10.2   WHAT IS A CONTROL ROOM?

Control rooms are specifically designed work environments used to monitor mission critical applications. For the most part, control rooms are 24/7 operations that require continuous monitoring and input of information essential to an operation or process. Although not generally accessible or familiar to the general public, control rooms are widely used in private companies, as well as public and government applications. Typical examples of control rooms include air traffic control towers, National Aeronautics and Space Administration's mission control rooms, 911 emergency response centers, electrical generation and transmission centers, traffic management centers, and similar facilities. These are examples of core 24/7 applications. However, as technology, globalization, and worldwide businesses continue to evolve, companies are starting to utilize control rooms in some less obvious applications. Examples of this include global supply chain and servicing centers, banks (transaction monitoring and security), customer support centers (large technology companies), security and monitoring (home security companies, internet providers), and other similar operations.

There are several factors that make control rooms unique environments.

First is the 24/7 nature of the operation. Depending on the application, the team that is monitoring a mission critical operation is required to continuously focus on the task at hand. Unlike an office environment, control room operators typically cannot leave their position during a given shift without a specific procedure in place to avoid compromising the mission or missing critical information.

Second, control rooms incorporate large amounts of technology that require significant interaction from the operator. While a typical office environment would require the employee to interact with one or two display screens, a control room operator may be required to monitor up to 16 different monitors.

Lastly, since each mission has its own unique workflows, technology, mission objectives, space constraints, and people, each control room is a unique environment and requires significant planning and design considerations to ensure an optimal HMI design, which will ultimately result in a safe and efficient operation.

10.3 HISTORY OF ERGONOMICS IN CONTROL ROOMS

Control rooms started to become commonplace with industrialization and started to appear in limited industries in the early 1900s. As manufacturing, power generation, and use of telephones became widespread, the need for 24/7 operations began to arise. Early on, the function of the control room was the primary focus of the control room design. The reason was twofold; there was no real understanding of the impact of ergonomics and the other factors on the operation (operators), and the technology at the time was what could be defined as single interface. This refers to the fact that most operations were directly tied to a single input by an operator. For example, if an operator was monitoring the steam flow, the regulation of each flow control valve was likely a dedicated interface (button) on a control panel. Technology such as monitors and multiple screen-based interfaces did not exist. As a result, each critical control feature had to be placed somewhere in the control room as shown in Figure 10.1. With so many inputs, the design focus was placed more on fitting everything in the physical space rather than considering the implications this had for the operator.

As technology started to evolve, visual displays became more commonplace in control rooms. This created not only new efficiencies, but also challenges when it came to the quality of the overall environment. The cathode ray tube (CRT)-based displays of the time were large, nonadjustable and, compared to today, had poor resolution and were susceptible to significant screen glare and flicker. This, combined with the widespread use of direct fluorescent lighting, in some aspects, created a more challenging operator environment than before.

The real shift to the application of ergonomics in control rooms came in the 1990s. The introduction of flat screen technology created a whole new set of opportunities when it came to display position and adjustment. In parallel, the use of keyboards and the development of interactive graphical user interfaces allowed for the consolidation of the control functions into a single display screen, controlled by a keyboard and a mouse. Lastly, linear drive mechanisms were introduced to the market, which ultimately allowed for the adoption of sit–stand technology in control rooms.

FIGURE 10.1 Single-interface control configuration. (Courtesy of Acroterion, http://commons.wikimedia .org/wiki/File:NS_Savannah_control_room_MD1.jpg#/media/File:NS_Savannah_control_room_MD1.jpg.)

At the same time the technology transition was taking place, there was also a shift in thinking toward ergonomics and overall workplace safety and productivity. Control room designers and ergonomists were starting to show tangible results of how proper ergonomic considerations in the design process were creating a better overall work environment, productivity, and safety. A catch phrase soon emerged in the late 1990s: "Good ergonomics equals good economics." In a 1996 presidential address to the Human Factors and Ergonomics Society's 40th Annual Meeting, Hendrick presented a paper titled "The Ergonomics of Economics is the Economics of Ergonomics" (1996). Citing several examples, Dr. Hendrick emphasized the emerging role of ergonomics in multiple industries and organizations. The main premise of the argument was that the proper consideration for ergonomics greatly contributes to the decrease of repetitive strain injuries, which, in turn, promotes performance and productivity, reduces errors, and greatly reduces injury-based claims and costs for organizations. At the same time, it sends a positive message to the employees regarding the importance of their overall health, safety, and job satisfaction.

This thought process started to gradually proliferate through various industries and applications, including control rooms. One of the best early examples of ergonomics playing a pivotal role in the transformational thinking and design approach was in the 911 control room environment; 911 centers are, to say the least, intense and high-stress operations. As a result, the health and safety of the operators, along with the public and the first responders, is always a primary concern. For 911 operators, their environment presents multiple challenges when it comes to human factors. In one sense, their tasks are highly repetitive. Incoming 911 calls are answered and processed by the operator, who will in turn engage the appropriate response teams, such as fire, police, or emergency medical services. In larger centers, there are call taker positions that answer incoming calls and then distribute them to the appropriate dispatcher who will engage the required frontline responders. This process can be repeated hundreds of times for a single operator in a single shift. Since performing these tasks requires interaction with both screens and dispatching equipment, such as radios and keyboards, reach zones and visual sightlines are very important. In addition, due to the intense nature of the tasks, having the opportunity to proactively change posture becomes challenging. In the 1990s, 911 centers were typically designed with static workstations, whereby CRT displays were freestanding on the desktop. Communication equipment and other accessories were typically fixed in place on the desktop and under the monitors as shown in Figure 10.2.

FIGURE 10.2 Fixed CRT-based equipment configuration. (Courtesy of Seattle Municipal Archives, Item 114076, Seattle, WA.)

As a result, none of the desired ergonomic features that are commonplace today were attainable or even considered. During that time, the instances of work-related repetitive strain injuries were on the rise. Employee demand, along with the public nature of 911 centers, spurred municipalities to start looking for ways to reduce these injuries. At the turn of the century, ergonomists began to get involved with addressing some of these issues in a significant way. Their research and recommendations started to find their way into specifications, which in turn drove architects, control room designers, and console manufacturers to create designs and product solutions geared toward addressing many of these key issues. In a very short time, design specifications started to include features such as full sit–stand capability, full adjustment of monitor arms and other accessories, streamlined user input controls, white noise control, and personal environmental units. In fact, today it is highly unlikely that a specification for the procurement of a new 911 center would be issued without these features being mandatory.

Figure 10.3 shows the Polk County Sherriff Emergency communication center. It utilizes a very open concept design with full adjustability of both the operator work surfaces (heights) and electronic adjustment of the displays (height, tilt, focal depth). This, combined with an innovative room design (layout, lighting, common audiovisual [AV] displays), creates an effective, collaborative, and satisfying work environment. It is an example of an operator-focused design, whereby the management and the design team considered operator comfort, and the associated ergonomic considerations, as a vital element of the design process.

What helps facilitate an innovative room design today is the continued evolution of various technologies. This, in turn, helps better reflect the operational needs of the environment. Lighting is an example of whereby lighting companies are investing in significant, ongoing research and development. With the evolution of LED-based technology, products are now available that allow for the consideration of indirect lighting fixtures and use of lighting color tones to address fatigue and circadian rhythms to name a few. The flexibility of control and dimming features allow designers to create not only specific lighting zones, but also overall room lighting modes. This could mean a lower lighting level during normal operations and a full illumination mode during an upset scenario.

Similarly, AV systems continue to evolve whereby common information can be effectively shared either thorough common viewing structures (large seamless display walls) or through localized (group) displays using large format displays to bring situational awareness to a group of operators and/or supervisors. The ability to toggle the desired information depending on the actual

FIGURE 10.3 Fully adjustable console design with sit–stand and electronic display positioning. (Courtesy of Evans Archive Images, Polk County Sheriff Emergency Communication Center, Polk County, Florida.)

operational scenario allows for more flexibility in the room layout and also provides more flexibility to adapt to changing technology or operational needs.

However, not all industries with control rooms have made such a significant transition. Control rooms by their nature are task-oriented operations, meaning that the incorporation of ergonomic considerations into the design is a balancing act between a multitude of operational constraints and technology. The good news is, ergonomics is firmly on the map in terms of being at the very least a consideration, if not a driving factor, in the design of these facilities.

10.4   FACTORS TO CONSIDER IN CONTROL ROOM DESIGN

What makes control rooms such an interesting environment is the multitude of different operational needs, the amount of technology (HMI), and the reactive nature of the environment. Control room operators make real-time decisions that have substantial implications on organizations and processes. As a result, ergonomics is just one of the factors that influences the design of a control room.

As shown in Figure 10.4, in order to adequately design a control room, all factors (and key stakeholders) must be considered. In general, not only is the control room itself a complex operation, but so is the process or the operation that it monitors. For example, a refinery control room is the heart of the operation in a complex and high-risk environment. The input and the output of information within the control room require not only a significant consideration of technology and equipment, but also critical infrastructure such as power distribution and process control system accommodations and similar infrastructures.

From a building infrastructure, the room is typically blast proof in order to ensure the safety and the continuity of the operations during a major incident. The allocated space for the control room is often limited due to either existing building constraints or limitations based on the overall plant layout. As a result of these factors, control rooms have unique needs that are typically not required in a normal office environment.

- Hardware—Each operator position in a control room must house a significant amount of equipment during a normal operation. This includes processors, keyboard-video-mouse switches, multiple keyboards, communication equipment, displays, alarm management systems, and similar equipment. Although the processors are now commonly remotely

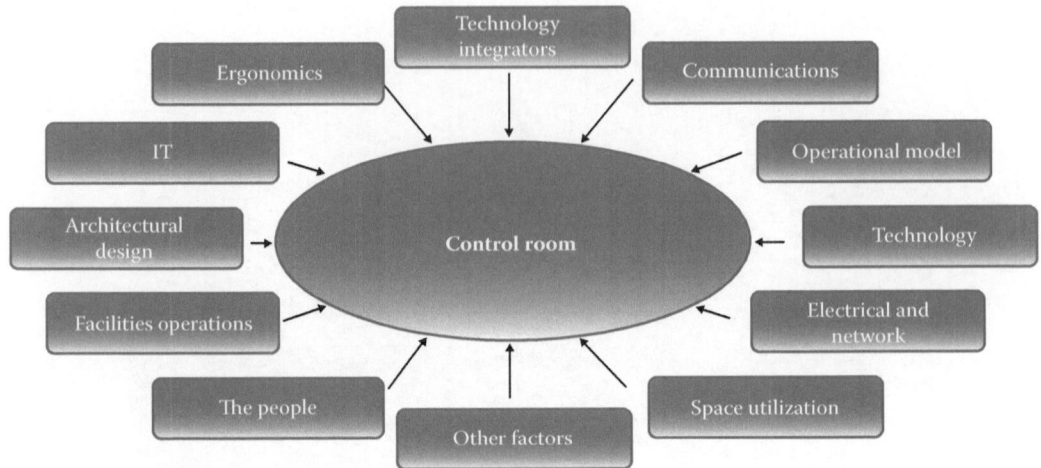

FIGURE 10.4 Factors that influence the design of a control room. (Courtesy of Evans Archive Images, Calgary, AB, Canada.)

stored (in an adjacent server room), a lot of control rooms still require local processors at the operator position.

- Electrical and data distribution—Due to the critical nature of the operation, the power distribution within a control room utilizes multiple uninterrupted power supplies to run critical equipment. In a typical scenario, a control room will have at a minimum three dedicated power circuits running to each operator position. Similarly, the data distribution is an important design factor, as it requires adequate consideration given the amount of information that needs to be transferred between the operator position and the rest of the process being monitored. To add to this, facilities that deal with intelligence tasks where the sensitivity of information is critical will often require dedicated data distribution channels and secure hardware housing. This adds to the complexity of the design while posing a layer of constraints that must be taken into account.

- Architectural/design standards—These standards have various requirements depending on the specific operation. For example, many control rooms do not have any windows, particularly if there are blast-proof requirements within the operation. In other cases, the operation itself may provide an architectural or a design constraint. A typical example of this is an air traffic control tower (ATCT). The core operational requirement for an ATCT is to provide a visual interface between the air traffic controller and the runway (or an incoming aircraft). As a result, ATCTs are built in a carefully selected location within an airport. The height of the ATCT is typically determined based on the overall airport location, the positioning of runways, the location of gate areas, etc. Finally, the number of controllers in a tower is determined by current and future operational needs of the airport. In terms of physical constraints, an ATCT is a structure that requires a long, low profile that supports the cab itself at the required height. As a result, the cab often has significant space constraints that need to be considered in the design process.

These are just a few examples of the factors that influence the operational realities of control room environments. The key takeaway is that designing a control room is about flexibility and balance.

As Figure 10.5 illustrates, the successful design of a control room is largely predicated on achieving the optimum balance between all the different factors. Placing too much emphasis on any one factor can greatly deteriorate the overall efficiency of the space. This is evident when looking at the

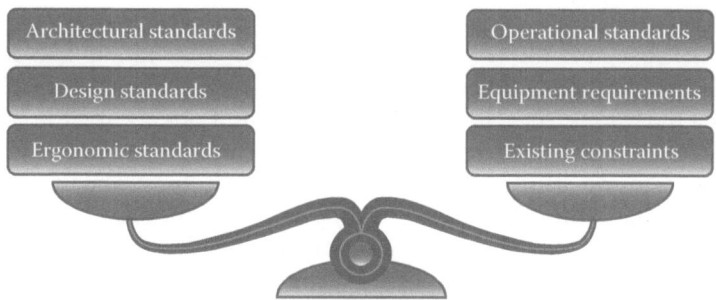

FIGURE 10.5 Balance and consideration of multiple factors influencing the control room design. (Courtesy of Evans Archive Images, Calgary, AB, Canada.)

history of control rooms, whereby placing a heavy emphasis on equipment and operational requirements adversely tipped the scale against ergonomics and other design considerations. Today, ergonomic considerations are firmly established as one of the key factors influencing design.

Balance is one of the biggest challenges that ergonomists face when it comes to control rooms. Ergonomic standards are often in direct conflict with operational and equipment needs. While most standards are written assuming a clean slate when it comes to creating the immediate operator environment, this is not the case in control rooms.

For example, say that a control room is an existing operation requiring a remodel after 25 years of operation. The control room is in an existing building that cannot be modified in terms of footprint, so the space is fairly limited. For the five operator positions that are needed, each operator needs to monitor four screens, and each position needs to have three central processing units (CPUs) mounted at the base of the console. The operators have used, up until now, static consoles at a height of 762 mm to view the CRT monitors.

In speaking with the operators, the first recommendation is to use a sit–stand console as opposed to a fixed-height desk.

The designers take all of the cabling and equipment needs and incorporate this into a console design as shown in Figure 10.6.

The base frame of the console requires approximately 75 mm for lower and upper cable raceways (each) as well as 458 mm of height clearance for a CPU. In addition, the computer

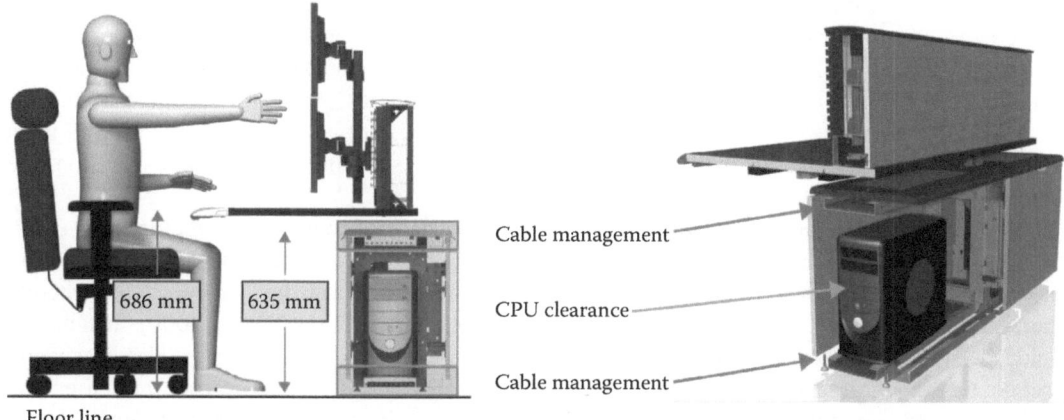

Cable management

CPU clearance

Cable management

FIGURE 10.6 Space constraints for equipment integration. (Courtesy of Evans Archive Images, Calgary, AB, Canada.)

equipment manufacturer specifies that spaces are required above and below the processor for adequate airflow. This makes the base cavity height at 635 mm. Taking into account the safety clearance and adding the work surface on height-adjustable columns (25 mm thick); this brings the total height of the work surface to 686 mm off the floor. Looking at the ergonomic standards for height-adjustable workstations, you determined that the work surface must be lowered to a height of 610 mm. This, of course, is not possible given the constraints mentioned earlier. So now the question is what is the compromise? On the one hand, the customer wants to follow your advice and create a better environment for the operators. On the other hand, the equipment and space constraints are dictating the minimum footprint and base cavity height. As an ergonomist, the only way that you can fall within the guidelines of the existing standards is to recommend a static work surface, in which case the recommended work surface height is 725 mm. This would satisfy both the standards and the equipment needs. However, this would leave the operator with a fixed-height console solution. The compromise approach would look at combining both features and standards. For a static work surface, the height of 725 mm assumes that the 95th percentile male can sit and still be comfortable in a working position. The 5th percentile female, however, would have to utilize an adjustable chair and a foot rest in order to be comfortably seated on the fixed height. Recall, in the example used, that the minimum height that the work surface can be lowered to is 686 mm. This would mean that if a height-adjustable chair and a footrest were used for shorter operators, all could still benefit from the sit–stand features of the console position.

Although this is a simple example, there are numerous other scenarios that may arise during the design process. Other examples include avoiding the visual sightline constraints of double-stacked monitors while viewing a common video display wall, providing the adequate level of acoustic isolation for operators versus allowing for optimal communication between positions, and balancing the overall room lighting levels versus the individual position lighting to name a few. In order to achieve this, the ergonomic requirements must be considered and communicated throughout the design process. A control room is often the final step in designing a complex facility or operation. When designing a new facility (operation), the amount of investment and time dedication to the core areas of the process is significant and typically goes through a long design and construction process. With so much time and capital invested in the core operational areas, the control room itself often gets temporarily pushed aside to focus on the rest of the project. While the control room is typically considered a placeholder in the overall design process, there is no specific focus on the detailed design of the space until much later in the design process. Once the focus shifts back to the control room design, it is often already constrained by the design decisions for other areas of the overall facility. Some examples of the constraints include inadequate space planning, no consideration for support functions (engineering rooms, information technology (IT) support, equipment rooms), heating, ventilating, and air-conditioning and structural limitations, accessibility, and similar constraints. All of these constraints have an impact on the quality of the control room design. There is often no way to get around these constraints, since the overall facility design may already be completed or the construction may have already begun.

In response to this, companies are starting to identify control room design requirements early on in the planning process to ensure that all the requirements of the control room are captured and incorporated into the overall facility planning. The process has also been formalized, whereby there are now control room design standards available, such as ISO 11064 (2000), which help guide designers and key stakeholders through the control room design process. Unlike specific ergonomic standards, ISO 11064 is a much broader guideline that focuses on the recommended steps and considerations that should be taken into account during the design process. However, the design standard is a broad guideline that attempts to identify the general steps and is intended to be used across different industry applications. It is up to the specific design teams to ensure that they are adequately evaluating their specific application and applying the appropriate design standards (ergonomic or other) that may apply.

10.5 THE INSIDE-OUT APPROACH

The early consideration of ergonomics in the design process and the introduction of control room design standards are now allowing for the control room to be designed by the operator and for the operator. This is important since the one-size-fits-all approach is not effective in a control room environment for a long term. A typical life cycle of a control room is anywhere from 15 to 25 years. In this time, the control room may go through several technology upgrades, but the core operation (workstations, placement, and similar operations) will remain the same. As a result, it is important to tailor the workstation and the overall room layout around the needs of the operator as opposed to having them conform to predesigned, off-the-shelf solutions. Hence, ergonomists and control room designers are now using the inside-out approach when it comes to control room design.

As Figure 10.7 illustrates, this approach starts with the operator. Once the specific equipment needs are identified through a discovery process with other stakeholders (IT, operations, control system integrators, etc.), they are placed in the ideal functional position in relation to the operator (step 1). At this point, the considerations such as optimum sightline viewing cones, primary versus secondary displays, reach zones, and similar considerations are incorporated in the design. Once this is complete, the workstation is then designed around the ideal equipment placement, and the rest of the hardware, such as personal computer cable management, environmental controls, etc., is incorporated in the position design (step 2). Next, the functional features of the position are designed. In this case, it is a sit–stand workstation with an independently adjustable keyboard platform (step 3). From here, the rest of the positions are placed in the room based on the interaction and communication needs of the specific control room. This includes concentration functional areas, location of supervisors, and similar needs (step 4). This process continues to expand to other areas of the operation such as examining operational adjacencies and support functions.

This approach is considered a luxury in the design process, particularly in situations where the control room design is done late in the process or there are preexisting site constraints. However,

FIGURE 10.7 The inside-out operator-focused design approach. (Courtesy of Evans Archive Images, Calgary, AB, Canada.)

more and more operations are adopting this approach. Often, this is because the benefits and the value proposition of early design consideration are effectively communicated by not only the ergonomists and the control room designers, but also the other stakeholders involved in the process.

10.6   AREAS OF ERGONOMICS IN CONTROL ROOMS

The International Ergonomics Association categorizes ergonomics as being broken into three primary domains: physical, cognitive, and organizational (Figure 10.8).

In a lot of industries and work environments, one or two domains are the primary focus for ergonomists. For example, if evaluating an assembly operation process, the main focus is on the physical tasks that the assembly worker performs. Things like reach zones, repetitive motion, and frequency of tasks are all considered. The physical environment (heat, light, ventilation, noise, and vibration) is also important. Software designers, on the other hand, will rely more on the cognitive ergonomics in order to design the most effective user interface, say on a smartphone touch screen. Organizational ergonomics often takes a broad look at the social aspects of an operation, such as coordinated teamwork, and often complements (or are used as a guideline) for more detailed processes covered by either cognitive or physical ergonomics.

When it comes to control rooms, all three domains are equally present. A good comparison would be the cockpit of a large aircraft. In larger aircraft, there can be up to three crew members in the cockpit as well as support functions such as flight attendants and service directors. When we think of a flight operation from an ergonomic standpoint, it starts with the individual positions within the cockpit. The physical placement of the flight controls and the engineering/navigation functions come to mind, and these need to be operable in turbulence or during the day or night. Then there is the extensive application of cognitive ergonomics in terms of monitoring a multitude of communication and alarm systems. Finally, there is the focus on the organizational aspect of crew interaction, both in terms of communication and specific task assignments during the operation.

Control rooms are structured in a similar fashion. They are, in effect, their own microorganizations that operate within a larger entity. Although control room operators may not be operating an aircraft, they are still responsible for monitoring and reacting to critical processes, which can have catastrophic consequences if not properly managed. Adequately considering all three domains in the control room environment is probably one of the biggest areas of opportunity for improvement in the industry today. While some industries do a very good job considering all of these aspects, others place a heavy emphasis on the physical ergonomics while not focusing on the cognitive and organizational aspects. Similarly, while some organizations do have formal organizational and cognitive guidelines, they are often developed as a generic approach and are not looked from the standpoint of a specific site-operation point of view. This in itself can result in constraints or, even worse, risks to the operation.

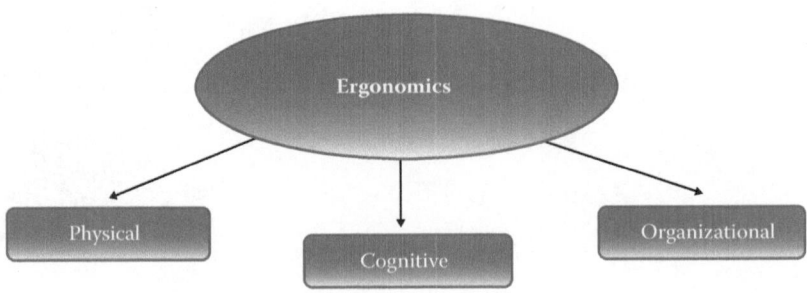

FIGURE 10.8   Categorization of ergonomics. (Courtesy of International Ergonomics Association, http://www.iea.cc/whats/.)

10.7 PHYSICAL ERGONOMICS

When we look at control rooms, most of us associate the environment as closely resembling that of a typical office setting. In both applications, an operator sits at a desk and interacts with technology, typically through a keyboard, a mouse, and a display screen. The difference in a control room is the scale and the focus on technology interaction.

Another comparison that can be made is to a manufacturing environment, whereby continuous repetitive motion is present, like a production assembly line. Ergonomists will look at these applications and will attempt to minimize the amount of repetitive stress injuries by examining the motions of the operator, looking at reach zones, incorporating layouts that will allow for a change of posture, and similar attempts. A control room equivalent of this would be a call taker in a 911 center. The operators will perform the same motions, like answering a phone call or activating a radio system, hundreds of time per shift. What is unique about the control room environment is the scale and the variability of these tasks. There is not only a greater amount of technology present, but also the functions between different roles within a control room that varies as well. A shift supervisor or a geographical area manager may have a completely different set of tasks than an incoming call talker or task dispatcher.

Figure 10.9 shows a typical process control application and equipment configuration. At first glance, it appears that the amount of equipment placed on the console is simply too much for one operator to monitor and is evidently in contradiction with ergonomic standards. However, this is not the case. It is an example of the compromise and the balance between operational needs and ergonomic design. The three lower central monitors are considered as the primary viewing screens. This is the area that the operator spends the most time focusing on. As a result, they are placed in the appropriate viewing cone. The upper monitors are used to monitor infrequent process alarms and system notifications. They are not assumed to be continuous viewing monitors and, as such, are placed in the secondary viewing area. The monitors that are to the left or the right of the three central screens are separate processes that require interaction when needed. They do not require continuous viewing and are often triggered by specific alarms or task requirements. If an operator is required to interact with these monitors, they will simply slide over to that area of the work surface and will, in effect, establish a new set of primary and secondary screens. The position is designed to allow this motion without obstruction while the operator remains seated by rolling their chair along the console work surface. Another key feature of this design is that the entire unit, including all the display screens and keyboards, is height adjustable thus allowing the operator to change posture during the operation. This flexibility, coupled with the fact that all displays and task lights are mounted on adjustable arms, which allow for forward/back, tilt, or side-to-side adjustments, give the operator(s) the flexibility to change the equipment placement.

FIGURE 10.9 Typical process control equipment configuration. (Courtesy of Evans Archive Images, Calgary, AB, Canada.)

There are a number of physical ergonomic standards that are being used in the control room industry. In general, they are the same standards that are used in office environments such as Business and Institutional Furniture Industry (BIFMA) G1 (2013) or ISO 9241-400 (2007). Outside of North America, different standards are used, depending on the specific anthropometric data for the region. In most cases, the standards are used as a guideline given the amount of other operational requirements. As a result, many companies that have a large number of control rooms will often develop their own ergonomic standards, which are put together by either in-house ergonomists or specialized consultants.

In addition, physical environments are also affected by temperature, lighting, acoustics, and similar factors. In general, most of these guidelines are covered as part of the ISO 11064 (2000) control room design standard. They describe the general recommended requirements. However, more detailed designs are often performed for specific applications in collaboration with subject matter experts in the respective areas.

These company specifications often combine multiple standards and the operational needs of a company. This often provides a very clear design direction and avoids the issue of conflicting priorities during the design process.

Given all of the different constraints that exist in this environment, a good general approach to physical ergonomics is to try and incorporate as much adjustability in the design as possible. While it may not be possible to avoid some of the equipment and processes needed for the operation, giving the operator the ability to adjust their environment will go a long way in ensuring comfort and safety.

10.8 COGNITIVE ERGONOMICS

Cognitive ergonomics plays a key role in the control room environment. When discussing physical ergonomics, the importance of creating the correct physical environment for the operator to perform his or her tasks was discussed. Those physical tasks are in most cases triggered by some cognitive-based decision (or corrective action). Control room operators monitor multiple processes and must, at the appropriate time, react to the changes in a standard process or be able to interpret variables to complete a common task. For example, a control room operator monitoring the operational process of a nuclear power generation facility must continuously monitor the temperature of the system in order to ensure that there is no overheating or pressure build up. While they may be simply monitoring the majority of the time, at a given point, they may need to react to an alarm or an indicator and take a corrective action. In this case, fatigue, alertness, and ability to recognize alarms are key abilities. On the opposite end of the spectrum, is an air traffic controller in a terminal radar approach facility (TRACON) monitoring air traffic at 30,000 ft. Unlike ATCT controllers, they do not have a visual of the aircraft. Instead, they depend on a screen to monitor multiple aircraft travelling at different altitudes, speeds, and directions while looking at two-dimensional images. They continuously need to process information, make ongoing, real-time decisions, and then communicate those to multiple contacts (aircraft).

The focus on the cognitive ergonomics is very strong in a number of 24/7 applications, typically those where there is a significant operational risk present or where there exists a considerable opportunity for loss of life. These industries have invested a significant amount of effort over the last several decades in the research and the development of solutions to ensure the cognitive aspects of their operations. One of the most common examples of cognitive ergonomics in control rooms is alarm management. It is present in almost any process control-related environment: electrical generation and transmission, refineries, pipeline control rooms, and similar environments. Alarm management is a particularly challenging aspect of process control, because the quality of the response to an automatically triggered alarm is only as good as the individual's ability to recognize and interpret it. In a control room, operators must have the ability to monitor multiple processes and be able to identify and disassociate what each visual or audible alarm means as well as the type of action that is required to correct it. This challenge is compounded by two common operational scenarios that occur: feast or famine. The *feast* refers to the alarms that occur often and are, ultimately, shut off or ignored by the operator because they are annoying. The *famine* refers to a very infrequent alarm that is triggered by a significant event in the system that requires immediate and critically important action. Both scenarios can have lasting consequences if ignored or improperly managed. Lastly, there is the scenario of too many alarms. This may be unavoidable in some control rooms, in which case the objective is to properly categorize and distinguish the various types of alarms, so the interpretations is second nature to the operator.

This is an aspect of control rooms where ergonomics and ergonomists have played a significant role for a number of decades. And this is not just end users. Companies that develop process control systems and alarm management devices have invested significant resources in better understanding how operators identify, process, and react to alarms. In some industries, and over time, end users, industry suppliers, and regulators have formed more formal processes and regulations to systematically monitor the development and the training in terms of alarm management and other aspects of control room operations. A good example of this is the pipeline industry. The American Petroleum Institute (API) (2015a) has developed a guideline for both pipeline control room management (API 2015b) and pipeline supervisory control and data acquisition alarm management (API 2007). These standards have been implemented by both end users in control rooms and industry suppliers of alarm systems. The auditing of the operations, which is based on the same standards, is performed by the U.S. Department of Transportation. This cohesive effort helps elevate both the quality of operations and the safety standards within the industry.

Not all control room applications have evolved to such a level of standards and formal review processes. In these industries, ergonomists and designers continue to work on educating the end users to the benefits of these approaches and how it would impact their specific operation.

10.9 ORGANIZATIONAL ERGONOMICS

Organizational ergonomics is the overlay structure that holds the operation of a control room together. Unfortunately, it is also one of the least applied processes when it comes to the design of control rooms. As mentioned earlier, most organizations that have control rooms as part of their process have, for the most part, well-defined organizational structures and processes. However, the control room is often its own ecosystem that requires a separate organizational structure. This is particularly true of, and exponentially important in, facilities that have multiple control rooms and/or support functions. A common oversight that end users make during the design process is not considering the various functional areas as a common team operating under an expanded, common physical area. While they focus on designing the needs for individual control rooms (or separate functions within the same control room), they fail to consider the elements required for the adequate interaction between these functions. As these functional areas continue down separate paths during the course of the design process, each path makes certain assumptions that may be in direct conflict with the other. At the point in the process where they are looked at together, it is often too late to make significant changes. The result can then be that there is a very well-designed central control room or a functional area within a consolidated room, but with poorly designed support functions such as engineering, break rooms, security access, or server rooms. This can also affect the traffic flows within the space. The incorrect placement of functional areas can often result in the congestion of areas that can pose a distraction or, in extreme cases, a risk to the operation (for example, during emergencies).

There are, of course, companies that do it right and make this planning part of the design process. Processes and guidelines such as the ISO 11064 (2000) help provide not only a template, but also a validation that organizational ergonomics are being considered in the design process.

A good example of incorporating the elements of organizational ergonomics is the recent design approach for ATCTs within the Federal Aviation Administration (FAA).

ATCTs are very complex working environments, and there are numerous considerations that need to be taken into account to achieve a safe and efficient operation. Aside from the direct interaction of controllers with the various equipment, there are challenges such as visual sightlines (look-up/

look down angles to the airspace and runway), visual sightlines across the tower cab, communication between controllers/supervisors, maintenance access with minimum disruption, and operational distractions (visual, acoustical, traffic flow). Adding to the challenge is the sheer volume of aircraft that need to be managed. For example, at O'Hare International Airport, Chicago, there were 74,017 aircraft operations in the month of March of 2015 (Chicago Department of Aviation 2015). This means that, on average, there is an arriving or a departing aircraft approximately every 40 s.

The design of the ATCT had not changed for the last 60 years. While there have been technology changes along the way, the configuration and the basic operation of the tower remained unchanged. The FAA was facing several challenges. The most important were the increased amount of air traffic and the resulting expansion of airport runways and volume in general. This required more controller positions within the ATCT environment. The rate of the technology change continues to put pressure on the current controller position structures, as they were not designed for flexible technology changes, but rather for fixed mount, analog equipment. Making any changes to the technology was very costly and highly disruptive. These existing fixed mount technology integration solutions did not offer the controllers any flexibility for equipment adjustment.

To overcome this challenge, the FAA assembled a broad internal team and worked with various industry experts to find a solution. But rather than simply looking to resolve the main issue, which was increasing the available space within the tower cab, they took a comprehensive look at how the solution could address the multiple aspects of the ATCT operation. They also involved all the internal stakeholders, which in turn provided crucial feedback about their aspects of the operation. This included air traffic controllers, supervisors, engineers, technical operator, managers, and other members of the organization. The result was substantial. The FAA had created a new solution that addressed the main objective; it significantly reduced the footprint of the tower consoles and therefore created expansion capacity. As a result, the existing ATCT infrastructure could be used rather than building a new one. But the solution went beyond that. Through the design process, the FAA was able to address and greatly optimize several key operational issues such as ease of deployment, quick technology replacement and reconfiguration, and parts commonality/flexibility as well as national deployment process development.

Figure 10.10 shows the recently renovated ATCT at Midway Airport. The interesting aspect of the design from a controller standpoint is that it allows for the configuration of operational positions based on the needs of each controller or function.

FIGURE 10.10   Modernized ATCT at Midway Airport. (Courtesy of FAA, Washington, DC.)

As shown in the image, one controller is standing (left), one is in a seated position (middle), and one is sitting in a high chair. The base structure is designed to be low off the floor and any equipment that is nonessential is mounted low to the ground. The main piece of equipment, the flight strip trays and the communication system, are mounted on a height-adjustable mechanism, thus allowing the individual controller to accordingly adjust the height. Unlike other control room applications, it is just this single piece of equipment that is adjustable, rather than the entire work surface. The reason is that this minimizes the visual obstruction for other controllers who may need to look across the cab to look at a runway. The low profile of the console and the equipment mounting allows for the controller to move significantly closer to the cab glass resulting in an improved look-up/look-down angle to the runway.

This is an example of how considering organizational ergonomics can greatly impact the quality of the control room environment.

10.10   NORMAL VERSUS UPSET SCENARIOS

One of the most critical considerations in a control room environment is how the environment will perform in an upset scenario. In most control rooms, the objective is to monitor an operation. Under normal circumstances, the design addresses issues such as operator fatigue, repetitive tasks, and alarm management. The question that must be asked is how the control room will respond to an emergency situation. In an electrical transmission control room, this may be a cascading power failure, similar to the blackout that affected the northeast of the United States and Canada in 2003. In a refinery, it may be a major explosion. One of the most well-known examples of an upset scenario is the events of 9/11 and the impact they had on control rooms of various types.

In an upset scenario, the entire dynamics and the operation of the control room changes. This includes everything from operator inputs, communication, traffic flows, amount of people in a control room, and similar factors. It is important to consider all these possible scenarios during the planning and design phase. From an ergonomics standpoint, this can impact the physical, cognitive, and organizational considerations at various levels.

One example of a changing control room environment is the events that took place at the FAA system command center (ATCSCC), in Herndon, Virginia, (Figure 10.11) on the morning of September 11, 2001.

FIGURE 10.11   FAA system command center in Herdon, Virginia. (Courtesy of NYC Aviation, http://www .nycaviation.com/2014/02/waking-issues-pilot-fatigue-part-two/#.VVDFo_lVhBd.)

During normal operations, the ATCSCC manages the national air space (NAS) for the United States, which is the busiest, largest, and most complex air space in the world. The center communicates with various FAA operational partners such as ATCTs, TRACON facilities, and air route traffic control centers. It also communicates with airline partners across the country. As part of its normal operations, the ATCSCC manages the NAS ensuring the safety and the operational efficiency of commercial, military, and general aviation aircraft. In terms of numbers, almost 15,000 air traffic controllers manage approximately 50,000 flights a day (FAA 2011). As part of the normal operations, the ATCSCC manages events such as severe weather, runway closures, or equipment outages.

On the morning of September 11, as the gravity of the situation started to become clear to the leadership within the FAA, the decision was made to clear the U.S. airspace. The sequence of events was described by Levin, Adams, and Morrison (2002). At 9:45 a.m., the order was issued to land 3949 aircraft. Within the hour, all the aircraft had been landed, and the air space was clear. More importantly, all planes were safely landed. Undoubtedly, the dynamics of the ATCSCC drastically changed on that day. The control room underwent an operational transformation to respond to the emergency. This included different individual tasks, different internal/external communication, increased number of people within the control room space, a significantly elevated stress level, and changing traffic patterns within the room. While this is an extreme example, it illustrates the importance of considering upset scenarios within the control room environment.

Although it is a rare occurrence, it is, in a way, exactly what a control room is built for. Some of the questions that should be asked are how does the operation change during an upset scenario, who should have access to the room, should there be additional space or spare unit provisions, what are the specific operational tasks expected, and what are the anticipated traffic patterns within the rooms or the adjacent rooms. This will, of course, vary by the operation, but it is a key aspect of the design that should not be overlooked.

In the case of Leon County in Tallahassee, Florida, shown in Figure 10.12, the thought process of considering both normal and upset operations was present from the very early design stage. The facility was designed in such a way that it consolidated the county traffic management center (TMC),

FIGURE 10.12 Leon County Consolidated dispatch center. (Courtesy of Evans Archive Images, Calgary, AB, Canada.)

the 911 center, and the emergency operations center (EOC). Besides the benefit of having a consolidated facility, whereby operational efficiencies were achieved, it also created a common space for effectively handling emergencies. In Florida, a common scenario would be a hurricane. Having the TMC, the EOC, and the 911 control rooms under one roof allows for the most efficient coordination of response. As Figure 10.12 shows, the TMC is positioned on the main level. Above it, to the left and the right, are the 911 and EOC rooms. This allows for direct visual sightlines and communication between the three spaces. In this case, the architectural design was able to capture the operational requirements in the base design of the building, which resulted in an efficient and collaborative environment for various agencies and stakeholders. This design approach is now being considered by other municipalities in terms of facility planning and design.

10.11   THE FUTURE OF ERGONOMICS IN CONTROL ROOMS

Control rooms will continue to push the limits of technology into the future, and the ergonomics of these spaces will be increasingly impacted. There is no doubt that ergonomics is firmly established as a key element of the design process. However, as the envelope is being pushed, so will the need to adapt and modify existing thoughts, processes, and standards when it comes to the human factors and the HMI.

Control room operators are no longer being viewed in the context of simply identifying issues to a given process and taking the appropriate corrective action. Operators are now being viewed as key contributors to the overall operation (or business) success, and their inputs and actions can have a direct impact on the efficiency, the cost structure, the safety, and the effectiveness of an operation. To facilitate this, organizations are placing an increasing emphasis on creating a work environment that will not only address issues such as fatigue, repetitive stress injuries, and productivity but also create a peripheral environment, whereby the operator will be in a position to provide a greater impact to the organization.

What is evident even today is the difference between established control room operators and those entering the workforce having just graduated from college or university. While the established operators are used to having a single display for each function and often separate keyboards to operate them, newer generations of employees are much more adept to using a single display to manipulate multiple functions. They will routinely utilize a single interface (a smartphone or a laptop) to perform multiple functions like reading the news, listening to music on a separate link, actively having a conversation on social media, and possibly taking a phone call. For teenagers and individuals in their early twenties, this has become the norm. This is why so many interfaces today allow for a multitiled display. In addition, communication today is not necessarily in person or just by e-mail. Real-time online collaboration is routine. This is already starting to have a substantial impact on the way we do things in everyday life. It is also starting to change the way we think about control rooms.

Today, there are companies who are looking into the future and changing the way we think of approaching 24/7 monitoring and response. An example of the adaptation of new technology and how it can translate into better working environments for individuals is the Honeywell Experion® Orion Console. In researching and being active in the control room industry, Honeywell realized that, in order to respond to the industry needs, they needed to create not just a process control system, but an overall solution that will give the operators and their organization the most effective and agnostic tool to address their operational challenges.

Figure 10.13 shows the new Experion Orion Console system (right) compared to the legacy Icon (left). The new system utilizes single, high-resolution displays that have the ability to tile various information inputs into a single screen. It is a single touch screen interface, which allows for the consolidation of the input functions into a single interface. This is a disruptive change compared to historical practices. What makes this design interesting is that it creates a new technical solution to address efficiency and accuracy, and it also places a significant design focus on the HMI and ergonomics aspects of the solution. It addresses issues such as operator effectiveness (improved situational awareness and control across the process), health and safety considerations (meets health and

FIGURE 10.13 Honeywell existing Icon and new Orion platforms. (Courtesy of Honeywell Process Solutions, Houston, Texas.)

safety requirements and reduces fatigue-related absences), and comfort and alertness of operators (reduction of fatigue-related incidents: see Honeywell [2015]). The system also allows for employers to attract a younger generation of employees by offering a solution that is in line with new, broader consumer-based technology.

Aside from the radically changed interactive interface, the system also changes the way alarm management is done. As previously discussed, alarm system management is a very important aspect of the cognitive considerations for the operator. The approach that the Orion system is taking is to move from a one-to-one alarm management system to more of a dashboard approach. This provides an at-a-glance approach to viewing the entire system on the display. The visual alarm display is a glass panel mounted behind the monitor that can be illuminated in various colors, thus providing an alarm indication to not only the operator, but also the rest of the control room.

From an organizational perspective, the system is changing the way we think about control room management and operations moving forward. It is designed to utilize a collaborative approach within the control room and move away from having the operator focus on purely monitoring (transactional) functions and more toward a broader contribution to the overall operation of the facility.

There are two technology features that allow this to happen. The first is an interactive collaboration screen that is typically placed between two operating positions as shown in Figure 10.14. This allows

FIGURE 10.14 Collaborative control approach. (Courtesy of Honeywell Process Solutions, Houston, Texas.)

for real-time inputs and collaborations with different areas of the facility, communication with remote experts, and a shared view of real-time information. The second is a handheld tablet that the operator can take with them. This tablet replicates all of the information that is displayed at the control station including the system overview and/or any possible alarms. The operator is now able to physically leave their station to work on collaborative tasks without compromising their primary (monitoring) function. This is a significant new capability, since it will allow the designers to look at the entire control room space in a different manner; the fact that the operator is no longer necessarily bound to a physical control station within the control room adds a degree of flexibility to the overall control room design that has not been available until now. These new capabilities will have an impact on traffic flows, sightlines, communication, real estate, and a number of other factors that influence the design of a control room.

Of course, there are numerous questions that inevitably arise with the implementation of new operating principles and technology. One potential concern is the intensive use of touch screen interfaces and the physical effects on the operator. Another example is the fact that the touch screen interface may inadvertently allow a user to select a response (similar to accidentally pressing a touch key on your phone). As these issues are being raised, there are also solutions being developed to address them. Like any new technology, it will likely take time for it to be widespread and fully adopted, but this is an example of a trend that is likely here to stay.

For ergonomists, this type of technology will change the physical, cognitive, and organizational thought processes when it comes to control room design. While this is currently mainly focused on the process control industry, similar technology-based advances are almost certain in other control room applications in the near future. As a result, ergonomic standards and design practices will have to accordingly adapt.

This is just one example of how the future of control rooms could evolve. One thing is for certain; ergonomics as a science has a very well-established presence in the design process. The inclusion of human-based design has now gone beyond simply resolving issues surrounding the physical tasks of the operator. The established relationship between ergonomics and overall productivity and value-add to organizations will continue to drive the need for further strengthening the link between ergonomics and the important role it plays in the control room environment. The environment will continue to be unique and will therefore continue to present challenges in bridging the gap between ergonomic standards found in typical office environments and those needed in a control room. As such, ergonomists will need to work with other stakeholders to find the right balance for each specific environment. This is an investment that is already proving worthwhile and will continue to provide value in the future.

REFERENCES

API (American Petroleum Institute). 2007. API 1167: Pipeline SCADA alarm management. Second edn. Washington, DC: API.

API. 2015a. API pipeline standards. From http://www.api.org/publications-standards-and-statistics/standards /annual-standards-plan/standards%20plan%20segments/pipeline.

API. 2015b. API 1168: Pipeline control room management. Second edn. Washington, DC: API.

BIFMA. 2013. BIFMA G1: Ergonomics guideline for furniture used in office spaces designed for computer use. Grand Rapids, MI: BIFMA.

Chicago Department of Aviation. 2015. Monthly operations, passengers, cargo summary by class. From http:// www.flychicago.com/SiteCollectionDocuments/OHare/AboutUs/Facts%20and%20Figures/Air%20 Traffic%20Data/0315%20ORD%20SUMMARY.pdf.

FAA (Federal Aviation Administration). 2011. Fact sheet. From https://www.faa.gov/news/press_releases /news_story.cfm?newsId=12903.

Hendrick, H. W. 1996. The ergonomics of economics is the economics of ergonomics. *Hum Fac Erg Soc P.* 40(1):1–10.

Honeywell Control Monitoring and Safety Systems. 2015. Control, monitoring and safety systems. From https://www.honeywellprocess.com/en-US/explore/products/control-monitoring-and-safety-systems /pages/default.aspx.

International Standards Organization (ISO). 2000. ISO 11064-1: Ergonomic design of control centers. First edn. Geneva: ISO.

ISO. 2007. ISO 9241-400: Ergonomics of human-system interaction—Part 400: Principles and requirements for physical input device.

Levin, A., M. Adams, and B. Morrison. 2002. Part I: Terror attacks brought drastic decision: Clear the skies. *USA Today*. From http://usatoday30.usatoday.com/news/sept11/2002-08-12-clearskies_x.htm.

11 Schools, Health, and Productivity

Leon Straker and Erin Howie

CONTENTS

11.1 INTRODUCTION

This chapter presents a systems model of schools as a workplace and provides a summary of ergonomics research on school playgrounds, classrooms, furniture, and IT used by children prior to outlining some future challenges.

11.2 SYSTEMS MODEL OF SCHOOLS AS A WORKPLACE

Schools are a complex system involving multiple groups of people, performing different tasks in different physical environments, and using different furniture, equipment, and information technologies. Figure 11.1 provides a visual representation of the typical elements in this system.

Child students are usually the largest group of people in schools, and in Australia, the ages typically range from 4 or 5 for kindergarten, 5–12 for primary/elementary classes, and 13–18 for secondary/high school classes. Children younger than 4 years old often attend child care centers, and although these centers share many features with schools, they are not specifically covered in this chapter, as their physical and organizational settings vary widely. There is also a large group of children with physical and mental conditions, which create special needs for the school system to effectively support their learning. The particular issues related to children with special needs will not be covered in this chapter. The adults in schools can be broadly classified as workers, including teachers, administrators, cleaners, gardeners, laboratory assistants, health service providers,

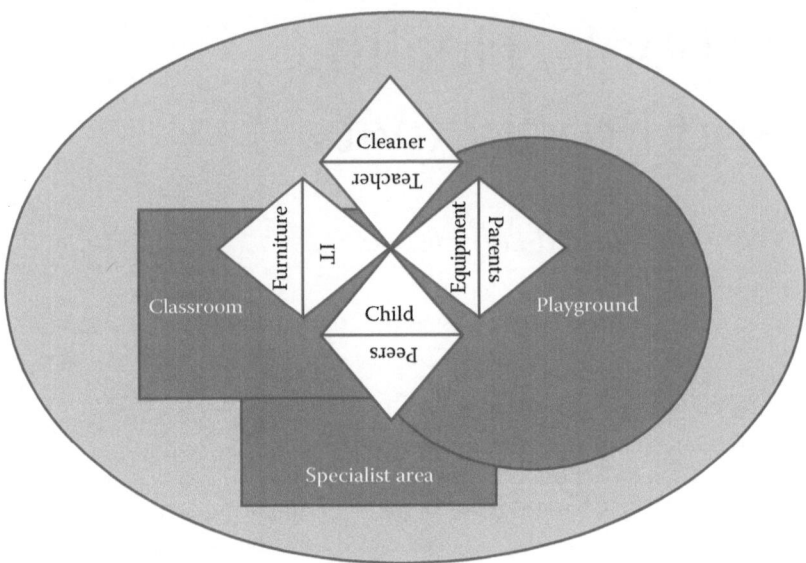

FIGURE 11.1   A systems model of school as a workplace.

and visitors (typically parents). While the focus of this chapter is on child students, some content on teacher and other worker issues with higher education facilities are covered in another chapter.

The primary task for the students at school is formal learning, although they are also involved in playing, socializing, and self-caring tasks. Children spend a large proportion of their time at school not only in classroom environments, but also in indoor and outdoor playground environments, and in specialist areas such as library, music room, and science laboratory. The furniture and equipment in classrooms typically include desks and chairs; in playgrounds, they typically include seating and play equipment; and children usually transport school-related materials in schoolbags. Information technology (IT) is an important element in schools, ranging from traditional paper-based reading and writing through blackboards/whiteboards to electronic-based desktop, laptop, and tablet computers and smart boards. There is also a growing awareness of the importance of the neighborhood to the school system. For example, the walkability of the surrounds impacts the proportion of children walking or cycling to school and thus the overall physical activity levels of children.

A number of health concerns have been raised for children in schools, including musculoskeletal disorders such as neck, back, or upper limb discomfort or pain related to awkward or sustained postures and repetitive motions, vision problems related to prolonged close vision work and computer use, mental health problems related to social interactions and mood, and sedentary behavior–related issues such as obesity and diabetes from sustained sedentariness and a lack of activity. The productivity for students relates to effective learning, with a large body of research exploring the methods to improve learning.

11.3   ERGONOMICS RESEARCH ON SCHOOL PLAYGROUNDS

The design of school playgrounds contributes to the student's health and learning by encouraging gross motor activities and minimizing injuries. School playgrounds provide opportunities for children to be physically active before, after, and between classes. The playground design features such as surface, fixed and movable equipment, and ground markings can encourage or inhibit the physical activity of children. Playgrounds are also important environments for gross motor skill development, bone and muscle strengthening activities, and social development. Children exhibit more motor skills and locomotor patterns, giving them more practice, on the playground compared to physical education class (Myers 1985). Additionally, contemporary (multipurpose-linked structures), adventure (movable tools and materials for children to create their own play area), and natural playgrounds may be even more effective at promoting positive child development than traditional playgrounds with metal equipment (Barbour 1999). Recent research has found that children have higher levels of activity on swings, fixed play equipment areas, and basketball courts, and less activity in field areas without equipment and hard-surface areas (Anthamatten et al. 2014). Simple interventions, such as adding lines to the playground and providing movable equipment (i.e., balls, jump ropes, hula hoops), can be used to encourage activity (Janssen et al. 2013). Physical activity guidelines for children emphasize the need for a variety of activities to promote both cardiovascular fitness and muscle and bone strength, such as jumping and climbing (USDHHS 2008). Playgrounds can provide an ideal opportunity for these types of activities.

Aside from encouraging participation in physical activity, the primary concerns of playground design have been injury prevention and playground safety. Common injuries requiring medical care include fractures, contusions, abrasions, and lacerations (Vollman et al. 2009). Three out of every four playground injuries are the result of falls, with impacting/striking and cutting/pinching/crushing causing the majority of the remaining injuries (Vollman et al. 2009). As the majority of injuries on playgrounds are due to falls, research has focused on the surface and equipment features. While dirt and grass surfaces were typical in the first half of the twentieth century, they became replaced with asphalt midcentury, as this was seen as more durable and did not become any more hard-packed and rocky with high use (Frost 1995). Artificial turf is now widely used for specialist sports fields.

Surface absorption is especially important for surfaces around fixed equipment where falls from a height are likely. Several research studies in the late twentieth century compared injury rates between playground surfaces underneath fixed equipment, finding those with rubber surfaces had the lowest risk of injury compared to those with wooden bark and concrete (Mott et al. 1997). Specific recommendations on the impact absorption of playground surfaces are now available (ASTM F1292-04 2004). Other factors that are important in playground safety and injury prevention include the depth and area of the surfaces, the height of the equipment, sharp edges, entrapment dangers, tripping hazards including elevated surfaces, the condition of the playground, the age appropriateness, and supervision.

Research has also studied which type of equipment may contribute to playground injuries. Falls from higher equipment, compared to standing height, are associated with greater risk of and severity of injuries, with the risk increasing significantly above 1.5 m (~5 ft) (Mott et al. 1997). Certain equipment such as swings, slides, seesaws, and monkey bars are more frequently associated with injuries (Petridou et al. 2002; Loder 2008). Climbing equipment, swings, and slides are the location for over 80% of injuries (Vollman et al. 2009). Swings are a particularly high risk for traumatic brain injuries (Loder 2008). Monkey bars have been linked to a high risk of fractures (Mott et al. 1997). Several recommendations have been made to remove these high-risk equipment from the playgrounds. With the removal of such equipment, the injury trends have shifted from severe head injuries to upper limb fractures (Mitchell et al. 2007). However, reducing the injury risk needs to be balanced against the important functions of the playground equipment in providing physically challenging tasks to aid gross motor skill and capacity development (Vredenburgh and Zackowitz 2008).

11.4   ERGONOMICS RESEARCH ON SCHOOL CLASSROOMS

The general classroom design and ongoing environment contribute to the student's health and learning. The important features of "the basic structural unit of our educational system" (Talton and Simpson 1987) include temperature and air quality, noise, light, color, and furniture arrangement.

A report for the United Kingdom Design Council for Building Schools for the Future found that poor air quality, noise, and poor lighting were detrimental to student health (Higgins et al. 2005). For example, poor air quality, including higher concentrations of nitrogen dioxide, dampness, and microbiologic pollutants, may be related to asthma and allergies and has been linked to increased absences; high noise environments have been linked to raised blood pressures (Cohen et al. 1980); and suboptimal lighting has been linked to a wide range of health outcomes including eye strain, headaches, fatigue, and even weight gain and dental cavities (Higgins et al. 2005).

A study of schools in the UK found that six built environment factors of the classroom, including light, color, choice (characteristics offering a sense of ownership), connection (clear and orientating corridors), complexity (school provides appropriate diversity), and flexibility (room allows for varied learning methods) explained 51% of the variability in the learning improvements of students (Barrett et al. 2013) demonstrating that classroom environment features are important for productivity in addition to health. While there has been limited research on the effects of temperature and humidity on cognitive performance, and the majority of research has been in adults, the limited evidence suggests that higher temperatures may have negative impacts on the speed of work and more complex cognitive tasks (Mendell and Heath 2005). Chronic noise, such as aircraft noise from a school located in a flight path near an airport (equal to 16 h of outdoor equivalent continuous noise level [leq] > 66 dBA), resulted in poorer reading comprehension than matched controls (Haines et al. 2001). However, research has also focused on the internal classroom noise including the noises from teaching equipment (such as computers and projectors), building services (such as air-conditioning), students and teachers in the classroom, and noise transmitted through the walls from other areas and classrooms in the school. A review on the effect of noise on students found several negative effects of increasing noise levels from both internal and external sources (Shield and Dockrell 2003). A primary effect of noise on students is the reduction in speech intelligibility, which limits the ability of the students to understand the teacher and the other students. Classroom noise has been associated with detrimental effects on reading, word intelligibility, letter and number recognition, and standardized test scores (Shield and Dockrell 2003). While classroom acoustics such as reverberation also contribute to speech intelligibility, noise is the most important contributing factor (Bradley, Reich, and Norcross 1999). It is recommended that the speech-to-noise ratio should be greater than 15 dB throughout the classroom, and ideally, it should be 25 dB. The WHO's guidelines for the maximum noise levels in classrooms is 35 dB L_{Aeq} (Berglund, Lindvall, and Schwela 1999).

Different colors have been proposed as supporting aspects of learning (Samuels and Stephens 1997); however, there is contradictory research demonstrating the improved learning with different colors (Higgins et al. 2005). The effect of colors on learning may depend on age, gender, and personal preference. For example, some authors recommend primary colors for younger children and paler colors for adolescents, while others suggest cool colors may enhance concentration (Higgins et al. 2005). Apart from color, the amount and quality of classroom lighting is important. While it has been reported that the natural sunlight may increase student learning through the stimulation of cortisol (Küller and Lindsten 1992), adequate levels of artificial light are important when natural sunlight is not available. Generally, higher illuminance levels are related to higher visual acuity (Berman et al. 2006), with 500 lux (~50 fc) as a standard recommendation for typical classrooms. However, this recommendation has not been updated for the growing use of backlit computer and electronic screens, which may have a different ideal lighting. Cool white fluorescent lighting is recommended for reading and focus, while warm white light may help collaborative work (Küller and Lindsten 1992; Mott et al. 2012). However, the glare and the flicker from fluorescent lighting may have detrimental effects on the visual and cognitive performances. Dynamic lighting, which alters the illuminance and the correlated color temperature across the day for different activity types, may improve student concentration (Sleegers et al. 2013).

How the classroom is laid out may also affect student learning, as different desk arrangements may facilitate different types of learning activities. Nonlinear semicircles and U-shaped formations increase face-to-face communication for interactive, collaborative learning activities, while rows of desks increase on-task behavior and emphasize individual student roles (Fernandes, Huang, and Rinaldo 2011). Additionally, other decorations in the classroom may influence productivity. For example, kindergarten students were more off-task and learned less when walls were highly decorated than when decorations were removed (Fisher, Godwin, and Seltman 2014).

Research from the 1970s and the 1980s on open classroom design, where physical walls are removed between classrooms, found no large effects on student outcomes (Hattie 2013). The overall classroom design may change the student and teacher behaviors and thus the health and productivity outcomes, although the combined architectural and pedagogical changes may have a greater effect (Woolner et al. 2007; Hattie 2013). The evaluations of classrooms in Australia and Canada found links between physical and psychosocial classroom environments and student cohesion, autonomy, involvement, task orientation, cooperation, and satisfaction, reinforcing the importance of good classroom design (Zandvliet and Straker 2001).

It is difficult to separate the effects of environmental factors from other social factors. For example, children who live in neighborhoods with chronically high noise levels (peak sound level of 74 dB) had impaired incidental memory compared to those in low-noise neighborhoods (57 dbB) (Lercher, Evans, and Meis 2003). However, schools with high levels of noise and poor air quality are often located in neighborhoods with lower socioeconomic status, which also has been negatively associated with student achievement. Further classroom environment research, particularly on light, color, noise, and air quality, is required to keep pace with new building construction technology, new school designs and IT use, and evolving teaching strategies.

11.5 ERGONOMICS RESEARCH ON CLASSROOM FURNITURE AND EQUIPMENT

Classroom furniture is designed to support the common child tasks of reading and writing on a desk and attending to the teacher or the other students presenting in the class. Evidence from observational studies in classrooms suggests that children spend a large proportion of their classroom time sitting, with flexed lumbar and cervical postures common (Murphy, Buckle, and Stubbs 2004; Geldhof et al. 2007). There is concern that the prolonged sitting, often in poor postures, could increase the risk of neck and back pains in particular. Children spend a large proportion of the day in the classroom, and there are several classroom factors that can contribute to musculoskeletal pain in children (Pollock and Straker 2008). Musculoskeletal pain in children is common, affects their ability to function, tracks into adolescence and potentially adulthood, and can interfere with learning (Pollock and Straker 2008). Some of the classroom factors that contribute to musculoskeletal pain in children include extended sitting in static postures and mismatched chairs (Hedge and Lueder 2008). More recently, there has been a concern that the high proportion of sedentary time at school may contribute to increased risk of cardiometabolic disorders (Abbott, Straker, and Mathiassen 2013). It has also been suggested that the chair/desk design and the resultant posture may influence on-task behavior and learning (Knight and Noyes 1999). Given these important potential implications, a focus of school ergonomics research has been on chair and desk designs.

11.5.1 CLASSROOM CHAIRS

The vast majority of ergonomics research on seating assumes sitting in chairs, despite sitting, kneeling, and squatting on the floor being common sitting postures in many non-Western cultures. Three chair-sitting postural options have been advocated in the literature: upright, reclined, and forward tilt (Figure 11.2).

The upright posture is typically defined by a vertical trunk with a 90° angles between the trunk and the thighs, between the thighs and the lower legs, and between the lower legs and the floor, with the upper arms vertical and the forearms horizontal. This posture provides a highly functional posture for many seated tasks such as typing. The potential concerns with this posture include the requirement for the lumbar extensor muscle activity to maintain lumbar lordosis.

The reclined posture is typically defined by a trunk reclined from the vertical such that there is a 110° angle between the trunk and the horizontal thighs. The reported advantages of the reclined posture include reduced lumbar extensor muscle activity and reduced intradiskal pressure (Grandjean and

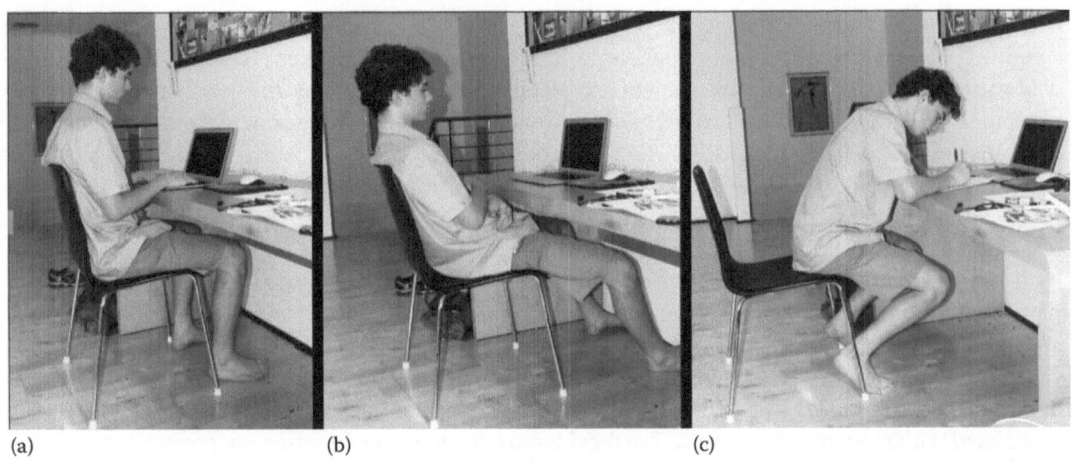

(a)   (b)   (c)

FIGURE 11.2   Three chair-sitting postural options: (a) upright, (b) reclined, and (c) forward tilt.

Hünting 1977). However, there are concerns about increased shoulder and neck muscle and joint loads through sustained shoulder flexion and neck flexion when working with a material on a desk, making this posture more potentially suitable for listening and watching presentations at the front of the class.

The forward tilt posture is typically defined by the thighs tilted forward below the horizontal. A reported advantage of creating a trunk thigh angle of around 135° is the promotion of a lumbar lordosis typical of the erect standing posture, through balanced muscle tension (Mandal 1976). This posture may be the most suitable for writing on and reading from a material flat on the desk.

In adult seating research, a backrest to support the lumbar lordosis in sitting, similar to the lordosis seen during standing, has been widely promoted based on the lower levels of back muscle activity and intervertebral disk pressure (Corlett and Eklund 1984). However, alternative lumbar loading research suggests that some lumbar flexion may be desirable (Dolan, Adams, and Mutton 1988). Chair backrests may also be needed to accommodate developing lumbar curvatures, with the typical adult lumbar curve reported to not develop until adolescence and into adulthood (Hedge and Lueder 2008).

The ergonomics evaluations of chair design and seating in the second half of the twentieth century tended to focus on comfort, reduction of muscle tension and activity, and reduction in intervertebral disk pressure. However, reducing physical stresses may not be the appropriate paradigm given the sedentariness of many children. It has been argued that for adults, work should provide the physical stresses necessary to improve the physical function, and that the absence of sufficient physical stress leads to the deterioration in bone and muscle strengths and capacity (Straker and Mathiassen 2009). Thus, selecting the chair and desk designs for school based on least muscle activity may not be appropriate. Other developmental changes such as a high prevalence of kyphosis, cervical vertebrae shape, and support for growth plates may also need to be considered in furniture design for children (Hedge and Lueder 2008).

A design paradox has been noted by Knight and Noyes (1999) that the classroom furniture is designed to "ensure that children stay in one place" to "facilitate monitoring" and reduce "distracting interactions," and ensure comfort, and yet "unnatural" long periods of "postural immobilization" may lead to discomfort. Indeed, Geldhof et al. (2007) observed that primary school children sat still for 85% of the class time, and Murphy et al. (2004) reported the reduced trunk movement during classroom sitting was related to neck and upper back pain in primary school children.

One design solution to this paradox is to encourage active sitting—trunk movement without getting up from sitting. The importance of the opportunities for postural change has been well recognized in ergonomics for some time (Corlett and Eklund 1984). Postural variety can be gained by seat designs which encourage alternating between the three advocated types of sitting posture, for example, alternating between forward leaning and reclined sitting postures when suited to the types of tasks performed (Grandjean and Hünting 1977). Seat designs can also encourage frequent small trunk movements and muscle contractions within a posture option by providing an unstable or tilting seat pan (Mandal 1976). For example, O'Sullivan et al. (2006) found that the adults sitting on a stool with an unstable cushion surface had more small movements of the pelvis than when sitting on a more stable stool surface.

Aside from trying to determine the seat design that encourages optimal postures and active sitting movements, the major focus of the ergonomics research on school seating has been to match the size of the seat to the anthropometry of children (Molenbroek, Kroon-Ramaekers, and Snijders 2003; Pollock and Straker 2008). Research based on the upright sitting philosophy has typically compared the inside knee (popliteal) height of children with the seat pan (horizontal seat surface) height, along with buttock–inside knee length with seat pan depth (Milanese and Grimmer 2004). Seat pan dimensions that are greater than child dimensions are likely to result in pressure on the thighs or the calves and discomfort. Panagiotopoulou et al. (2004) and Chung and Wong (2007) reported that mismatches were common. However, there is no convincing evidence of any increase in spinal pain risk with the mismatch of the chair and child anthropometry (Milanese and Grimmer 2004; Chung and Wong 2007; Skoffer 2007). Authors have recommended various ranges on the seat pan heights for different class ages to attempt to accommodate the differences in the anthropometry within and between the age groups. While office ergonomics research has resulted in most office workplaces providing height-adjustable chairs for workers, schools mainly provide a range of fixed-height chairs, reportedly for economic reasons. However, some studies report that some students had access to height- and backrest-adjustable and chairs with titling seats and height- and angle-adjustable tables (Skoffer 2007). Linton et al. (1994) found that when adjustable features were provided, children needed an instruction to appropriately use those features.

11.5.2 CLASSROOM DESKS

The influence of the desk height and angle on student upper body posture has long been a concern (Mandal 1982). Horizontal desk surfaces have been criticized as encouraging excessive trunk and neck flexion to read and write, with recommendations that desk surfaces be slightly tilted to reduce this effect (Mandal 1976; Freudenthal et al. 1991). While more upright trunk and head postures have been observed with tilted desks, there are practical problems with pens and paper materials rolling or sliding off the sloping desks. The desk height has been traditionally recommended to be around sitting elbow height (Molenbroek, Kroon-Ramaekers, and Snijders 2003); however, Mandal (1982) reported preferences for higher seats and desks, as did Aagaard-Hansen and Storr-Paulsen (1995). The incorporation of a foot rail into the desk to encourage postural variety has also been reported.

The majority of research on school chairs and desks is now several decades old. Since that time, there have been changes in the physical characteristics for many populations of children (increases in height, weight, and adiposity; decreases in muscle strength, and aerobic fitness), along with the changes in the teaching practices and IT used. The desks now must often accommodate both pencil-and-paper and computer or mobile technology tasks (Pollock and Straker 2008). Despite the lack of convincing evidence for back and neck pain being related to school desk and chair features (Troussier 1999; Skoffer 2007), the importance of providing chairs and desks which support reasonable postures, encourage movement, and are matched to the size of the child appears compelling (Hedge and Lueder 2008), to assist with physical health and development and attentive learning.

11.5.3 OTHER EQUIPMENT—SCHOOLBAGS

Carrying excessive load in poorly designed schoolbags has long been seen as a risk for back and neck pain in children (Malhotra and Gupta 1965), and there is a common belief that greater school-bag weight increases the risk of spinal pain for children. While Grimmer and Williams (2000) did

FIGURE 11.3 Comparison of school bag styles: large and small two-strap backpacks, wheeled bag, gym bag.

find an association between higher bag weight and back pain, van Ghent et al. (2003) actually found a lower level of back pain in children whose bag weighed >18% of their body weight. Neither of these cross-sectional studies could account for the reverse causality, i.e., children with spinal pain may choose lower bag weights. However, Jones et al. (2003) reported from a prospective study that there was no association between schoolbag weight and back pain one year later. Laboratory studies have clearly demonstrated the acute posture changes in response to the load magnitude and position (Pascoe et al. 1997; Korovessis et al. 2005), and that the energy expenditure increases with greater load magnitude (Hong et al. 2000) as would be expected, although these changes are not necessarily harmful. Thus, there is ongoing debate over what magnitude of load is appropriate for different aged children, with common recommendations suggesting that 10–15% of the child's body weight should be the upper limit (Negrini, Carabalona, and Sibilla 1999; Grimmer and Williams 2000; Moore, White, and Moore 2007). While many studies imply or directly conclude that higher loads are undesirable, an alternative view has been proposed that some degree of loading is desirable and could help children develop strength and fitness. Haselgrove et al. (2008) found an interaction between the duration of schoolbag carriage and the method of transport to school, with children walking or cycling to school appearing to gain some protection from back and neck pain from this regular physical activity/exercise. Interestingly, the perception of schoolbag heaviness or related fatigue has been frequently related to experienced back and neck pain, suggesting that it may not be the absolute weight carried, but the mismatch between the student strength and fitness and the load carriage requirements which leads to pain (Negrini and Carabalona 2002).

In addition to the bag weight, Grimmer and Williams (2000) also reported a cross-sectional relationship between increased bag carriage duration and back pain, suggesting that other loading aspects may be important. The current practice for many children is now to wear a two-strap back-pack; however, other bag styles are also used including suitcase, gym bag, one-strap back-pack, and airline wheeled bag (Figure 11.3). Some studies have suggested that carrying a schoolbag on one shoulder may increase the risk of musculoskeletal pain (Skoffer 2007), but others have not found such a relationship (Murphy, Buckle, and Stubbs 2007). One of the limitations of much of the available evidence is that the studies have often not considered that children may be carrying other sports bags, musical instruments, etc., in addition to a normal schoolbag.

11.6   ERGONOMICS RESEARCH ON SCHOOL IT USE

Schools have a very long tradition of using paper for individual reading and individual writing, with a pen/pencil. The use of blackboards for teacher or group reading and writing has also a long history. During the last couple of decades of the twentieth century, there was a rapid introduction of electronic computing technology into schools. Initially this usually took the form of specialist computer laboratories with rows of desktop computers. Subsequently, shared desktop computers in normal classrooms became prevalent, with some schools using individual laptop computers. Currently there is a rapid influx of tablet computers, along with smart boards, which provide interactive internet-enabled interfaces (Figure 11.4). Concerns have been raised about the health and productivity impacts of IT use in schools (Pollock and Straker 2008). The health concerns have mainly been about the impact on posture and muscle activity and thus the risk of musculoskeletal discomfort and disorders, but also include the concerns about vision, sedentariness, and mental health (Straker and Pollock 2005; Bener et al. 2011; Pegrum, Oakley, and Faulkner 2013; Saunders et al. 2013; Hamilton 2012). The productivity issues have included not only a positive potential for greater student engagement, but also a negative potential for greater student distraction (Huang et al. 2012; Kanala, Nousiainen, and Kankaanranta 2013; Pamuk et al. 2013).

The early research on computer use in schools built on research conducted on office workers in the 1980s, and examined the impact on posture (Laeser, Maxwell, and Hedge 1998; Oates, Evans, and Hedge 1999), the discomfort related to laptop computer use (Harris and Straker 2000), the importance of matching to the anthropometry of the individual (Straker, Briggs, and Greig 2002), and the impact of the screen position and forearm support on posture and muscle activity (Briggs, Straker, and Greig 2004; Greig, Straker, and Briggs 2005).

The spinal muscle activity and the posture of children using paper, desktop, laptop, and tablet computers have been compared in a number of laboratory studies (Straker, Burgess-Limerick et al. 2008; Straker, Coleman et al. 2008; Straker, Burgess-Limerick et al. 2009; Maslen and Straker 2009; Straker, Maslen et al. 2009) and field studies (Laeser, Maxwell, and Hedge 1998; Oates, Evans, and Hedge 1999; Ciccarelli et al. 2011). Desktop computer use has tended to result in more upright and more symmetrical trunk and head postures, associated with the higher, central visual target, and in lower average levels of postural muscle activity. Working with paper flat on a desk has been associated with considerable spinal flexion and increased upper trapezius muscle activity. Laptop computers, with a visual target between the desktop and the paper tasks, have been associated with intermediate spinal postures. For older (taller) children, the connected screen and keyboard of laptop computers create problems with either greater neck flexion (if the keyboard position is set for arms) or greater arm and hand elevation (if the screen is set for head posture). The single laboratory study on tablet computer use by children used now-outdated technology, but found postures and muscle activities similar to paper-based tasks. While the mean muscle activity and the postures were more favorable for desktop computers than those for paper tasks, there was a higher level of monotony. Thus, the risk of musculoskeletal discomfort may be similar for different electronic and paper technologies, but via different mechanisms.

FIGURE 11.4   Comparison of IT used in schools: desktop, laptop, tablet, and smart phone.

Epidemiological studies have provided some evidence for an association between increased computer use and neck and back pain (Auvinen et al. 2007), although other studies have found no association (Briggs et al. 2009; Brink et al. 2009). The relationship between computer use, posture, and neck pain is confounded by gender effects (Straker et al. 2011).

The majority of research on technology in schools relates to education and productivity outcomes. However, this research has been limited by case study or nonrandomized designs and small sample sizes. The research on children in one elementary school classroom found that a mobile application enhanced the children's motivation for writing (Kanala, Nousiainen, and Kankaanranta 2013), but other studies found that 10% of the students reported distraction when using e-books (Huang et al. 2012), and teachers perceived that tablet computers distracted students from the lesson (Pamuk et al. 2013). An evaluation of a large wireless mobile device implementation project in the UK found that the participating schools had an 8% increase in retention, 9.7% increase in achievement, and other qualitative improvements (Attewell, Savill-Smith, and Douch 2009). A similar project in Texas found that students in the participating schools became more technologically proficient, interacted more with peers in small groups, had fewer disciplinary actions, and improved standardized math test scores. A meta-analysis of the effects of educational technology on reading found a small overall effect of 0.16, with the greatest effects in secondary schools (Cheung and Slavin 2012). A nonrandomized study of primary school students with laptops found that these students had higher grade point averages, end-of-year grades, and state test scores, and made greater improvements in academic achievement over time (Cengiz Gulek and Demirtas 2005). Several mobile learning technology interventions have shown improvements in math performance. Elementary school boys using an interactive computerized tabletop improved their math performance (Jackson et al. 2013). An iPod touch intervention improved the multiplication math skills compared a normal lesson (Kiger, Herro, and Prunty 2012). A math application on an iPad improved the fraction knowledge in elementary school students (Riconscente 2013). Additionally, students found that e-books were more acceptable and had an equal reading rate compared to traditional books (Huang et al. 2012). However, like several physical factors in the classroom environment, it is the quality of the technology and how it is used by the students and the teachers, and not the technology itself, that are thought to contribute most to the student learning outcomes (Lei 2010).

11.7 TEACHERS

When considering the ergonomics of a classroom, it is important to take an ergonomic systems perspective which includes pedagogy, curriculum, environments, and multiple users (Legg and Jacobs 2008). One often-overlooked factor in schools is the teachers. The teachers, however, have the ability to influence the children's interaction with the technology, furniture, and spaces at school, and are in turn also affected by the classroom environment.

11.7.1 TEACHERS INFLUENCING STUDENT ERGONOMICS

Teachers often have little training in ergonomics, for either themselves or their students. For example, a survey found that only 12% of teachers had received information on computer ergonomics, while many teachers had concerns about the students' postures and discomfort while using the computers (Williams, Cook, and Zigler 2000). Additionally, over 90% of the teachers in another survey wanted more information on computer ergonomics provided through a training course or printed format (Dockrell et al. 2007).

Teachers also have the ability to control several factors of the environment that have been previously described. For example, many principals allow the teachers to select the furniture or arrange the furniture in the classroom. Teachers can also control some of the classroom noise levels such as through classroom rules and their own voice (which also contributes to noise levels). There is a large variation in how teachers control student noise as shown in the variation in the teacher voice-to-noise ratios (Lindstrom et al. 2011). Similarly, teachers can select video and interactive software for school and home learning which matches the developmental level of their class (Hana 2008).

11.7.2 Teacher Ergonomics

Teachers are also physically affected by the environment they work in. Teachers report a high prevalence of musculoskeletal pain. One study found a high prevalence of low back pain in primary and high school teachers, and that this was positively associated with prolonged sitting and standing, working hours with the computer, and correcting exam papers, and negatively associated with rest and physical activity (Mohseni Bandpei et al. 2014). Another study examined neck/shoulder pain and low back pain in school teachers and found that the neck/shoulder pain was associated with prolonged standing and sitting (static postures), while the low back pain was associated with twisting, uncomfortable back support, as well as static postures (Yue, Liu, and Li 2012).

There are many factors, which are unique to the teaching profession, that may influence these levels of musculoskeletal pain (Bennett, Woodcock, and Diane 2006). Teachers work in an environment that is predominantly designed for students with student-sized desks and chairs. Teachers therefore frequently stoop or crouch to interact with the child at the child's eye level. While teachers have more movement and light activity during school time compared to office workers (Parry 2014), the after-class time involving lesson preparation and marking student work can involve prolonged bouts of static and awkward postures.

Teachers, like students and many other workers, also spend increasingly greater time working with computers. While some of this teacher–computer interaction may be in workstations designed for adults, teachers may also use computers in workstations designed for children and in locations not designed for computer interaction.

11.8 THE FUTURE OF SCHOOL ERGONOMICS

The available evidence supports the importance of many aspects of the school workplace in terms of the potential impact on student health and productivity. However, the current evidence is patchy, with limited research activity in some areas in recent years. While the general principles of ergonomics are supported, including the importance of involving students and other stakeholders in the design (Horton et al. 2009), the specific details of the appropriate designs are often unknown. The substantial diversity globally, in physical and technological school environments (ranging from bush hut with no furniture and limited IT to resource-rich classrooms with each student having a powerful Internet-connected mobile computing device) and in teaching philosophies and practices (ranging from rigid curriculum teacher delivery and rote learning to student-centered investigative learning) further limits the capacity of the current research evidence to inform good practice. Substantial cultural differences also exist in school systems within and between countries, with the available evidence, and thus the information presented in this chapter is biased toward the English language and often affluent Western country research. The changes in school populations and administrative systems are going to require new applications of the ergonomics principles. School population numbers are increasing with increasing population levels. In many cultures, children are also attending formal or semiformal school systems from earlier ages and extra tutorial sessions before or after formal schooling. Children with disabilities are increasingly being integrated into mainstream schools creating new challenges. Some school systems are increasing time spent in school by children, and others are restructuring the age groups serviced by particular school levels, for example, to create a middle school between elementary school and high school.

Thus, a challenge for the ergonomics of schools is to provide evidence for the current diversity of school workplaces, in addition to keeping up with the interrelated changes to pedagogy, technology, and physical environments.

11.8.1 Changes in Pedagogy

The pedagogical changes over the last few decades in Western countries have moved from teacher-directed to more independent student learning. For example, some current pedagogical initiatives are lifelong learning, problem-based learning, collaborative learning, and online learning (Gulland and Phillips 2008). Such initiatives often have goals related to global competencies with an emphasis on skills in critical thinking, active learning, problem solving, communicating, contextualizing knowledge, and fluencies in technology, information, and media. While school systems, which encourage rote learning and massed practice, currently lead the international literacy and numeracy rankings, there is a realization that this approach may limit the children's ability to succeed in a global environment requiring flexibility and creative problem solving.

Supporting new ways of teaching may require changes in school designs to support new tasks and ways of working. For example, a focus on more collaborative student-driven group work may need the support from the changes in classroom furniture arrangements (from rows all facing the front of the class to clusters). Similarly, flipped classrooms where students are expected to do the background concept reading at home with experiential learning at school may need more extensive family involvement and technology availability at home. The adult office design is currently embracing activity-based work, where individual staff no longer has a set office but rather the workplace provides a range of different physical work environments, and workers move between work spaces during the day depending on the task they are performing. School classrooms are also exploring task-based spaces, which include not only sitting desk clusters, but also standing individual and group work spaces, couches, floor mats and "caves." The potential advantages of the variety of workspaces include support for a wider range of postures and more encouragement and opportunity to move during class time—both within a task space and between task spaces.

11.8.2 Changes in Technology

The shift from laboratories with desktop computers to individual laptop and tablet computers also provides the potential for more movement during class time. A study in the early 2000s found that a "moving school" was able to improve the children's sitting habits and physical activity. The moving school provided work spaces that were customized for specific learning tasks (including dynamic seating), and children could move between them to enable learning (Cardon et al. 2004). Mobile learning technology may also afford opportunities for movement, as students are not restricted to being at their desk or even in the classroom to utilize computers (Richardson et al. 2013). The students have the opportunity to easily take their "desk" (e.g., tablet computer or laptop computer) wherever they are working, including active and standing workstations, if they are available. Additionally, mobile apps and interactive smart boards may be used as teacher aids for leading active lessons and activity breaks in the classroom. Changes in the computer input to systems requiring whole-body movement enable movement while learning which may be particularly useful (Maddison 2013).

11.8.3 Changes in the Physical Environment

New trends in building materials and architectural designs are likely to influence future schools. For example, a school design based around a learning street involving a spacious and well-lit, double-loaded corridor as the social artery of a school has been recently promoted (Stevenson 2010). There is an increasing recognition of the importance of exposure to outdoor natural environments. A review of over 150 studies on outdoor education found that fieldwork and outdoor visits, outdoor adventure education, and school grounds/community projects positively benefitted students (Rickinson, Council, and Britain 2004). Some of the benefits included improved long-term memory from the fieldwork, improved interpersonal skills and self-perceptions from the outdoor adventure experiences, and gains in specific curricula areas such as science from school grounds projects. Changes in mobile IT are creating new opportunities for combining technology and nature. For example, one classroom in China was able to use mobile tablet computers combined with an outdoor ecological pool to create a science activity for students that was engaging and increased the students' science knowledge through student inquiry (Liu et al. 2009). Large-scale environmental changes such as increasing pollution and climate change may also have an impact on the future school.

11.9 CONCLUSION

Schools are the occupational workplace for up to 12 years for most people, and those years are when people are most responsive and vulnerable, physically and mentally. Thus, no other workplace should have a higher priority for ergonomics research—to ensure school workplaces are optimally designed and managed to support the health and productivity of children.

REFERENCES

Aagaard, J., and A. Storr-Paulsen. 1995. A comparative study of three different kinds of school furniture. *Ergonomics*. 38(5):1025–35.

Abbott, R. A., L. M. Straker, and S. E. Mathiassen. 2013. Patterning of children's sedentary time at and away from school. *Obesity (Silver Spring)*. 21(1):E131–3.

Anthamatten, P., L. Brink, B. Kingston, E. Kutchman, S. Lampe, and C. Nigg. 2014. An assessment of schoolyard features and behavior patterns in children's utilization and physical activity. *J Phys Act Health*. 11(3):564–73.

ASTM (American Society for Testing and Materials) F1292-04. 2004. *Standard Specification for Impact Attenuation of Surfacing Materials within the Use Zone of Playground Equipment*. West Conshohocken, PA: ASTM.

Attewell, J., C. Savill-Smith, and R. Douch. 2009. *The Impact of Mobile Learning: Examining What It Means for Teaching and Learning*. London: Learning and Skills Network.

Auvinen, J., T. Tammelin, S. Taimela, P. Zitting, and J. Karppinen. 2007. Neck and shoulder pains in relation to physical activity and sedentary activities in adolescence. *Spine*. 32(9):1038–44.

Barbour, A. C. 1999. The impact of playground design on the play behaviors of children with differing levels of physical competence. *Early Child Res Q*. 14(1):75–98.

Barrett, P., Y. Zhang, J. Moffat, and K. Kobbacy. 2013. A holistic, multi-level analysis identifying the impact of classroom design on pupils' learning. *Build Environ*. 59:678–89.

Bener, A., H. S. Al-Mahdi, A. I. Ali, M. Al-Nufal, P. J. Vachhani, and I. Tewfik. 2011. Obesity and low vision as a result of excessive Internet use and television viewing. *Int J Food Sci Nutr*. 62(1):60–2.

Bennett, C., A. Woodcock, and T. Diane. 2006. Ergonomics for students and staff. In *Safe and Healthy School Environments*. Frumkin, H., R. Geller, L. Rubin and J. Nodvin (eds.). New York: Oxford University Press.

Berglund, B., T. Lindvall, and D. H. Schwela. 1999. "Guidelines for community noise." Geneva: Department of the Protection of the Human Environment, Occupational and Environmental Health World Health Organization. Available at http://www.who.int/docstore/peh/noise/Comnoise-1.pdf.

Berman, S., M. Navvab, M. Martin, J. Sheedy, and W. Tithof. 2006. A comparison of traditional and high colour temperature lighting on the near acuity of elementary school children. *Light Res Technol*. 38(1):41–9.

Bradley, J. S., R. D. Reich, and S. G. Norcross. 1999. On the combined effects of signal-to-noise ratio and room acoustics on speech intelligibility. *J Acoust Soc Am*. 106(4 Pt 1):1820–8.

Briggs, A., L. Straker, and A. Greig. 2004. Upper quadrant postural changes of school children in response to interaction with different information technologies. *Ergonomics*. 47(7):790–819.

Briggs, A. M., L. M. Straker, N. L. Bear, and A. J. Smith. 2009. Neck/shoulder pain in adolescents is not related to the level or nature of self-reported physical activity or type of sedentary activity in an Australian pregnancy cohort. *BMC Musculoskel Disord*. 10(1):87.

Brink, Y., L. C. Crous, Q. A. Louw, K. Grimmer-Somers, and K. Schreve. 2009. The association between postural alignment and psychosocial factors to upper quadrant pain in high school students: A prospective study. *Man Ther*. 14(6):647–53.

Cardon, G., D. De Clercq, I. De Bourdeaudhuij, and D. Breithecker. 2004. Sitting habits in elementary school-children: A traditional versus a "moving school." *Patient Educ Couns*. 54(2):133–42.

Cengiz Gulek, J., and H. Demirtas. 2005. Learning with technology: The impact of laptop use on student achievement. *J Technol Learn Assess*. 3(2).

Cheung, A. C., and R. E. Slavin. 2012. How features of educational technology applications affect student reading outcomes: A meta-analysis. *Educ Res Rev*. 7(3):198–215.

Chung, J. W., and T. K. Wong. 2007. Anthropometric evaluation for primary school furniture design. *Ergonomics*. 50(3):323–34.

Ciccarelli, M., L. Straker, S. E. Mathiassen, and C. Pollock. 2011. ITKids part II: Variation of postures and muscle activity in children using different information and communication technologies. *Work*. 3(4):413–27.

Cohen, S., G. W. Evans, D. S. Krantz, and D. Stokols. 1980. Physiological, motivational, and cognitive effects of aircraft noise on children: Moving from the laboratory to the field. *Am Psychol*. 35(3):231–43.

Corlett, E., and J. Eklund. 1984. How does a backrest work? *Appl Ergon*. 15(2):111–4.

Dockrell, S., E. Fallon, M. Kelly, B. Masterson, and N. Shields. 2007. School children's use of computers and teachers' education in computer ergonomics. *Ergonomics*. 50(10):1657–67.

Dolan, P., M. Adams, and W. Mutton. 1988. Commonly adopted postures and their effect on the lumbar spine. *Spine*. 13(2):197–201.

Fernandes, A. C., J. Huang, and V. Rinaldo. 2011. Does where a student sits really matter? The impact of seating locations on student classroom learning. *Int J App Educ Studies*. 10(1):66–77.

Fisher, A. V., K. E. Godwin, and H. Seltman. 2014. Visual environment, attention allocation, and learning in young children when too much of a good thing may be bad. *Psychol Sci*. 25(7):1362–70.

Freudenthal, A., M. van Riel, J. Molenbroek, and C. Snijders. 1991. The effect on sitting posture of a desk with a ten-degree inclination using an adjustable chair and table. *Appl Ergon*. 22(5):329–36.

Frost, J. L. 1995. History of playground safety in america. *Child Environ Q*. 2:13–23.

Geldhof, E., D. De Clercq, I. De Bourdeaudhuij, and G. Cardon. 2007. Classroom postures of 8–12 year old children. *Ergonomics*. 50(10):1571–81.

Grandjean, E., and W. Hünting. 1977. Ergonomics of posture—Review of various problems of standing and sitting posture. *Appl Ergon*. 8(3):135–40.

Greig, A. M., L. M. Straker, and A. M. Briggs. 2005. Cervical erector spinae and upper trapezius muscle activity in children using different information technologies. *Physiotherapy*. 9(2):119–26.

Grimmer, K., and M. Williams. 2000. Gender-age environmental associates of adolescent low back pain. *Appl Ergon*. 31(4):343–60.

Gulland, D., and J. Phillips. 2008. New directions in learning. In *Ergonomics for Children: Designing Products and Places for Toddlers to Teens*. Lueder, R., and V. J. Berg Rice (eds.). Boca Raton, FL: Taylor & Francis.

Haines, M. M., S. A. Stansfeld, R. S. Job, B. Berglund, and J. Head. 2001. Chronic aircraft noise exposure, stress responses, mental health and cognitive performance in school children. *Psychol Med*. 31(02):26577.

Hamilton, J. R. 2012. The electronic lonely crowd-patterns and effects of electronic media usage among contemporary adolescents. *Int J Arts Sci*. 5(3):165–74.

Hana, L. 2008. Designing electronic media for children. In *Ergonomics for Children: Designing Products and Places for Toddlers to Teens*. Lueder, R., and V. J. Berg Rice (eds.). Boca Raton, FL: Taylor & Francis.

Harris, C., and L. Straker. 2000. Survey of physical ergonomics issues associated with school childrens' use of laptop computers. *Int J Ind Ergon*. 26(3):337–46.

Haselgrove, C., L. Straker, A. Smith, P. O'Sullivan, M. Perry, and N. Sloan. 2008. Perceived school bag load, duration of carriage, and method of transport to school are associated with spinal pain in adolescents: An observational study. *Aust J Physiother*. 54(3):193–200.

Hattie, J. 2013. *Visible Learning: A Synthesis of over 800 Meta-Analyses relating to Achievement*. London: Routledge.

Hedge, A., and R. Lueder. 2008. Classroom furniture. In *Ergonomics for Children: Designing Products and Places for Toddlers to Teens*. Lueder, R., and V. J. Berg Rice (eds.). Boca Raton, FL: Taylor & Francis.

Higgins, S., E. Hall, K. Wall, P. Woolner, and C. McCaughey. 2005. The impact of school environments: A literature review. New South Wales: The Centre for Learning and Teaching, School of Education, Communication and Language Science, University of Newcastle.

Hong, Y., J. X. Li, A. S. K. Wong, and P. D. Robinson. 2000. Effects of load carriage on heart rate, blood pressure and energy expenditure in children. *Ergonomics*. 43(6):717–27.

Horton, J., P. Kraftl, A. Woodcock, M. Newman, M. Kinross, P. Adey, and O. den Besten. 2009. Involving pupils in school design: A guide for schools. Northampton: University of Northampton. Available at http://nectar.northampton.ac.uk/2505/1/Horton20092505.pdf.

Huang, Y.-M., T.-H. Liang, Y.-N. Su, and N.-S. Chen. 2012. Empowering personalized learning with an interactive e-book learning system for elementary school students. *Educ Technol Res Dev*. 60(4):703–22.

Jackson, A. T., B. J. Brummel, C. L. Pollet, and D. D. Greer. 2013. An evaluation of interactive tabletops in elementary mathematics education. *Educ Technol Res Dev*. 61(2):311–32.

Janssen, M., J. W. Twisk, H. M. Toussaint, W. van Mechelen, and E. A. Verhagen. 2013. Effectiveness of the PLAYgrounds programme on PA levels during recess in 6-year-old to 12-year-old children. *Br J Sports Med*. 49(4):259–64.

Jones, G. T., K. D. Watson, A. J. Silman, D .P. Symmons, and G. J. Macfarlane. 2003. Predictors of low back pain in British schoolchildren: A population-based prospective cohort study. *Pediatrics*. 111(4):822–28.

Kanala, S., T. Nousiainen, and M. Kankaanranta. 2013. Using a mobile application to support children's writing motivation. *Interact Technol Smart Educ*. 10(1):4–14.

Kiger, D., D. Herro, and D. Prunty. 2012. Examining the influence of a mobile learning intervention on third grade math achievement. *J Res Technol Educ*. 45(1):61–82.

Knight, G., and J. Noyes. 1999. Children's behaviour and the design of school furniture. *Ergonomics*. 42(5):747–60.

Korovessis, P., G. Koureas, S. Zacharatos, and Z. Papazisis. 2005. Backpacks, back pain, sagittal spinal curves and trunk alignment in adolescents: A logistic and multinomial logistic analysis. *Spine*. 30(2):247–55.

Küller, R., and C. Lindsten. 1992. Health and behavior of children in classrooms with and without windows. *J Environ Psychol*. 12(4):305–17.

Laeser, K. L., L. E. Maxwell, and A. Hedge. 1998. The effect of computer workstation design on student posture. *J Res Computing Educ*. 31(2):173–88.

Legg, S., and K. Jacobs. 2008. Ergonomics for schools. *Work*. 31(4):489–93.

Lei, J. 2010. Quantity versus quality: A new approach to examine the relationship between technology use and student outcomes. *Brit J Educ Technol*. 41(3):455–72.

Lercher, P., G. W. Evans, and M. Meis. 2003. Ambient noise and cognitive processes among primary schoolchildren. *Environ Behav*. 35(6):725–35.

Lindstrom, F., K. P. Waye, M. Södersten, A. McAllister, and S. Ternström. 2011. Observations of the relationship between noise exposure and preschool teacher voice usage in day-care center environments. *J Voice*. 25(2):166–72.

Linton, S. J., A.-L. Hellsing, T. Halme, and K. Åkerstedt. 1994. The effects of ergonomically designed school furniture on pupils' attitudes, symptoms and behaviour. *Appl Ergon*. 25(5):299–304.

Liu, T.-C., H. Peng, W.-H. Wu, and M.-S. Lin. 2009. The effects of mobile natural-science learning based on the 5E learning cycle: A case study. *Educ Technol Soc*. 12(4):344–58.

Loder, R. T. 2008. The demographics of playground equipment injuries in children. *J Pediatr Surg*. 43(4):691–9.

Malhotra, M., and J. S. Gupta. 1965. Carrying of school bags by children. *Ergonomics*. 8(1):55–60.

Mandal, A. C. 1976. Work-chair with tilting seat. *Ergonomics*. 19(2):157–64.

Mandal, A. 1982. The correct height of school furniture. *Human Factors*. 24(3):257–69.

Maslen, B., and L. Straker. 2009. A comparison of posture and muscle activity means and variation amongst young children, older children and young adults while working with computers. *Work*. 32(3):311–20.

Mendell, M. J., and G. A. Heath. 2005. Do indoor pollutants and thermal conditions in schools influence student performance? A critical review of the literature. *Indoor Air*. 15(1):27–52.

Milanese, S., and K. Grimmer. 2004. School furniture and the user population: An anthropometric perspective. *Ergonomics*. 47(4):416–26.

Mitchell, R., S. Sherker, M. Cavanagh, and D. Eager. 2007. Falls from playground equipment: Will the new Australian playground safety standard make a difference and how will we tell? *Health Promot J Austr*. 18(2):98–104.

Mohseni Bandpei, M. A., F. Ehsani, H. Behtash, and M. Ghanipour. 2014. Occupational low back pain in primary and high school teachers: Prevalence and associated factors. *J Manipulative Physiol Ther*. 37(9):702–8.

Molenbroek, J. F., Y. M. Kroon-Ramaekers, and C. J. Snijders. 2003. Revision of the design of a standard for the dimensions of school furniture. *Ergonomics*. 46(7):681–94.

Moore, M. J., G. L. White, and D. L. Moore. 2007. Association of relative backpack weight with reported pain, pain sites, medical utilization, and lost school time in children and adolescents. *J Sch Health*. 77(5):232–9.

Mott, A., K. Rolfe, R. James, R. Evans, A. Kemp, F. Dunstan, K. Kemp, and J. Sibert. 1997. Safety of surfaces and equipment for children in playgrounds. *Lancet*. 349(9069):1874–6.

Mott, M. S., D. H. Robinson, A. Walden, J. Burnette, and A. S. Rutherford. 2012. Illuminating the effects of dynamic lighting on student learning. *Sage Open*. 2(2):2158244012445585.

Murphy, S., P. Buckle, and D. Stubbs. 2004. Classroom posture and self-reported back and neck pain in schoolchildren. *Appl Ergon.* 35(2):113–20.

Murphy, S., P. Buckle, and D. Stubbs. 2007. A cross-sectional study of self-reported back and neck pain among English schoolchildren and associated physical and psychological risk factors. *Appl Ergon.* 38(6):797–804.

Myers, G. D. 1985. Motor behavior of kindergartners during physical education and free play. In *When Children Play.* Frost, J. L., and S. Sunderlin (eds.), pp. 151–155. Wheaton, MD: Association for Children Education International.

Negrini, S., and R. Carabalona. 2002. Backpacks on! Schoolchildren's perceptions of load, associations with back pain and factors determining the load. *Spine.* 27(2):187–95.

Negrini, S., R. Carabalona, and P. Sibilla. 1999. Backpack as a daily load for schoolchildren. *The Lancet.* 354(9194):1974.

O'Sullivan, P., W. Dankaerts, A. Burnett, L. Straker, G. Bargon, N. Moloney, M. Perry, and S. Tsang. 2006. Lumbopelvic kinematics and trunk muscle activity during sitting on stable and unstable surfaces. *J Orthop Sports Phys Ther.* 36(1):19–25.

Oates, S., G. W. Evans, and A. Hedge. 1999. An anthropometric and postural risk assessment of children's school computer work environments. *Comput Schools.* 14(3–4):55–63.

Pamuk, S., R. Çakır, M. Ergun, H. Yılmaz, and C. Ayas. 2013. The use of tablet PC and interactive board from the perspectives of teachers and students: Evaluation of the FATİH Project. *Educ Sci: Theory Pract.* 13(3):1815–22.

Panagiotopoulou, G., K. Christoulas, A. Papanckolaou, and K. Mandroukas. 2004. Classroom furniture dimensions and anthropometric measures in primary school. *Appl Ergon.* 35(2):121–8.

Parry, S. 2014. *Sedentary Time and Physical Activity Exposure Patterns and Musculoskeletal Symptoms of Australian Office Workers.* Western Australia: School of Physiotherapy and Exercise Science, Curtin University.

Pascoe, D. D., D. E. Pascoe, Y. T. Wang, D.-M. Shim, and C. K. Kim. 1997. Influence of carrying book bags on gait cycle and posture of youths. *Ergonomics.* 40(6):631–40.

Pegrum, M., G. Oakley, and R. Faulkner. 2013. Schools going mobile: A study of the adoption of mobile hand-held technologies in Western Australian independent schools. *Austral J Educ Technol.* 29(1):66–81.

Petridou, E., J. Sibert, X. Dedoukou, I. Skalkidis, and D. Trichopoulos. 2002. Injuries in public and private playgrounds: The relative contribution of structural, equipment and human factors. *Acta Paediatr.* 91(6):691–7.

Pollock, C., and L. Straker. 2008. Information and communication technology in schools. In *Ergonomics for Children: Designing Products and Places for Toddlers to Teens.* Lueder, R., and V. J. Berg Rice (eds.). Boca Raton, FL: Taylor & Francis.

Richardson, J. W., S. McLeod, K. Flora, N. J. Sauers, S. Kannan, and M. Sincar. 2013. Large-scale 1:1 computing initiatives: An open access database. *Int J Educ Dec Using Inf Commun Technol.* 9(1):4–18.

Rickinson, M., F. S. Council, and G. Britain. 2004. *A Review of Research on Outdoor Learning.* London: National Foundation for Educational Research and King's College.

Riconscente, M. M. 2013. Results from a controlled study of the iPad fractions game motion math. *Games Culture.* 8(4):186–214.

Samuels, R., and H. Stephens. 1997. Colour and light in schools: Theoretical and empirical background. Sydney, Australia: University of New South Wales. Available at http://www.urbanclimateresearch .org/pdfDocuments/light/Robert%20Samuels%20and%20Harry%20Stephens%20Light%20and%20 Colour%20at%20School%201997.pdf.

Saunders, T. J., M. S. Tremblay, M. E. Mathieu, M. Henderson, J. O'Loughlin, A. Tremblay, and J. P. Chaput. 2013. Associations of sedentary behavior, sedentary bouts and breaks in sedentary time with cardiometabolic risk in children with a family history of obesity. *PLoS One.* 8(11):e79143.

Shield, B. M., and J. E. Dockrell. 2003. The effects of noise on children at school: A review. *Build Acoust.* 10(2):97–116.

Skoffer, B. 2007. Low back pain in 15- to 16-year-old children in relation to school furniture and carrying of the school bag. *Spine.* 32(24):E713–7.

Sleegers, P., N. Moolenaar, M. Galetzka, A. Pruyn, B. Sarroukh, and B. van der Zande. 2013. Lighting affects students' concentration positively: Findings from three Dutch studies. *Light Res Technol.* 45(2):159–75.

Stevenson, K. R. 2010. *Educational Trends Shaping School Planning, Design, Construction, Funding and Operation.* Washington, DC: National Clearinghouse for Educational Facilities.

Straker, L., and S. E. Mathiassen. 2009. Increased physical work loads in modern work—A necessity for better health and performance? *Ergonomics.* 52(10):1215–25.

Straker, L., and C. Pollock. 2005. Optimizing the interaction of children with information and communication technologies. *Ergonomics*. 48(5):506–21.

Straker, L., A. Briggs, and A. Greig. 2002. The effect of individually adjusted workstations on upper quadrant posture and muscle activity in school children. *Work*. 18(3):239–48.

Straker, L., R. Burgess-Limerick, C. Pollock, J. Coleman, R. Skoss, and B. Maslen. 2008. Children's posture and muscle activity at different computer display heights and during paper information technology use. *Hum Factors*. 50(1):49–61.

Straker, L., R. Burgess-Limerick, C. Pollock, and B. Maslen. 2009. The effect of forearm support on children's head, neck and upper limb posture and muscle activity during computer use. *J Electromyogr Kinesiol*. 19(5):965–74.

Straker, L. M., J. Coleman, R. Skoss, B. A. Maslen, R. Burgess-Limerick, and C. M. Pollock. 2008. A comparison of posture and muscle activity during tablet computer, desktop computer and paper use by young children. *Ergonomics*. 51(4):540–55.

Straker, L., B. Maslen, R. Burgess-Limerick, and C. Pollock. 2009. Children have less variable postures and muscle activities when using new electronic information technology compared with old paper-based information technology. *J Electromyogr Kinesiol*. 19(2):e132–43.

Straker, L. M., A. J. Smith, N. Bear, P. B. O'Sullivan, and N. H. de Klerk. 2011. Neck/shoulder pain, habitual spinal posture and computer use in adolescents: The importance of gender. *Ergonomics*. 54(6):539–46.

Talton, E. L., and R. D. Simpson. 1987. Relationships of attitude toward classroom environment with attitude toward and achievement in science among tenth grade biology students. *J Res Sci Teach*. 24(6):507–25.

Troussier, B. 1999. Comparative study of two different kinds of school furniture among children. *Ergonomics*. 42(3):516–26.

USDHHS (U.S. Department of Health and Human Services). 2008. *2008 Physical Activity Guidelines for Americans*. Washington, DC: USDHHS.

van Gent, C., J. J. Dols, M. Carolien, R. A. H. Sing, and H. C. de Vet. 2003. The weight of schoolbags and the occurrence of neck, shoulder, and back pain in young adolescents. *Spine*. 28(9):916–21.

Vollman, D., R. Witsaman, R. D. Comstock, and G. A. Smith. 2009. Epidemiology of playground equipment-related injuries to children in the United States, 1996–2005. *Clin Pediatr (Phila)*. 48(1):66–71.

Vredenburgh, A. G., and I. B. Zackowitz. 2008. Playground safety and ergonomics. In *Ergonomics for Children: Designing Products and Places for Toddlers to Teens*. Lueder, R., and V. J. Berg Rice (eds.). Boca Raton, FL: Taylor & Francis.

Williams, I. M., T. Cook, and T. Zigler. 2000. Computer ergonomics for teachers and students. *Hum Fac Erg Soc P*. 44(9):2–91–2–93.

Woolner, P., E. Hall, S. Higgins, C. McCaughey, and K. Wall. 2007. A sound foundation? What we know about the impact of environments on learning and the implications for building schools for the future. *Ox Rev Educ*. 33(1):47–70.

Yue, P., F. Liu, and L. Li. 2012. Neck/shoulder pain and low back pain among school teachers in China, prevalence and risk factors. *BMC Pub Health*. 12:789.

Zandvliet, D. B., and L. M. Straker. 2001. Physical and psychosocial aspects of the learning environment in information technology rich classrooms. *Ergonomics*. 44(9):838–57.

12 Ergonomic Concerns in Universities and Colleges

Danny S. Nou

CONTENTS

12.1 INTRODUCTION

This chapter demonstrates the various issues that universities have as a workplace and provides a summary of unique university environments such as laboratory, dining services, animal, and custodial ergonomics. Some universities also have medical centers, which require safe patient-handling ergonomics. The particular issues related to patient handling or computer ergonomics will not be covered in this chapter.

12.2 INTRODUCTION TO THE AMERICAN UNIVERSITY SYSTEM

In 2015, there were over 4600 universities in the United States with over 2800 four-year institutions and 1800 two-year institutions. American universities developed independent accreditation organizations to vouch for the quality of their degrees. The accreditation agencies rate the universities and the colleges on criteria such as the academic quality, the quality of their libraries, the publishing records of their faculty, and the degrees that their faculty hold.

In terms of student population size, universities that are considered small have fewer than 5000 students. These are typically private colleges like Hobart, Colgate, etc. Many American universities fall into the medium-sized category, between 5000 and 15,000 students. Some examples are Yale, Brown, Howard, Duke, Stanford, etc. The largest universities usually mean more than 15,000 students. The University of Southern California, New York University, and the University of Pennsylvania qualify as large on the private side, the University of California at Davis, Michigan State, and with the University of Texas at Austin on the public side. Some larger universities are both public and private, such as Cornell.

Finally, the faculty (professors, lecturers, researchers) and staff (human resources, health and safety, custodial, etc.) populations of the schools are significant populations as well. University faculty sizes range from 200 to more than 2000 faculty depending on the student population and the research facilities. The staff on the campus is usually at least double to ten times the amount of the faculty population.

12.3 THE STRUCTURE OF AMERICAN UNIVERSITIES

Universities are usually spilt in two main groups: educational (faculty, students, and researchers) and university staff (custodians, groundskeepers, technicians, etc.). Unlike other organizations or companies, the business of the university is education, and the staff members provide the support to the functions of the university. Each of these groups has various types of workers that perform different tasks in various environments.

The educational group represents unique ergonomic situations in the workplace. Many faculty members conduct various research projects (arts, engineering, sciences) that use a diverse plethora of laboratory tools, furniture, and IT. One faculty member might only have computers in his laboratory as compared to an animal science faculty, which can handle large animals and work outdoors. There are also issues with each faculty member usually being in charge of their own budgets, which can create difficulties for an ergonomic program to effectively support all faculty. Issues such as procuring ergonomic products, applying administrative solutions, and obtaining furniture can be difficult without appropriate funds from the university or the faculty member.

The university staff represents the various functions that help run the day-to-day operations of the university. The staff members in universities can include various jobs such as lab technicians, administrators, custodians, gardeners, health and safety, university police, and health service providers. Some ergonomic issues that the university staff handle are animal and material handling, dining services, various pieces IT, and laboratory management. The focus of this chapter is to break down the various ergonomic issues on campus and the unique issues the university has to handling them.

12.3.1 FACULTY AND STUDENT ERGONOMICS

12.3.1.1 Office Ergonomics

University environments entail a great variety of physical and mental activities. Often, the core activities of any office job take place at the desk or at the workstation. Many people consider that workstations from the home office or the corporate office should not change much in the educational world, but there are some slight differences. In addition to the physical dimensions of the workspace and the furniture, other features should also be considered in any design, reorganization, or relocation of faculty workstations.

First, faculty members need to communicate with people in private that have various levels of confidentiality. Faculty members are often meeting with other faculty members, students, guests, and university staff to discuss various research- or university-related projects. As a result, the workstation needs to facilitate interpersonal contact and group discussion. The spaces need to be intimate enough to conduct one-on-one meetings and large enough to encourage tutorial-like discussions. These spaces should allow for easy movement, accommodating visitors where necessary, and storage (Figure 12.1). Table 12.1 provides some ranges.

Second, the faculty workstation needs to provide a level of seclusion that does not interfere with their concentration. The faculty members need to conduct many solitary job functions such as preparing lesson plans, grading student work, reading academic journals, writing manuscripts, reports, and grant proposals, and various other independent tasks. As a result, it is recommended that these offices have fixed or permanent walls to isolate for sound and outside noises that may interfere with

FIGURE 12.1 Faculty office.

TABLE 12.1
Ranges for Workstations in the University

Employee Type	Group	Workstation Type	Recommended Area: ft² (m²)
Faculty	Dean	Office	240 (~22.3)
	Faculty (full/part time)	Office	80–160 (~7.4–14.8)
	Emeriti (active/inactive)	Office/shared office	80–160 (~7.4–14.8)
	Lecturers, visiting faculty	Shared office or cube	80 (~7.4)
Staff	Program director	Office	64–100 (~6–9.3)
	Full time	Small office, shared office or cube	64–100 (~6–9.3)
	Contractors or temporary	Shared office or cube	64–80 (~6–7.4)
	Research associates	Small office, shared office or cube	64–100 (~6–9.3)
	Part time	Shared office or cube	36–80 (~3.3–7.4)
	Student workers	Cube	36–64 (~3.3–6)
Students	Graduate students	Cube	30–64 (~3.3–6)
	Teacher's assistants	Cube	30–48 (~3.3–4.5)

Source: Stanford University, *Space and Furniture Planning Guidelines*, Stanford, CA, 2009. With permission.

their concentration and to protect confidentiality (Figure 12.2). Faculty workstations should not be shared spaces if possible to allow for visual privacy and storage of documents, books, personal assets, and confidential documents essential to the job function (Figure 12.3). In addition, unlike many shared corporate environments, most faculty spaces are personalized to some degree and each then has their own unique ergonomic challenges. Some faculty members like unorganized spaces that create the simulated comforts of a home office. As an ergonomist, discretion must be taken when recommending ergonomic equipment or changes to furniture in these offices.

When recommending office spaces for faculty or students, several factors can affect the adoption rate. The major aspects include the following:

- Type of work—If most of the workday is spent on field assignments, meetings, site visits, and consultations, a smaller office space may be quite satisfactory. However, for faculty members who perform their job at their workstation most of the time, a small space may

FIGURE 12.2 Faculty workstation.

FIGURE 12.3 Graduate student shared space.

create discomfort due to feelings of confinement. Some job functions that, for example, include frequent meetings in their office space or require the use of multiple sources of material for consultation, research, writing, confidential grading, etc., mean that a faculty member should be assigned more work space.

- Academic climate—The perception of a designated personal space is a matter of comparison. Faculty members generally accept the fact that those at the higher levels in seniority or funding may have larger offices. As an ergonomist, you must be careful of your recommendations to various faculty members, because if the space is not equal to their peers, then the space will likely be deemed too small.
- Individual perception—The amount of space available can have a profound psychological meaning. It is natural for people to strive to occupy more space, for this perception may signify importance, respect, and more authority or power. In the academic setting, the amount of our personal space is often linked with ones status within the university.
- Anthropometrics—The actual office space requirements depend on the size and the shape of people, simply because an office has to accommodate them, enable them to move safely and unhindered in the workspace, and allow them to complete their jobs.
- Equipment—Some faculty may require more equipment than others. For example, some faculty can use a laptop computer, while others require workstations with multiple large screens. Some faculty may have microscopes or other equipment items in their office.

12.3.1.2 Lecture Halls and Classrooms

Lecture halls and classrooms are evolving as universities combine ergonomics in order to create an effective learning environment. The traditional paper-based lecture halls with their rows of desks facing forward all focused on the educators are slowly being phased out. The traditional chalk-and-talk classrooms with blackboards are being replaced by electronic media capabilities—computers, screens, and electronic projectors. Many classrooms are slowly transforming from formal lecture formats into more informal spaces, like coffee shops and residence halls. Understanding physical and cognitive ergonomics is important in the design of the lecture halls and the classrooms in order to improve comfort and stimulate learning. The effective allocation of classroom space in the university depends on several factors such as the following:

- Classroom space allocation and utilization—Classroom space allocation and utilization is the analysis of the adequate classroom numbers and sizes of classrooms, in the appropriate locations, to serve the academic needs. The analysis contains key pieces of information such as the amount of large/medium/small classrooms in a building and their availability at certain times.
- Classroom space for each seat—Classroom space for each seat is the estimation of correctly sized spaces per seat within any given classroom. This is to ensure that the classrooms intended for 100 students are in fact adequately sized to seat 100 students with the appropriate furniture.
- Classroom technology support—Classroom technology support ensures that the technological capabilities of the classroom support the teaching needs of the faculty. Items such as projectors, computers, monitors, and sound and video systems are some examples of these tools.
- Flexibility of classroom space—Flexibility is a key factor in the design of classrooms. The configuration of the room and the furniture layout should have the ability to change as the educational style evolves, and the classroom designs should reflect this.

The physical aspects of a space can promote or discourage certain kinds of learning and learning activities (Van Note Chism 2006). Modern lecture halls feature rows of comfortable, flexible chairs, desks, or linear work surfaces that support technology such as wireless Internet access and screen sharing, document cameras with projection systems, multiple screens for optimal viewing, temperature controls, stimulating wall colors, and much more. These learning spaces should incorporate certain elements including flexibility, comfort, and sensory stimulation.

Ergonomics in the classroom should focus on sensory comfort so that students can see and hear well (Gee 2006). The room dimensions and the physical layout may need to be flexible to accommodate different learning activities, such as collaborative group work, provide a sense of personal space, and encourage social interaction and communication (McVey 1996). As technology evolves, the classroom needs to accommodate the implementation of traditional educational material such as blackboards and whiteboards, as well as electronic technologies such as personal computers, projectors, etc. Ergonomics ensures that these tools are implemented in a way that does not interfere with basic human sensory comfort. So the noise from a computer, the glare from windows or lights, or the overbearing heat from equipment should not interfere with the learning and the teaching that occur in the space (Figure 12.4).

Table 12.2 provides a range of guidelines for different types of classroom spaces.

12.3.1.3 Dormitories and Residence Halls

Most colleges and universities provide single- or multiple-occupancy rooms for their students, usually at a cost. These buildings consist of many such rooms, like an apartment building, and the number of rooms varies quite widely from just a few to hundreds. Most residence halls are much

FIGURE 12.4   Several examples of modern classrooms.

TABLE 12.2
Classroom Space Guidelines

Room Type	Capacity (No. of Stations)	Movable Desks and Chairs: ft²/Station (m²/Station)	Fixed Pedestal Desks and Chair: ft²/Station (m²/Station)	Auditorium Seating: ft²/Station (m²/Station)
Seminar/small class	0–25	16–26 (~1.5–2.5)	20–22 (~1.8–2)	N/A
Classrooms/lecture hall	26–49	16–26 (~1.5–2.5)	18–20 (~1.7–1.8)	N/A
Classrooms/lecture hall	50–99	16–22 (~1.5–2)	18–20 (~1.7–1.8)	14–17 (~1.3–1.6)
Classrooms/lecture hall	100–149	16–22 (~1.5–2)	18–20 (~1.7–1.8)	12–15 (~1.1–1.4)
Classrooms/lecture hall	150–299	16–22 (~1.5–2)	17–19 (~1.6–1.7)	10–14 (~0.9–1.3)
Classrooms/lecture hall	300+	16–22 (~1.5–2)	16–18 (~1.5–1.7)	10–14 (~0.9–1.3)

Source: Cornell University, *Space Planning Guidelines*, Ithaca, NY, 1994. With permission.

closer to campus than comparable private housing such as apartment buildings. This convenience is a major factor in the choice of where to live, since living physically closer to the classrooms is often preferred, particularly for first-year students who may not be permitted to park vehicles on campus. However, many dormitories present unique ergonomic hazards mainly due to space considerations. The rooms often have multiple occupancies ranging from two to six occupants in a single room. Dormitories usually provide each person a bed or bunk bed, a table with a standard chair, and a closet space (Figure 12.5). These pieces of furniture are usually not flexible, unable to be moved

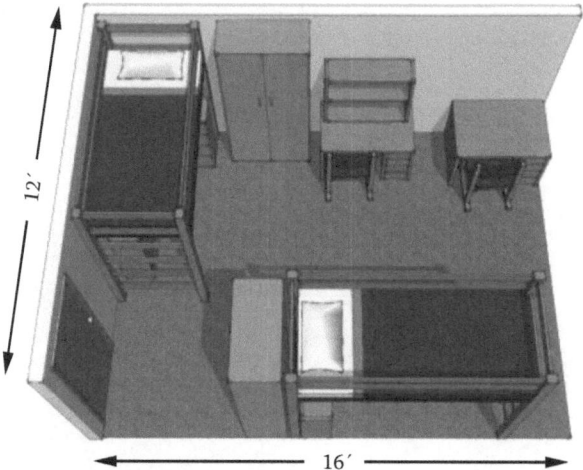

FIGURE 12.5   Dormitory room example.

easily, and heavy, which can force students into awkward postures. Many pieces of furniture do not provide sufficient ergonomic features to be recommended for long periods of work (Menéndez et al. 2009).

Most office/desk chairs feature wheels on the bottom that make it easy to maneuver the chair in and around the desk. Dormitory chairs, however, often stand on planks of wood connecting the two legs at each side, which makes it very difficult to slide around or scoot closer to the desk while someone is occupying it. The seat cushions do not often have enough support for prolonged sitting and can be uncomfortable for larger students. The backrest does not usually have any cushioning or lumbar support that causes the students to slouch or lean forward in order to sit comfortably (Figure 12.6).

Dormitory furniture should accommodate the various sizes of the users that live in the university residence. Universities should procure beds, closets, desks, and chairs with casters to allow for easy movement and cleaning of the room. Desks that are adjustable by height make it possible for individuals of different sizes to use the desk to do schoolwork without discomfort. Chairs should have the basic ergonomic features such as casters, lumbar support, and height adjustment. Beds sizes should accommodate for larger populations in order to be usable for most of the student population. Dormitories are where students spend a majority of their time resting and studying; the furniture in these spaces should accommodate these various activities without putting the students' bodies in awkward postures.

FIGURE 12.6   Standard college dormitory chair.

FIGURE 12.7   Librarian performing a book-lifting task.

12.3.1.4 Library Ergonomics

Libraries are a university's principal repository of information and accumulated knowledge. Although computers continually grow in use, much of the information is still retained in a physical format, i.e., newspaper, pamphlets, books, videos, film reels, slides, and photographs. Moving these materials is a large part of the ergonomic exposures faced in library work. People working in libraries perform numerous manual-handling tasks, such as shelving books and maneuvering book carts (Figure 12.7). These tasks can put stress on the back, the shoulders, the arms, the hands and the wrists, and can increase the risk of repetitive stress injuries (Adeyemi 2010).

When employees shelve heavy materials, strategies such as footstools to stand on or power grips to hold books should be used to reduce the stress on shoulders, arms, and upper back (Sornam et al. 2011). In addition, when employees are maneuvering book carts, there should be a focus on training to educate on the proper pushing rather than the pulling of the carts to reduce the strains on the upper extremities (Chandra et al. 2009). Lifting and carrying are often strenuous, and back injuries are a common issue for librarians (Gavgani et al. 2013). Libraries should have elevators to convey library materials from one location to the other. The workers who will be affected by the ergonomic changes must be involved in the discussions before the changes are implemented. Their input can help determine the necessary and appropriate changes. The goal of ergonomics is to look for ways to make the job fit the worker, instead of forcing the worker to conform to the job. Proper positioning of the computers is crucial to prevent pain and injury (Chengalur et al. 2004). Make sure that the computers are situated directly in front of the workers to avoid uncomfortable positions. Place computers perpendicular to the light sources and consider buying screen protectors to cut down on glare.

12.3.1.5 Laboratory Ergonomics

A university often has many researchers, students, and faculty who participate in science experiments, and each of these workplaces has various ergonomic risks. Tasks such as pipetting, microscope work, fume hood activities, and data analysis are all associated with high risk to the upper extremities. In some university labs, it is general practice to have the same person perform the experiments or the tests from start to finish. These tests can be quite lengthy and fatiguing. It is important to know how long to work and how long to rest, stretch, or perform a task that encourages the off-loading of the stressed muscles. Some of the most high-risk tasks involve pipetting and micromanipulation activities such as using the microscope or fume hood. Pipetting has typically been a source of hand and shoulder problems due to its repetitiveness and force output (Asundi et al. 2005).

The microscope work can be a source of eye strain, and neck, lower back, shoulder, and arm problems, if conducted for long periods without breaks (Darragh et al. 2008). The forward head and extended arm positions during hood work are difficult to avoid, but the strain caused by them can be minimized. The issues with typical laboratory tasks are summarized in the following. For a more extensive review of these issues, see Chapter 13 of this book.

- Pipetting—Pipettes are important tools for transferring small and precise volumes of liquids in university laboratories. Researchers often pipette for several hours a day, placing them at a high risk of discomfort or injury. Research has shown an increase in hand and shoulder injuries when pipetting for >300 h per year (David and Buckle 1997). The traditional plunger forces can reach 4 kg during dispensing and tip ejection. This is above the dynamic peak force levels of 3 kg (6.6 lbs) and 2.1 kg (4.6 lbs) that are recommended for male and female workers, respectively, by Kroemer (1989). Awkward elbow and wrist postures can reduce the available grip strength from 25% to 60% (Figure 12.8). Rempel et al. (1998) reported high pressure levels on the median nerve when working in the forearm rotation and the wrist flexion. When high precision is required, the workload on the muscles is even greater (Asundi et al. 2005), further increasing the risk of injury. The best defenses against pipetting injuries include using well-designed low-force pipettes, working in neutral postures, and minimizing repetition. Electronic pipettes reduce force but can be inaccurate and heavy, primarily depending on the battery weight and the amount of fluid being transferred. Multichannel pipettes are faster, decrease repetition, but may have high plunger and tip forces, and are generally heavier, and the tips may seal unevenly. There are some pipettes with innovative designs that reduce awkward postures and force. James and Glascock (2005) compared traditional pipettes with the Ovation BioNatural pipette (Figure 12.9). The users in a clinical lab preferred the Ovation for comfort, accuracy, and general use. Hand forces, shoulder elevation, and wrist deviation, flexion, and extension were decreased, but the forearm rotation was increased, although this is considered less of a risk factor than the awkward wrist postures.
- Microscope Ergonomics—The human body is not adapted to sitting or standing, bent over a microscope eyepiece for prolonged periods (James 1995). Microscope work requires the head and the arms to be flexed in a forward position and inclined toward the microscope with abducted shoulders, a posture that can irritate soft tissues, such as muscles, ligaments, and disks (Kreczy et al. 1999). Workers often place their feet on ring-style footrests that

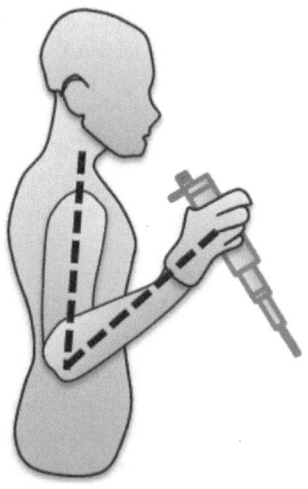

FIGURE 12.8 Typical pipetting posture.

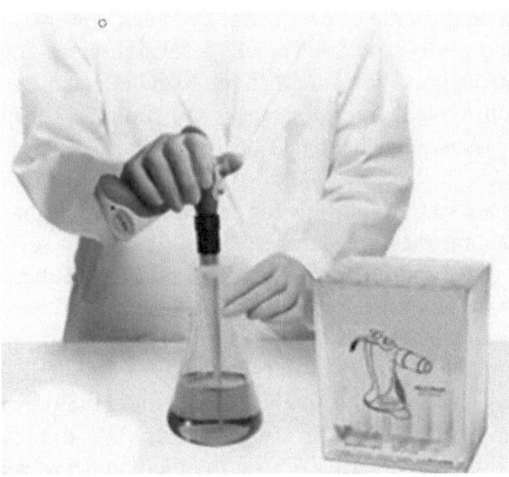

FIGURE 12.9 Ovation BioNatural pipetting posture. (Courtesy of Esslab, http://www.esslab.com/Vistalab /images/MacroPage.jpg, Essex, London.)

are common to many lab stools, which places the person in a posture that is further exaggerated. Some risk factors from microscope work include head inclinations of up to 45° and upper back inclination at angles of up to 30°, awkward posture of the upper extremities, and repetitive motions that increases muscular fatigue (James 1995). The major issue with using conventional microscopes is that viewing the specimens requires the users to maintain a flexed neck posture while the hands are in a relatively fixed position. The biomechanical posture of maintaining an incline of 30° from the vertical can produce significant muscle contractions, muscle fatigue, and in some cases pain (Kreczy et al. 1999). The repetitive motions of the hands and the contact stress of the arms resting on a hard surface can cause pain and nerve injury, leading to repetitive stress injuries and/or carpal tunnel syndrome. A recommendation is to promote a neutral erect working posture. The optical path (distance from the ocular lenses to the specimen being viewed) should range between 45 and 55 cm (18–21.5 in.). The eyepieces should be no more than 30° above the horizontal plane of the desk surface (Helander and Prabhu 1991). A majority of older microscopes, however, have much shorter optical path dimensions (25–30 cm or 10–12 in.) with the eyepieces angled at 60° above horizontal (Haines and McAtamney 1993). This recommendation can be simplified in Figure 12.10 to minimize fatigue. The goal in any ergonomic posture is to raise the microscope high enough to reduce the neck flexion while allowing the forearms to be parallel to the floor.

• Laboratory Micromanipulation—The work at hoods and biosafety cabinets is especially awkward, often without legroom, height adjustability, or appropriate arm support (Figure 12.11). Many university laboratories are designed for specific purposes, and the accommodation of workers is usually not a priority over the scientific discoveries. One of the main issues that compound a micromanipulation task is the work surface height. The counters are typically not adjustable and are set at 0.9–1.0 m (35.4–39.4 in.) high: too low for standing and too high for sitting. Linear laboratory counters do not often afford sufficient space and increase the reach distances as the instruments are laterally spread out. In addition, there is usually insufficient legroom under these linear counters, and that prevents the employees from sitting closer to their work. The legroom space is often filled with cabinetry or improvised storage, further compromising the legroom and increasing the reach distances. There is frequently an inadequate planning of the laboratory space for material and information flows, including the location of computers.

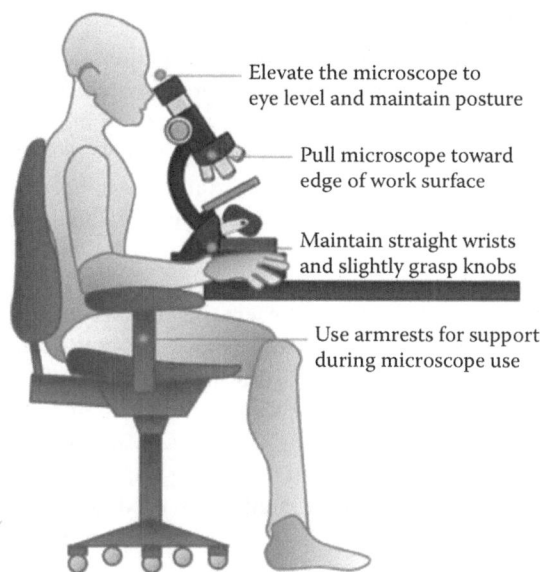

Elevate the microscope to
eye level and maintain posture

Pull microscope toward
edge of work surface

Maintain straight wrists
and slightly grasp knobs

Use armrests for support
during microscope use

FIGURE 12.10   Typical microscope posture.

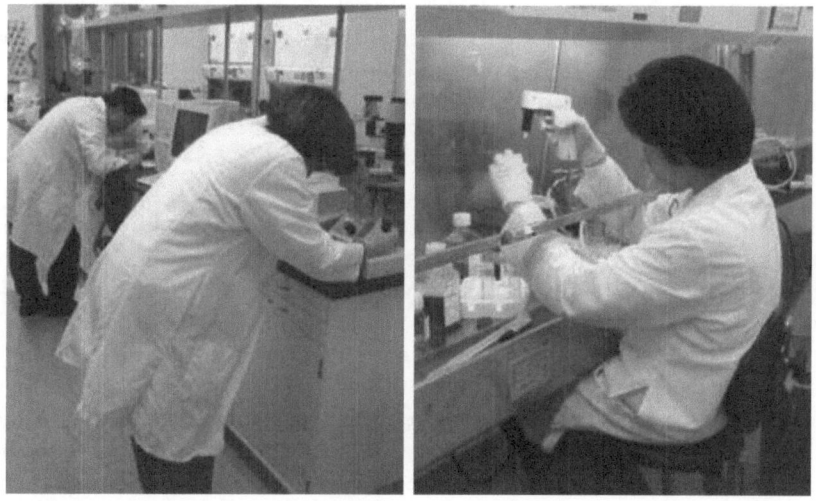

FIGURE 12.11   Typical micromanipulation postures.

The technological changes in the university laboratory are often made in an incremental manner when funding is available to purchase equipment. If well planned, these can be opportunities
for ergonomics improvement rather than for the introduction of new injury risk factors. Modular
furniture and freestanding, adjustable work surfaces from 0.8 to 1.1 m (27–34 in.) are more adaptive than fixed-height laboratory benches, allowing height adjustments to take task requirements
and equipment scale into account. Instruments of different sizes can have appropriate work surface
heights, so that users can reach the areas of operation without awkward postures. Incremental and
periodic changes in equipment and technology can be easily and inexpensively accommodated by
adjusting specific workstations. For more information on laboratory ergonomics, see Chapter 13
of this book.

12.3.1.6 Animal-Handling Ergonomics

At universities, animal care employees play a critical role in maintaining the health and well-being of animal test subjects and supporting ongoing university research. To perform these job functions, the workers can be exposed to ergonomic risks such as repetitive motion, strain, and awkward postures. Some of the most strenuous tasks involve handling water bottles, cleaning cages, feeding, and changing the bedding for various small animals (for example, see Project Woodchuck [1997]). This section will describe these issues and the various ergonomic solutions. For this section, we will only discuss small animal handling.

- Water bottle handling—Water bottles represent the most frequently handled item throughout the small animal cage-cleaning process. The water bottle handling process includes cleaning, filling, capping and uncapping, transporting to and from the cage room, placing on and removing from the cages, emptying, and sanitizing. The bottles are placed in racks and filled before transport, making them heavy and difficult to handle. The average weights of these tray can range from 9 to 18 kg (20–40 lb) depending on the amount of water and the types of water bottle containers used (glass, plastic, or metal bottles [Figure 12.12]) (Kerst 2003). Several issues can arise with the repeated handling of the water bottles including
  - Repetitive motion and forceful pinching/gripping associated with capping and uncapping bottles
  - Forceful exertions and awkward postures when grasping and tipping the racks to dump the water
  - Repetitive and awkward postures when lifting, carrying, loading, and unloading the heavy racks
  - Sustained forceful exertion while pushing and pulling the heavy racks of water bottles on carts
  - Awkward postures when removing the bottle racks from the tunnel washer

  The best ergonomic solution that can reduce water bottle handling is having the universities switch to automatic watering systems or prepackaged water (Hydropac) solutions, which can provide sterile water to an animal for 2 weeks at a time, thereby reducing the labor costs and the repetitive motions (see Figure 12.13). The prepackaged water is lighter compared to glass or plastic bottles, reducing the overall lifting and cleaning of these watering systems. These solutions reduce the amount of pinching associated with pinching and the overall repetitive motions of refilling the water bottles. If this solution is cost prohibitive, administrative solutions such as switching hands or arranging the bottle rack within close proximity can reduce the repetitive task strains.

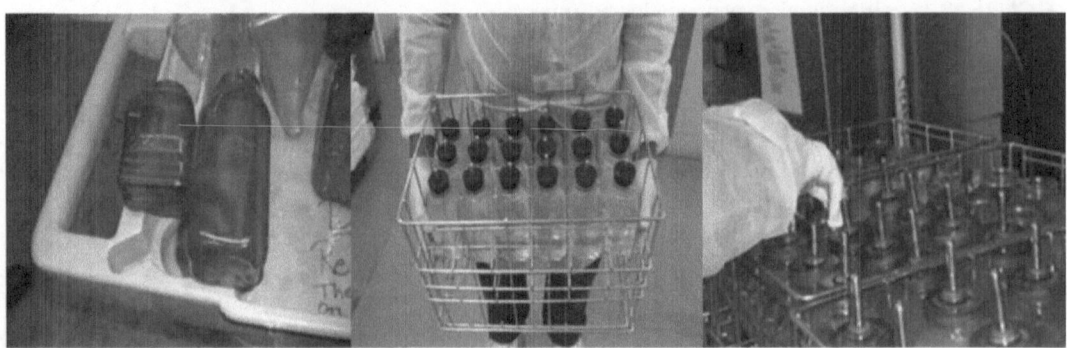

FIGURE 12.12 Examples of animal water bottles.

FIGURE 12.13 Examples of Hydropac animal watering system. (Courtesy of Hydropac, http://www.hydropac .net/storage/download/Hydropac_Product_Guide.pdf, Seaford, DE.)

- Cage changing—Laboratory animal care staff is an occupational group that is at a high risk for injury throughout a university system (Kerst 2003). This is a highly regulated work environment that requires a specialized skill set to provide care for the animal subjects. The physical nature of the work involves repetitive tasks, and the staff use awkward postures and forceful exertions in areas crowded with cages and equipment (ILAR 2011). The process for cleaning many small animals (rats, mice, rabbits) is relatively simple but has high amounts of repetitive motion. The supplies for cage changing (clean rabbit cage racks, prefilled rodent boxes, fresh feed, water bottles, etc.) are transported to the animal room. The racks are repositioned in the room, and the animals are transferred into the clean caging. The trash is removed; the floors are swept and mopped; and all surfaces are wiped down with a disinfectant (Figure 12.14). The ergonomic risk factors associated with these tasks include
  - Repetitive pinching, gripping, and awkward wrist postures when handling the cage lids and the rodents
  - Static postures when holding the cage lids during rodent transfers
  - Bending and lifting while obtaining the supplies and performing the cleaning tasks
  - Bending or kneeling to remove the cages from the racks
  - Overhead reaching to access the cages on the top areas of the racks
  - Prolonged standing on hard floors

FIGURE 12.14 Typical cage cleaning task.

The best ergonomic solution that can reduce ergonomic risk is to place the work surfaces at appropriate heights for each worker. The height range from 34 to 37 in. (85–110 cm) serve as a good baseline to establish these workstations (Chengalur et al. 2004). This solution will reduce the need for unnecessary bending and reaching when scooping feed, refilling water, and emptying bedding (UCEPT 2014).

- Cage cleaning and bedding dispensing—Universities that conduct animal research usually have facilities where they transport their dirty cages to be cleaned and rebedded for new animals. The dirty cages are transported to a dedicated emptying and cleaning area and typically dumped by manually tipping, banging, and scraping the dirty bedding into an open container. Once the bedding is emptied, the cages are turned upside down on a tunnel washer or placed on racks for the cage washer. Once the cages are cleaned, the bedding is manually scooped or occasionally automatically dispensed into the clean cages. Some of the ergonomic risk factors for these job tasks include
  - Awkward gripping postures while grasping multiple cages and manually scooping the bedding
  - Bending forward at the waist while manually scooping the bedding into cages placed low on the carts
  - Repetitive and forceful motions while banging the cages to remove the dirty bedding
  - Awkward shoulder postures while holding the cage to scrape the contents

Figure 12.15 shows some solutions to cage cleaning which include the automation of these work processes with robotics in order to reduce the risk of injury and improve the efficiency and the air quality (Hesler 2009). Recognizing that this equipment can be cost prohibitive, providing maneuverable carts and a tiered racking system can reduce the amount of repetitive lifting and reaching associated with cage bedding (UCEPT 2014).

- Cart moving and rack—Animal care tasks involve frequent pushing, pulling, or maneuvering large carts and cage racks (Figure 12.16). This includes maneuvering the cage racks into and out of the cage washer and moving these racks within the rooms or from room to room. It also involves maneuvering the loaded carts (or similar material-handling equipment) to transport equipment, supplies, large cages, etc., within the facility. The challenges reported with these push/pull tasks include limited space, poor cart/rack maneuverability, inappropriate and poorly maintained casters, steep ramps, and heavy loads. The ergonomic risk factors for these tasks include
  - Awkward shoulder, hand, and wrist postures while grasping the handles
  - Lower back, neck, and shoulder strain to maneuver the heavy loads or push up the steep ramps
  - Forceful grip on the handles

FIGURE 12.15   Typical bedding scoop cage-cleaning task.

FIGURE 12.16 Examples of different animal cage racks.

Ramps leading into the cage washers should have a gradual slope to minimize the forces required to push the racks into the washer. The casters and the wheels on the cage racks should be carefully evaluated and selected, so there are minimal push and pull forces required to move the racks. Position the racks close to the body, with the elbows slightly bent, to make it easier to maneuver the racks. The loaded cart height should not exceed 56 in. to the top of the cart to ensure that the shorter handlers are able to see over the load during the operation. Using information from the Liberty Mutual tables, the recommended load should not exceed 225 N (50 lbf or 23 kgf) while using the whole body or 110 N (24 lbf or 11 kgf) while using only the upper extremities (Eastman Kodak 2004).

12.3.1.7 Dining Services Ergonomics

At many large universities, dining services plays a critical role in providing food for thousands of students, guests, staff, and faculty. The workers are often required to prepare, cook, and dispose of large quantities of food that have high ergonomic risk associated with them (Grant and Daniel 1997). Some of the most strenuous tasks involve food preparation, material handling in the kitchen, and transportation/storage of food. This section will describe these issues and the various ergonomic solutions.

- Food preparation—Food preparation are tasks that involve cutting, mixing, lifting, and moving large amounts of food. Universities usually order large quantities of bulk food to feed their campuses, and the workers must be able to handle large quantities rarely seen in normal restaurants. The food is usually not presliced or peeled, which requires the workers to prepare the materials for cooking. These workers often perform cutting and slicing tasks for hours of preparation (Figure 12.17). Many of the injuries that often occur in the dining services involve cutting or slicing hands and/or fingers and increased strain in the wrists (Grant and Daniel 1997).

  The best ergonomic solution for food preparation in the kitchen is having universities switch to precut or prepeeled foods. This solution reduces the amount of awkward wrist postures associated with slicing and overall repetitive motions of cutting food (Marsot 2007). In addition, the differences between a dull and a sharp knife (20 N; 4.5 lbf; 2 kgf) can increase three-fold (60 N; 13.5 lbf; 6 kgf) by repeated use and improper upkeep of knives. Meat slicing machines can eliminate the need to manually cut meats; ensure that the machines are placed at the correct height to minimize awkward wrist and shoulder postures, which for sitting is 28 in. (71 cm) and for standing, 37 in. (94 cm) Also ensure that the workers follow the appropriate meat slicing safety procedures, including using guards and proper cleaning techniques. If this solution is cost prohibitive, solutions such as automation or completing food preparation during slow periods can reduce fatigue during the peak hours. In addition, providing knives with various handle sizes for larger and smaller hands can reduce the force required to cut food. Ensure that the workers have appropriate

FIGURE 12.17 Typical standing food preparation work surface.

knives to choose from, and that they use the appropriate knife for the task. The steel in the blade should be easy to sharpen and should be regularly maintained to ensure that the cutting takes minimal force and repetition (McGorry et al. 2003).

- Material handling in the kitchen—Material handling in dining commons is probably at the highest risk for serious injury. University employees handle heavy (over 66 lb; 20 kg) boxes (Figure 12.18) or oddly shaped packages (meat, vegetables, fruits) which vary on a daily basis (UC Ergonomics Project Team [UCEPT] 2012). For example, baking bread for a university population requires kneading and mixing large amounts of dough in a short period. Studies show that the prolonged stoop posture and high can lead to serious injuries during times when the demand for food is high (Nou 2012). In addition, these high stress times do not allow for adequate rest for the employees to recover from muscle fatigue, putting them at further risk for injury.

Some solutions to these issues include installing height-adjustable work surfaces that will improve productivity and comfort. Installing these in strategic locations to accommodate employee height differences and to make heavy tasks (i.e., using meat slicers and cheese graters) and light work (i.e., slicing, peeling, and cleaning foods) easier to perform. Such work surfaces should have a height range of at least 28–44 in. (71–112 cm).

FIGURE 12.18 Typical food package handling task.

12.3.1.8 Custodial Services

Custodial services are responsible for maintaining building interiors and ensuring that the campus remains clean. Custodial staff generally work in buildings that are planned for other workers and not typically designed to accommodate cleaning, and this can cause problems (e.g., location of taps, storage facilities, access, and unsuitable floor materials). Custodial workers are at risk for repetitive motion injuries during routine tasks such as using a broom or a mop, moving furniture, using a vacuum cleaner, cleaning restrooms, and moving trash. In addition, custodial workers may find themselves frequently bending at the waist and lifting awkward objects.

- Mopping—Mopping is done in various areas in the university such as laboratories, hallways, and dining commons (Figure 12.19). The mop pails used by the custodians usually have a bisected water bucket, one side for clean, soapy water and the other for dirty mop water, and a removable roller system to squeeze the water out of the mop (Hagner and Hagberg 1989). In order to activate the rollers, the custodian must bend over and push or pull on a short lever arm with considerable force. At the same time, the bucket must be prevented from rolling by either holding it with another hand or placing a foot in front of the wheels. The potential risks from frequent mopping include back and shoulder flexion when mopping hard to reach areas (Chang et al. 2012). Some solutions include using automated floor cleaning equipment that can work in a variety of locations and will reduce the physical risks associated with manual mopping (Haslam and Williams 1999). For larger areas, no-touch cleaning systems and automatic scrubbers can significantly reduce the ergonomic risks and provide a higher level of cleaning, especially for larger areas (Figure 12.20). If these solutions are cost prohibited, providing adjustable handles and lightweight mop heads will reduce the amount of strain placed on the workers.
- Vacuuming—Many buildings may need a combination of vacuums to safely clean all areas. It is best to identify the most efficient and practical vacuum for each area to be cleaned. Establish and enforce a regular maintenance program for all vacuums. The physical demands of vacuuming tasks exhibit risk due to repetitive bending of the back and shoulder motion (Koerber 2006). There are two main types of vacuum cleaning machines: backpack and canister/barrel machines. Some solutions when choosing the appropriate vacuum include finding the best wheels or casters for the job so that the equipment easily rolls. Consider the best size and material of the wheels or the casters as well as the type

FIGURE 12.19 Typical manual mopping task.

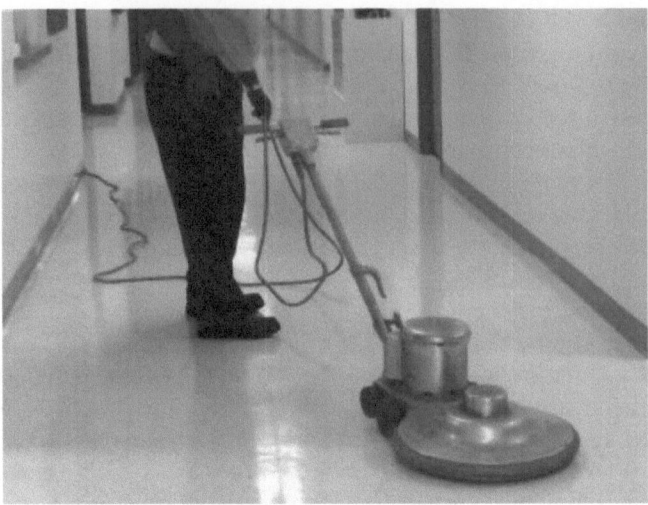

FIGURE 12.20 Typical powered floor scrubbing and polishing task.

of floor and the work environment. Canister/barrel vacuums should be used in hallways, offices, residence halls, and small-to-medium spaces (Figure 12.21). Backpack vacuum cleaners should be used to clean hard-to-reach areas or where upright vacuums are not practical for use, such as stairs, chandeliers, and windowsills (Figure 12.22). The use of backpack vacuums in large areas should be avoided, as this is inefficient and creates excessive physical load to the worker (Koerber 2006). Lighter weight models represent a trade-off: less weight for less power with smaller bags and less capacity.

- Moving furniture—A university usually has various types of buildings (dormitories, laboratories, offices, etc.) that have different pieces of furniture that have to be cleaned or moved. Moving and lifting heavy furniture represent a significant risk to the custodial workers (Figure 12.23). The most obvious injury risks are associated with the lifting and moving of heavy objects. These exposures are heightened by the repetitive movement, the awkward shapes of the objects moved, and the need to walk with and carry these loads from place to place (Paskiewicz et al. 2003). This section provides the moving and the heavy custodial staff with information about how to safely work and to reduce the risk

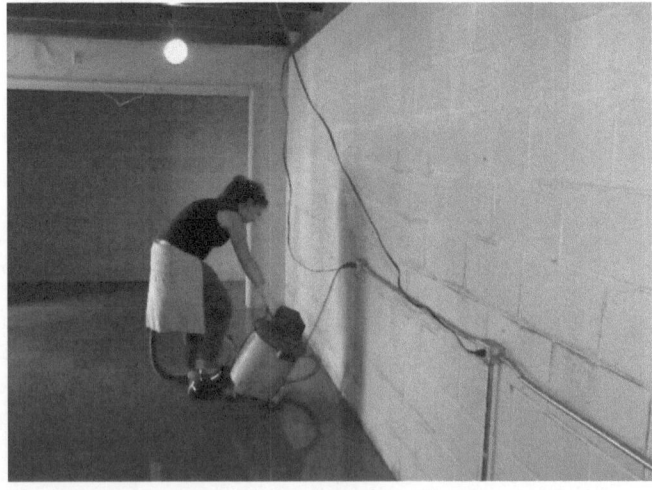

FIGURE 12.21 Typical vacuuming with a floor canister vacuum cleaner.

FIGURE 12.22 Typical backpack vacuum cleaner use.

of ergonomic injury. Team lift policies should be established and proper moving equipment should be provided. All lift teams should have a minimum of two people handling any piece of furniture to have the maximum power of each lifter (Sharp et al. 1997). The setting-up and tearing-down of furniture to accommodate various events demand the frequent moving of furniture specifically designed for this use. This type of furniture should be lightweight, easy to move, and easy to stack and store.

- Trash disposal—The custodial staff at a large university has the enormous task of cleaning not only classrooms and office spaces, but also student residential areas and on-site event locations. The number of trips is hard enough, but the most painful part could be holding the dumpster lid open with one hand while throwing in the garbage bags (Goggin 2006). Trash cans and barrels that do not have handles introduce a pinch grip injury risk, as the worker grips the top edge of the trash can to lift it. Repeated gripping of potentially heavy

FIGURE 12.23 Typical furniture lift task.

FIGURE 12.24 Typical garbage lift.

loads leads to hand injuries and strains (Boisclair 2010). Not only is this method dirty and less effective, but also throwing items with one hand and over the shoulder could potentially create injuries. Pulling the bags out of larger trash cans requires hand gripping the bag to extend over the shoulder height in order for the bag to be removed from the trash can (Figure 12.24). When the trash is transferred from trash cans to larger dumpsters, the workers must lift the large, heavy bags of debris up and over the side of the dumpster. Even small dumpsters require the worker to extend their arms up and over the shoulder height.

FIGURE 12.25 Dumpster with a lid lift.

Some solutions to these issues are replacing taller dumpster with shorter and wider dumpsters to reduce the overhead reaching while maintaining the storage volume (UCEPT 2012). Installing these dumpsters in various locations to accommodate employees that often access the dumpsters can reduce the overall strain. In addition, having tools that can prop a dumpster lid open (see Figure 12.25) or a stepping stool can reduce the overhead lifting.

12.4 CONCLUSIONS

Universities are microcommunities that have various services in education and the support system that runs the university. The ergonomic issues of the university are as unique as the research being conducted in the various laboratories. In order to handle the ergonomic issues of the university, an ergonomist must be aware of the staff and faculty job risks. The obstacles to ergonomic implementations such as faculty funding, departmental fiefdoms, and bureaucratic processes are part of the university environment. However, this should not deter the ergonomist from trying to emphasize the importance of ergonomics in the university and applying administrative and engineering solutions to the work place.

REFERENCES

Adeyemi, A. O. 2010. ICT facilities: Ergonomic effects on academic library staff. *Lib Philos Pract*. Paper 343. From http://digitalcommons.unl.edu/cgi/viewcontent.cgi?article=1351&context=libphilprac.

Asundi, K. R., J. M. Bach, and D. M. Rempel. 2005. Thumb force and muscle loads are influenced by the design of a mechanical pipette and by pipetting tasks. *Human Factors*. 47:67–76.

Boisclair, L. 2010. Recycling made easy: An ergonomic approach to home recycling and trash containers. From http://digitalcommons.calpoly.edu/imesp/44/.

Chandra, A. M., S. Gosh, S. Barman, and D. P. Chakravarti. 2009. Ergonomic issues in academic libraries in Kolkata, West Bengal: A pilot study. *Lib Philos Pract*. June.

Chang, J.-H., J.-D. Wu, C.-Y. Liu, and D.-J. Hsu. 2012. Prevalence of musculoskeletal disorders and ergonomic assessments of cleaners. *Amer J Indust Med*. 55(7):593–604.

Chengalur, S. N., Rodgers, S. H., and T. E. Bernard. 2004. *Kodak's Ergonomic Design for People at Work*. Secon edn. Hoboken, NJ: Wiley.

Cornell University. 1994. *Space Planning Guidelines*. Ithaca, NY: Cornell University.

Darragh, A. R., H. Harrison, and S. Kenny. 2008. Effect of an ergonomics intervention on workstations of microscope workers. *Amer J Occup Therapy*. 62:61–9.

David, G., and P. Buckle. 1997. A questionnaire survey of the ergonomic problems associated with pipettes and their usage with specific reference to work-related upper limb disorders. *Appl Ergon*. 28(4):257–62.

Eastman Kodak Company. 2004. *Kodak's Ergonomic Design for People at Work*. Hoboken, NJ: Wiley.

Gavgani, V. Z., J. Nazari, M. A. Jafarabadi, and F. Rastegari. 2013. Is librarians' health affected by ergonomic factors at the work place? *Lib Phil Pract. (e-journal)*. Paper 893.

Gee, L. 2006. Human-centered design guidelines. In Oblinger, D. (ed.) *Learning Spaces*. From http://www.educause.edu/learningspacesch10.

Goggins, R. 2006. Ergonomic solutions for cleaning workers. *Proc 2006 ASSE Prof Develop Conf*. Park Ridge, IL: American Society of Safety Engineers.

Grant, K. A., and J. H. Daniel. 1997. An electromyographic study of strength and upper extremity muscle activity in simulated meat cutting tasks. *Appl Ergon*. 28(2):129–37.

Hagner, I.-M., and M. Hagberg. Evaluation of two floor-mopping work methods by measurement of load. *Ergonomics*. 32(4):401–8.

Haines, H. and McAtamney, L. 1993. Applying ergonomics to improve microscopy work. *Microscopy and Analysis*. 17–19.

Haslam, R. A., and H. J. Williams. 1999. Ergonomics considerations in the design and use of single disc floor cleaning machines. *Appl Ergon*. 30(5):391–9.

Helander, M. and P. Prabhu. 1991. Planning and implementation of microscope work. *Appl Ergon*. 22:36–42.

ILAR (Institute of Laboratory Animal Resources). 2011. *Guide for the Care and Use of Laboratory Animals*. Eight edn. Washington DC: National Academies Press.

James, T. 1995. Microscope use. *Hum Fac Erg Soc P*. 39(10):573–7.

James, T., and N. Glascock. 2005. Comparison of traditional and alternative pipettes—Use and preference. *Appl Ergon Conf.* New Orleans, LA.

Kerst, J. 2003. An ergonomics process for the care and use of research animals. *ILAR J.* 44(1):3–12.

Koerber, J. A. 2006. *Ergonomic Comparison of Upright and Backpack Vacuum Cleaners.* Diss. Davis, CA: University of California.

Kreczy, A., M. Kofler, and A. Gschwendtner. 1999. Underestimated health hazard: Proposal for an ergonomic microscope workstation. *Lancet.* 354:1701–2.

Kroemer, K. H. E. 1989. Cumulative trauma disorders: Their recognition and ergonomics measures to avoid them. *Appl Ergon.* 20:274–80.

Marsot, J., L. Claudon, and M. Jacqmin. 2007. Assessment of knife sharpness by means of a cutting force measuring system. *Appl Ergon.* 38(1):83–9.

McGorry, R. W., P. C. Dowd, and P. G. Dempsey. 2003. Cutting moments and grip forces in meat cutting operations and the effect of knife sharpness. *Appl Ergon.* 34(4):375–82.

McVey, G. 1996. Ergonomics and the learning environment. In Jonassen, D. (ed.) *Handbook of Research for Education Communications and Technology.* New York: Macmillan.

Menéndez, C. C., B. C. Amick, C.-H. Chang, R. B. Harrist, M. Jenkins, M. Robertson, I. Janowitz, D. M. Rempel, J. N. Katz, and J. T. Dennerlein. 2009. The epidemiology of upper extremity musculoskeletal symptoms on a college campus. *Work.* 34(4):401–8.

Nou, D., B. J. Miller, and F. A. Fathallah. 2012. Low back muscle fatigue measurements of cyclic and prolonged stooped work. *Hum Fac Erg Soc P.* 56(1):1196–1200.

Paskiewicz, K. J., and F. A. Fathallah. 2003. *Effects of the GRIPSystem (TM) on Muscle Activity and Kinematics of the Lower Back.* Diss. Davis, CA: University of California.

Project Woodchuck. 1997. http://ergo.human.cornell.edu/ergoprojects/woodchuck/intro.htm. Ithaca, NY: Cornell University.

Rempel, D., J. Bach, L. Gordon, and Y. So. 1998. Effects of forearm pronation/supination on carpal tunnel pressure. *J Hand Surg.* 23A:38–42.

Sharp, M. A., V. J. Rice, B. C. Nindl, and T. L. Williamson. 1997. Effects of team size on the maximum weight bar lifting strength of military personnel. *Human Factors.* 39(3):481–8.

Sornam, A. et al. 2011. Ergonomics and library professionals: A study. *SRELS J Info Manage.* 48(6):625–40.

Stanford University. 2009. *Space and Furniture Planning Guidelines.* Stanford, CA: Stanford University.

UCEPT (University of California Ergonomics Project Team). 2012. *Ergonomics Study of Dining Services Positions at the University of California.*

UCEPT. 2014. *Ergonomics Study of Animal Care Positions at the University of California.*

Van Note Chism, N. 2006. Challenging traditional assumptions and rethinking learning spaces. In Oblinger, D. (ed.) *Learning Spaces.* From http://www.educause.edu/learningspaces.

ADDITIONAL SUGGESTED READING

Aickin, C. 1997. Ergonomic assessment (manual handling) of cleaning work. *Proc—Int Workplace Health Safety Forum.* Queensland.

Andrew, M., Y. Bhambhani, and J. Wessel. 1998. Physiological and perceptual responses during household activities performed by healthy women. *Amer J Indust Med.* 39:180–93.

Balogh, I., P. Oerbaek, K. Ohlsson, C. Nordander, J. Unge, and J. Winkel. 2004. Self-assessed and directly measured occupational physical activities—Influence of musculoskeletal complaints, age and gender. *Appl Ergon.* 35:49–56.

Brown, M. 2006. Trends in learning space design. In Oblinger, D. (ed.) *Learning Spaces.* From http://www.educause.edu/learningspacesch9.

Cornell, P. 2006. The impact of changes in teaching and learning on furniture and the learning environment. *New Dir Teach Learn.* 92(winter):33–43. From http://www.eric.ed.gov/ERICWebPortal.

Gaudry, B. 1998. *Manual Material Handling Prevention for Cleaning Contractors and Their Employees in State Government Schools.* New South Wales: WorkCover.

Gober, J. 2011. Get a grip: Three things that are wrong with many handles. From http://www.humantech.com/blog/.

Gunn, S. M., A. G. Brooks, R. T. Withers, C. J. Gore, N. Owen, M. L. Booth, and A. E. Bauman. 2002. Determining energy expenditure during some household and garden tasks. *Med Sc Sports & Exerc.* 34(5):895–902.

Hesler, J. and N. Lehner (eds.). 2009. *Planning and Designing Research Animal Facilities.* London: Elsevier.

Huerkamp, M. J., M. A. Gladle, M. P. Mottet, and K. Forde. 2009. Ergonomic considerations and allergen management. In Hessler, J. R., and N. D. M. Lehner (eds.) *Planning and Designing Research Animal Facilities.* San Diego, CA: Elsevier. 115–28.

Humantech. 2009. *The Handbook of Ergonomic Design Guidelines.* Second edn. Ann Arbor, MI: Humantech, Inc.

Joint Information Systems Committee (JISC). 2006. Designing spaces for effective learning. From JISC e-learning Programme, http://www.jisc.ac.uk (accessed July 27, 2010).

LeBleanc, S., J. Percifield, M. Huerkamp, and M. McGarry. 2010. Application of ergonomics to animal facility operations. *ALN Magazine.* October 1 issue. From http://www.alnmag.com/articles/2010/10 /applicationergonomicsanimal-facility-operations.

Loruso, A., S. Bruno, and N. L'Abbate. 2009. Musculoskeletal disorders among university student computer users [in Italian]. *Med Lav.* 100(1):29–34.

National Occupational Health and Safety Commission. 2004. National code of practice for the prevention of musculoskeletal disorders (MSD) from manual handling at work. *NOHSC: 2005.* Canberra: AGPS.

Norman, J. F., J. A. Kautz, H. D. Wengler et al. 2003. Physical demands of vacuuming in women using different models of vacuum cleaners. *Med Sci Sports Exercise.* 35(2):364–9.

Panero, J., and M. Zelnik. 1979. *Human Dimension & Interior Space.* New York: Whitney Library of Design.

Paver, C., J. Crosbie, R. Lee, and G. Paver. 1997. Analysis of wet mopping in the cleaning industry. Report to WorkCover. New South Wales: WorkCover.

Putz-Anderson, V. 1988. *Cumulative Trauma Disorders: A Manual for Musculoskeletal Disease of the Upper Limbs.* Bristol, PA: Taylor & Francis.

Village, J., M. Koehoorn, S. Hossain, and A. Ostry. 2009. Quantifying tasks, ergonomic exposures and injury rates among school custodial workers. *Ergonomics.* 52(6):723–34.

Woods, V. and P. Buckle. 2004. An investigation into the design and use of workplace cleaning equipment. *Int J Indust Ergon.* 35:247–66.

Woods, V., P. Buckle, and M. Haisman. 1999. Musculoskeletal Health of Cleaners. No. 215. *Robens Centre for Health and Ergonomics.*

13 Biotechnology Laboratories
Promoting Health and Productivity with Ergonomics

Cynthia M. Burt

CONTENTS

13.1 INTRODUCTION

The nature of biotechnology work has changed as laboratory processes and equipment have evolved. Automated systems such as robotic high-throughput systems and image analysis have improved the workflow while increasing the precision of laboratory work. Unfortunately, the interface of the biotechnology worker with the improved tools and equipment remains a challenge. Many tasks continue to be completed with traditional microscopes, pipettes, and tools such as forceps that require prolonged awkward postures and repetitive motions.

Numerous studies have shown that many laboratory workers suffer from prolonged exposure to force, repetition, awkward postures, static postures, and contact stress. More than 85% of cytotechnologists have musculoskeletal discomfort (MSD) associated with poor workplace ergonomics (Thompson et al. 2003). Laboratory personnel who work longer than 16 years have significantly higher risks of developing MSDs (Haile et al. 2012). A dose–response effect for hand and shoulder problems has been associated with manual pipette use (Bjorksten et al. 1994; David and Buckle 1997). Differences in body size, strength, and skill level affect the physical capabilities of laboratory workers, contributing to discomfort and injuries. Ergonomics can minimize the risks by matching laboratory tasks, equipment, and furniture to individual parameters. By making the routine and repetitive tasks more comfortable and easier to do, ergonomics reduces physical and psychological stresses, lowers fatigue, and, as a result, reduces human error. Ergonomics in laboratory design and practice should be as important to consider as chemical safety or biological safety.

13.2 ERGONOMICS ANALYSIS OF LABORATORIES

High-risk jobs associated with MSDs can be initially identified through a review of work-related injury data. This review should be followed-up with employee and management interviews. The workers can provide valuable information about their jobs, including the identification of the most difficult or fatiguing tasks completed. These tasks frequently have high ergonomic risks and should be targeted for further evaluation.

Employee interviews can include a symptom survey or a visual analog scale to measure the location, the type, and/or the level of pain (McCormack et al. 1988). The Borg scale (Borg 1970) can be used to measure the perceived exertion. The Standardized Nordic Questionnaire (Kuorinka et al. 1987) is frequently used for the analysis of musculoskeletal symptoms in occupational settings. Standard operating procedures and job safety analyses outline the specific steps and timelines required to complete the tasks, and they can provide the estimates of the expected exposure times. The potential high-risk tasks that expose workers to prolonged or static awkward postures, repetition, and forceful exertions should be further evaluated.

A variety of tools has been developed to identify and prioritize workplace ergonomic concerns. They include checklists and observational surveys. Table 13.1 provides an example of a laboratory ergonomics checklist.

Although they are easy to use, risk factor checklists have limitations. They are qualitative in nature, usually limited to yes and no answers, and can be overly sensitive to hazards. More specific information can be obtained with an observational survey to gather quantitative information to determine the relative risk of the tasks identified as problematic using a checklist.

The tools that evaluate the relative risk level of work are called *semiquantitative*. They require more effort than a simple checklist to collect data and determine the risk, but are not sophisticated enough to predict injury with any degree of certainty. They do provide a method to identify hot spots or tasks that should be prioritized.

One of the first widely recognized semiquantitative tools was developed by Suzanne Rodgers who proposed the muscle fatigue analysis (Chengalur et al. 2003) to assess the total body risk. This tool assesses the amount of fatigue that accumulates in the muscles during various work patterns within 5 min of work. It is most appropriate for assessing tasks requiring frequent exertions and awkward postures that are performed for an hour or more. The ANSI Z365 checklist (National Safety Council 2002) is based on the work of Suzanne Rodgers and provides a tool to assess the risks associated with work postures, force, work rate, work layout, tool design, and flexibility of the workstations. The rapid upper limb assessment (RULA) (McAtamney and Corlett 1993) and the rapid entire body assessment (REBA) (Hignett and McAtamney 2000) are screening tools that evaluate the risk factors of posture, static muscle contraction, repetitive motion, and force relative to a specific job. The RULA evaluates the upper extremity risk, and the REBA assesses the neck, trunk, and leg risk factors in addition to the upper extremities. Each risk factor contributes to a score that determines the degree of risk exposure and recommends the level of follow-up action that should be taken. RULA has not done well in the validation arena (Bao et al. 2007).

Quantitative assessments require greater expertise and involve computations to measure or predict the mechanical forces acting on or within the body. Useful examples of the tools to evaluate high repetition tasks common to the laboratory include the Moore and Garg strain index (Moore and Garg 1995) and the hand activity level (HAL) threshold limit values (TLVs) (American Conference of Governmental Industrial Hygienists [ACGIH] 2014). The strain index provides a method to quantify the risk factors associated with developing an MSD in a hand-intensive task. The ACGIH HAL TLV uses hand activity levels and peak hand forces to evaluate the risk factors for developing disorders in the hand, the wrist, or the forearm when completing single task jobs with 4 or more hours of repetitive work. Both the HAL and the strain index have been validated by independent studies, with the strain index having the best validation of the two tools (Spielholz et al. 2008; Harris et al. 2011; Kapellusch et al. 2013).

The National Institute for Occupational Safety and Health (NIOSH) revised lifting equation (Waters et al. 1994) and the Liberty Mutual Manual Materials Handling Snook tables (Snook and Ciriello 1991) can be used to assess the laboratory manual material-handling tasks. The NIOSH revised lifting equation calculates the recommended weight limit of objects that lab workers may

TABLE 13.1
Laboratory Ergonomics Checklist

	Yes	No
Laboratory Benches		

Is the workbench height appropriate for the work performed? (precision work = 1–2 in. above the elbow
  height; light work = 2–4 in. below the elbow height; heavy work = 8 in. or more below the elbow height)
Is there adequate clearance for the legs and the knees to sit at the bench for seated tasks?
Is there sufficient surface space for the tasks completed?
Is antifatigue matting available where prolonged standing tasks are completed?

	Yes	No
Laboratory Chairs/Stools		

Do chairs have seat height, tilt, and depth adjustments?
If chairs have armrests, can they be adjusted to allow the close access to the bench?
Are the worker's feet supported by the floor, a footrest, or a foot ring?

	Yes	No
Microscopes		

Are the microscopes positioned at or close to the front edge of the bench over a cutout to allow for sitting?
Is the bench, chair, or microscope height adjustable to allow the workers to maintain neutral back, head,
  and neck postures? (no more than 25° of forward bending)
Can the forearms be supported on the bench or arm supports when using microscope controls?
Does the worker take hourly breaks from microscope work?

	Yes	No
Pipettes		

Is a variety of pipettes available and appropriate for the tasks completed? (manual, electronic,
  multichannel)
Are pipettes and pipette supplies positioned within close reach of the worker? (within 18–22 in. distance)
Is the bench or the chair height adjustable to allow the workers to maintain neutral shoulder, arm, and
  back postures? (no more than 25° of forward bending)
Are pipettes calibrated and serviced as recommended by the manufacturer?
Do workers hold pipettes using light force?
Do workers insert pipette tips using light force?

	Yes	No
Fume Hoods/BSCs/Glove Boxes (Containment Devices)		

Is the containment device set at sitting (29–30 in.) or standing (36 in.) height as appropriate for work
  completed?
Can the workers adjust the chair or the containment device to work using neutral back, head, and neck
  postures? (no more than 25° of forward bending)
Are the shoulders and the arms in neutral positions (below the chest height and close to the sides) when
  working in the containment device?
Are the work supplies within easy reach (18–22 in.)?
Are sit–stand chairs or adjustable stools available to use when working at containment devices?

	Yes	No
Computer Workstations		

Are the computers located at cutouts with leg and knee room available?
Are the feet supported by the floor, the footrest, or the foot ring?
Are the input devices located at seated or standing elbow height and within easy reach?
Is the monitor at least an arm's length distance and slightly below the seated eye level?
If the source documents are frequently used, is there a document holder?

	Yes	No
Miscellaneous		

Do the workers rotate tasks and take breaks after 20 min of repetitious work?
Do carboys and containers have handles and pumps to dispense liquids?
Can smaller aliquots be dispensed to prevent lifting heavy loads?
Are heavy items stored on shelves between the knee (or the knuckle) and waist levels?

safely handle. The Liberty Mutual Snook tables are convenient to assess lifting, lowering, pushing, pulling, and carrying tasks. Both assessment methods are effective to identify the tasks associated with a higher risk of back injury claims, but the Liberty Mutual tables have better specificity and are less likely to identify non-risk and low-risk tasks as if they were high-risk tasks (Marras et al. 1999).

13.3   MICROSCOPE ERGONOMICS

Due to their design, microscopes require the operators to use prolonged awkward postures and repetitive motions that can result in discomfort and permanent injury. The OSHA reported that "microscope work is straining both to the visual system and the musculoskeletal system. Operators are forced into an unusual exacting position with little possibility to move the head or body … such as the head bent over the eye tubes, the upper part of the body bent forward, the hand reaching high up for a focusing control, or with the wrists bent in an unnatural position." Unfortunately, for microscope users, the human body works best when it is moving and changing positions. Prolonged static, awkward postures and repetitive wrist and hand motions required to use a microscope take a high toll on the musculoskeletal and visual systems. Over 50% of microscope users report discomfort or injury to the neck, shoulders, back, arms, wrists, hands, and eyes (Kalavar and Hunting 1996; Thompson et al. 2003; Lorusso et al. 2007; Jain and Shetty 2014). Around 80% of microscope users have reported musculoskeletal discomfort with 20% reporting lost work time (Haines and McAtamney 1993; Kalavar and Hunting 1996).

Microscope work requires the head and the arms to be held in a forward position with the shoulders rounded and the head and the neck tilted or inclined toward the microscope as shown in Figure 13.1.

Putz-Anderson (1995) identified the typical postures associated with microscope work to include head inclinations of up to 45° and upper back inclinations of up to 30°. From a biomechanical perspective, head inclinations of 30° from the vertical can produce significant muscle contractions, fatigue, and pain (Chaffin and Andersson 1984). These awkward postures held for prolonged periods result in discomfort or injury to the neck, shoulders, back, arms, and wrists.

To view the specimens through a microscope lens, the microscope operators must maintain a static flexed neck posture with the hands in a relatively fixed position. The optical path (the distance from the ocular lenses to the specimen being viewed) depends on the microscope type and optics. The distances should range between 18 and 21.5 in.; 45.7–54.6 cm) with the eyepieces no more than 30° above the horizontal plane of the work surface. Many older dissecting microscopes have

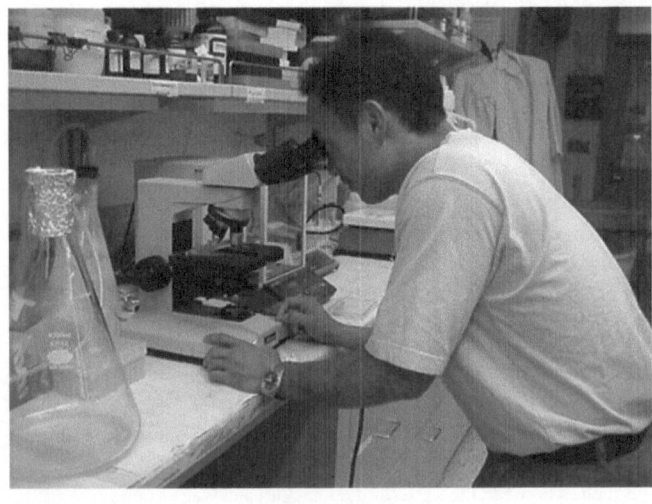

FIGURE 13.1   Microscope posture.

shorter optical paths (10–12 in.; 24.5–30.5 cm) with the eyepiece angles 60° above horizontal. This traditional design forces the user into either neck flexion or awkward wrist postures. Most workers try to find a balance and end up with discomfort in their neck, shoulders, back, wrist, and hands.

Eye discomfort and fatigue are common complaints associated with microscope work (Holper et al. 2005). These problems are compounded when the user has farsightedness (presbyopia) or astigmatism. Even though the diopter adjustment on the microscope eyepieces can compensate for minor focus problems, microscopists with significant visual issues should wear corrective eyeglasses. Specialized high eye point eyepieces are available from microscope manufacturers to provide a longer eye point necessary for observation with eyeglasses.

Video camera systems that display specimens on a computer monitor or a television screen can help prevent the visual strain when using a microscope. Unfortunately, not all microscope work can currently be completed with a video system. As technology evolves, future microscope designs will be available to replace the traditional observation tubes and eyepieces with charge-coupled device or complementary metal oxide semiconductor image sensors.

Microscopes are precision instruments and must be kept clean and aligned. Proper alignment of the microscope light source and the optical pathway produces bright and sharp images. Expanded view fields and eyepieces with larger field diagrams can also increase the visibility of specimens to reduce eye fatigue.

Aftermarket adapters can modify conventional microscopes to improve the user fit. Height extenders elevate the height of the eyepiece to adjust the position of the observation tube. Optical wedges provide adjustment of the angle of the eyepiece between 30° and 80° to reduce the awkward head and neck postures when looking through the eyepiece. Extended eye tubes increase the distance from the eyepiece to the binocular head, and can help to reduce the awkward head, neck, and back postures when using a microscope in a confined or hard-to-access space such as a biosafety cabinet (BSC). The microscope itself can be elevated and/or tilted with aftermarket stands, adjustable tables, or even books and risers. Arm pads, as shown in Figure 13.2, support the forearms and

FIGURE 13.2 Microscope arm supports.

position the hands, so the controls can be reached and operated in neutral postures. Extended stage controls allow the user to control the stage with the arms supported on the work surface.

Fortunately, manufacturers have recognized the value of the user-friendly design in marketing their products. As a result, many of today's microscopes incorporate ergonomics to reduce discomfort and increase productivity. Cytotechnologists at Duke University Medical Center (James et al. 2000) had significantly less discomfort in the neck and shoulder regions after their conventional Zeiss microscope was replaced with a Nikon Eclipse E400 microscope with ergonomic features. These features included a tilting and telescoping head, riser tubes, single-hand focus controls, and in-line focusing. Although the sample size was small ($n = 5$), the study provides support for the use of ergonomically designed microscopes to reduce user discomfort. The researchers suggested that the introduction of ergonomically designed workstations and ergonomics training would further improve the comfort levels.

The use of adjustable laboratory furniture and benches can increase the comfort of microscopists. Sillanpää et al. (2003) evaluated the effectiveness of an adjustable microscope table compared to a standard, nonadjustable laboratory table. The adjustable table was height adjustable with a tilting middle section supporting the microscope. The left and right sections were independently height adjustable and designed for arm support. The front of the table had a cutout allowing the user to place their forearms on the side sections. The new table allowed the microscope to be used with the head in an upright position with the forearms supported by the table. The subjects reported that the effort required to work at the new table was "lighter or much lighter" than at the original table. The measured levels of trapezius and deltoid activity were significantly lower, indicating the use of more relaxed neck and arm positions.

Sillanpää and Nyberg (2010) completed a follow-up study making further improvements to the design of the adjustable microscope table and again found reduced muscle load in the neck–shoulder and upper extremities. Since an adjustable table is less costly than adapting or replacing a microscope,

---

TABLE 13.2

Microscope Guidelines

Work Position/Posture

Sit close to the work surface at the bench cutout.

If a cutout is not available, use a sit–stand chair to position the hips at 60° to compensate for limited leg and knee spaces.

Avoid forward-leaning postures to look through the eyepiece. Adjust chair, bench, or microscope position to work with neutral or upright head, neck, and back postures.

Microscope Position

Position the microscope close to the front edge, or even beyond front edge of the counter.

Adjust the height so the eyepiece is slightly below the eye level with a downward gaze of 30°–45° below the horizontal.

Position the eyepiece so the neck flexion (downward bend) is no greater than 10°–15° below the horizontal.

Microscope Design/Accessories

Select microscopes with adjustable height, tilt, and length eyepieces, and low stages that are easy to reach with the forearms supported on table surface.

Consider the use of aftermarket adjustable eyepieces, optic wedges, and tilting stands if you do not have an adjustable microscope.

Use pads and arm supports to reduce the workload on the shoulders and the contact stress on the forearms.

Work Habits

Implement task rotation and spread work throughout day whenever possible.

Take frequent rest breaks. Move and stretch whenever possible.

Rest your eyes by focusing on distant objects during breaks.

Keep scopes clean and aligned.

---

the replacement of a traditional laboratory bench with an adjustable table can be an economical way to reduce the risk level.

Research demonstrates that incorporating ergonomics into the design of the microscopes and the microscope tables increases user comfort and improves productivity. Administrative controls such as work–rest schedules and task rotation are also important to implement into the laboratory routine. The users must be trained to properly use a microscope and avoid prolonged awkward postures. Table 13.2 provides the basic guidelines for microscope users.

13.4 PIPETTING ERGONOMICS

A pipette is used in biotechnology laboratories to transfer a precise volume of liquid. Pipetting work demands concentration, accuracy, and precision, frequently under time constraints to achieve the optimal results. Laboratory workers often pipette for several hours a day, placing themselves at risk of discomfort or injury. Bjorksten et al. (1994) found an increase in hand and shoulder injuries when pipetting for 300 h per year. David and Buckle (1997) reported hand injuries after using pipettes for 220 h per year, with increased discomfort occurring after 60 min of continuous pipetting. The British Columbia Institute of Technology (Occupational Health and Safety Agency for Healthcare [OHSAH] 2005a) reported that between 60% and 80% of workers in six healthcare sites had hand, back, shoulder, or neck injuries associated with pipetting.

The risk factors associated with micropipetting include repetition, force, and posture. McGlothin and Hales (1997) reported that workers in production labs completed between 6000 and 11,700 repetitive motions per day with the pipette. Traditional micropipette plunger forces can reach 4 kg (8.8 lb) during the dispensing and the tip ejection (Bjorksten et al. 1994; Fredriksson 1995). This is above the dynamic peak force levels of 3 kg (6.6 lb) and 2.1 kg (4.6 lb) recommended for male and female workers. respectively (Kroemer 1989). Exceeding the recommended force levels significantly affects the pinch strength. Females are more susceptible to this effect than males. In a study by Lin and Chen (2009), females experienced a greater reduction in the maximum pinch force (31.5%) than males (14.8%) after pipetting for an 8 h workday.

The use of a pipette requires lifting the upper arm in front of the body (shoulder flexion) between 45° and 90°. Transporting and transferring the liquids require frequent lifting of the elbow out to the side of the body (shoulder abduction) as well as repetitive twisting of the forearm between the palm-up and palm-down positions (pronation and supination). Rempel et al. (1998) reported high pressure levels on the median nerve when working in forearm rotation and wrist flexion. When high precision is required, the workload on the muscles is even greater, increasing the risk for tendon damage (Visser et al. 2004; Asundi et al. 2005). The addition of high viscosity fluids further increases thumb force, cycle time, and risk exposure (Asundi et al. 2005).

Pipetting in a containment device, such as a BSC, promotes awkward head and neck postures as well as arm extension into the hood as seen in Figure 13.3. This stresses the neck and the shoulder and compounds awkward hand and thumb postures required to activate the pipette plunger.

The awkward elbow and wrist postures reduce the grip strength from 25% to 60% (Grandjean 1988; Hallbeck and McMullin 1993; Kattel et al. 1996). The lateral pinch strength is reduced when the wrist is bent down or back or deviated to the side (Mital and Kumar 1998). Workers exposed to high-intensity and high-precision pipetting who use nonneutral arm, wrist, and hand postures are at a high risk of developing musculoskeletal disorders. Because pipetting is a highly technique-dependent activity, the aches and the pains associated with these disorders can reduce the precision and the accuracy and negatively affect productivity. A summary of musculoskeletal injuries and symptoms associated with pipetting is presented in Table 13.3.

Today's ergonomically designed pipette systems that use motors, latch-modes, or magnetic assists to significantly reduce force and repetition should be sought out as a priority for pipette risk control. Pipette manufacturers have developed these technologies to reduce the plunger forces without sacrificing precision and accuracy. Magnetic assist pipettes reduce the force needed to aspirate

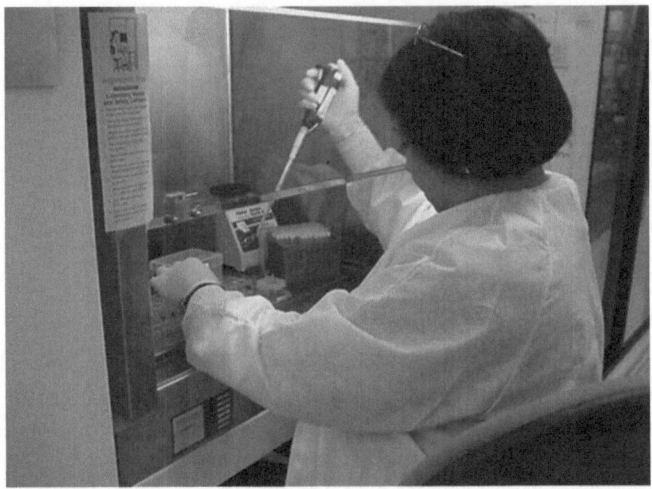

FIGURE 13.3   Pipetting in fume hood awkward postures.

TABLE 13.3
Pipetting Injuries

Activity	Symptoms	Injury
Tightly gripping the pipette	Numbness or tingling in the fingers or the hand Hand weakness Pain in the thumb and the radial side of the wrist	DeQuervain's tenosynovitis—Inflammation of the sheath that surrounds the tendons of the thumb Carpometacarpal joint deterioration Lateral epicondylitis—Inflammation of the tendon inserting into the elbow
Inserting the pipette tip and moving the pipette away from the body to access the tips	Pain and swelling in the wrist and elbow—outer part of the elbow becomes sore	Lateral epicondylitis (tennis elbow)—Inflammation of the tendon inserting into the elbow
Pushing the plunger to aspirate or blow out liquid	Numbness or tingling in the fingers or the hand Hand weakness Pain in the thumb and the radial side of the wrist Radiating pain from the hand up to the elbow or the neck Reduced thumb and finger movements, locking up	DeQuervain's tenosynovitis—Inflammation of the sheath that surrounds the tendons of the thumb Trigger finger—Inflammation of the sheath around the tendon in the fingers and the thumbs
Turning the wrist to transfer liquids and ejecting the tip	Numbness or tingling in the palmar side of the index, the middle, and half the ring fingers Hand weakness Pain from the hand to the elbow or the neck	Carpal tunnel syndrome—Compression of the median nerve in the wrist
Leaning on or resting the elbow on a hard surface	Numbness or pain in ring and little fingers Hand weakness Pain when elbow is touched	Cubital tunnel syndrome—Compression of the ulnar nerve at the elbow
Elevating the arm to transfer liquids and eject tips	Pain in the shoulder and the neck	Neck and shoulder nerve compressions Back strain
Leaning forward and bending the neck and the back	Pain in neck, shoulders, and back	Neck and shoulder nerve compressions Back strain

and blowout with the use of weaker aspiration and blowout springs. A magnet integrated into the plunger motion helps the user to find and hold the first stop. Latch-mode pipettes eliminate the blowout spring and divide dispensing and aspirating between the thumb and the fingers. However, Asundi et al. (2005) identified unexpected static thumb muscle loading due to users holding their thumb in an extended position after plunger dispensing while using a latch-mode pipette. The study recommended the addition of thumb rests to the latch-mode pipette and user training to avoid awkward thumb postures.

Rainin introduced the first pipette to use an electronic motor to control the plunger in 1984. These electronic pipettes have repeater functions to eliminate repetitive motions required to draw up liquids. Today's electronic pipettes as pictured in Figure 13.4 are lightweight, have comfortable designs, and provide high accuracy and precision. Nevala-Puranen and Lintula (2001) tested the ergonomics and the usability of an electronic pipette compared to two mechanical pipettes. The muscular strain of the upper extremity and the perceived strain of the thumb were lower with the use of the electronic pipette. Electronic pipettes were preferred for volume adjustability, ease of viewing, volume display, and time cycle in a study completed by Lichty et al. (2011). The repeating function of these pipettes significantly reduces repetitive motion.

Well-designed, lightweight, low-force pipettes reduce the risk exposure when pipetting. Finger hooks allow the user to rest the hand before, during, and after a pipette cycle. Pipette diameters between 1 and 1.5 in. (2.54 and 3.81 cm) are preferred by most users. Grant et al. (1992) reported that the forearm muscle activity is increased when the handle diameter is as little as 0.39 in. (1 cm) smaller than the user's inside grip diameter. Their study recommended that manufacturers provide a variety of handle sizes to match the size of the user's grip. In a pipette usability study completed by Lintula and Nevala (2006), the users preferred the shortest and lightest pipette with three different handle sizes that could be matched to the user's hand size.

Significant force is needed to load and eject the tips when pipetting with traditional pipettes. These forces are intensified when using multichannel pipettes. The tip ejection force of a traditional pipette averages 4 kg (8.9 lb), twice the recommended maximum force of about 2 kg (4.45 lb). New pipettes with ergonomic designs use spring-loaded tip attachment and ejection systems to reduce the tip ejection force levels below the recommended maximum force levels.

FIGURE 13.4   Axial electronic pipettes.

The highest level of thumb muscle activity associated with pipetting occurs when changing the volume adjustment dial while pipetting (Asundi et al. 2005). Both the location and the design of the pipette dials affect the workload on the thumb. The volume adjustment dials should be easy to read, reach, and grip, operate with low force, and be large enough to require minimal rotations to adjust the volumes.

Pipettes with finger-actuated triggers reduce the thumb strain and increase the precision when pipetting. Lee and Jiang (1999) compared two pipettes with thumb-operated plungers with a modified pipette that had a multifinger-operated plunger. The modified pipette reduced muscular stress, increased performance, and had better user ratings.

Although micropipettes with thumb-operated plungers comprise the large majority of the pipettes in laboratories, micropipettes with finger-operated plungers are available. The Finnpipette Novus electronic multichannel pipette has trigger action buttons that allow dispensing with the index fingers to reduce the thumb motion. The Biohit Proline XL macrovolume electronic pipette has finger-operated controls and is gripped low on the shaft of the pipette to reduce the shoulder flexion.

Traditional axial pipettes produce biomechanical loads on the shoulders, arms, wrists, and hands of the users (Bjorksten et al. 1994; Fredriksson 1995; David and Buckle 1997; Lee and Jiang 1999; Asundi et al. 2005; Lintula and Nevala 2006; Lu and Sudharkaran 2005). Sormunen and Nevala (2013) reported that pipettes with lighter weight, shorter length, and smaller circumference were preferred by users. The general recommendations for axial pipettes included a length of 22–23 cm (8.7–9 in.), a weight of 68–76 g (2.4–2.7 oz), and a rotating grip to allow for grip variation and ambidextrous work.

Ovation designed a nonaxial pipette to reduce awkward shoulder and upper extremity postures. James and Glascock (2005) and Lu and Sudhakaran (2005) compared traditional pipettes with the Ovation model which is shown in Figure 13.5. Clinical laboratory workers preferred the Ovation for comfort, accuracy, and general use. Hand forces, shoulder elevation, and wrist deviation, flexion, and extension were decreased. The forearm pronation was increased with this pipette; however, this motion is considered less of a risk factor than forearm supination and awkward wrist postures.

Postural awareness, use of proper pipetting techniques, and ergonomic workstation setup and design are important to control the risk exposure. The laboratory workbench height should be at the pipette tip level when the pipette is held in a standing position with the hand at elbow height (Park and Buckholtz 2013). If the workbench is not adjustable, the lab stool seat height should be sufficiently adjustable to attain this position. The pipetting should be done while sitting or standing at a cutout or unobstructed open space under the workbench to allow the worker to get close to the

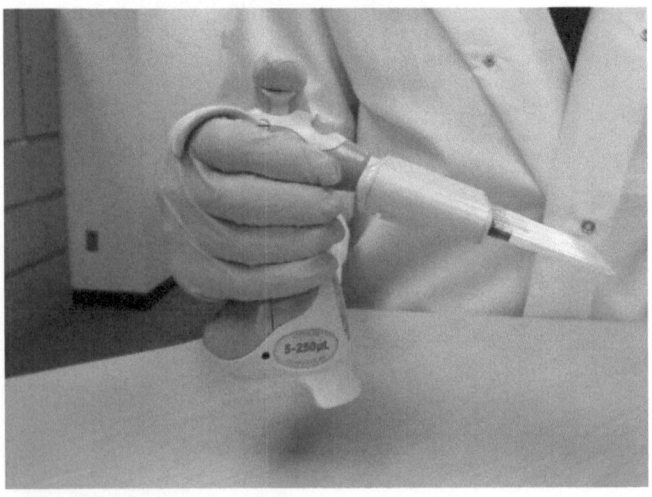

FIGURE 13.5 Ovation nonaxial pipette.

work surface. The use of supportive shoes with inserts, foot rails, and floor mats can help reduce the fatigue when standing.

Rempel et al. (2010) reported that the arm supports are effective in reducing the muscle loading in the neck and the shoulder during pipetting. Both a counterbalanced arm support and a gel pad supporting the elbow reduced deltoid and trapezius muscle activities. However, the use of the gel pad promoted wrist flexion to reach the tubes close to the elbow. Additional research is needed to measure the effect of the arm supports on lower arm and hand muscles.

The location of work tools and accessories is important when pipetting. Place the frequently used items within a 15 in. (38.1 cm) distance for easy reach when sitting and 20 in. (50.8 cm) when standing. Position vials, tips, and receptacles at heights and angles that are easy to reach. Use shorter length pipettes, tips, tubes, and containers whenever possible. Turntables, central or recessed waste receptacles, and work platforms can reduce reach and improve workflow. A prototype pipetting workstation pictured in Figure 13.6 incorporates the use of cutouts, recessed wells, and height adjustability to keep the work tools within close reach.

If work is completed in a fume hood, use a height-adjustable cabinet with angled sashes that have both vertical and horizontal adjustments. Adjust the sashes as much as possible to provide easy access into the hoods while maintaining the proper airflow. Set the chair height to position the airfoil at seated elbow height and pull up close to the hood. Sit–stand chairs can help achieve this position by allowing the worker to sit closer to the bench with the hips and knees at about 60° rather than at a 90° angle. Turntables and tilt stands are especially important to position the materials within close reach. Multiwall plate stands allow the user to angle the tubes and the plates to view using neutral head and neck postures while dispensing.

Specialty pipettes designed with a pistol grip, such as the Pipette Aid seen in Figure 13.7, angle the pipette to decrease the shoulder elevation and the wrist extension. This style of pipette is especially useful when working in a BSC. Many of these pistol grip pipettes have finger trigger grips to reduce thumb force and repetition required to pipette.

Task rotation and breaks reduce the risk exposure. Limit continuous pipetting to 20–30 min periods when possible. Take 3–5 min microbreaks every 20–30 min, if pipetting must be completed for longer periods. Complete gentle hand, arm, and shoulder stretches during these breaks.

Laboratory workers who are technically skilled work faster with less effort. Tip immersion depth and angle, cadence, tip position, and force affect both accuracy and efficiency. Pipettes must be kept calibrated and in good working order to maximize productivity and ensure accuracy.

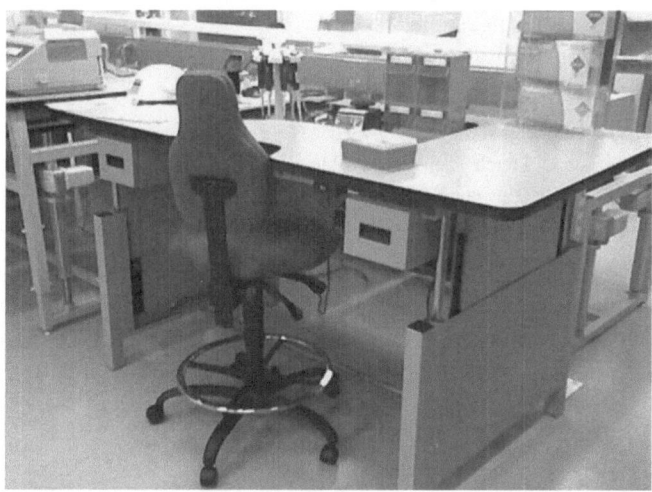

FIGURE 13.6 Modified DNA pipetting station. (Courtesy of Ira Janowitz, PT, Lawrence Berkeley National Laboratories.)

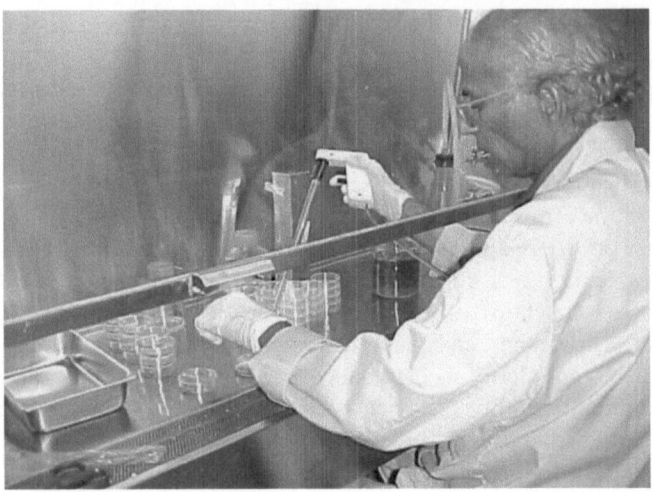

FIGURE 13.7   Pistol grip pipette.

TABLE 13.4
Pipetting Guidelines

Work Position/Posture
Sit or stand close to the work surface at a bench cutout.
If a bench cutout is not available, consider standing rather than sitting, If sitting is necessary, use a sit–stand chair to
   position the hips at 60° to compensate for limited leg and knee spaces.
Avoid elevating the shoulders and the arms to pipette. Adjust the height of the chair or the work surface to keep the hands
   at chest height.

Pipette and Accessory Placement
Avoid leaning and reaching postures. Place supplies within close reach on the bench top.
Use accessories such as turntables, pipette holders, and angled dispenser containers to reduce reach.

Pipette Design/Accessories
Select the right pipette for the job. Consider electronic pipettes and multichannel pipettes to reduce repetition and force.
Select pipettes with reduced trigger and tip ejection forces.
Use shorter length pipettes and tips when possible.

Work Habits
Implement task rotation and spread work throughout day whenever possible.
Take frequent rest breaks. Move and stretch whenever possible.
Keep pipettes clean and in proper working condition.

Pipetting work is challenging. Follow the general guidelines presented in Table 13.4 to reduce
the exposure to musculoskeletal disorders. Set up the bench to avoid forward leaning and reaching.
Limit the static and repetitive work by rotating tasks, taking breaks, and moving around. Select the
best tools for the job and keep them in proper working order. When pipetting, an ounce of prevention is worth a pound of cure.

13.5   BSCs AND FUME HOODS

Ventilated workstations such as BSCs and fume hoods are designed to protect the user and the environment from pathogens. Unfortunately, this protection brings unique structural requirements that

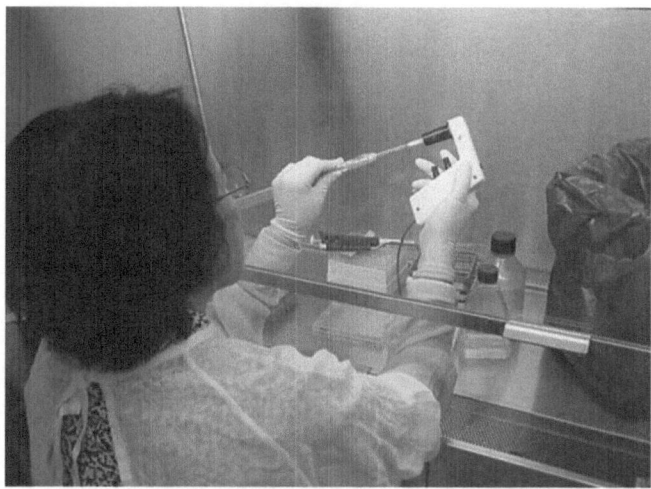

FIGURE 13.8   Awkward postures in a cabinet.

limit the incorporation of ergonomics into their design. Most are at a fixed height, and have limited legroom, hard and sharp metal surfaces, and wide downdrafts that force the user to work with forward reaching and elevated arms as seen in Figure 13.8. The presence of a viewing sash limits both the abilities to comfortably see and reach into the cabinet or hood.

Many laboratory workers use BSCs and hoods on a daily basis for many hours. Fortunately, many modern manufacturers have begun to recognize the importance of incorporating ergonomics into the design and the operation of BSCs (Jones and Eagleson 2001). BSCs and hoods are available with options such as adjustable height, angled front sashes, grills providing forearm support, increased leg clearance, and improved lighting systems.

Whenever possible, select BSCs and hoods with a height-adjustable base stand to improve the sitting or the standing position and maximize foot and arm supports. Before selecting an adjustable base, consult with an engineer or a biosafety specialist to ensure that the containment device height can be adjusted based on the ventilation system and the work performed in the device. Ducted BSCs (such as thimble-connected class II A1 or A2 or hard-ducted class II B1 or B2) may not allow for adjustable height due to safety concerns. Remove any supplies, equipment, drawers, or cabinet doors under the BSC or the hood to maximize knee clearance.

The size and shape of the plenum under the work surface affects thigh and knee clearances and impacts the position of the upper extremities when working. Wide plenums are especially problematic for smaller stature workers. The wider the plenum, the higher the worker has to elevate their arms to work in the hood, increasing the strain on the shoulders. The use of a sit–stand chair, as seen in Figure 13.9, can reduce the hip angle from 90° to about 60° to allow the worker to get closer to the BSC or the hood and work with more neutral neck, shoulder, and arm postures.

Fume hoods with large combination horizontal and vertical sashes provide greater clearance to access the materials inside the cabinet. Sash openings larger than 30 in. (76.2 cm) provide unobstructed viewing into the enclosure. Sashes that operate with single counterbalanced weights limit the force required to lift and are preferred to sash designs that fold up rather than slide. Fold-up sashes have limited flexibility and often have frames that interfere with the visual sightlines to look inside the hood.

The front sashes of the BSCs and the fume hoods with a 10° slope promote neutral head and neck postures and reduce glare. Modified sashes are available from manufacturers to use a microscope in the BSC.

BSCs and hoods with reduced front grill depth allow the work to be positioned closer to the front of the hood to reduce reach. Low profile airfoils enable the users to lift the materials over the airfoil with minimum shoulder elevation. Wide-access openings provide forearm support; sloped

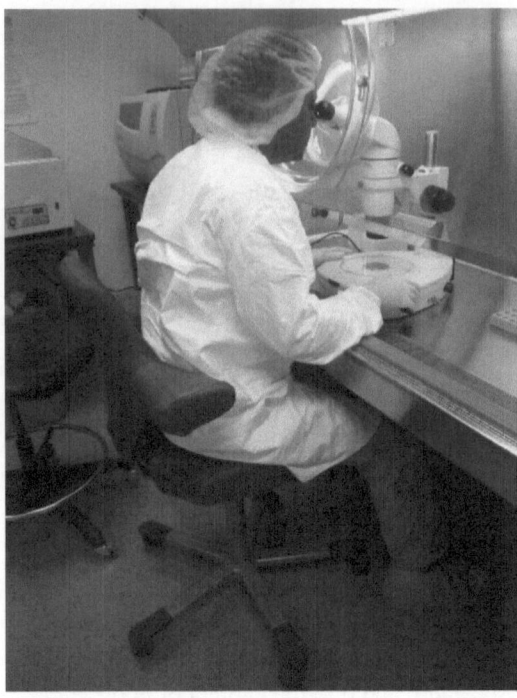

FIGURE 13.9   Sit–stand chair (Capisco) at a BSC.

armrest extensions allow the forearm and the elbow to be supported when working with items such as beakers or plates inside the BSC. Several manufacturers provide foam armrest pads and elbow rests for padding and support of the arms while keeping them off the front air grill. Any pads and armrests used on the grill must not interfere with the airflow and must stand up to the disinfection procedures. An example of an elbow rest available from Nuaire is pictured in Figure 13.10.

To limit the reach, place the materials as close as possible to the front of the cabinet or the hood while complying with safety and sterility requirements. Use carousels or turntables to transport supplies and equipment closer to the worker such as pictured in Figure 13.11.

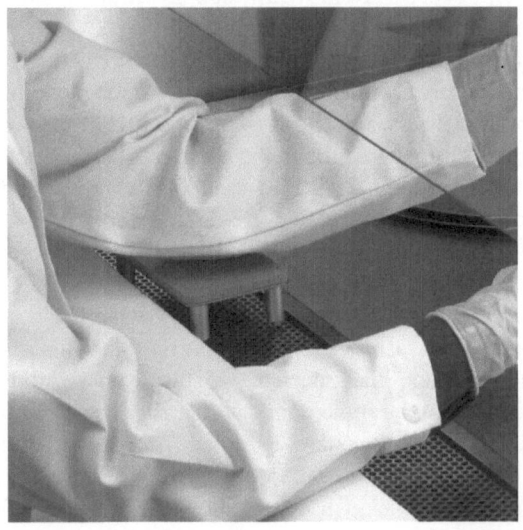

FIGURE 13.10   Elbowrest model. (Courtesy of Nuaire Manufacturing Company.)

FIGURE 13.11   BSC overhead carousels. (Courtesy of Nuaire Manufacturing Company.)

Look for cabinets and hoods with lower surfaces or wells to reduce the height of tall containers to limit the shoulder elevation. Tilt bins and bottles or place horizontally on stands within the hoods or the BSCs for easier access.

Proper lighting and nonreflective matte surfaces are important to minimize the glare in a containment device. The light source at the top front of the work area should be flicker-free and angled down and away from the worker. Energy-saving fluorescent canopy lighting with solid-state ballasts reduces flicker and minimizes heat. Lab personnel can also wear dark-colored lab coats to reduce the reflections and the glare on the sash. Manufacturers are currently developing LED lighting systems that improve the light levels, reduce the reflections and the glare while meeting the biosafety requirements.

Cleaning and disinfecting the containment cabinets before and after use can be challenging. Use cleaning devices with handles such as a Swiffer to reach the inside the cabinet. Keep face shields clean and streak-free for viewing while working.

Eye level controls that face down toward the user provide greater visibility and easier access. Electrical outlets should be located toward the front of the cabinet and within arm's reach for the convenience of the user and the maintenance department.

The noise generated by BSCs can be distracting and irritating. Cabinets should be located away from high-use work areas and have high performance airflow systems to reduce noise levels and improve efficiency. Locate cabinets away from doors, windows, and trafficked areas to ensure the integrity of the air curtain at the face, which is vital to ensure adequate ventilation. The size of the work surface should allow for all the materials for the experiment and the waste containers to be placed inside the BSC prior to initiation. This ensures the proper use of the BSC for safety (breaking the air curtain impedes the ascetic environment and safety) and reduces twisting to reach the materials outside the cabinet.

Few studies have been completed regarding the ergonomics of BSCs. A participatory study was completed by Nevala and Toivonen (2007) to identify BSC ergonomic concepts. After determining the optimal slope of the sash (10° sitting and 11° standing), they assessed the usability of a variety of arm supports, chairs, a forehead support, and a pipette holder placed in an electrical height-adjustable cabinet. A height-adjustable arm support beam, arm pads integrated into a lab coat, and segmental support plates were all found to reduce fatigue and discomfort with the supporting beam ranked as best. The users preferred a saddle chair over a perching stool or a Capisco stool. However, the users were familiar with the saddle chair prior to the study, and this may have affected the results. The pipette holder was useful with a suggestion made to provide a magnetic pipette holder fastened on the cabinet wall to save space. The electrically adjustable height of the cabinet was preferred, but a greater range in adjustability was needed to optimize flexibility and improve working postures. The forehead support was most suitable for standing postures.

BSCs present significant ergonomic challenges to the users. Few scientific studies have been completed to integrate ergonomics into their design to improve usability. Further research is needed to assist the manufacturers to develop improved products.

13.6   GLOVE BOXES

A glove box is a sealed container with gloves built into the sides designed to allow workers to see and manipulate objects in an isolated atmosphere. Different types of glove boxes exist: one allows a person to work with hazardous substances such as radioactive materials or chemicals; another is designed for infectious disease agents; and the last allows the manipulation of substances that must be contained within an inert atmosphere, such as argon or nitrogen. This type of glove box is pictured in Figure 13.12.

Glove box users experience awkward neck, shoulder, arm, wrist, and hand postures due to the accessibility limitations into the box through the glove ports, and the restrictions in moving the materials to pass through chambers. Thick gloves promote the overcompensation of the grip force to manipulate the objects and can create discomfort, especially when they are bulky and inflexible.

Since postural discomfort and visual strain are more likely when obstructions interfere with clear views of the materials, the glove boxes should have full-view glass chambers. Viewing glass windows with 15° slopes promote neutral head, neck, and back postures and decrease the reach distances.

To minimize the restricted movement created by working through a glove port, the workers should stand rather than sit when using a glove box. The average reach distance is between 6 in. (15.2 cm) and 16 in. (40 cm) when seated, and up to 20 in. (50.8 cm) when standing. Select glove boxes with height-adjustable legs to accommodate the anthropometric range of the users. Set the glove box height based on the type of the work completed. (Fine work = 1–2 in. [2.54–5 cm] above the elbow height; light work = 2–4 in. [5–10 cm] below the elbow height; heavy work = 8 in. [20.3 cm] or more below the elbow height.)

Typical glove box ports are circular and between 5 in. (12.7 cm) and 10 in. (25.4 cm) in diameter. This limited size opening restricts movement requiring the use of static and nonneutral postures to work. Manufacturers will provide larger custom-sized and custom-shaped ports if requested.

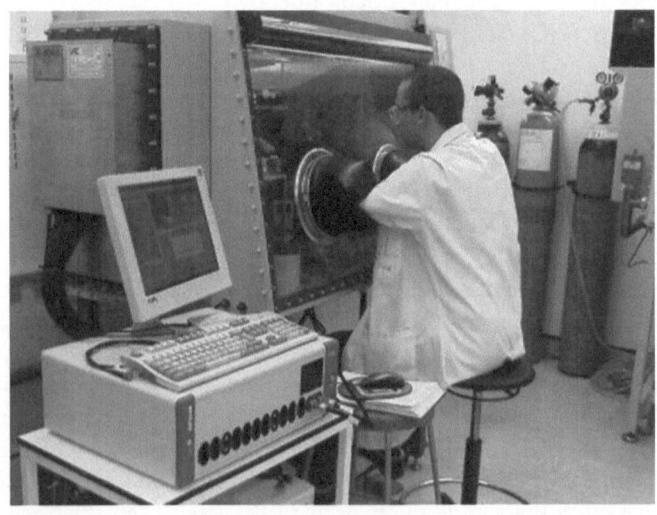

FIGURE 13.12   Glove box.

TABLE 13.5
BSC and Glove Box Guidelines

Work Position/Posture

Sit rather than stand when working at a glove box. The reach distance is increased when standing as compared to sitting.

If sitting is necessary, adjust the chair to sit close to the cabinet. Consider the use of a sit–stand chair to position the hips at 60° rather than at 90°.

Avoid elevating the shoulders and the arms to work in the cabinet or the glove box. Adjust the height of the chair, the cabinet, or the glove box to work at standing or seated elbow height.

Remove drawers, equipment, and supplies from under workbenches, cabinets, and glove boxes to increase the leg clearance.

Accessories

Place the supplies within close reach in the hood or the glove box.

Use accessories such as turntables, forearm supports and elevated platforms, conveyors, and motorized pass-through trays.

If you sit to work, select an adjustable chair and use a nonskid adjustable footrest.

Use antifatigue shoe inserts or matting when standing for prolonged periods.

Hood/Glove Box Design

Select cabinets with height-adjustable bases and low profile cabinet design to maximize knee/thigh clearance.

Select cabinets with frameless windows, angled sashes, and cool white lighting to increase visibility.

Work Habits

Avoid leaning on hard surfaces and edges. Use pads or pad front edges of the cabinet.

Implement task rotation and spread work throughout day whenever possible.

Take frequent rest breaks. Move and stretch whenever possible.

Keep cabinets in proper working condition and make sure lighting is flicker-free.

Whenever possible, specify the oval glove box ports that are between 15 in. (38 cm) to 18 in. (45.7 cm) in diameter to provide a greater range of working postures and movement.

Shoulder and arm strains can result from handling materials in the glove box. To reduce the strain, place frequently used and heavy materials close to the front of the box. Use motorized pass-through trays and conveyors to eliminate or reduce manual manipulations and reaching. Place varied height work platforms inside the box to position the materials and the equipment within closer reach. Air pumps and bottom-dispensing carboys can be used to transfer liquids. As with any task requiring constrained and awkward postures, limit the exposure and provide frequent breaks, stretches, and task rotation. The general guidelines for workers who must use BSCs and glove boxes are presented in Table 13.5.

13.7 MICROTOME AND CRYOSTAT WORK

The operation of a manual rotary microtome requires prolonged repetitive motion of the upper extremity. It is common for a laboratory technician to turn the microtome wheel over 1000 times per day. If the workbench is too high, awkward shoulder postures compound the high repetition to increase the risk exposure.

The best way to reduce the risks associated with the operation of a manual rotary microtome is to replace it with an automated unit. If it is not possible to use an automated microtome, select a manual model with foot pedals to augment the hand wheels to reduce the hand and arm motions. Place the microtome at a height and a position to minimize the shoulder flexion and the arm abduction when turning the wheel, while keeping the foot pedals accessible. If the workbench is not height adjustable, provide an adjustable chair with an adequate height range and a lower back support. A large

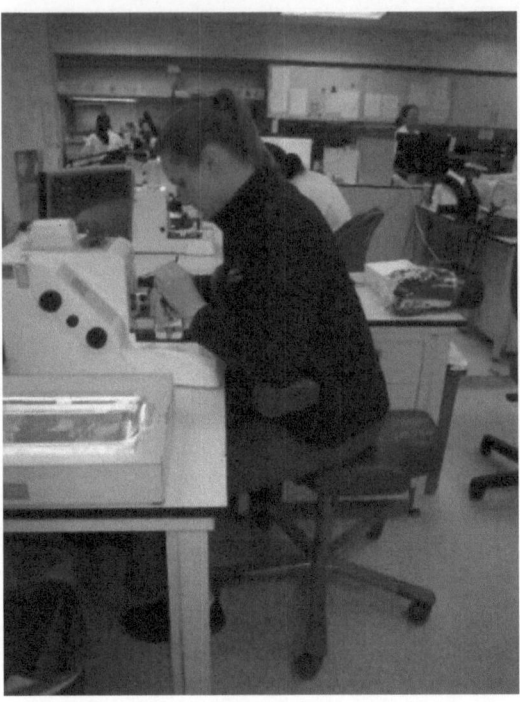

FIGURE 13.13   Microtome.

foot ring or a sturdy footrest may be necessary to provide the foot support if the operator's feet do not reach the floor. Specialty sit–stand chairs such as the Hag Capisco pictured in Figure 13.13 can provide the upper body support when working in a forward-reaching posture. The padding can be added to the work surface to eliminate the sharp edges and to avoid forearm and elbow compressions that can result from leaning and reaching postures.

Traditional wheel handles require repetitive wrist extension and flexion to turn the hand wheel. Adjust the feed wheel position or add an adapter to the existing handle to place the operator's hand in a neutral pistol grip position to reduce the force required to operate the hand wheel. Consider adding angled mirrors above the microtome to reduce awkward head and neck postures to view the specimens.

A cryostat is basically a microtome placed inside of a freezer to maintain a temperature of about −20°C (−4°F). The ergonomic concerns are similar to those of a microtome with the added issues of increased downward gaze into the unit and working in a confined space in cold temperatures. Working in a freezer promotes neck and lower back flexions and extended arm reaches. The contact stress on the forearms combined with the cold temperature decreases the blood flow and increases the risk of developing an MSD.

Cold exposure should be limited as much as possible. Always keep tools such as forceps outside of the cryostat when not being used. Take breaks that are long enough for the hands to return to normal temperatures. Light cardio exercise such as walking during breaks can facilitate the blood flow to your hands to help warm them up.

13.8   MICROMANIPULATION

Handling specimen plates, vials, and tubes, removing caps and screw-off lids, and using small hand tools such as forceps require the use of small muscle groups to pinch and grip. Tools and supplies designed to reduce pinch and grip forces and repetitive motions can reduce injury risk. Plastic vials with fewer threads on the lids reduce the number of twisting motions required to cap and uncap. (Note: Reducing the number of threads can lead to leakage, so be cautious when implementing this

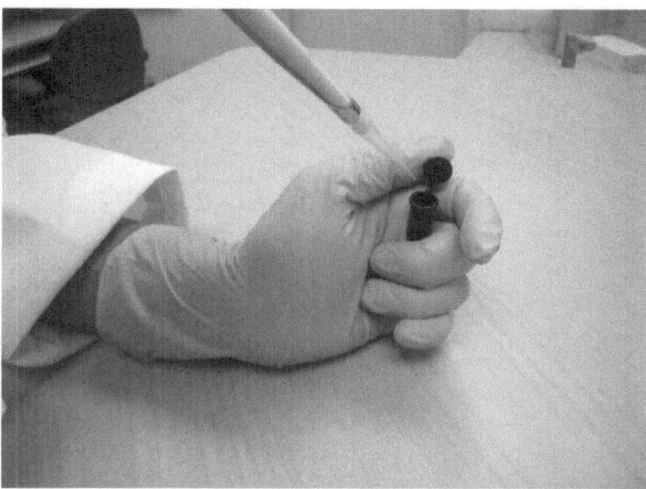

FIGURE 13.14   Back spring design centrifuge tube.

control.) Use jar openers to remove jar and bottle lids. Centrifuge tubes with a back spring design pictured in Figure 13.14 are easier to open than tubes that open by flipping up the lip on the front of the cap. (Note: If used with biohazards, this design can present a risk due to potential spray.) Manual and automated decapping tools reduce or eliminate the finger force required to open tubes, and a freestanding manual decapper is available to allow one-handed uncapping.

Precision hand tools are commercially available with molded and foam grips. Hand tools and small pipettes can be built up with foam or corks to transfer a finger pinch into a handgrip. Negative action, or self-closing tweezers, and clamping rings eliminate the need to apply constant finger pressure to the grip. Low-force forceps and tweezers also reduce the pinch force. The distribution of force can be accomplished by alternating holding the forceps between the index and the middle finger rather than the thumb and the index finger. Storage bins and beakers can be tilted or angled to reduce the reach when accessing or disposing of vials, tubes, caps, and samples.

Take frequent breaks and implement task rotation schedules when completing micromanipulation tasks. Select alternative tasks that use larger muscle groups to rest overused muscles in the fingers and the thumbs when rotating tasks.

13.9   VOLUMETRIC AND MATERIALS HANDLING

Laboratory work involves the handling of large volumes of fluids in beakers, bottles, carboys, drums, and other containers, many made of heavy glass such as the beakers in Figure 13.15. Five gallon (18.9 L) containers weigh over 40 lb (18.1 kg), and 55 gallon (208.2 liters) drums, more than 200 lb (90.9 kg). Transferring and transporting liquids can involve awkward postures and excessive force, especially when the containers are stored in hard-to-access areas.

Handles make it easier to hold, transfer, and pour liquids from the containers. Consider the use of beakers with handles or dispensing bottle jackets. Bottom-dispensing carboys eliminate the need to twist off the caps and the lids and to lift the bottle to pour out the liquid. Bottle-top dispensers fit on the top of bottle and pull up the liquid into the syringe to dispense through the nozzle by depressing a plunger. Self-closing caps allow one-handed pouring of the liquids and reduce the force required to open and close the containers.

Use burettes to dispense small volumes of chemicals from large containers. These clamping devices provide the controlled release of the liquids by droplet or stream to replace the pouring from cylinders or bottles. When transferring wet or dry materials between drums and containers, use drum pumps to avoid lifting and tilting the drums.

FIGURE 13.15   Dispensing liquids.

Transport the drums with hand trucks or drum handlers to avoid lifting or rolling. Both manual and powered drum trucks are available. Battery-operated stair-climbing hand trucks can be used to lift drums upstairs. Powered cart pushers motorize the pushing tasks to reduce the force and the awkward postures associated with transporting heavy loads. Lift and tilt carts, load-leveling carts and tables, and conveyors provide adjustable heights and positions to eliminate awkward back postures and reduce reaching for materials.

13.10   ERGONOMICS DESIGN AND BIOTECHNOLOGY LABORATORIES

Today's laboratory is a maze of interdependent and sequential tasks requiring ongoing interaction and collaboration of the workers. As a result, the physical design and layout can significantly affect performance, health, product quality, and productivity. Due to the complexity of a laboratory, a cross-functional design team should be composed of laboratory researchers, health and safety personnel, an ergonomist, an architect/engineer, and maintenance personnel. It is important to consider the perspectives and the knowledge of these other subject matter experts. Something ergonomically recommended may not be appropriate or feasible from an engineering, biosafety, or chemical safety perspective.

An understanding of the planned use for the laboratory is necessary to provide the foundation for the functional design and layout of the space. A review of the user needs, the performance the expectations, the equipment requirements, and the regulatory constraints is needed to develop this understanding. Work spaces should be tailored to match the workers, human anthropometry (body measurements) to minimize awkward postures and improve task efficiency. The laboratory design should minimize the distances the workers must cover to retrieve materials and lab equipment, minimize the awkward reaching and bending postures for equipment and supplies, and position the workstation surfaces at heights that promote neutral postures. Sufficient storage space must be provided, and the workflow must be optimized to eliminate redundancies. The tasks must be defined to determine whether the workers should sit, stand, or be mobile. Sitting is better when visually intensive or precise work is required, the activities are repetitive, and the tasks are completed for longer than 5 min. However, if the workers must reach more than 15 in. (38 cm) past the front edge of the workstation, standing is recommended (Sengupta and Das 2000).

Traditional lab benches are typically not adjustable and are set at between 35 in. (88.9 cm) and 39 in. (99 cm) high. The counters are traditionally linear, limiting the work space, and increasing the reach distances for instruments and supplies. Figure 13.16 provides an example of a traditional laboratory. The insufficient space due to the placement of cabinetry, equipment, and supplies under the counter further compromises the legroom and increases the reach distances and the awkward postures.

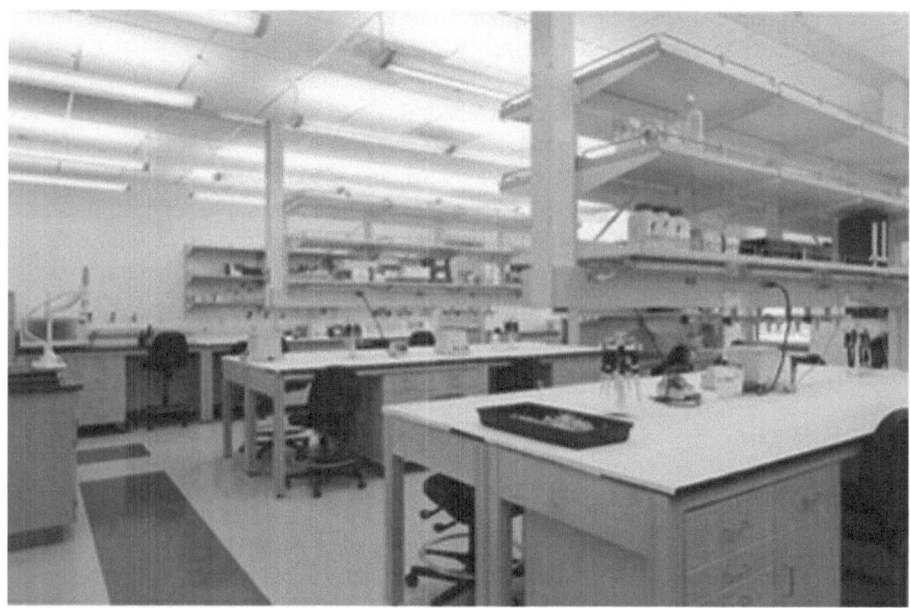

FIGURE 13.16   Typical laboratory design.

The basic layout of the laboratory space should be flexible and nonlinear to optimize worker interactions and reduce reach distances for supplies and instruments (Garikes 2004; Janowitz et al. 2009). The traditional linear surface requires extended reach distances for work supplies and materials. Angled workstations and cockpits provide increased work surface within closer reach of the user as demonstrated in Figure 13.17. The OHSAH in British Columbia (2005b) completed a redesign of microbiology and core laboratories that demonstrated the effectiveness of substituting nonlinear workstations for linear designs.

Flexibility is a critical design element to incorporate into the laboratory construction or modification (Garikes 2004). The ability to expand, modify, or change the footprint of the lab is necessary to keep pace with the technological improvements and the process changes to ensure an ergonomic workplace. Modular furniture and freestanding adjustable work surfaces should be specified in place of fixed-height lab benches whenever possible to allow the work surfaces to be set at a variety

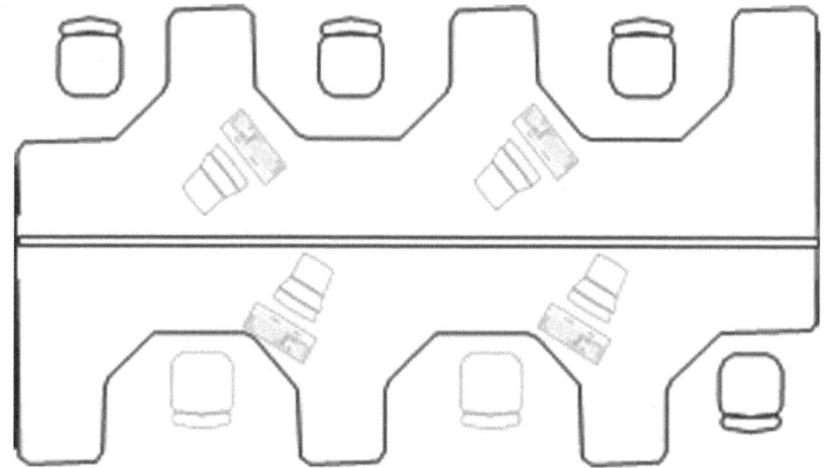

FIGURE 13.17   Angled laboratory workstations.

of locations and heights. This provides the opportunity to adjust the benches and the workstations as the changes are made to the equipment and the work processes.

Don Prowler noted at the Laboratory Safety and Environmental Management National Conference held in Alexandria, Virginia, in 2001: "What the cathedral was to the 14th century, and the office building was to the 20th century, the laboratory is to the 21st century" (Grant 2009). Both religion and business have made tremendous impacts on how humans live and work. The biotechnical advances made in today's laboratories are currently making even greater impacts on human health and well-being. Ergonomics has played a critical role in promoting the health and the productivity of workers in these laboratories by optimizing the physical design of the laboratories and how researchers work within them.

REFERENCES

ACGIH (American Conference of Governmental Industrial Hygienists). 2014. *2014 TLVs and BEIs: Threshold Limit Values for Chemical Substances and Physical Agents and Biological Exposure Indices.* Cincinnati, OH: Signature Publications.

Asundi, K. R., J. M. Bach, and D. M. Rempel. 2005. Thumb force and muscle loads are influenced by the design of a mechanical pipette and by pipetting tasks. *Human Factors.* 47:67–76.

Bao, S., N. Howard, P. Spielhotz, and B. Silverstein. 2007. Two posture analysis approaches and their application in a modified rapid upper limb assessment evaluation. *Ergonomics.* 50:2118–36.

Bjorksten, M. G., B. Almby, and E. S. Jansson. 1994. Hand and shoulder ailments among laboratory technicians using modern plunger-operated pipettes. *Appl Ergon.* 25:88–94.

Borg, G. 1970. Perceived exertion as an indicator of somatic stress. *Scand J Rehab Med.* 2:92–8.

Chaffin, D. B., and G. B. J. Andersson. 1984. *Occupational Biomechanics.* New York: John Wiley & Sons.

Chengalur, S. N., S. H. Rodgers, and T. E. Bernard. 2003. *Kodak's Ergonomic Design for People at Work,* 2nd edn. New York: John Wiley & Sons.

David, G., and P. Buckle. 1997. A questionnaire survey of the ergonomic problems associated with pipettes and their usage with specific reference to work-related upper limb disorders. *App Ergon.* 28:257–62.

Fredriksson, K. 1995. Laboratory work with automatic pipettes: A study on how pipetting affects the thumb. *Ergonomics.* 38(5):1067–73.

Garikes, R. W. 2004. Lean lab design. *Med Lab Obs.* 36:20–2, 34.

Grandjean, E. 1988. *Fitting the Task to the Man.* Philadelphia, PA: Taylor & Francis.

Grant, B. 2009. Can Labs go green? *The Scientist,* June 1. http://www.the-scientist.com/?articles.view/articleNo/25097/title/Can-Labs-Go-Green-/

Grant, K. A., D. J. Habes, and L. L. Steward. 1992. An analysis of handle designs for reducing manual effort: The influence of grip diameter. *Int J Indust Ergon.* 10:199–206.

Haile, E. L., B. Taye, and F. Hussen. 2012. Ergonomic workstations and work-related musculoskeletal disorders in the clinical laboratory. *Lab Med.* 43:11–9.

Haines, H., and L. McAtamney. 1993. Applying ergonomics to improve microscopy. *Microscop Anal.* 1:17–9.

Hallbeck, M. S., and D. L. McMullin. 1993. Maximal power grasp and three-jaw chuck pinch force as a function of wrist position, age and glove type. *Int J Indust Ergon.* 11:195–206.

Harris, C., E. A. Eisen, R. Goldberg, N. Krause, and D. Rempel. 2011. Workplace and individual factors in wrist tendinosis among blue-collar workers—The San Francisco study (1st place, PREMUS best paper competition). *Scand J Work and Environ Health.* 37:85–98.

Hignett, S., and L. McAtamney. 2000. Rapid entire body Assessment (REBA). *Appl Ergon.* 2:201–5.

Holper, L., C. Scutaru, and G. Schacke. 2005. Ergonomics at the microscope workstation: Prevalence of work-related musculoskeletal and visual complaints. *Zentralblatt fur Arbeitsmedizin, Arbeitsschutz und Ergonomie.* 55:186–96.

Jain, G., and P. Shetty. 2014. Occupational concerns associated with regular use of a microscope. *Int J Occup Med Environ Health.* 27:591–98.

James, T., and N. Glascock. 2005. Comparison of traditional and alternative pipettes—Use and preference. Paper presented at the annual Applied Ergonomics Conference, New Orleans, LA, March 21–25.

James, T., S. Lamar, T. Marker, and L. Frederick. 2000. An intervention study comparing traditional and ergonomic microscopes. *Hum Fac Erg Soc P.* 44:6–31.

Janowitz, I. 2009. Introduction to ergonomics in the biosciences. Presentation at 22nd Annual Occupational Safety and Health Institute Biotech and Laboratory Ergonomics, Center for Occupational and Environmental Health, San Francisco, July 30–31.

Jones, R. L., and D. Eagleson. 2001. Ergonomic considerations in the development of a class II, type A/B3 biological safety cabinet. *Amer Clin Lab.* 4:37–42.

Kalavar, S. S., and K. L. Hunting. 1996. Musculoskeletal symptoms among cytotechnologists. *Lab Med.* 27: 765–9.

Kapellusch, J. M., A. Garg, S. S. Bao, B. A. Silverstein, S. E. Burt, A. M. Dale, B. A. Evanoff et al. 2013. Pooling job physical exposure data from multiple independent studies in a consortium study of carpal tunnel syndrome. *Ergonomics.* 56(6):1021–37.

Kattel, B. P., T. K. Fredericks, J. E. Fernandez, and D. C. Lee. 1996. The effect of upper extremity posture on maximal grip strength. *Int J Indust Ergon.* 18:423–9.

Kroemer, K. H. E. 1989. Cumulative trauma disorders: Their recognition and ergonomics measures to avoid them. *Appl Ergon.* 20:274–80.

Kuorinka, B., A. Jonsson, H. Vinterberg, F. Biering-Sorensen, G. Andersson, and K. Jorgensen. 1987. Standardized Nordic questionnaires for the analysis of musculoskeletal symptoms. *Appl Ergon.* 18: 233–7.

Lee, Y., and M. Jiang. 1999. An ergonomic design and performance evaluation of pipettes. *App Ergon.* 30: 487–93.

Lichty, M. G., I. L. Janowitz, and D. M. Rempel. 2011. Ergonomic evaluation of ten single channel pipettes. *Work.* 39:177–85.

Lin, Y. H., and C. Y. Chen. 2009. Effects of syringe size and hand condition on thumb loading and muscle activity during pipetting tasks. *J Chinese Inst Indust Eng.* 26:493–8.

Lintula, M., and N. Nevala. 2006. Ergonomics and the usability of mechanical single-channel liquid dosage pipettes. *Int J Indust Ergon.* 20:257–63.

Lorusso, A., S. Bruno, F. Caputo, and N. L'Abbate. 2007. Risk factors for musculoskeletal complaints among microscope workers. *G Ital Med Lav Ergon.* 29:932–7.

Lu, M. L., and S. Sudharkaran. 2005. Evaluation of effectiveness of ergonomic pipettes. Presented at Applied Ergonomics Conference, New Orleans, LA, March 21–25.

Marras, W., L. J. Fine, S. A. Ferguson, and T. R. Waters. 1999. The effectiveness of commonly used lifting assessment methods to identify industrial jobs associated with elevated risk of low-back disorders. *Ergonomics.* 42:229–45.

McAtamney, L., and E. N. Corlett. 1993. RULA: A survey method for the investigation of work-related upper limb disorders. *Appl Ergon.* 24:91–9.

McCormack, H. M., D. J. Horne, and S. Sheather. 1988. Clinical applications of visual analogue scales: A critical review. *Psychol Med.* 18:1007–19.

McGlothin, J., and T. Hales. 1997. Ergonomic and epidemiologic evaluation of a biological laboratory. *Proc Amer Indust Hyg Conf Expo.* 1997:54–5.

Mital, S., and S. Kumar. 1998. Human muscle strength definitions, measurements and usage. *Int J Indust Ergon.* 22:101–21.

Moore, J. S., and A. Garg. 1995. The strain index: A proposed method to analyze jobs for risk of distal upper extremity disorders. *Amer Indust Hyg Assoc J.* 56:443–58.

National Safety Council. 2002. Management of Work-Related Musculoskeletal Disorders Accredited Standards Committee Z365 Working Draft. Washington, DC: National Safety Council.

Nevala-Puranen, N., and M. Lintula. 2001. Ergonomics and usability of Finnpipettes. In *Proc NES 2001. Promotion of Health through Ergonomic Working and Living Conditions. Outcomes and Methods of Research and Practice.* Tampere: University of Tampere 7:230–33.

Nevala, N., and R. Toivonen. 2007. Ergonomic Product Development of Biological Safety Cabinets. Report made to Finnish Institute of Occupational Health. Kuopio: Finnish Institute of Occupational Health, March 8.

OHSAH (Occupational Health & Safety Agency for Healthcare in British Columbia). 2005a. Evaluation of Best Practices for Alleviating and Preventing Cumulative Trauma Disorder Amongst Healthcare Laboratory Technologists Involved in Pipetting Work. Final Report. Vancouver, B.C. February 1.

OHSAH. 2005b. Ergonomic Assessments in Microbiology and Core Laboratories. Final report. Vancouver, B.C. July 8.

Park, J. K., and B. Buckholtz. 2013. Effects of work surface height on muscle activity and posture of the upper extremity during simulated pipetting. *Ergonomics.* 56:1147–58.

Putz-Anderson, V. 1995. Microscope use. Paper presented at the Howard Hughes Medical Institute. EH&S Conference, Washington, DC, April.

Rempel, D., J. Bach, L. Gordon, and Y. So. 1998. Effects of forearm pronation/supination on carpal tunnel pressure. *J Hand Surg.* 23A:38–42.

Rempel, P., I. Janowitz, M. Alexandre, D. Lee, and D. Rempel. 2010. The effect of two alternative arm supports on shoulder and upper back muscle loading during pipetting. *Work*. 39:195–200.

Sengupta, A. K., and B. Das. 2000. Maximum reach envelope for the seated and standing male and female for industrial workstation design. *Ergonomics*. 43:1390–1404.

Sillanpää, J., and M. Nyberg. 2010. Science, technology applications and research science. *Amer J Clin Pathol*. 4:543–8.

Sillanpää, J., M. Nyberg, and P. Laippala. 2003. A new table for work with a microscope—A solution to ergonomic problems. *Appl Ergon*. 34:621–8.

Snook, S. H., and V. M. Ciriello. 1991. The design of manual handling tasks: Revised tables of maximum acceptable weights and forces. *Ergonomics*. 34:1197–1213.

Sormunen, E., and N. Nevala. 2013. User-oriented evaluation of mechanical single-channel axial pipettes. *Appl Ergon*. 44:785–91.

Spielholz, P., S. Bao, B. Silverstein, J. Fan, C. Smith, and C. Salazar. 2008. Reliability and validity assessment of the hand activity level threshold limit value and strain index using expert ratings of mono-task jobs. *J Occup Environ Hyg*. 5:250–7.

Thompson, S. K., E. Mason, and S. Dukes. 2003. Ergonomics and cytotechnologists: Reported musculoskeletal discomfort. *Diagnos Cytopathol*. 29:364–7.

Visser, B., M. De Looze, M. De Graaff, and J. Van Dieen. 2004. Effects of precision demands and mental pressure on muscle activation and hand forces in computer mouse tasks. *Ergonomics*. 47:202–17.

Waters, T. R., V. Putz-Anderson, and A. Garg. 1994. *Applications Manual for the Revised Lifting Equation*. Cincinnati, OH: NIOSH.

14 Work Design and Health for Hospitality Workers

Laura Punnett, Pamela Vossenas,
W. Gary Allread, and Noor Nahar Sheikh

CONTENTS

14.1 INTRODUCTION

Internationally, the hospitality sector is a significant contributor to local, regional, and global economies. The travel and tourism industry accounts for 9.5% of the global GDP and over 250 million jobs, with an expected growth of 3.9% in 2015 (Ernst & Young Global [EYG] 2015). In 2014, it was reported that over 1.3 million hotel rooms are part of a global pipeline; China, the United States, Brazil, India, and Indonesia are the five countries with the most hotels under construction (Haussman 2014). Given the significance of the hospitality sector in the global economy and its projected growth in the global workforce, the ergonomic hazards in the hospitality industry and their potential impact on the workers' health deserve serious attention.

14.2 WORKFORCE OVERVIEW

In the United States, the accommodations subsector of the hospitality industry employed 1.8 million workers in 2012, with an expected 10% increase of 181,000 employees by 2022 (Bureau of Labor Statistics [BLS] 2013a). The corresponding numbers for the largest occupational groups within the accommodations subsection are housekeeping cleaners (428,000 workers; 9.8% increase by 2022); food preparation and serving, combined (453,100 workers; 9% increase by 2022); and hotel, motel, and desk clerks (221,100 workers; 13.6% increase by 2022) (BLS 2013a). Representing a smaller but essential workforce, the occupation of baggage porters, bellhops, and concierges (23,400 in 2012) is expected to grow by 9.8% by 2022 (BLS 2013a).

In a large, representative study of unionized U.S. hotel workers, housekeepers comprised the largest single occupational group at about 25%; they were overwhelmingly female and nonwhite (Buchanan et al. 2010). Next, the aggregated departments of hotel banquet, kitchen, and dishwashing together accounted for approximately 25% of the study population; the remaining 50% were dispersed over a broad range of other jobs.

On a global level, women are 55% and men 45% of the tourism workforce, although these proportions vary by job classification within sectors and subsectors, as well as geographically between countries and even among regions and urban centers of the same country (Baum 2013). Similarly, the distribution of employees by industry subsectors differs by country. One common element worldwide is the significant role of migrant labor in the hotel industry (Baum 2012).

14.3 HOTEL WORKER INJURY RATES

According to the U.S. BLS, the incidence rate of nonfatal occupational injuries in the accommodations subsector is 61% higher than the rate for all private industry: 5.0 cases versus 3.1 per 100 workers (BLS 2013b). The two types of lodging establishments that account for the majority of the hotel employees are hotels and motels, with 5.2 cases per 100 workers, and casino hotels with 4.3 cases per 100 workers (68% and 39% higher, respectively, than all private industry).

The records of hotel and restaurant employees (all jobs combined) in Denmark indicated an increased standardized hospitalization ratio for musculoskeletal diseases (Hannerz et al. 2002). The lower back/waist was the body region with the highest pain intensity. The study authors speculated that the risks were likely underestimated for this sector due to certain employment factors such as high turnover rate and high percentage of immigrant labor that would produce job misclassification.

The risk of musculoskeletal disorders increases with occupational exposure to physical ergonomic stressors. Both repetitive motions and forceful exertions are well-established risk factors for upper extremity disorders such as tendonitis, while heavy lifting, repeated bending, and twisting are implicated with lower back disorders (LBDs) (National Research Council and the Institute of Medicine [NRC/IOM] 2001). These job features exist in all of the major hospitality occupations, with varying frequency and intensity, as detailed later in this chapter. Many hotel workers spend a large part of the work shift on their feet; static standing in particular is associated with leg and back pains (Messing et al. 2008; Tissot et al. 2009). Psychosocial stressors, especially low decision-making opportunity, are prevalent and may interact with the physical load in the development of musculoskeletal disorders and other health effects. In addition, many hospitality jobs involve evening and night works, which entail a wide range of health effects such as disrupted sleep, digestive processes, and immune responses (Herichova 2013). Shift work may also cause social and psychological stresses, such as when family life is disrupted (Chiang et al. 2010).

In addition to the physical pain and suffering of occupational injuries and musculoskeletal disorders, medical and productivity costs are substantial. For lower back pain, in particular, the human toll is significant, with an estimated 818,000 disability-adjusted life years lost annually worldwide (Punnett et al. 2005). In a recent study of 65 low-wage occupations for all U.S. industries in 2010, housekeeping cleaners ranked third with medical and productivity costs totaling $3 billion, and combined food preparation and serving workers, including fast food, ranked fifth with $2 billion in total of medical and productivity costs (Leigh 2012). These two occupation groups are the largest occupation groups within the accommodations subsector. The other job groups mentioned earlier were also cited among the low-wage occupations: hotel, motel, and desk clerks with $72 million in total of medical and productivity costs; and baggage porters and bellhops with $140 million in total of medical and productivity costs (Leigh 2012). Interestingly, despite the low wages paid in these jobs, productivity costs exceeded medical costs.

14.3.1 HOTEL BUSINESS TRENDS AND THEIR IMPACT ON ERGONOMIC JOB FEATURES

The hotel industry competition for guests has led to profound changes in guest room accommodations, with corresponding changes in the workload and the work organization for hotel housekeepers. The hospitality industry's Battle of the Bed Wars began in the late 1990s with the launching of luxury bedding by leading U.S. hotel companies as a way to brand themselves and be differentiated from their competitors (Bauman 2006). With the introduction of more pillows, triple sheeting, duvets, and 250-thread count sheets, the weight of the linen per guest room increased (Alexander 2006). However, extra comfort for the guests creates extra work for the hotel housekeepers. The room quota for daily cleaning was typically established for traditional bedding and may not take the recent changes into account, such as the size and the weight of the bed (king versus double) and the linens; the number of beds in a room (one bed versus two double beds, known as *double doubles*); the size of the room (such as suites with pull-out sofa beds and kitchenettes); and the increased effort to maneuver housekeeping carts.

Other business trends indirectly affect occupational health and safety, by reducing job security and/or potentially obscuring the source of decision-making. One trend is hiring contracted labor to clean the hotel rooms. Relevant consequences of the contingent nature of this employment include no paid sick days, labor conditions resulting in greater work intensification, and higher risk of psychological distress (Sanon 2014). The ownership of hotels by public or private real estate owners that sign franchise agreements with leading hotel companies to fly the corporate flag and brand, then contract with a hospitality management company to hire the employees and manage the property, is an increasingly common business practice in the United States. In 2013, the top 15 management companies in the United States each had more than 8000 hotel rooms in their portfolio. The next surge in this hospitality management model is expected to take place in Europe and Latin America (EYG 2015). With the hospitality management company as the employer, it is difficult to identify who makes the decisions about the workload, the productivity demands, the safety practices, and the control of workplace hazards. Also, the arrangements between large hotel brand owners and management companies are described as "increasingly detached," with the owner focus on the return on investment instead of the investment in longer-term career development of the employees; the latter point particularly affects women workers who may seek more flexible work (Baum 2013).

The majority of new hotel development is driven by global hotel companies, often external to the countries where the hotels are being built. Thus, there is a concern about exporting ergonomic hazards to developing countries that may be (1) unaware of the magnitude of the hazards in this industry, (2) ill-equipped to mitigate the negative health impacts, and (3) lack the resources to make prevention-through-design adjustments to the hospitality workplaces after the construction or the renovation has taken place and the investment dollars may no longer exist. These concerns, although not identical, are similar to those raised in the past by public health experts about the exportation of pesticides, asbestos, and other workplace and environmental hazards to developing countries (Castleman and Navarro 1987; Smith and Root 1999). These concerns deserve serious attention as opportunities to protect large numbers of potentially affected workers.

14.4   HOTEL JOBS AND THEIR ERGONOMIC HAZARDS

The five aforementioned job classifications—hotel housekeepers, kitchen workers, banquet servers, bellhops and front desk clerks—have been prioritized here for the identification of the ergonomic risk factors and the potential intervention measures because they reflect the largest job groups in the accommodations industry.

14.4.1   Hotel Room Cleaner

14.4.1.1   Description of Job Tasks

Hotel rooms cleaners, also called housekeepers or room attendants, clean hotel guest rooms. They lift mattresses to change sheets (Figure 14.1), replace pillowcases and duvets, and remove soiled linens (BLS 2015a). They clean room surfaces, polish furniture, dust, and vacuum. They remove dirty sheets and towels, empty wastebaskets, and transport these items to the proper area. In the guest bathrooms, hotel housekeepers change towels, restock toilet paper, soaps, and other amenities, and clean bathroom floors, shower walls, and fixtures. They also move linens, towels, toiletries, cleaning supplies, and vacuum cleaners in and out of the rooms using wheeled carts. As part of their routine tasks, hotel housekeepers may lift heavy furniture or work in stooped or bent postures, and they are on their feet for most of the work day (O*NET 2015a).

14.4.1.2   Work-Related Health Outcomes

Hotel housekeepers have high rates of overexertion and musculoskeletal disorders. In a large three-year cohort study of the employees in 50 hotels, representing 55,327 worker-years of observation,

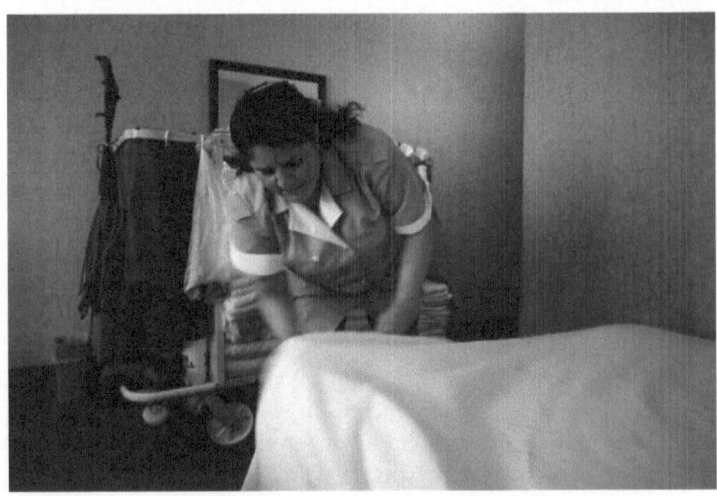

FIGURE 14.1   Example of a typical housekeeper task, making a bed using items from a full cart of linens. (Courtesy of Photodisc/Thinkstock, Charlotte, NC.)

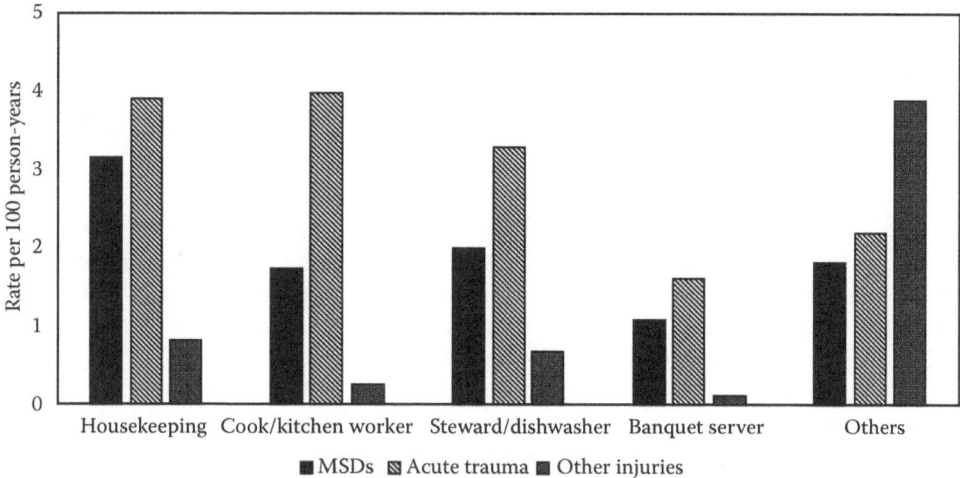

FIGURE 14.2 Injury rates for selected groups of hotel employees. (Data from Buchanan, S., P. Vossenas, N. Krause et al.: Occupational injury disparities in the U.S. hotel industry. *Am J Ind Med.* 2010. 53. 116–25. Copyright Wiley-VCH Verlag GmbH & Co. KGaA. Reproduced with permission.)

housekeepers had the highest injury rate of 7.9 per 100 person-years, compared to an overall injury rate of 5.2 injuries for all hotel workers studied (Buchanan et al. 2010). By nature of injury, housekeepers had the highest rate of musculoskeletal disorders of 3.16, compared to 2.0 for all jobs studied, and the second highest acute trauma injury rate (3.9), after kitchen workers, compared to 2.7 for all jobs (Figure 14.2). In separate analyses of 4230 hotel housekeeper injuries logged as work-related in 102 hotels, 44% were strains and sprains and 21% were contusions. Of all injuries, 32% occurred in the upper extremities and 22% on the trunk (Frumin et al. 2006).

The 2013 incidence rate of lost work time, nonfatal injuries, and illnesses for "maids and housekeepers" was 2.5 times higher than for all private industry: 277.3 versus 109.4 per 10,000 workers (BLS 2013b). By nature of injury, the incidence rates for housekeepers compared to all private industry were at least three times higher for sprains and strains (119.8 versus 40.2), non-specific pain (58.8 versus 19.1), and bruises and contusions (30.1 versus 8.9). The rates for carpal tunnel (1.4 versus 0.7) and tendonitis (0.6 versus 0.3) were twice as high for housekeepers, and four times higher for multiple traumas with sprains and strains (5.2 versus 1.3). The official statistics may underrepresent the problem. Among hotel housekeepers stating that they had work-related pain, 50% (Lee and Krause 2002) to 67% (Scherzer et al. 2005) had not reported it to the management.

Among almost 1000 hotel room cleaners, 63% had severe lower back pain, while 59% had upper back pain and 43% neck pain (Krause et al. 2005). Those in the highest quartile of physical workload and ergonomics problems were three to five times more likely to report severe pain than room cleaners in the lowest exposure category. In a separate U.S. study of hotel room cleaners, 56% reported severe or very severe pain in the shoulder (Burgel et al. 2010). Similarly, hotel room cleaners in Brazil ranked shoulder pain second only to back pain (Silva Jr. et al. 2012). Finally, 60% of female housekeepers self-reported poor or fair health, almost five times higher than the general female population (12.6%) (Krause et al. 2010).

14.4.1.3 Ergonomic Exposures

Hotel room cleaners have particularly repetitive and strenuous work. The ergonomic risk factors such as awkward postures, repetitive actions, high force exertions, standing for long periods, pushing heavy carts loaded with linens and cleaning supplies, and lack of sufficient physical recovery time are cited for this workforce across many countries (WorkSafeBC 2001a; Division of Occupational Safety and Health [Cal/OSHA] Consultation Service 2005; European Agency for Safety and Health

2009; NIOSH 2011; Workplace Safety and Health Council 2013). Cleaning bathrooms, dusting, and cleaning ceilings with a duster or a vacuum represent a high risk of joint overload because of bending and reaching requirements (European Agency for Safety and Health 2009); repetitively using a cloth or a sponge to clean large surfaces was identified as a moderate risk. Room upgrades such as increased use of chrome, large mirrors, and floor-to-ceiling glass shower walls have resulted in more cleaning time required, repetitive polishing motions, and extended reaches. Other specific tasks deemed especially problematic include changing the linens on larger beds with heavier mattresses or adding extra sheets or heavy bedspreads; cleaning large mirrors, dark furniture, surfaces made of porous materials, and added amenities; pushing heavy supply/linen carts or vacuum cleaners; and moving furniture (Krause et al. 2005; Seifert and Messing 2006). Additional ergonomic problems result from equipment and tools in need of repair, such as vacuum cleaners, and a lack of mops (32% of room cleaners) and other equipment.

With specific regard to bed making, Milburn and Barrett (1999) found that many tasks like lifting sections of the mattress and placing and removing bedding produced spinal loads above the recommended safe lifting limits, a result later confirmed by Orr (2004) for lifting the corners of king-sized mattresses. Beds positioned close to walls and other furniture result in awkward body postures during bed making (Barrett and Milburn 1997). Lumbosacral compression forces increased with bed size and when the beds were pushed; the latter resulted from the combination of larger beds and the amount of twisting required. Beds positioned low to the floor further increased the physical stress on housekeepers. Both studies by these authors concluded that the industry's trend of installing larger or heavier mattresses, and positioned closer to the floor, will likely lead to further increases in these spinal loads.

The physical demands of room cleaning load the upper body, as well. Bed linens have markedly changed in the hospitality industry over the last two decades. The duvets can now weigh as much as 14 lb and require eight snapping movements of the shoulder to fluff them for placement on the beds (Orr 2007).

How a facility organizes the housekeeping activities further impacts the job's physical demands. Hotel room cleaners are often required to perform their jobs quickly, which is related to the daily work quota established by the facility and the lack of advance notice of how much cleaning each room will require. The workload is further impacted by the mix of stay-over rooms (for existing guests) and those that require more cleaning effort, e.g., checkouts (for departing guests) and rooms with two beds. Other work organization factors that affect the pace of the housekeeping work include the amount of traveling (between floors and between rooms and the linen supply closet) required to complete the daily cleaning assignment (Lee and Krause 2002; Vossenas 2007); the number of scheduled "housemen" to help clean the extra dirty rooms or handle the dirty linens (Lee and Krause 2002; Vossenas 2007); and the number, the length, and the timing of rest breaks (Cal/OSHA Consultation Service 2005).

Rapid work pace is an important risk factor for musculoskeletal disorders (NRC/IOM 2001). The dynamic movements of the back have long been known to increase the trunk loading (Marras et al. 1987). Milburn and Barrett (1999) estimated that the lumbar load in numerous bed-making tasks, such as lifting mattresses to remove and replace the bed linens, was up to five times higher when the dynamic motion aspect of these activities was included than when assessed only with the static modeling. In an evaluation of hotel housekeeping tasks using the lumbar motion monitor, not one of the 16 room cleaning tasks measured was categorized as low risk, and the majority were clustered near the high risk benchmark value (Figure 14.3) (Allread et al. 2013).

Female floor cleaners, who performed repetitive activities for at least half of the workday, developed more damage to their median nerves than women in office jobs involving keyboard operation and other manual tasks (Bekkelund et al. 2001). Weigall et al. (2005) evaluated the upper body postures and the movements of a predominantly female population of janitors, custodians, and hotel/motel room attendants as they vacuumed and cleaned toilets. They determined that the exposures

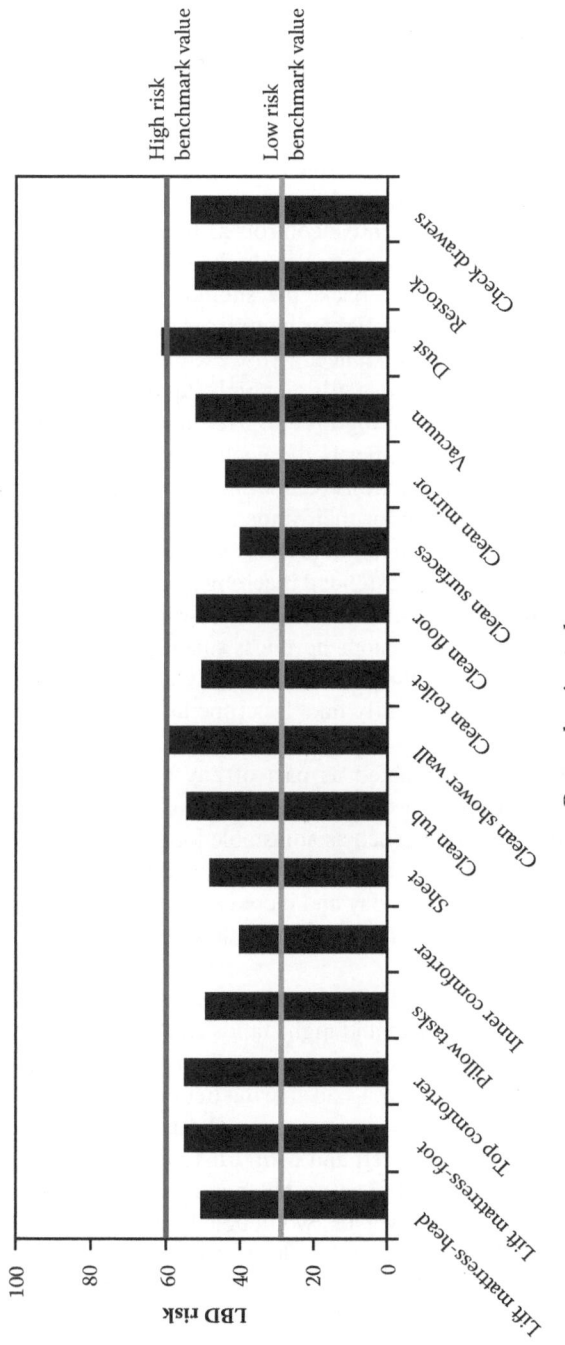

FIGURE 14.3 Summary results of LBD risk values for several common housekeeping tasks, estimated using the lumbar motion monitor. (Data from Allread, G., S. A. Ferguson, and W. S. Marras, Using the lumbar motion monitor to assess housekeeper room-cleaning tasks, Paper presented at Cal/OSHA Advisory Committee Meeting, Housekeepers in the Hotel and Hospitability Industry, Los Angeles, CA, http://www.dir.ca.gov/dosh/doshreg/Housekeeper_OSU_LMM_Summaries.pdf, 2013. With permission.)

during vacuuming reached a level of moderate risk, while cleaning toilets was judged to necessitate immediate intervention efforts.

14.4.1.4 Recommendations for Prevention

Because of the multiple related elements that influence the level of effort required by hotel house-keepers to perform their work, a systems approach should be used to identify interventions for each task. These elements include the physical hotel environment itself, the way the housekeeping job is organized, the tools and the equipment used, and the methods that the employers require the house-keepers to follow. The hotel room is a housekeeper's primary workplace, followed by the hallway where the cart is used throughout the shift and the linen supply closet. As always, the interventions should follow the standard hierarchy of hazard elimination whenever possible, followed by engineering-based improvements, with administrative controls as the last option.

The hotel room design is rarely considered with attention to its substantial impact on the house-keeper. To reduce the physical demands on the back, the shoulders, and other musculoskeletal regions during bed making, it is preferable to use lighter mattresses and to have the beds raised far enough from the floor that the housekeeper does not need to use extreme back postures. The beds on fixed platforms should have ample clearance from walls and nightstands, to provide the housekeep-ers with more postural flexibility in the bed-making task (Barrett and Milburn 1997). More space gives the housekeepers greater ability to use proper body mechanics. Fitted bottom sheets reduce bending, twisting, and a number of repetitive motions (Cal/OSHA Consultation Service 2005). The upgrades should be selected with consideration for their impact on the cleaning task requirements: matte surfaces rather than chrome, smaller mirrors and glass surfaces will avoid the detrimental increases in repetitive and forceful motions and extended reaching and bending. Light-stained wood furniture and patterned fabric on upholstered furniture show less dirt, thus taking less time and effort to clean. Wall-mounted (flat-screen) televisions have less surface area to be cleaned. In bath-rooms, the fixture design choices could similarly beneficially impact the housekeepers. Smooth, nonporous surfaces that do not show dirt as readily take less time and effort to clean, as do toilets that are suspended from the wall.

Hotel renovation upgrades should be evaluated as part of any ergonomic hazard assessment of the daily cleaning tasks, and they may necessitate accompanying measures for ergonomic risk reduction: provision of appropriate equipment, such as adjustable long-handled cleaning tools (Cal/OSHA Consultation Service 2005); a decrease in the workload by reducing the number of cleaning tasks per room, the number of rooms cleaned per day and increased staffing of housemen for heavier tasks (Casey and Rosskam 2009); and training the staff about safe work practices that avoid high-risk postures (Institute for Ergonomics 2011a).

The decisions about how housekeeping work is organized have ergonomic implications. Avoiding unnecessary daily polishing of wooden armoires and nightstands can reduce the repetitive motions. Søgaard et al. (2006) stressed the importance of the cleaners doing tasks that substantially vary which part of the body is biomechanically loaded. Load sharing (team cleaning) allows more physi-cally demanding tasks to be alternated between housekeepers and can reduce the time and the effort to perform specific tasks, e.g., bed making (Barrett and Milburn 1997; Milburn and Barrett 1999). Reviewing the housekeepers' daily assignment sheet to balance the number of different types of rooms, e.g., stay over versus check out or king versus two double beds, will help reduce the work-load. Adjusting how the work is scheduled to allow time for restocking carts as needed, instead of loading an entire day's supply at the start of a shift, results in fewer items stocked and less force to push the cart (Silva Jr. et al. 2012). In short, many feasible changes in the housekeeping workload and the work organization exist for hazard reduction and injury prevention.

Increasingly, product suppliers are recognizing the benefits that housekeepers gain from using cleaning tools and equipment that have ergonomic design features. Longer or telescopic handles are recommended for use in this industry (Cal/OSHA Consultation Service 2005) and are commer-cially available on many mops, sponges, toilet brushes, and dusters, with the objective of allowing the housekeepers to clean with less deviated arm, shoulder, and back postures. Compared to traditional bathtub/shower cleaning with a sponge, housecleaners who used a long-handled tool to perform the task could do so using neutral postures of the wrist, the shoulder, and the trunk for more of the time, and with more stable footing (Janowitz et al. 2005). They also subjectively felt more comfortable with this new method. These types of handles can reduce the risk of acute trauma (e.g., slips, falls). Also, the handles on these tools that provide a better contour to the hand (i.e., rounded, with no sharp edges) and are slip resistant not only reduce the amount of grip strength needed during their use but can also improve the cleaning quality and reduce the cleaning time too. Similarly, vacuum cleaner handles impact the ease or the difficulty with which the product is used. In addition to a slip-resistant surface, a full or a partially looped handle provides more user control and reduces the amount of muscle strength needed to grip it.

Finally, because housekeepers frequently move their supply carts, appropriate design features can reduce this effort. Obviously, motorized carts substantially reduce or eliminate the required physical effort compared to manual carts. Although retrofitting carts with motors might be cost prohibitive, Intilli (1999) reported that low-cost modifications to existing carts improved their handling to the satisfaction of housekeepers and management. For example, in one hotel, the cart pull forces dropped by more than 20% when the cartwheels made of ribbed rubber with sleeve bearings were replaced with solid, hard plastic ones with full ball bearings. The feedback from the housekeepers that the carts were easier to use, a reduced number of reported cart-related injuries and management expectation that the wheel replacement costs would be repaid by reduced injury costs, resulted in the hotel replacing the wheels on the hotel's entire fleet of carts. Any focus on improving the housekeeping carts (or other equipment) should include a plan for routine inspection and repair, since workable but damaged equipment requires more effort and time to use.

Training room cleaners and managers about safer work methods and body positioning during cleaning is a key element of a systems approach to ergonomic hazard reduction and hotel employer's injury prevention program. Ergonomic training procedures that can be taught for bed making include bending the knees to minimize the amount of back bending during the task; temporarily moving the bed away from the wall, if possible; and walking around the bed rather than reaching across it to remove the bedding. In addition, hotel room cleaners should be taught how to use length-adjustable cleaning tools to reduce taxing body postures (Figure 14.4). In the bathroom, ergonomic cleaning practices include walking along the wall, the floor, and the mirror surfaces to minimize overreaching and bending; this principle also applies to dusting and vacuuming. In

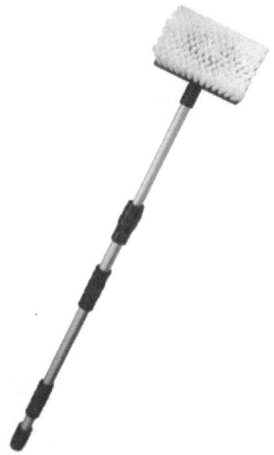

FIGURE 14.4 Length-adjustable brush which allows a housekeeper to adapt the tool to her/his body size and to the task being performed. (Courtesy of Le Do/Shutterstock.com, New York.)

one study, cleaners who were instructed to use ergonomics guidelines during their work generated significantly less muscle load and cardiovascular effort than those in a control group who were told to disregard previous ergonomics training (Samani et al. 2012). The importance of pushing supply carts should be stressed, since pulling them results in loading similar to lifting, twisting the spine, and asymmetrically loading the body. Training the managers (Institute for Ergonomics 2011b) and the housekeepers (Institute for Ergonomics 2011a) about the safety and health benefits derived from taking time to adapt current cleaning methods to include these safe work practices is crucial to the success of a workplace injury prevention program.

Ergonomics training should be part of a larger effort to improve the cleaning tasks, by involving the affected employees in decisions related to the job design and the selection of tools and equipment. Such a participatory approach was shown to improve the quality of the job risk analyses performed with cleaners in offices and the quality of resulting ergonomics changes (Kumar et al. 2005). Similarly, the involvement of cleaners in hospitals was found to markedly lower injury rates, workers' compensation costs, and lost work hours (Carrivick et al. 2005).

14.4.2 Kitchen Worker

14.4.2.1 Description of Job Tasks

The kitchen staff in the hospitality industry includes cooks, chefs, food preparation workers, dishwashers, and those who perform other tasks, which may vary with the type of culinary service provided and the size of the food establishment (BLS 2015b,c). In hotel kitchens, the cooks may bake, steam, grill, broil, or fry an assortment of foods (Figure 14.5). The food preparers assist the cooks and the chefs by washing, peeling and cutting vegetables and fruit, making cold food items, and readying meats and other dishes. The kitchen workers unload and store food supplies, and weigh, measure, and mix ingredients. Among kitchen staff, the dishwashers clean the food preparation areas and wash utensils, pots, pans, glasses, and tableware; they carry supplies to dining areas or use a cart or a hand truck to do so. After the cooking process, the kitchen workers store food in containers, put it in storage areas, and monitor the temperatures of the food, the coolers, and the refrigerators to prevent spoilage. The kitchen staff may operate commercial equipment such as dishwashing or pot-washing machines, broilers, steamers, ovens, slicers, and grinders, as well as clean these equipment and task-specific areas of the kitchen. These kitchen occupations involve extensive standing, walking, handling loads, and working in moderately close proximity to others at a hectic pace (O*NET 2015b).

FIGURE 14.5 Examples of typical kitchen worker tasks, preparing food at a variety of locations. (Courtesy of Ingram Publishing/Thinkstock, Charlotte, NC.)

14.4.2.2   Work-Related Health Outcomes

Kitchen work has been moderately well studied, although few investigations have specifically addressed food preparation workers in the hotel setting. In general, cooks have a high prevalence of musculoskeletal disorders in the neck, the lower back, the shoulder, the elbow, and the wrist. Exposure to heat and humidity is common, especially during the hot months of the year, and can substantially increase the risk of fatigue. Acute traumatic injuries may result from burns, knives, slips and falls, fire, and electrical hazards. The median number of lost workdays for injury to cooks and stewards exceeded the national median of 7.0 days for all industries in 2004 (BLS 2005). For both jobs, the median number of days of restricted activity averaged 10 days, indicating a marked impact of these injuries on employee health and productivity (Vossenas et al. 2010).

In a large study of unionized hotel workers, cooks/kitchen workers ("cooks") and stewards/dishwashers ("stewards") each had higher injury rates than all other hotel job groups (6.0 versus 5.2 cases), due to the notably higher rates of acute traumatic injury (Buchanan et al. 2010). Musculoskeletal injuries were about one-third of all recorded injuries in both groups. Two-thirds of all overexertion injuries to cooks and one-half to stewards were due to heavy lifting (Vossenas et al. 2010).

In a study of 905 hotel restaurant workers (no job titles specified) in Taiwan, 84% reported experiencing a work-related musculoskeletal disorder in the previous month, with the highest prevalence (over 50%) for the shoulder and the lower back (Chyuan et al. 2004). Among 495 female kitchen workers in Finland (71% employed in schools), the 3-month prevalence of musculoskeletal pain was 87%, with the neck, the lower back, and the forearm/hands as the leading sites (Haukka et al. 2006). Among all the subjects, 73% had concurrent musculoskeletal pain in at least two sites and 36% in four or more sites.

14.4.2.3   Ergonomic Exposures

Again, kitchen work has mostly been studied in other settings but cooking tasks are broadly similar whether performed in restaurant or hotel kitchens, although the meal style and the type of cuisine may have a meaningful influence on the physical workload characteristics (Xu et al. 2013). The most pronounced risk factors are awkward postures (especially neck and shoulder flexions), heavy lifting, and other applied forces, repeated movements during food preparation, such as chopping and stirring, and prolonged standing. Kitchen workers often need to reach above the shoulder height to access the supplies, extend their arms, bend across counters and prep tables to plate food, and reach and twist to wash dishes (WorkSafeBC 2001b; Workplace Safety and Health Council 2013). The fast-paced environments of hotel kitchens often exacerbate the effects of the numerous ergonomic stressors.

Heavy lifting is a common feature of kitchen work. Kitchens that operate on a large scale involve food preparation with ingredients purchased and handled in bulk (WorkSafeBC 2001a). It is not uncommon for staples such as flour, sugar, and rice to be received in 50 lb containers. Also, dishwashers and dining room attendants handle stacked dinnerware and pots and pans that can be quite heavy. The handling of these items, especially if they are stored below the knee level or above the shoulder height, results in bending, extreme reaches, and high loading on the spine (WorkSafeBC 2001b).

Ergonomic factors are also potential contributors to acute injuries. For example, the objects stacked on high shelves may fall when being placed or retrieved, and in a kitchen, many of these items are hot, heavy, and/or sharp. Such hazards are especially important when workers are moving quickly and in crowded spaces.

14.4.2.4   Recommendations for Prevention

Many awkward postures and lifting tasks in the kitchen work can be reduced or eliminated by giving attention to item storage locations. Store heavy items on lower shelves to avoid straining to lift them from above the shoulder height, and keep heavier items in the power zone between the knee and chest levels (WorkSafeBC 2001a, 2009). Stacked items must be stable to avoid the risk of workers being struck by falling objects.

Awkward work postures can also be reduced by choosing a work height appropriate for the task. The activities that require more use of force, such as chopping vegetables or cutting meat, should be done at waist level, as this will provide more upper body strength (WorkSafeBC 2001b) and reduce the strain on the musculoskeletal system (Pekkarinen and Anttonen 1988). In contrast, performing detail-oriented tasks (e.g., icing cupcakes) at or near elbow height improves the visual acuity and provides more neutral postures (Workplace Safety and Health Council 2013). Height-adjustable tables can accommodate both types of work (Figure 14.6). For fixed-height tables, adjustable feet on the cutting boards provide a way to make the table height appropriate for the task and the worker. Lowering the rinse nozzle to midbody level helps the stewards reduce their reach while washing dishes (WorkSafeBC 2001c).

Several approaches can be taken to deal with the high amounts of weight that may be handled. Where possible, the bulk items should be purchased in lower-weight containers (e.g., 25 instead of 50 lb) (Health and Safety Executive 2012), and the workers should be educated about the detrimental impact of attempting to lift too much weight (e.g., a large stack of plates) at any one time. If the bulk items are not purchased in lower-weight packages, then they should be transferred into smaller storage containers (e.g., two 25 lb containers instead of the original 50 lb container). A similar approach for handling liquids is to purchase pails that hold fewer gallons, e.g., a 3 gal pail instead of the more common 5 gal ones. Dishwashers can fill the racks only halfway with the dishes to reduce the weight (WorkSafeBC 2001c). These simple approaches could help to reduce the cumulative amount of weight lifted daily.

The use of carts to transport heavier items, compared to manual carrying, would also reduce the physical demands of the kitchen work (WorkSafeBC 2001a; Workplace Safety and Health Council 2013). The carts should be the same height as the tables and the counters used in the work area, so that the items can be slid from one surface to the next rather than being lifted. Similar to cart use in

FIGURE 14.6 Table with vertical adjustment, which can be modified according to the task and the worker height. (Courtesy of UNITE HERE! International Union, New York.)

other hotel occupations, appropriate handle height and wheel size and regular cart maintenance will reduce forceful pushing and related strain. The training of food workers and maintenance staff on safe postures and maintenance of carts is an important element of a systems-wide approach to injury prevention in the hospitality industry (Workplace Safety and Health Council 2013).

While well-designed carts may be more costly, the employer may find them to be quite cost effective. For example, in grocery store shelf stocking, height-adjustable carts were studied in common tasks such as loading and unloading the cart with products such as soup cans. The adjustable carts showed a benefit on the back in terms of biomechanical loading, as well as a significant decrease in the overall task time, compared to a traditional cart. The participants favorably rated the height-adjustable cart on "ease of use" and "more productive" (Davis and Orta-Anes 2014).

There are several opportunities to reduce the demands of prolonged standing. Comfortable footwear is an obvious solution, but floor matting aimed to prevent slips and falls that is also compressible can reduce back and leg fatigue (WorkSafeBC 2001a,b,c; 2002). Placing the mats at least 8 in. under the workstation will provide a smooth work surface, and beveled edges on the mats will reduce trip hazards. Mats should cover the entire work area where foot movements take place and be replaced when become worn (Ramsey et al. 2014). Muscular fatigue can also be reduced by making a step or a rail available to those who perform stationary tasks such as chopping and dishwashing (WorkSafeBC 2001b,c). This allows the workers to alternate their standing postures and provide periods of rest to their legs and feet.

It is important to recognize the opportunities to involve the workers in reducing the hazards common to kitchen work, as they know their jobs and have ideas about solutions (WorkSafeBC 2002). A participatory ergonomics study with mostly municipal school kitchen workers, performing similar tasks, although not in the hotel setting, achieved high participation rates (60–83%) and found a statistically significant increase in the knowledge of ergonomics and in the confidence to solve ergonomic problems (e.g., apply new knowledge to purchase tools and equipment), and a perceived decrease in workload and musculoskeletal pain (Pehkonen et al. 2009).

14.4.3 BANQUET SERVER

14.4.3.1 Description of Job Tasks

Banquet servers set tables with linens, dishes, glassware, utensils, and water pitchers before meals and remove and replenish as needed during and after the meals (Figure 14.7). They serve food and

FIGURE 14.7   Example of a typical banquet server task, carrying beverages to a reception. (Courtesy of youri/Shutterstock.com, New York.)

drinks to customers seated around large tables in ballrooms and banquet halls. Banquet servers are also responsible for cleaning up and returning dirty tabletop items to the kitchen. They often use large trays and tall banquet carts with multiple shelves to transport dishes, utensils, and other supplies between the kitchen and the dining hall. Banquet serving involves almost constant standing and walking, continual contact with others, working in a team, and regular trunk, hand, arm, and leg movements while handling loads and placing or moving objects (O*NET 2015c,d).

14.4.3.2   Work-Related Health Outcomes

Much of the scant literature on food service addresses caterers or restaurant wait staff. While it is likely relevant, there is too little research specifically about hotel banquet serving to cite firm estimates of injury or musculoskeletal disorder rates. However, there is evidence that the wait staff disproportionately suffer from work-related musculoskeletal disorders in the lower back, the shoulder, the elbow, and the foot (Xu et al. 2013). Among hotel food servers, lower back pain was associated with frequent bending while moving/lifting heavy objects (Chyuan et al. 2004). Shoulder pain was associated with frequent, prolonged moving of heavy objects and bending while moving/lifting heavy objects; and finger/wrist pain was associated with continual twisting and frequent vigorous wrist actions.

14.4.3.3   Ergonomic Exposures

Lifting heavy loads is an important and well-recognized ergonomic hazard. The primary opportunities for musculoskeletal stress among the banquet staff involve carrying loaded trays, especially those that are unbalanced, and serving beverages from coffee pots and pitchers (WorkSafeBC 2001d); walking long distances from supply stations external to the banquet hall and then carrying the heavy loads long distances within the banquet hall; and pushing large, tall carts laden with heavy loads. Carrying trays and heavy pitchers often results in bent wrists that are loaded from the weight of the food and drinks (WorkSafeBC 2001e), which causes high biomechanical load. Meal servers push and pull heavily loaded carts on a daily basis. Pushing and pulling motions are less widely appreciated as hazardous, despite their association with musculoskeletal problems affecting the lower back and the shoulder (Jansen et al. 2002).

Awkward postures, especially neck and shoulder flexions, are other important risk factors. The serving staff may repeatedly bend their bodies and reach as they set tables, serve diners, and then clear tables at the conclusion of the function (Workplace Safety and Health Council 2013). Setting up the function space itself can be physically stressful when carrying heavy tables, chairs, and other equipment. Finally, as with other hospitality workers, the banquet staff must often stand for long periods.

The physical exposures are multiplied by long working hours and long walking distances. The psychosocial stressors are notable as well but have been little examined, especially in their interaction with the physical stressors. Chiang et al. (2010) noted that food service workers in hotels and restaurants "perform repetitive tasks, require fewer skills and use of cognitive capacity, and enjoy limited autonomy, control and flexibility over how they perform and arrange their work." They also reported that the lack of job control and family-unfriendly work practices intensified the effects of work demands on the job stress.

14.4.3.4   Recommendations for Prevention

Allowing the servers to use their shoulders and hands in a power grip for support can reduce the effort to carry large trays, as can balancing the items on it with the heaviest items in the center (Figure 14.8) (Workplace Safety and Health Council 2013). Training servers on item placement, safe tray-carrying methods, and, where possible, alternating the sides of the body used for the task will help reduce the physical demands (WorkSafeBC 2001e). Providing smaller banquet trays, lighter dinnerware, and shorter carrying distances will further reduce the load (WorkSafeBC 2001a).

FIGURE 14.8 Good work practice of keeping a serving tray close to the body and balancing it with hands in a power grip. (Courtesy of UNITE HERE! International Union, New York.)

The function space itself can be set up to reduce the physical demands. Increasing the numbers of supply stations in the hallways leading to the banquet hall will shorten the travel distances between the kitchen and where the trays are restocked, reducing the effort of carrying the trays laden with dirty dishes, glasses, empty water pitchers, and coffee pots. Using trolleys to transport tables and chairs to and from storage will reduce or eliminate the need to carry these items (Workplace Safety and Health Council 2013). In addition, the fewer stacked chairs that are lifted or moved at any one time will also lessen the muscular loading. In terms of placement, tables spaced further apart, and having fewer seats per table, will allow the servers to move around more easily and do their work with less bending and reaching (WorkSafeBC 2001d).

Regarding psychosocial aspects, the worker latitude in decision-making can ameliorate the potential stresses. Chiang et al. (2010) recommended organizational support for food service employees to exercise "discretionary power when serving their customers and enhances their ability to make appropriate decisions on the spot thereby reducing stress that may arise from increased work demands. It is therefore important that the employees have the opportunity to reduce the job demands (i.e., high job control) and the support to do so (i.e., work life balance practices)."

14.4.4 Bellhop

14.4.4.1 Description of Job Tasks

Bellhops greet hotel guests upon arrival and transport the guests' personal items into the hotel and to their room using a luggage cart (Figure 14.9), and upon departure, escort them to the lobby for check out. They also check and retrieve the guests' luggage for short-term storage. When tour groups stay at a hotel, the bellhops will handle large quantities of luggage in short periods. Bellhops are trained to help guests with special needs. The bellhop tasks involve almost constant standing, walking, stooping, lifting loads, and working in very close proximity of others (O*NET 2015e).

FIGURE 14.9 Example of a typical bellhop task, loading a luggage cart from an incoming guest. (Courtesy of Andrea Chu/DigitalVision/Thinkstock, Charlotte, NC.)

14.4.4.2 Work-Related Health Outcomes

There is a dearth of epidemiological literature on workplace injuries occurring to bellhops and porters in the hospitality sector despite the governmental guidelines that identify the ergonomic risk factors for these occupations (O*NET 2015e) and the potential risk of musculoskeletal strain injuries (WorkSafeBC 2001f). In a review of risk assessments from a regional government surveillance program of manual loads in certain hotel occupations, hotel porters had a range of NIOSH lifting index scores that included the highest risk level (0.77–3.75) for musculoskeletal disorders (Fontani et al. 2010). The national-level workers' compensation data linked to hospital discharge data yielded the finding that out of 10 occupations, the category of elementary occupations—to which hotel porters belong—ranked third for the highest rates of serious threat to life injuries (40.1 per 100,000 worker years) (Cryer et al. 2014).

14.4.4.3 Ergonomic Exposures

Luggage and other guest belongings considerably vary in terms of size, shape, and ease of transport, and they rarely provide any indication as to weight. Often large numbers of bags must be moved in a short period; in addition, the storage areas may become overfilled and disordered when large groups are checking in or out at the same time. This can result in bellhops needing to reach and lift heavy or difficult-to-handle items and rapidly maneuver them and in tight spaces, possibly resulting in overexertion.

Bellhops also spend a significant amount of time moving guest belongings on luggage carts. While the use of carts potentially reduces the lifting demands, by allowing the luggage to be transferred more easily (e.g., hotel lobby to room), they also pose their own set of risks. For example, the friction of the cartwheels considerably increases when moved on carpet, as compared to smoother surfaces such as concrete, tile, or even asphalt (Al-Eisawi et al. 1999). In addition, the carts may have to be moved over uneven surfaces, such as door thresholds or elevators not stopped level with floors. Individual subjects' trunk motions were found to fluctuate markedly when pushing a cart that suddenly stopped (Lee et al. 2011); such incidents repeated over time could increase the risk of developing an LBD.

14.4.4.4 Recommendations for Prevention

Reducing the physical demands of luggage handling can be achieved using a three-pronged approach: providing bellhops with sufficient handling equipment, having adequate luggage storage space, and ensuring that the employees know and are able to use appropriate lifting practices.

First, there should be enough luggage carts so that one is available any time as needed by a bellhop (WorkSafeBC 2001f). Similar to housekeeping and foodservice carts, the wheel design is critical, as larger wheel diameters have been found to lower the minimum amount of force to push or pull a cart (Al-Eisawi et al. 1999). A proactive maintenance program should ensure that the carts are kept in good working order, as damaged or misaligned tires increase the force required to move them along the floor.

Second, an important element to assess is the size and configuration of the luggage storage area provided in the hotel. This should allow all smaller bags to be stored on shelves near the waist level, all larger bags to be stored upright on the floor or on low shelves, and sufficient space to easily walk around and access all bags (Figure 14.10). Thus, sufficient storage space can reduce the physical effort to handle luggage and provide more opportunities for bellhops to use safer lifting techniques.

Third, training bellhops on proper luggage-handling methods will provide them with the knowledge to perform their job in ways that reduce its physical demands. The recommended strategies include pulling the bag toward the body before lifting it, alternating the side of the body used to lift, and pushing rather than pulling luggage carts (WorkSafeBC 2001f). The training should emphasize the importance of using bag handles, as lifting objects with handles significantly reduces spinal loads (Davis et al. 1998), oxygen consumption, heart rate, and perceived exertion (Jung and Jung 2003). The handles also increase the maximum amount of weight that the individuals were

FIGURE 14.10 Hotel area with multiple shelves that allow for the appropriate storage of luggage according to size and weight. (Courtesy of UNITE HERE! International Union, New York.)

willing to lift (Garg and Saxena 1980). In addition, the training should address the importance of luggage storage techniques; for example, tipping the bags on end, compared to stacking them, significantly reduced spinal loads and shoulder muscle activity (Korkmaz et al. 2006).

14.4.5 FRONT DESK CLERK

14.4.5.1 Description of Job Tasks

Front desk clerks greet guests upon arrival at the registration desk, locate the reservation, obtain guest information, designate a guest room, and provide room keys. During the guest's stay, they assist the guest with any changes to the reservation, problems with the guest room, guest complaints, and answer guests' questions. Upon departure, the front desk clerks provide a copy of the guest's bill and obtain the payment for the hotel stay. Behind the scenes, the front desk clerk manages the reservations, keeps track of which rooms are occupied, and notifies the housekeeping department when the guests have checked out. The work activities include direct communicating with guests, using computers and telephone systems, resolving conflicts, and communicating with guests and hotel staff (O*NET 2015f).

14.4.5.2 Work-Related Health Outcomes

There appears to be no specific literature on the health of front desk clerks, so it is unknown whether or not they experience work-related musculoskeletal or other health problems. However, ergonomic exposures that have been identified in this job (see Section 14.5.3) deserve attention due to their known effects in other occupations.

14.4.5.3 Ergonomic Exposures

Many front desk clerks stand in one place for prolonged periods as they greet, check in, and check out guests. This can produce lower back and leg fatigue (WorkSafeBC 2001g). In addition, the locations of the equipment used by these employees (e.g., telephone, printer, and key card system) may cause them to bend or reach past their normal range of motion (Figure 14.11). The clerk's use of a computer may cause bending of the wrists and/or the neck, depending on the placement of the keyboard, the mouse, and the monitor (WorkSafeBC 2001g). This, too, can result in muscular discomfort. The neck and the shoulder can also become strained if a clerk cradles the handset there while on the phone (Workplace Safety and Health Council 2013).

FIGURE 14.11 Example of a typical front desk clerk work space requiring work at varying heights. (Courtesy of kzenon/iStock/Thinkstock, Charlotte, NC.)

14.4.5.4 Recommendations for Prevention

Because front desk clerks frequently use computer workstations, standard ergonomics guidelines for computer work apply. For example, positioning the keyboard at approximately elbow height, and the mouse at this same level next to the keyboard, reduces the strain on the arms and the wrists (WorkSafeBC 2001g). The demands on the neck are minimized when the top of the computer monitor is at approximately eye level (Workplace Safety and Health Council 2013) and in-line with the keyboard and mouse. Allowing the employees to sit while using the computer also reduces the time that the neck is flexed forward (WorkSafeBC 2001g). The front desk areas often have multiple computers, and numerous employees may work on these at any given time. Therefore, having each keyboard, mouse, and monitor on an adjustable, articulating arm will provide clerks of varying body sizes the ability to appropriately position the equipment. The clerks will benefit as well from having other frequently used items within arm's reach of their primary work location, as this minimizes the potential for repeated overextension (WorkSafeBC 2001g).

Muscular fatigue that may develop due to prolonged standing can be addressed by providing the ability for front desk clerks to alternate between standing and sitting postures (Figure 14.12). If the clerks currently stand, sitting is an appropriate option only if the design of the front desk area allows for leg clearance and, thus, a comfortable seated posture. Any (traditional or sit–stand) chair should allow clerks to maintain the same neck, arm, and wrist postures during computer use as when standing. Allowing the employees to sit when not greeting guests while performing other tasks, such as answering calls, would reduce the time that the front desk clerks are on their feet and the resulting muscular fatigue and contact stress.

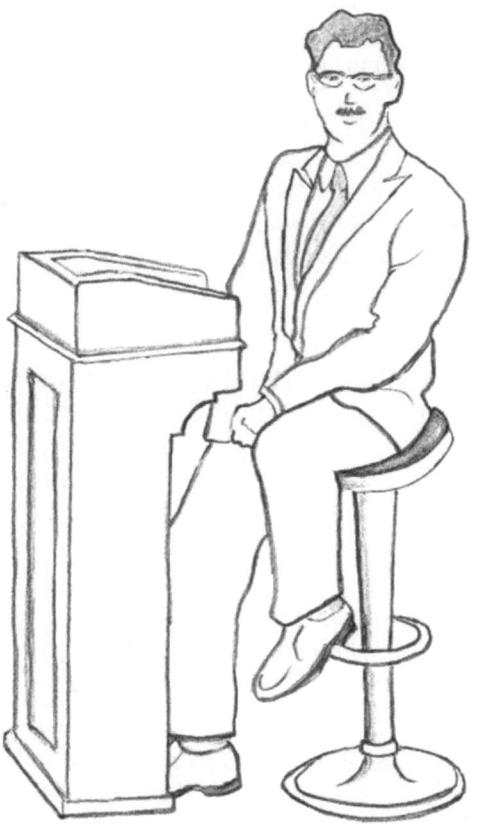

FIGURE 14.12 Front desk work area that allows the clerks to alternate between sitting and standing. (Courtesy of UNITE HERE! International Union, New York.)

For front desk work performed standing, a step or a rail placed on the floor below the counter will allow the clerks to alternate their foot positions more easily and reduce back and lower leg stresses (WorkSafe BC 2001g). Antifatigue matting along the front desk area and cushioned shoes may also help to lower static muscle tension and strain (Workplace Safety and Health Council 2013).

14.5 FUTURE RESEARCH AND DEVELOPMENT NEEDS

Despite the global economic importance of the hospitality sector, a number of jobs, including bellhop and front desk clerk, have been inadequately studied to date with regard to their ergonomic features and potential for musculoskeletal disorder or other health risks, as noted earlier.

Another research gap is particularly salient in light of the high proportion of women in the cleaning workforce in general, and hotel housekeeping in particular. There appears to be no research to date specifically on the gender differences in the musculoskeletal demands in cleaning work; a limited number of studies on painting and other construction tasks suggest that women may use different muscle recruitment strategies than men and experience different loading patterns, even after taking anthropometry into account (e.g., Rosati et al. 2014). The prevalence of upper extremity and shoulder pain in female housekeepers underscores the need for data on the musculoskeletal loading related to specific tasks and the development of better tool and equipment design for the (female-dominated) workforce in cleaning occupations, in the hospitality and other industries. The female anthropometry should be addressed in intervention research for the design of cleaning tools and equipment such as long-handled tools, carts, and vacuum cleaners. A study comparing work rates, energy expenditure, and perceived exertion by cleaners using a backpack versus an upright vacuum cleaner found that female cleaners had higher ratings of perceived exertion than males for both types of vacuums (Mengelkoch and Clark 2006). Additional studies are needed to understand the potential gender differences in the impact of ergonomic exposures and the effectiveness of potential interventions.

Work organization issues represent another important domain for intervention research. Barebones staffing results in faster work pace for many hotel occupations, increasing the repetitiveness of the work and hence the musculoskeletal strain from ergonomic exposures. This combination of factors may also contribute to the risk of acute traumatic injuries, another topic that deserves far more research than has been conducted to date.

ACKNOWLEDGMENTS

The authors thank the following UNITE HERE! International Union staff: Virginia Blaisdell for sketches, Diana Colon-Guzman for proofreading the manuscript, and Jeanette Rivera for clerical assistance.

REFERENCES

Al-Eisawi, K. W., C. J. Kerk, J. J. Congleton et al. 1999. Factors affecting minimum push and pull forces of manual carts. *Appl Ergon*. 30:235–45.

Alexander, M. 2006. Luxury hospitality linen—The bed wars. *Textile Rental*. 34–40. From http://www.laundry experts.biz/BedWars_Alexander_0806.pdf (accessed July 28, 2015).

Allread, G., S. A. Ferguson, and W. S. Marras. 2013. Using the lumbar motion monitor to assess housekeeper room-cleaning tasks. Paper presented at Cal/OSHA Advisory Committee Meeting, Housekeepers in the Hotel and Hospitality Industry, Los Angeles, CA. From http://www.dir.ca.gov/dosh/doshreg/Housekeeper_OSU_LMM_Summaries.pdf (accessed August 27, 2015).

Barrett, R., and P. Milburn. 1997. Lumbar loads in occupational bedmaking: A static planar analysis. *J Occup Health Safety*. 13:35–46.

Baum, T. 2012. Migrant workers in the international hotel industry. International Migration Paper No. 112. Geneva: Sectoral Activities Department, International Migration Branch, International Labour Office.

Baum, T. 2013. International perspectives on women and work in hotels, catering and tourism. Working Paper No. 289. Geneva: International Labour Office.

Baumann, G. 2006. The bed race: Hotel companies' everlasting pursuit of differentiation. *Hospitality Trends.* From http://www.hospitalitynet.org/news/4027006/html (accessed August 4, 2015).

Bekkelund, S. I., T. Torbergsen, A. K. Rom et al. 2001. Increased risk of median nerve dysfunction in floor cleaners: A controlled clinical and neurophysiological study. *Scand J Plast Reconstr Hand Surg.* 35: 317–21.

Buchanan, S., P. Vossenas, N. Krause et al. 2010. Occupational injury disparities in the U.S. hotel industry. *Am J Ind Med.* 53:116–25.

BLS (Bureau of Labor Statistics). 2005. The editor's desk: Distribution of days away from work due to work-place injuries and illnesses, 2004. Washington, DC: U.S. Bureau of Labor Statistics, U.S. Department of Labor. From http://www.bls.gov/opub/ted/2005/dec/wk3/art02.htm (accessed September 18, 2014).

BLS. 2013a. Employment by industry, occupation, and percent distribution: 2012 and projected 2022: 721100 Traveler accommodation. Employment Projections Program. Washington, DC: U.S. Bureau of Labor Statistics, U. S. Department of Labor.

BLS. 2013b. Table 15: Incidence rates for nonfatal occupational injuries and illnesses involving days away from work per 10,000 full-time workers by selected worker occupation and nature of injury or illness, private industry, state government, and local government, 2013. Washington, DC: U.S. Bureau of Labor Statistics, U. S. Department of Labor.

BLS. 2015a. Maids and housekeeping cleaners. In *Occupational Outlook Handbook*, 2014–15 edn. Washington, DC: U.S. Bureau of Labor Statistics, U. S. Department of Labor. From http://www.bls.gov/ooh/building-and-grounds-cleaning/maids-and-housekeeping-cleaners.htm (accessed July 21, 2015).

BLS. 2015b. Cooks. In *Occupational Outlook Handbook*, 2014–15 edn. Washington, DC: U.S. Bureau of Labor Statistics, U. S. Department of Labor. From http://www.bls.gov/ooh/food-preparation-and-serving/cooks.htm (accessed May 21, 2015).

BLS. 2015c. Data for occupations not covered in detail: Dishwashers. In *Occupational Outlook Handbook*, 2014–15 edn. Washington, DC: U.S. Bureau of Labor Statistics, U. S. Department of Labor. From http://www.bls.gov/ooh/food-preparation-and-serving/food-preparation-workers.htm (accessed May 21, 2015).

Burgel, B. J., M. C. White, M. Gillen et al. 2010. Psychosocial work factors and shoulder pain in hotel room cleaners. *Am J Ind Med.* 53:743–56.

Cal/OSHA (Division of Occupational Safety and Health) Consultation Service Research and Education Unit. 2005. *Working Safer and Easier for Janitors, Custodians, and Housekeepers.* Oakland, CA: California Department of Industrial Relations.

Carrivick, P. J. W., A. H. Lee, K. K. W. Yau, and M. R. Stevenson. 2005. Evaluating the effectiveness of a participatory ergonomics approach in reducing the risk and severity of injuries from manual handling. *Ergonomics.* 48:907–14.

Casey, M., and E. Rosskam. 2009. Organizing to reduce hotel workers' injuries. In Schnall, P., Dobson, M., and Rosskam, E. (eds.) *Unhealthy Work: Causes, Consequences, Cures.* Amityville, NY: Baywood Publishing Company.

Castleman, B. I., and V. Navarro. 1987. International mobility of hazardous products, industries and wastes. *Int J Health Serv.* 17:617–33.

Chiang, F. F. T., T. A. Birtch, and H. K. Kwan 2010. The moderating roles of job control and work-life balance practices on employee stress in the hotel and catering industry. *Int J Hosp Manage.* 29:25–32.

Chyuan, J. Y., C. L. Du, W. Y. Yeh et al. 2004. Musculoskeletal disorders in hotel restaurant workers. *Occup Med.* 54:55–7.

Cryer, C., A. Samaranayaka, and J. D. Langley. 2014. The epidemiology of life-threatening work-related injury: A demonstration paper. *Am J Ind Med.* 57:425–37.

Davis, K. G., and L. Orta-Anes. 2014. Potential of adjustable height carts in reducing the risk of low back injury in grocery stockers. *Appl Ergon.* 45:285–92.

Davis, K. G., W. S. Marras, and T. R. Waters. 1998. Reduction of spinal loading through the use of handles. *Ergonomics.* 41:1155–68.

Ernst & Young Global (EYG). 2015. Global hospitality insights: Top thoughts for 2015. EYG #DF0196. From http://www.ey.com/GL/en/Industries/Real-Estate/ey-global-hospitality-insights-2015 (accessed August 4, 2015).

European Agency for Safety and Health at Work. 2009. Preventing harm to cleaning workers. Luxembourg: Office for Official Publications of the European Communities.

Fontani, S., I. Mercuri, R. Salicco et al. 2010. Manual handling of loads in the hotel trade: The experience of the Local Health Unit Milan. *La Medicina Del Lavoro.* 101(6):437–45.

Frumin, E., J. Moriarty, P. Vossenas et al. 2006. Workload-related musculoskeletal disorders among hotel housekeepers: Employer records reveal a growing national problem. Presented at NIOSH National Occupational Research Agenda (NORA) Symposium, Washington, DC.

Garg, A., and U. Saxena. 1980. Container characteristics and maximum acceptable weight of lift. *Human Factors.* 22:487–95.

Hannerz, H., F. Tüschsen, and T. S. Kristensen. 2002. Hospitalizations among employees in the Danish hotel and restaurant industry. *Eur J Pub Health.* 12:192–7.

Haukka, E., P. Leino-Arjas, S. Solovieva et al. 2006. Co-occurrence of musculoskeletal pain among female kitchen workers. *Int Arch Occup Environ Health.* 80:141–8.

Haussman, G. 2014. New hotel construction pipeline in global shift. *Hotel Interactive.* From http://www .hotelinteractive.com/article.aspx?articleid=32644 (accessed August 4, 2015).

Health and Safety Executive. 2012. Preventing back pain and other aches and pains to kitchen and food service staff. HSE Catering Information Sheet No. 24. From http://www.hse.gov.uk/pubns/cais24.htm (accessed August 7, 2015).

Herichova, I. 2013. Changes of physiological functions induced by shift work. *Endocrine Reg.* 47:159–70.

Institute for Ergonomics. 2011a. Housekeepers: Practices to improve health & safety using ergonomics. Columbus, OH: Ohio State University. From https://www.osha.gov/dte/grant_materials/fy10/sh-20998 -10/Housekeeper_Ergo_Handout-English.pdf (accessed July 28, 2015).

Institute for Ergonomics. 2011b. Housekeeper managers: Improving housekeeping work using ergonomics. Columbus, OH: Ohio State University. From https://www.osha.gov/dte/grant_materials/fy10/sh-20998 -10/Housekeeper_Manager_Ergo_Handout.pdf (accessed July 28, 2015).

Intilli, H. 1999. The effects of converting wheels on housekeeping carts in a large urban hotel. *Amer Assoc OHN J.* 47:466–9.

Janowitz, I., G. U. Goldner, and S. S. Scott. 2005. Ergonomics evaluation of the use of a handled shower-cleaning tool. *Proc HFES 49th Ann Meet.* 49:1724–8.

Jansen, J. P., M. J. M. Hoozemans, A. J. van der Beek et al. 2002. Evaluation of ergonomic adjustments of catering carts to reduce external pushing forces. *Appl Ergon.* 33:117–27.

Jung, H. S., and H. S. Jung. 2003. Development and ergonomic evaluation of polypropylene laminated bags with carrying handles. *Int J Industr Ergonom.* 31:223–34.

Korkmaz, S. V., J. A. Hoyle, and G. G. Knapik. 2006. Baggage handling in an airplane cargo hold: An ergonomic intervention study. *Int J Ind Ergonom.* 36:301–12.

Krause, N., R. Rugulies, and C. Maslach. 2010. Effort-reward imbalance at work and self-reported health of Las Vegas hotel room cleaners. *Am J Ind Med.* 53:372–86.

Krause, N., T. Scherzer, and R. Rugulies. 2005. Physical workload, work intensification, and prevalence of pain in low wage workers: Results from a participatory research project with hotel room cleaners in Las Vegas. *Am J Ind Med.* 48:326–37.

Kumar, R., M. Chaikumarn, and J. Lundberg. 2005. Participatory ergonomics and an evaluation of a low-cost improvement effect on cleaners' working posture. *Int J Occup Safety Ergon.* 11:203–10.

Lee, P. T., and N. Krause. 2002. The impact of a worker health study on working conditions. *J Public Health Policy.* 23:268–85.

Lee, Y. J., M. J. M. Hoozemans, and J. H. van Dieën. 2011. Control of trunk motion following sudden stop perturbations during cart pushing. *J Biomech.* 44:121–7.

Leigh, J. P. 2012. *Numbers and Costs of Occupational Injury and Illness in Low-Wage Occupations.* Davis, CA: Center for Poverty Research, and Center for Health Care Policy and Research, University of California Davis.

Marras, W. S., S. L. Rajulu, and P. E. Wongsam. 1987. Trunk force development during static and dynamic lifts. *Human Factors.* 29:19–29.

Mengelkoch, L. J., and K. Clark. 2006. Comparison of work rates, energy expenditure, and perceived exertion during a 1-h vacuuming task with a backpack vacuum cleaner and an upright vacuum cleaner. *Appl Ergon.* 37:159–65.

Messing, K., F. Tissot, and S. Stock. 2008. Distal lower-extremity pain and work postures in the Quebec population. *Am J Public Health.* 98:705–13.

Milburn, P. D., and R. S. Barrett. 1999. Lumbosacral loads in bedmaking. *Appl Ergon.* 30:263–73.

NIOSH. 2011. Services sector: Occupational safety and health needs for the next decade of NORA—Safety and health among hotel cleaners. Publication No 2011-194. Cincinnati, OH: National Institute of Occupational Safety and Health, U.S. Department of Health and Human Services.

NRC/IOM (National Research Council and the Institute of Medicine). 2001. Musculoskeletal disorders and the workplace: Low back and upper extremities. Washington, DC: Panel on Musculoskeletal Disorders and the Workplace, Commission on Behavioral and Social Sciences and Education, National Academy Press.

O*NET OnLine (2015a). 37-2012.00. Summary report for maids and housekeeping cleaners. National Center for O*NET Development. Washington, DC. From http://www.onetonline.org/link/summary/37-2012.00 (accessed July 21, 2015).

O*NET OnLine (2015b). 35-9021.00. Dishwashers. National Center for O*NET Development. Washington, DC. From http://www.onetonline.org/link/summary/35-9021.00 (accessed July 19, 2015).

O*NET OnLine (2015c). Summary report for 35-9011.00—Dining room and cafeteria attendants and bartender helpers. National Center for O*NET Development. Washington, DC. From http://www.onetonline.org/link/summary/35-9011.0 (accessed July 21, 2015).

O*NET OnLine (2015d). Summary report for 35-3041.00—Food servers, nonrestaurant. National Center for O*NET Development. Washington, DC. From http://www.onetonline.org/link/summary/35-3041.00 (accessed July 21, 2015).

O*NET OnLine (2015e). Summary report for 39-6011.00—Baggage porters and bellhops. National Center for O*NET Development. Washington, DC. From http://www.onetonline.org/link/summary/39-6011.00 (accessed July 20, 2015).

O*NET OnLine (2015f). Summary report for 43-4081.00—Hotel, motel, and resort desk clerks. National Center for O*NET Development. Washington, DC. From http://www.onetonline.org/link/summary/43-4081.00 (accessed July 20, 2015).

Orr, G. 2004. Ergonomic task analysis for hotel housekeeping. Report to UNITE HERE! Alexandria, VA: Orr Consulting.

Orr, G. 2007. NIOSH lifting index evaluation of luxury hotel beds. Paper presented in From Science to Negotiated Solutions: Resolving Biomechanical Hazards of Hotel Work. Roundtable No. 246. Philadelphia, PA: American Industrial Hygiene Annual Meeting.

Pehkonen, I., E. P. Takala, R. Ketola et al. 2009. Evaluation of a participatory ergonomic intervention process in kitchen work. *Appl Ergon*. 40:115–23.

Pekkarinen, A., and H. Anttonen. 1988. The effect of working height on the loading of the muscular and skeletal systems in the kitchens of workplace canteens. *Appl Ergon*. 19:306–8.

Punnett, L., A. Prüss-Ustün, D. I. Nelson et al. 2005. Estimating the global burden of low back pain attributable to combined occupational exposures. *Am J Ind Med*. 48:459–69.

Ramsey, J. G., Musolin, K., Ceballos, D., and Mead K. 2014. Evaluation of ergonomic risk factors, thermal exposures and job stress at an airline catering facility. NIOSH HHE Report No. 2011-0131-3221. Cincinnati, OH: National Institute of Occupational Safety and Health, Centers for Disease Control and Prevention, U.S. Department of Health and Human Services. From http://www.cdc.gov/niosh/hhe/reports/pdfs/2011-0131-3221.pdf (accessed August 6, 2015).

Rosati, P. M., J. N. Chopp, and C. R. Dickerson. 2014. Investigating shoulder muscle loading and exerted forces during wall painting tasks: Influence of gender, work height and paint tool design. *Appl Ergon*. 45:1133–9.

Samani, A., A. Holtermann, K. Søgaard et al. 2012. Following ergonomics guidelines decreases physical and cardiovascular workload during cleaning tasks. *Ergonomics*. 55:295–307.

Sanon, M. A. V. 2014. Agency-hired hotel housekeepers: An at-risk group for adverse health outcomes. *Workplace Health Saf*. 62(2):86.

Scherzer, T., R. Rugulies, and N. Krause. 2005. Work-related pain and injury and barriers to workers' compensation among Las Vegas hotel room cleaners. *Am J Public Health*. 95:483–88.

Seifert, A. M., and K. Messing. 2006. Cleaning up after globalization: An ergonomic analysis of work activity of hotel cleaners. *Antipode*. 38:557–578.

Silva, J. S., Jr., L. R. C. Correa, and L. C. Morrone. 2012. Evaluation of lumbar overload in hotel maids. *Work (Reading, Mass.)*. 41 Suppl 1:2496–8.

Smith, C., and D. Root. 1999. The export of pesticides: Shipments from U.S. ports: 1995–1996. *Int J Occup Env Heal*. 5:141–50.

Søgaard, K., A. K. Blangsted, A. Herod, and L. Finsen. 2006. Work design and the labouring body: Examining the impacts of work organization on Danish cleaners' health. *Antipode*. 38:579–602.

Tissot, F., K. Messing, and S. Stock. 2009. Studying the relationship between low back pain and working postures among those who stand and those who sit most of the working day. *Ergonomics*. 52:1402–18.

Vossenas, P. 2007. Negotiated solutions in collective bargaining agreements: A critical step in reducing ergonomic hazards of hotel housekeeping. Paper presented at the American Industrial Hygiene Annual Meeting, Philadelphia, PA.

Vossenas, P., D. Colon, P. Martin et al. 2010. The need for ergonomic interventions in the hospitality sector. Paper presented at the annual Applied Ergonomics Conference, Denver, CO.

Weigall, F., K. Simpson, A. F. Bell et al. 2005. *An Assessment of the Repetitive Manual Tasks of Cleaners.* Sydney: WorkCover New South Wales.

Workplace Safety and Health Council. 2013. Workplace safety and health guidelines: Hospitality and entertainment industries. Singapore: Workplace Safety and Health Council and Ministry of Manpower. From https://www.wshc.sg/files/wshc/upload/infostop/attachments/2014/IS2013120200766/WSH_Guidelines _Hospitality_and_Entertainment_Industries.pdf (accessed July 28, 2015).

WorkSafeBC (Workers' Compensation Board of British Columbia). 2001a. Preventing injuries to hotel and restaurant workers: Focus report. Richmond, BC: Workers' Compensation Board of British Columbia. From http://www.worksafebc.com/publications/reports/focus_reports/assets/pdf/focushotel.pdf (accessed September 7, 2015).

WorkSafeBC. 2001b. Preventing injuries to kitchen staff: Ergonomic tips for the hospitality industry. Richmond, BC: Workers' Compensation Board of British Columbia. From http://www.worksafebc.com/publica tions/health_and_safety/by_topic/assets/pdf/kitchenstaff.pdf (accessed July 29, 2015).

WorkSafeBC. 2001c. Preventing injuries to dishwashers: Ergonomic tips for the hospitality industry. Richmond, BC: Workers' Compensation Board of British Columbia. From http://www.worksafebc.com/publica tions/health_and_safety/by_topic/assets/pdf/dishwashing.pdf (accessed July 29, 2015).

WorkSafeBC. 2001d. Preventing injuries when serving banquets. Ergonomic tips for the hospitality industry. Richmond, BC: Workers' Compensation Board of British Columbia. From http://www.worksafebc.com /publications/health_and_safety/by_topic/assets/pdf/banquet.pdf (accessed July 29, 2015).

WorkSafeBC. 2001e. Preventing injuries to servers. Ergonomic tips for the hospitality industry. Richmond, BC: Workers' Compensation Board of British Columbia. From http://www.worksafebc.com/publications /health_and_safety/by_topic/assets/pdf/server.pdf (accessed July 29, 2015).

WorkSafeBC. 2001f. Preventing injuries when handling luggage. Ergonomic tips for the hospitality industry. Richmond, BC: Workers' Compensation Board of British Columbia. From http://www.worksafebc.com /publications/health_and_safety/by_topic/assets/pdf/luggage.pdf (accessed July 29, 2015).

WorkSafeBC. 2001g. Preventing injuries to front desk agents. Ergonomic tips for the hospitality industry. Richmond, BC: Workers' Compensation Board of British Columbia. From http://www.worksafebc.com /publications/health_and_safety/by_topic/assets/pdf/frontdesk.pdf (accessed July 29, 2015).

WorkSafeBC. 2002. Health and Safety for Hospitality Small Business. Richmond, BC: Workers' Compensation Board of British Columbia. From http://www.worksafebc.com/publications/health_and_safety/by_topic /assets/pdf/hosp_smbiz.pdf (accessed March 20, 2016).

WorkSafeBC. 2009. Health and safety for hospitality small business. Richmond, BC: Workers' Compensation Board of British Columbia. From http://www.worksafebc.com/publications/health_and_safety/by_topic /assets/pdf/hosp_smbiz.pdf (accessed September 18, 2014).

Xu, Y., A. S. K. Chen, and C. W. P. Li-Tsang. 2013. Prevalence and risk factors of work-related musculoskeletal disorders in the catering industry: A systematic review. *Work.* 44:107–116.

15 Managing the Safety and the Performance of Home-Based Teleworkers
A Macroergonomics Perspective

Michelle M. Robertson and Wayne S. Maynard

CONTENTS

15.1 INTRODUCTION

Regular home-based, non-self-employed populations have grown by 10.3% since 2005 and 6.5% in 2014 (Global Workplace Analytics 2015), and it was estimated that 100 million U.S. workers would be telecommuting by 2010 (The Telework Coalition [TelCoa] 2015). With new computer-based information communication technologies, coupled with environmental and economic issues, the patterns of office and computer works will be impacted, as teleworking is changing how office employees are working, commuting, and communicating (International Telework Association and Council 1999; Davis and Polonko 2001; Harrington and Walker 2004). The advances in IT are allowing selective employees to work anywhere and at anytime. The traditional work location for some employees is changing from the corporate office to the virtual work locations, such as home, hotel, airport, shared and satellite offices, client office, and vehicle. In other words, telework allows the employees and their tasks to be shared across settings away from a central place of business or a physical organizational location (Belanger et al. 2001; Davis and Polonko 2001; Karnowski and White 2002). This trend toward alternative work styles and the distributed workforce is a widespread practice that has steadily increased in the United States and abroad (Robertson and Vink 2012).

Over the last several years, both the private industry and the federal government have encouraged these alternative workplace changes, many of which have been observed to be beneficial to the economy, to the environment, and to the quality of family life (Apgar 1998). In fact, a 40 min commute equates to 8 working weeks per year as reported by the Colorado Telework Coalition (TelCoa 2015). More recently, the federal government, encouraged by the congressional legislation and the Office Program Management (OPM) telework directive, is further supporting this initiative by developing telework programs including flexible and alternative office workplaces, such as satellite offices or shared work spaces (OPM 2015).

While telework offers an attractive alternative to traditional work locations, it is not without challenges for the employers and the employees (Ilozor et al. 2001). Many of these challenges are not unique, as many have been anticipated for years (Galtiz 1984). These challenges include the ways to best manage the employees who work at home rather than at the corporate site, the implementation and support of the required IT, the lack of social and group interaction, the changes in job autonomy, the absence of mentoring and career development, the balancing of work and personal conflicts, the extended work hours and workload, and lack of sound risk management that addresses safety and health issues (Ilozor et al. 2001). All these challenges can have a cumulative impact on the employee morale, stress, and MSD such as lower back pain and upper extremity disorders. Given that telework is one of the fastest growing fields in the service sector, the manner in which these factors are addressed will significantly impact not only the safety and the health of the employee, but also the organizational effectiveness (Robertson and Vink 2012). As these work styles and virtual workplaces continue to emerge, understanding and designing effective work systems using a macroergonomics perspective is essential to achieve the benefits of telework.

This chapter will first provide a literature background that frames the issues and the challenges of telework. Second, a conceptual model is given that captures the macroergonomics or the work systems design issues based on the empirical data derived from the literature and the case studies for understanding the outcomes in a telework environment. This conceptual model provides a diagnostic tool for identifying problems and issues to improve safety and health, as well as organizational performance of teleworkers. Further, we will provide an outline for a macroergonomics process in managing the safety and the health of the teleworkers along with guidelines and tips to manage the teleworkers. This approach explains the impact of the organizational, physical, and psychosocial risk factors on the teleworker's safety and health, which is also applicable to the conventional office workstation design.

15.2   MACROERGONOMICS: WORK SYSTEMS DESIGN MODEL FOR TELEWORK

Understanding and designing effective and safe work systems using a macoerognomics or a work systems design approach is essential for achieving the benefits of telework. A systematic view of the emerging trend of telework is needed to better understand the complexities of telework and how to best manage and design a telework program. A macroergonomics or a work systems design perspective can provide a framework to conceptualize these various systems elements and issues. Macroergonomics is concerned with the design of the overall work systems involving the technological and personnel subsystems, the external environment, and their interrelationships. It is a top-down sociotechnical systems approach to the design of the work systems. Incorporating a human-centered design approach, macroergonomics focuses on the human–organization technology interface. This human-centered design systematically considers the worker's professional and psychosocial characteristics in designing the work system, and then the macroergonomics process carries this work systems design effort through the ergonomics design of specific jobs and related hardware and software interfaces. Integral to this design process is the joint optimization design of the technical and personnel subsystems, using a humanized task approach in allocating functions and tasks. A primary methodology of macroergonomics is participatory ergonomics, an approach that involves the employees at all organization levels in the design process and has been shown to be effective in creating viable, integrated safety and health solutions to solve workplace issues (Imada 2002; Robertson et al. 2013). Using systems analysis models and design processes, the micro, meso, and macroergonomics factors and their interactions can be identified throughout the work systems design process. Designing work systems that have congruent subsystems and alignment and consider the sociotechnical characteristics of the organization can lead to substantial benefits such as increased productivity, health, and safety. Figure 15.1 depicts a macroergonomics model that captures various hypothesized workplace and job designs, and technological, psychosocial, and organizational factors related to telework, based on earlier work by Robertson et al. (2003).

Incorporated into this model are the sociotechnical systems factors (Hendrick and Kleiner 2002) as well as the balance model proposed by Carayon and Smith (2000). This micro, meso, and macroergonomics model organizes and structures the identified risks and issues of teleworking into three levels (individual, group, and organizational) (Figure 15.1). Within each of these levels, the technological and personnel subsystems, the psychosocial factors, and the physical work environment factors are

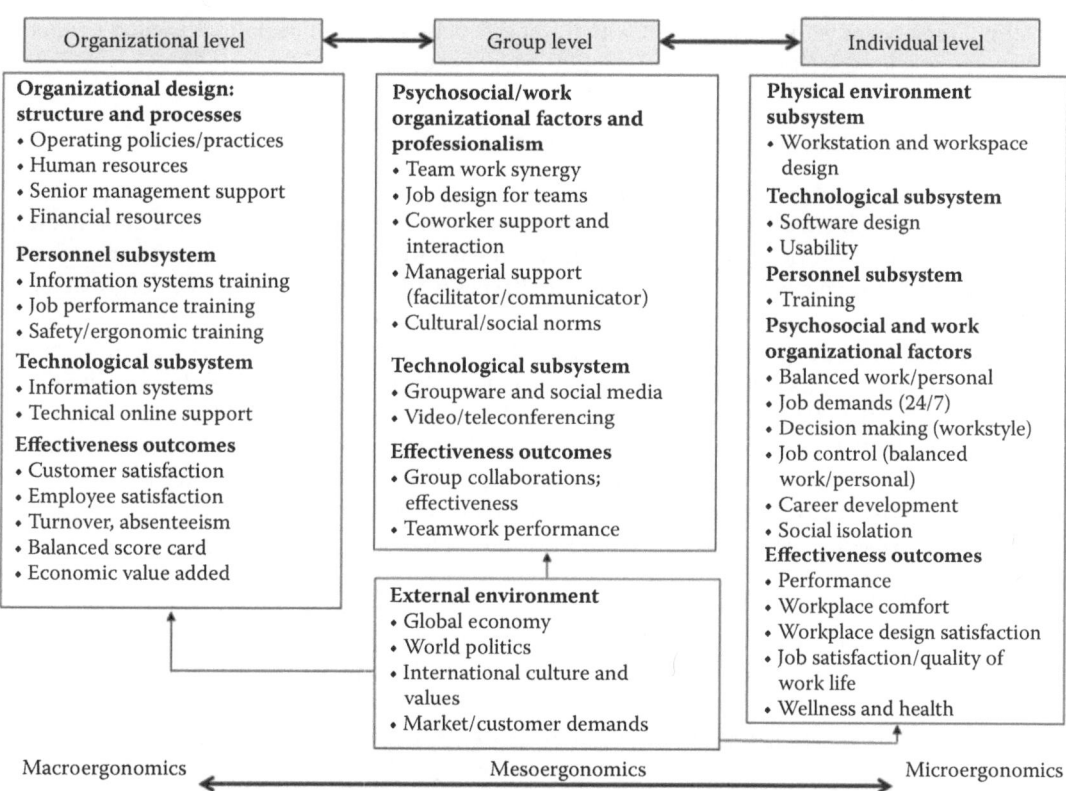

FIGURE 15.1 Telework: A macroergonomics work systems conceptual model. (Based on Robertson, M. M., W. S. Maynard, and J. R. Mcdevitt, *Prof Safety.*, 48, 30–6, 2003; Robertson, M. M., W. S. Maynard, Y. H. Huang, and J. R. McDevitt, *Hum Fac Erg Soc P.*, 46, 1330–4, 2004. With permission.)

identified along with the organizational design factors. Together, these subsystems, factors, and external environment comprise the overall work systems design for teleworkers. The potential effectiveness ratings and outcomes that could measure the success and the impact of a telework program are given at the bottom of the figure. A continuum is shown of macroergonomics issues to consider in the organizational design, mesoergonomics group level interactions to microergonomics physical and psychosocial individual design factors (Hendrick and Kleiner 2002; Karsh et al. 2014; Murphy et al. 2014). These factors were derived from literature and exemplary industry case studies. Several reported business case studies (Apgar 1998; Bailey and Kurland 2002) concluded that to have a successful telework program, the factors noted in the model should be systematically addressed at each of the levels and with the appropriate stakeholders.

15.3 DEFINING AND DESCRIBING TELEWORK AND PRACTICES

15.3.1 WHY TELEWORK?

Telework, defined as working for an employer at an alternative work location, such as the home with an electronic link, is common for millions of Americans (Hamilton 1987; Greengard 1998; Youngblood 1999). Organizations establish telework programs for several reasons: reduce corporate office space needs, attend to government regulations such as clean air standards, enhance employee retention, improve employee recruitment efforts, and increase productivity (Hamilton 1987; Hequet 1994; McGonegle 1996; Piskurich 1996). Employees view telework as the ability to gain more flexibility for time management and balancing work and personal (home) responsibilities, thus reducing stress. Other benefits are less time spent commuting and lower transportation costs. Moreover, the potential for a better work environment with fewer distractions and office politics is cited as another advantage of telework (Bailey and Kurland 2002; Grzywacz et al. 2007).

15.3.2 WAYS IN WHICH COMPANIES IMPLEMENT TELEWORK

There are a variety of companies implementing telework programs with each defining telework differently, yet each having common program elements. How these organizations define telework typically comprise a variety of work situations in which a worker performs job duties in a location other than the corporate office. This includes employees who regularly work in the corporate office but occasionally spend a day working elsewhere (home or while traveling) as well as employees who have telework agreements stating when and where they work. Other telework situations include mobile employees (e.g., sales staff and service technicians) whose job responsibilities require them to be out of the office (Hamilton 1987; Hequet 1994; Ellsion 1999; Wilde 2000; Dziak 2001; Bailey and Kurland 2002). Finally, a satellite office is where the employee works from a remote office, and there is a high concentration of employees, whereas a neighborhood work center provides office space for employees from different companies. At these work centers, the employee's company is usually responsible for the technical and administrative requirements.

15.3.2.1 Description of Telework Programs

A significant diversity exists in the telework programs' policies, structures, and evaluation processes. Some of the programs are highly structured with formalized procedures, such as written agreements containing job issues of performance standards, communication policies, and safety guidelines. The structured programs establish set working hours, especially for job positions requiring customer service. These programs also have defined specific working locations, equipment, and maintenance procedures. Human resource and development policies for selecting teleworkers and managers are highly formalized, and telework training is provided for each. The training prepares them to understand the different roles they would have, effectively communicate, define performance standards, and manage safety and ergonomics issues of the new workplace(s).

Smaller companies and jobs that are defined with a high level of professionalism and role/task variability typically have informal procedures regarding the role of a teleworker. Informal agreements are generally made between the employee and the manager, often at the initiation of the employee. These employees do not normally receive any formal training or standardized computer and workplace equipment. Informal programs have a low level of formalization regarding telework procedures, including nonstandard equipment, and a high level of schedule flexibility.

15.4 EMPIRICAL RESEARCH FINDINGS ON TELEWORKERS' PERCEPTIONS AND HEALTH

Telework studies have examined the effects of this type of alternative work style on the teleworkers' self-perceptions of performance, communication, job design, career development, job satisfaction, stress, health, and quality of work/home life. The majority of these field studies are cross-sectional in design where the primary study instruments are surveys, interviews, and descriptive case studies. These studies consist primarily of pilot studies, each addressing issues such as perceptions about the work arrangement, work attitudes, and work/family conflict (McCloskey and Igbaria 1998; Ellsion 1999).

A growing number of effects of teleworking have been observed, possibly due to the advances in information computer technology and the experiences in planning and implementing telework programs. Historically, study results from the 1980s showed an increase in productivity, flextime, job autonomy, and company loyalty. Also, an increase in social isolation and role ambiguity, along with a decrease in task significance, feedback, work group job enrichment, communication, and home autonomy, was reported in several studies (Olson and Primps 1984; Atkinson 1985; Hamilton 1987).

Subsequent research has revealed some similar results with slight shifts in teleworkers' perceptions of organizational and managerial supports, job control, work group communication and managerial interactions, and technology support (Durbin 1991; Trent et al. 1994). Durbin (1991) conducted a study where he compared job satisfaction and group productivity (transactions/h) between 34 in-house (conventional office) employees and 34 teleworkers. The teleworkers indicated higher satisfaction on items related to work conditions, coworker relationships, and scheduling. Productivity was estimated to be 30% higher for the teleworkers. Another investigation examining teleworkers, a one-year pilot study of 20 teleworkers, revealed that they expressed having more control of their time and reported being more productive when working at home (Miller 2009). However, they did report a lower level of job satisfaction. In two more recent studies, teleworkers reported high levels of social isolation, absence of professional help, and fewer opportunities for job promotion (Claes 2000; Belanger et al. 2001).

The communication patterns among and between teleworkers, their managers, and coworkers (in the office and out) has been a topic of several research studies, as reported in the meta-analysis by Belanger et al. (2001). Three variables of interest that were believed to impact the outcomes in the teleworker studies were (1) the availability of information systems (IS) technology, (2) the availability of communication technologies, and (3) the communication patterns of teleworkers within their work groups. The study results indicated that the technology variables positively impact productivity, performance, and satisfaction of teleworkers, while the interaction between the person and the technology variables was significant in predicting the perceived productivity. Furthermore, work group communication, as measured by the centrality of individuals, negatively affected the perceived productivity and performance (Belanger et al. 2001). Ilozor et al. (2001) surveyed 43 teleworkers and noted several aspects of communication strategies that had a significant and positive influence on job satisfaction. These were communicating job responsibilities, having goals and objectives, meeting deadlines and job expectations, communicating freely and regularly, providing appropriate equipment, receiving training and career development, and reviewing work and salary regularly. The study results also identified that cultural, economic, and social contexts may have a distinct part to play in the impact of the management communication strategies on the job satisfaction of the teleworkers. These observations are consistent with the conclusions reported by Harrington and Walker (2004), where a rational culture and a group culture, which emphasize human resources and member participation, facilitate the successful and effective implementation of teleworker programs.

The investigations concerned with the occupational health and safety issues of teleworkers indicate a minimal level of knowledge and awareness regarding the ergonomics and safety issues for the teleworkers (Lenckus 1997; Unsworth 1998; Healy 2000). Not surprisingly, a slow but continuous flow of ergonomics claims for teleworkers regarding MSDs have been reported (Lenckus 1997; Unsworth 1998).

One qualitative cross-sectional study noted that 60% of teleworkers reported that they were left to themselves to set up their home-based workstation with no ergonomics or technical advice (Montreuil and Lippel 2003). Home-based telework was generally reported by workers as having a positive effect on their health, although they also identified the potential problems arising from workstation design, long hours, and isolation (Montreuil and Lippel 2003). Given that home-based telework is usually performed with a computer as the principal tool, computer work is recognized for contributing to the development of musculoskeletal problems in the neck, the shoulders, the wrists, the hand, and the lumbar regions (Bergqvist et al. 1995; Marcus and Gerr 1996; Carayon et al. 1999). In their sample of 33 teleworkers, Montreuil and Lippel (2003) found that 54.5% complained of pain in their upper limbs, back, or neck, which they attributed to inadequate furnishings, static posture, computer use, and lifting heavy objects.

Given the aforementioned evidence, concerns have been raised that teleworkers are so committed to working at home that this may lead to less than optimal practices regarding health and safety. For example, teleworkers may commit to working long hours in order to ensure that their productivity meets or even surpasses expectations (NIOSH 2002). Moreover, having to respond to customers within a time-constraint context and using computer equipment that is poorly adapted to the employees' needs may result in a situation that is conducive to the onset of musculoskeletal disorders as shown by Sznelwar et al. (1999). Clearly, there is a need to address ergonomics issues and related musculoskeletal symptoms for home-based teleworkers. A manager's perception regarding how telework influences their managerial role is also essential to understand, especially in managing the safety and the health of remote, off-site workers. However, many telework studies focus on only the employees and not on the managers who are in a supervisory role. As reported by Ilozor et al. (2001) and Belanger et al. (2001), the majority of telework research focuses on the work management processes, such as job satisfaction and family and life balance issues, and these perceptions are typically gathered by cross-sectional surveys. For example, in a meta-analysis of 46 cross-sectional survey studies (Gajendran and Harrison 2007), the researchers found that there were minimal detrimental effects on the quality of the workplace relationships for teleworkers. However, if the duration of the telework was more than 2.5 days a week, it had negative effects on the coworkers' relationships. Conversely, the longer the telework duration, the more beneficial effects on work–family conflict were shown, along with job satisfaction, performance, turnover, and role stress.

15.4.1 Advantages and Disadvantages of Telework

Telework is not for everyone and not for an entire organizational workforce. Several of the advantages of telework, such as fewer distractions, are viewed as disadvantages by others—less social contact with coworkers, more isolation, less career development, and fewer opportunities for informal learning (Hamilton 1987; Piskurich 1996; McCloskey and Igbaria 1998). This double-edged sword of advantages/disadvantages of telework brings forward the potential misconceptions by both employees and managers. Some employees regard it as a substitute for day care services, whereas managers believe that they will lose control of their employees. These misconceptions can be reduced and the disadvantages limited when a program is systematically defined, planned, and managed.

There is insufficient research to determine whether telework is positive or negative for employees regarding the safety and health effects (Gajendran and Harrison 2007). Working at home may reduce stress and injury risk by harmonizing the work and family demands and minimizing the daily commutes. The positive factors frequently reported include the elimination of office stress, individual tailoring of work environments, greater accommodation of the disabled, reduced rates of sickness absenteeism, increased productivity, a better sense of control over the job and the workplace, and a higher level of job autonomy (Apgar 1998; Bailey and Kurland 2002; Grzywacz et al. 2007). Conversely, the negative issues often raised are social isolation, career stagnation, work–family conflict, and higher perceived workload levels (Baruch and Nicholson 1997; McCloskey and Igbaria 1998). The presumed benefits need to be balanced against the risks from loss of safety and ergonomics oversight, introduction of occupational hazards into the home working environment, blurring of work and family roles, social isolation from peers, and constant feeling of being linked to the workplace (NIOSH 2002).

It needs to be noted that there are several limitations to these reported studies regarding their research design and generalizability. Existing studies are limited due to differences in the definition of telework, small samples sizes, one-time data collection period, lack of a control group, and extraneous variables. Only a few studies have focused on stress intervention programs to improve the well-being of the teleworkers at remote workplaces, and these results report positive changes in developing stress-coping strategies while teleworking.

In general, teleworking appears to have both positive and negative aspects. The disadvantages of telework may be mitigated by having a high level of support from the organization and the managers to create a culture that embraces teleworking. Further, the study results suggest that the potential advantages gained from designing an effective telework program provide employees with a work environment and culture that can enhance not only individual and team performances, but also job satisfaction, quality of work life, and organizational effectiveness.

15.4.2 Successful Telework Program Elements

Planning for telework programs necessitates defining the technological and personnel subsystems components of the program. This is best accomplished by conducting a needs assessment which identifies the range and the boundaries of the program, and who is the target population (e.g., data entry, knowledge workers, customer service, software developers and engineers, copy editors). Successful telework programs share a common thread of spending sufficient time prior to implementing a telework program to thoroughly plan and systematically define the organization's telework program. The level of formalized policies and procedures is dependent on the organizational design and structure of the company (e.g., research and development company versus telecommunications company) as well as the size of the company and its products and services. Included in the planning phase is designing a telework pilot study implemented on a trial basis to evaluate the program effectiveness, identify areas for improvement, and enhance the positive aspects.

Part of defining a successful telework program is to clearly establish the program elements and processes. These include the following:

1. Legal issues regarding equipment and maintenance contracts, insurance (e.g., worker's compensation, home owners, disability), and union issues
2. Human resource and development issues, such as employee selection, skill and career development, training, communication patterns between employee and manager as well as within peer groups and nonteleworkers (training programs that address health, safety, and ergonomics issues for the workplace design and computing habits are essential, plus providing online assistance in these areas, and performance management standards and productivity expectations)

TABLE 15.1

Telework Safety Program Evaluation and Management Guide

Job Selection for Telework

1. Are the right job positions being identified as appropriate telework home-based assignments?
2. Are the jobs primarily independent work that is generally processed electronically and daily interactions with peers and supervisors are not critical for effective product delivery?
3. How will the performance of the teleworker be evaluated?
4. Do the employee and the manager have a sound relationship that is built on trust, respect, and mutual understanding of job expectations?

Maintain Open Communication Lines

1. Do work-at-home employees regularly communicate with their managers and peers, and are they kept current on the company happenings?
2. Are biweekly teleconference meetings being held between coworkers?
3. Are the employees encouraged to share their information and problem-solving ideas with all relevant coworkers (traditional versus teleworkers)?
4. Are informal performance discussions occurring at least every month to review the progress toward the objectives?

Home-Based Telework Guidelines and Recommendations for Setting Up Equipment

1. Are the guidelines for setting up a home office, including equipment and ergonomic accessories, offered, and do they provide general recommendations?
2. Is there a policy addressing what ergonomic accessories and office furniture will be paid for by the company?
3. Are the work surfaces approximately 26 in. high and at least 24 in. deep for the computer, the printer, and the optional fax machine?
4. Is there an ergonomically designed chair provided?
5. Are the file cabinets well designed and secured?

Design of the Home-Based Office

1. Is the location in a dedicated space, private, quiet, and secure?
2. Is the space large enough to accommodate the work activity, the equipment, and the furniture requirements? Is the area smaller than 6 ft × 6 ft?
3. Does the employee use a desktop or a laptop for computing? If a laptop computer is used, is a full-size monitor, an external detachable keyboard, and a docking station provided?
4. Is the heating, ventilation, air conditioning, and the utilities evaluated for employee health and safety, and property protection, compliant to the environmental standards?
5. Are the building structures sound, well maintained, and in compliance with the local codes? Have the employees verified compliance with the local codes and restrictions for home-based assignments?
6. Are improvements to the premises and the equipment, including doors, locks, and alarm systems necessary?
7. Has the employee consulted their company's risk management department regarding property and liability insurance requirements for home-based work?

Safety and Ergonomics Assessment of Home-Based Offices and Action Plans

1. Are there self-assessment surveys for ergonomics, computer workstations, and home hazards? If so, are these surveys administered online or by hard copy?
2. What is done with the telework safety and health surveys after they are received? What kind of follow-up exists to determine whether the hazards are corrected?
3. How is survey data collected, analyzed, and used for improving the safety at off-site environments?

Ergonomics and Safety Training for Home-Based Offices

1. Are there training programs for work-at-home workers that include understanding risk factors, ergonomic solutions, symptom recognition, and reporting process? If so, are the training programs administered via intranet, hard copy, or other means?
2. Is the training assessed to ensure that it is completed and that learning has taken place?

(Continued)

TABLE 15.1 (CONTINUED)

Telework Safety Program Evaluation and Management Guide

Early Reporting of Symptoms, Follow-Up Actions and Return to Work

1. How do home-based employees report symptoms and general health concerns when they feel that it is work related? What are the processes for reporting accidents and job-related illnesses and injuries? Do they feel they can do so without reprisal of job action? Is confidentiality of reports maintained?

2. Has the risk management department been notified of how to report the accidents of others if they are injured?

3. What are the policies of one's company's WC insurer providing specific site coding in their claims databases for identifying injuries that occur to at-home or off-site workers? Are these data used for determining safety and risk management priorities for off-site workers?

4. Is there a return-to-work strategy for disabled workers who work at home or off-site? Are the workers able to receive quality healthcare? What assurances are there that this is accomplished?

IT and Telephone Equipment Support and Service

1. Is there a procedure for reporting computer and systems problems that impact the work-at-home employee? Are these problems promptly resolved? What assurance or feedback occurs to ensure that the problem(s) are resolved?

2. Is the appropriate telephone service provided? Consider the frequency of calls and the desired capability of the need for high-speed internet service.

Source: Robertson, M. M., W. S. Maynard, and J. R. Mcdevitt, *Prof Safety.*, 48, 30–6, 2003.

3. IS and technology issues and support for the telecommuter including equipment and maintenance (IS training for employees on how to maintain their computers, get software upgrades, etc., needs to be provided)

4. Organizational and culture issues, financial and motivational support, and commitment from the senior managers of the program and continued improvement of the program elements acquired through programmatic evaluation processes

Designing effective communication strategies that allow the managers and the employees to define job responsibilities, set goals and job expectations, and regularly review work and performance are just some of the challenges that organizations must address to implement successful telework programs (Gajendran and Harrison 2007). Others include establishing policies and procedures regarding appropriate technology and equipment and training employees to manage these technologies (Ilozor et al. 2001). Table 15.1 provides a comprehensive framework and guide to evaluate a telework program summarizing many of the key elements discussed earlier.

15.5 MANAGING THE MACROERGONOMICS PROCESS FOR THE SAFETY AND THE HEALTH OF TELEWORKERS

Given the evidence provided by the earlier study results, it is critical that before managing the safety and health risks of teleworkers, assurance is made that the job position and the person is right for the a telework arrangement. Research has found that employees who are successful working at home are self-directed and motivated, with a history of solid job performance (Bailey and Kurland 2002). Additionally, optimizing the work environment of teleworkers, and reducing the risk of claims and injury costs, is critical to incorporating a safety process using a macroergonomics approach. Actively involved stakeholders are the key to the success of the telework programs. These key stakeholders are both within and external to the organization and include human resources, leadership personnel, safety and health professionals, engineering and maintenance, health care provider, rehab provider, the worker's compensation (WC) insurer and others. In the middle of the process is the teleworker where the communication between them and all stakeholders is necessary to promote and encourage safety (preinjury/prevention) and to enable (postinjury) return to work after an injury/illness (Robertson et al. 2003, 2004).

In the safety management process, surveillance, or worksite analysis, is essential. It is difficult to manage safety without detailed injury and hazard information obtained through the surveillance efforts. Obtaining accurate and complete injury data and hazard information to effectively manage the teleworker safety is a challenge. Still, part of designing a teleworker program is training the employees and the managers to provide and maintain a safe working environment. The managers should be trained in effective risk management concerning their employees and what policies and procedures to follow and implement. Further, the employees need to understand office ergonomics guidelines and other related safety issues and hazards and to be trained in applying these guidelines. Three safety and health surveillance approaches are recommended to assist managers and safety and health professionals in managing the risks associated with teleworking, (Robertson et al. 2003; Dainoff et al. 2012):

1. Job surveys—Employers should utilize checklists and surveys to understand the work hazards. Unless the worker voluntarily offers the information, the employers may not know what hazards exist in the home environment. Whether an employer should visit the home to do an inspection is debated and involves privacy issues (Mills et al. 2001; Von Bergen 2008; Jaakson and Kallaste 2010). From the OSHA viewpoint, U.S. employers are not responsible for the work performed at home unless it is unusually hazardous and do not have to conduct site visits (Mills et al. 2001). Employer liabilities, however, do apply to all work-related illnesses and injuries regardless of where they occur (Mills et al. 2001). Some companies have policies that prohibit employees from visiting other employees at home for this purpose (Jaakson and Kallaste 2010). The majority of companies rely on self-assessments of at-home workplaces (Healy 2000).

2. Safety records—Valuable information is provided in records such as WC claims reports and OSHA logs, since some teleworkers are concerned and have fears of reprisal in reporting and there could be a WC data quality issue. Furthermore, WC insurer claims that the databases may not accurately code for employee injuries occurring off-site, thus the detailed reports may not be available. Given the lack of quality data for the managers regarding trends and loss issues involving off-site workers, the managers may not have a clear assessment of their employees' safety. These claims cost data indicate health care quality issues and return-to-work issues impacting the length and the cost of the disability of off-site workers.

3. Employee reports—The reporting of hazards, injuries, or symptoms promptly to the employer is essential for treatment and prevention. Some teleworkers are reluctant to do so, fearing that reporting work-related hazards or injuries may result in the cancellation of the telework agreement. These teleworker's may visit their personal physician and rely on health insurance to pay the bill instead of reporting a work-related injury.

15.5.1 RECOMMENDED SAFETY AND ERGONOMICS GUIDELINES FOR WORKING AT HOME

For a home-based teleworker, there are several issues to consider in one's company policies to maintain a safe and comfortable work-at-home environment (e.g., Jaakson and Kallaste 2010). To evaluate a home-based teleworker program and its effectiveness, begin by answering the following questions shown in Table 15.1 (Robertson et al. 2003). If an answer is "no" or "I don't know," target that item for follow-up and improvement.

15.5.1.1 Preparing and Planning the Home-Based Office Work Space

There are several key design issues that should be incorporated when planning the work space for teleworkers (Robertson et al. 2003; Dainoff et al. 2012). These include the following:

1. Work space location—Teleworkers' should identify a location that provides a separate physical work space, preferably away from the flow of activity in their house. Interruptions by family members can be distracting. Ensure that there is a lockable door, so that the teleworker can control the entry into their work area. Teleworkers may try to have an understanding with the family members or the roommates that they will need privacy to conduct business in a professional manner. A good rule of thumb when planning one's workspace needs is to identify, at a minimum, a 6 ft × 6 ft (1.8 m × 1.8 m) space for the primary work area. The work area should have at least two means of egress. One way out can be a window, if there is a safe means of getting from the window to the ground.

2. Storage—Depending on what is needed for references or storage; expect the space requirements to grow. Measure the space to host both lateral and vertical files as each will provide proper storage space for reference materials. The lateral files have a footprint of 36 in. × 18 in. (90 cm × 45 cm), and vertical files are 15 in. × 18 in. (38 cm × 45 cm). Place bookcases and filing cabinets so that one needs to stand up to access them.

3. Movement—Employees need to be reminded to get up and move periodically, to not continuously sit throughout the day. Support and encourage them to vary their working posture and plan their movement into their office design. Additional space requirements will be necessary.

4. Office windows—Select an office in a small room without windows, if possible. With a closed-room design, two doors are necessary for life safety. A window helps to provide a view greater than 12 ft (3.7 m) away, providing the opportunity for the eye muscles to relax.

5. Computer display—The computer monitor should not be placed next to a window as the light can create glare on the screen. Ideally, it is best to find a space for the computer monitor on a north wall.

6. Cable and electrical management—Trip and fall hazards are present when extension cords and wirings cross the travel area. Fasten up and place all cables and extension cords out of the way, and select a location with access to sufficient electrical power outlets. If one has any questions about electrical supply, have a licensed electrician evaluate their needs and install additional outlets, if necessary. Residential-type extension cords are not a good choice; look for a cord with a minimum of 14 gauge wire. If a power strip is used, look for the types with surge and overload protections.

15.5.1.2 Selecting Home-Based Office and Computer Furniture

Determining the appropriate office furniture for home-based teleworkers needs to be done carefully and methodically, especially regarding one's desk and chair (Robertson et al. 2003; Dainoff et al. 2012). Depending on the company policies, if they provide specific office furniture for the home office, plan in advance where it will be placed to ensure that it will fit. If, according to the company telework policy, the employee is purchasing the furniture, then the manager or someone who is familiar with getting surplus furniture needs to guide the employee in the selection process.

15.5.1.3 Workstation Design, Monitor Placement, Keyboard Trays, and Storage

Desktop dimensions are important as they will need to accommodate one's computer, keyboard, phone, paper, references, stapler, sundry items like penholders and paper clips, and possibly a fax machine, CD/DVD drive, scanner, and printer. Be careful of the cheap office furniture in advertising fliers, as this furniture offers little flexibility in monitor placement and adjustment. Depending on the computer terminal size, it may impact the placement of other work tool accessories. Plan accordingly by measuring the work surface to ensure the proper placement of the monitor along with a document holder and other large reference book holders.

A work surface of at least 30 in. (76 cm) deep will be necessary if the computer monitor is large. If it is less than this, it will create problems. It is not unusual to find that the depth of the terminal combined with the depth of the keyboard exceeds 24 in. (60 cm). If this is the case, a keyboard tray may need to be installed. The keyboard should not be placed to the side of the monitor, as it will cause one to twist their neck in an awkward position, eventually leading to neck and shoulder pain.

An adjustable work surface is better than a fixed-height surface, as one will be able to set it at the correct height. Fixed-height desks or workstations are usually in the range of 28–29 in. (71–74 cm), and this height may be a problem for many people. An adjustable keyboard tray may be required in order to bring the keyboard down to a comfortable position, even when using a standard office chair adjusted to its highest point. Keyboard trays or holders should be at least 26 in. wide (66 cm) and at least 10 in. deep (25 cm) or more. The keyboard holders are generally not as stable as a desk-top and can be loose or bouncy. A keyboard tray tends to push the person further away from the main working surface, including their office accessories placed on the work desk. Consequently, the phone may be harder to reach, as one has to stretch out their arm and place themselves into awkward positions just to write as well as find themselves leaning and stretching out to read documents and other reference materials.

Workstation desks should consist of a solid, substantial desk that will not tip over when loaded up or when an overloaded drawer is pulled out. Inspect the edges of the workstation for any raised edges; work surfaces should have a smooth, matte finish. Ensure that the work area under the desk has enough leg clearance and is at least 17 in. deep at the knee. Center drawer desks are a poor choice as the drawer will not allow the keyboard tray to be adjusted to the correct working height and will impact one's leg clearance. Using a table with a cantilever work surface support, which is designed without impinging on one's leg room, is preferred. Ensuring that the filing cabinets are well secured to a wall or another cabinet in case of earthquakes or climbing children is critical.

15.5.1.4 Ergonomic Chair Design

A well-designed ergonomic office chair is a critical component of the teleworker's home office. Common household chairs will create problems in the long run due to the lack of adjustments and support for full-time office and computer work. Select a commercial office chair with height adjustability, back-tilt mechanism, lumbar support, and a seat pan that has the right width and length. Some companies make arrangements with furniture suppliers to allow the employees to wisely select one after trying it out for a specified time. Most office chairs adjust in the range of 16–21 in. (40–53 cm). Approximately 15% of the female population and 2% of the male population will need footrests, even at the 16 in. chair height. A chair should have a rounded or a waterfall front edge to the seat pan and a five- or six-point swivel pedestal base with casters appropriate for the surface (smooth floor or carpet). Heavily contoured seat pans are not recommended, as these may not fit some people. Armrest adjustability is preferable, and it should allow the person to raise or lower them accordingly to support a comfortable and optimal body posture. Lowering the armrests may be necessary to allow the person to move close to the keyboard. The armrests are also important for chair ingress and egress, especially for older people. The design of the chair backrest should not be so wide that the elbow bumps into the backrest and restricts movement.

15.5.1.5 Consider the Design of the Computing Environment

The ambient lighting around the screen should not exceed 500 lux (~50 fc), if there is a regular light-emissive terminal. With a flat panel display, the lighting can increase to around 750 lux (75 fc). Using bare light bulbs is not recommended because they can be a glare source and do not allow for a visually comfortable workstation environment. The use of indirect fluorescent or LED lighting or fluorescent or LED lighting with diffusers that can orient the light source directly downward are the best choice. Avoid having any bright light sources in one's immediate field of view. Preferably, the location of the light sources should be behind the person, over a shoulder at an angle, or at a right angle to the person in order to eliminate any reflections on the screen.

The office walls and the wall coverings should be made of nonreflective materials. Enamel paint or shiny wallpaper can be very reflective and should be avoided, if possible. Be cautious to not place framed artwork or photographs in one's immediate field of view because they tend to have a relatively high reflectance. Most noise at home will come from televisions, stereos, and conversation. Demanding complete quiet while a person works in the kitchen is unreasonable. Locating ones' office out of the mainstream of activity will allow the family or the roommates to conduct normal lives. The home office should have adequate ventilation. If the home has a forced hot air system or a central air, a duct should be in the work area. Most home carpeting and carpet pads are softer and less durable than commercial carpeting used in offices. The chairs will not roll well on carpeted surfaces, thus posing a challenge when trying to easily change one's position when performing tasks. A solid carpet protector can be helpful but can also be a problem if the chair rolls too easily. If the home office is below construction grade or is questionable given the age of the home, consider testing it for radon.

15.5.2 Laptop/Notebook Computer Use at Home

15.5.2.1 Laptops/Notebooks and Visual Demands

The design characteristics of many laptops/notebooks can create eye discomfort due to the lack of image clarity compared to a full-size computer monitor or flat panel. The effective solutions for those whose work requires a substantial amount of visual interaction with the screen is to use a docking system or to simply attach a full-size terminal. Additionally, a full-size keyboard and mouse or other pointing device should be used to reduce any possibility of awkward computing postures. To support the use of a laptop with a detached keyboard requires the work surface depth to be 24 in. (60 cm). The concerns regarding the onset of eye fatigue and strain when using a laptop at home are acknowledged, and several ergonomics solutions are proposed (Dainoff et al. 2012). First, following the 20-20-20 rule and integrating minibreaks into one's work habits is essential. Every 20 min, focusing on a distant object 20 ft away for 20 s can give one's eyes a break before continuing the work on the screen. Second, keep the screen clean at all times by using appropriate antistatic cleaning materials. Third, have a separate keyboard to ensure the proper positioning, or otherwise the display will have to be tipped backward. Fourth, reflective lighting may be a source of annoyance for laptop users. Use drapes, shades, or blinds to control the glare. Whenever possible, use indirect light while avoiding intense or uneven lighting in the field of vision. If the employees' are experiencing fatigue or visual discomfort after following these suggestions, have them consult an eye care specialist and inform them of their computer use.

15.5.2.2 Laptops and Computing Postures

While using a laptop, it is certainly a challenge to maintain comfortable hand and arm positions. Keeping one's head in a comfortable position, not overly turned or tilted, is essential. Adjust the screen brightness and the contrast levels to allow for a comfortable viewing of the screen. Change positions often to avoid discomfort and muscle fatigue. Stop and rest if one is beginning to exprience discomfort while computing. Taking periodic breaks and stretchings one's arms, hands, and fingers will help, along with the visual minibreaks noted earlier. Frequent, shorter breaks are of greater benefit than fewer, longer breaks. When typing, use a light touch and do not pound on the keys. There is no need to push harder on the keys than necessary. While keying or mousing, keep the wrists in a straight, nonrigid position. Never position the wrists in an exaggerated angle or in a position that causes tension in the wrists. The hands and the wrists should be free to move when typing and are not rested on a palm rest, table, or thighs while typing on a laptop. Keeping the fingers relaxed and nonrigid when operating a laptop or an input device is essential as well as paying close attention to the position of one's ring finger, pinkie finger, and thumb while keying. The fingers should not be tensely held up in the air or scrunched into the side of one's hand when using the input device or keying. Working with a laptop keyboard for long periods can be uncomfortable and fatiguing. For an employee who works a lot with numbers, the use of a laptop can be problematic. For these task-specific situations, consider getting a regular-sized configuration number pad as a peripheral input device.

15.5.2.3   Making a Good Ergonomic Fit

It is important to adjust the workstation to fit the employee once the furniture and the equipment are installed. Shown in Figure 15.2 are the essential measurements needed to obtain an ergonomic fit (Robertson et al. 2003; Dainoff et al. 2012). First, adjust the chair to the person's leg height (A in the drawing). Next, add the elbow measurement and the leg measurement (A + B) (Figure 15.2), and see if the work surface where one plans to put the keyboard is about 1 in. less (2.5 cm). If the chair is adjusted up to compensate, one will find that their heels are now off the floor. A footrest may be needed to allow the feet to rest flat on the floor or with the toes at a slight upward angle. It is not recommended to sit with one's feet extended for a long period.

The screen should be positioned for a moderate downward gaze angle, and between 20 in. (50 cm) and 30 in. (76 cm) from one's eyes. Those who are farsighted might even find the monitor comfortable at 40 in. (102 cm). If the employee is a hunt-and-peck typist, it might be easier for them to have the monitor closer and lower, so that they are not moving their head and neck up and down. The eyes move fairly easily through an arc of about 30°, so a fairly low monitor reduces A and A + B repeated neck motions. For those who touch type, a monitor at a higher position will probably be more comfortable.

One note regarding eyewear, a very common problem is presbyopia, the loss of the eye's ability to see close objects clearly. Presbyopia is usually corrected with bifocals or trifocals. If one is a touch typist or knowledgeable about the keyboard, they may need to only glance at the keyboard occasionally. Given this situation, it would benefit the employee to get monocular lenses to replace the bifocals while working. The strength of the monocular lens should be set for the distance from one's eyes to the screen. A good option for a person who is a typist is to specify the bifocal lenses for computer work, so that the top lens is set for the terminal distance, and the lower lens is for the keyboard.

Using a document holder can be a case of personal preference; however, if the work involves a lot of transcription from a printed document, it will be important to have a document holder. Document

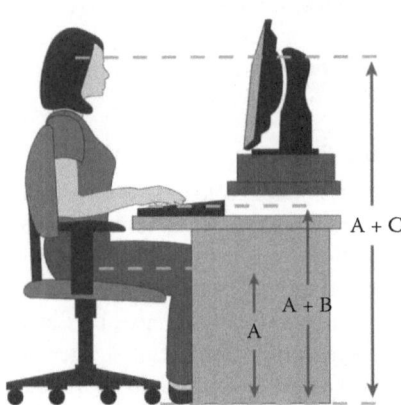

FIGURE 15.2   Achieving a good ergonomic fit. (Dainoff, M., W. Maynard, M. M. Robertson, and J. Andersen: *Handbook of Human Factors and Ergonomics.* 2012. Copyright Wiley-VCH Verlag GmbH & Co. KGaA. Reproduced with permission.)

holders are typically designed to be on the side of the terminal or between the terminal and the keyboard. The location of the document holder will be determined by whether one is a touch typist, whether they have presbyopia, and what type of display they are using.

If the work surface is the wrong height and one decides to buy a keyboard holder or tray, it is important to get one that is wide enough to hold the keyboard and the mouse/mouse pad or the track ball. This is usually about 28 in. (71 cm) unless one has a split keyboard or some other type that is wider than a standard expanded keyboard. Avoid setup situations that would necessitate reaching out for the mouse, especially with the shoulder raised. If one has multiple computers and terminals, it is best to use an A-B switch for the keyboard, so the desktop is not cluttered with keyboards and mice. The hands should be cupped and above the keyboard when typing, while the wrist is straight or very slightly extended. A wrist rest can, in some cases, be a problem, as it may encourage one to use it to support the hands while typing. Wrist rests, however, can allow for a soft place to relax the hands when not typing.

Figure 15.3 presents ergonomics tips to maximize the teleworker's comfort when computing. It is a flowchart of ergonomics recommendations for setting up a workstation, including chair, monitor, keyboard, input devices, work area, lighting, and accessories. Healthy computing habits are provided to guide the computer user on how to incorporate healthy ergo breaks and be aware of computing postures. Four reference postures, based on the recommended working postures from ANSI-HFES 100 (HFES 2007) are shown (upright sitting, reclined sitting, declined sitting, and standing). The benefits of incorporating the standing positions throughout the workday while computing are noted; however, it is still essential to be trained on how to properly adjust and use sit–stand workstations to avoid musculoskeletal and visual symptoms (Robertson et al. 2013; Karol and Robertson [in press]).

15.5.3 Telework Safety and Health Survey

An example of a telework safety and health survey is provided in Table 15.2. This survey was designed to identify the opportunities for improving the safety and the health of the employees' residential office work environment. It is to be completed by the employee, and the results are to be discussed with their manager or supervisor. It is important to conduct such a survey to gain feedback from the employees' regarding the ergonomics of their computer workstation. Agreed-upon action plans to improve the safety of the employee's residential office using the survey information should be accomplished. The employees complete the telework safety and health survey with the caveat that there will be no visit to their home as a result of the assessment. The survey information is to be solely used to improve the safety and the health of the employee's residential office. A no response to any survey question should alert the manager to discuss the issue(s) and to determine an agreeable solution.

15.6 CONCLUSION

Unique challenges and opportunities regarding telework are raised. Work style, workplace, and safety and health risks need to be considered when managing and implementing a telework program. To have a successful program, the organizations need to design and implement their programs with a systems-oriented, macroergonomics approach. Previous research identifies several key components necessary to implement an effective telework program. Understanding the critical systems elements of a teleworker program can lead to a successful teleworker program that will provide not only a safe and healthy work environment, but also a productive work arrangement.

Ergonomic tips to maximize your comfort when computing

Ergo-guide from Liberty Mutual Insurance

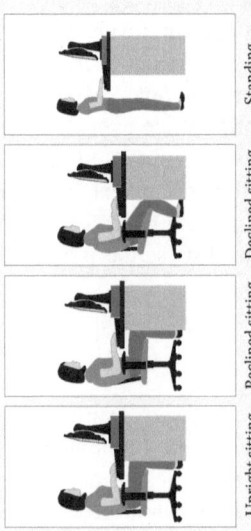

Upright sitting   Reclined sitting   Declined sitting   Standing

1. Chair and posture

- Use the backrest of the chair to provide full support to your lower back
- Make sure your chair allows clearance behind your knees when seated against the backrest

Maintain proper body posture by:

- Sitting with your hips and knees at a 90 degree or greater angle
- Keep your feet flat on the floor or on a footrest
- Keep your arms relaxed at your sides; ideally with elbows at 70–135 degrees
- Change posture frequently; common postures include upright sitting, reclined sitting, declined sitting and standing

2. Monitor

- Place the monitor directly in front of you—about an arm's length away
- Position the top of the monitor screen at, or below, eye level

3. Keyboard and input devices

- Adjust the keyboard or chair height to keep forearms, wrists and hands in a straight line
- Place mouse and other input devices near to and at the same height as your keyboard
- Keep your elbows close to your body

4. Work area and lighting

- Allow ample clearance to move your knees and legs under the keyboard and desk
- Avoid contact stress with the edge of the desk and keyboard

To reduce glare and shadows on your work surface:

- Adjust window shades or decrease overhead lighting
- Adjust the monitor screen or add an antiglare filter
- Add a task light to properly illuminate paper references

5. Accessories

- Get a headset if you regularly talk on the phone for extended periods of time. Use a lowered voice

Use an adjustable document holder to:

- Place reference materials as close to the computer screen as possible
- Keep materials at the same height and distance as your computer screen
- Use your ergonomic accessories to support body posture (e.g., lumbar support, arm rests, monitor blocks, external keyboard)

6. Healthy computing habits

- Use a softer touch when keying; relax your grip on the mouse
- Avoid working too long in one position
- Change your body posture frequently
- Take frequent breaks. Stretch periodically
- Give your eyes a visual break

Liberty Mutual.
INSURANCE

libertymutualgroup.com/riskcontrolservices   in   🐦 @LibertyB2B

The illustrations, instructions and principles contained in the material are general in scope and, to the best of our knowledge, current at the time of publication. No attempt has been made to interpret any referenced codes, standards or regulations. Please refer to the appropriate code-, standard-, or regulation-making authority for interpretation or clarification. Provided that you always reproduce our copyright notice and any other notice of rights, disclaimers, and limitations, and provided that no copy in whole or in part is transferred, sold, lent, or leased to any third party, you may make and distribute copies of this publication for your internal use.

© 2015 Liberty Mutual Insurance, 175 Berkeley Street, Boston, MA 02116.   RC 5334 R2   08/12

FIGURE 15.3   Ergo-guide: Ergonomics tips to maximize comfort when computing.

TABLE 15.2
Example of a Teleworker Safety and Health Survey

Instructions to employee: The survey below is intended to identify opportunities for improving the safety and health of your residential office work environment. We recognize the importance of comfort and productivity and request your feedback on the ergonomics of your computer workstation. Please take a few minute to complete the assessment including your computer set-up (if any). No visit to your home will result from this assessment. The information you provide will only be used to improve the safety and health of your residential office. A "No" response to any question below should be discussed with your manager or supervisor to determine an agreeable solution.

Name: _____

Organization: _____

Address: _____

City/State/Zip Code: _____

Business Telephone: _____

Briefly describe your designated work area:

_____

_____

_____

_____

_____

A. Workplace Environment

1.	Are the following environmental conditions adequate to maintain your normal level of job performance?	
	a. temperature,	Yes [ ]　No [ ] NA [ ]
	b. illumination,	Yes [ ]　No [ ] NA [ ]
	c. noise levels	Yes [ ]　No [ ] NA [ ]
1.	Does the work area have adequate fresh air ventilation and is it free of unusual odors?	Yes [ ]　No [ ] NA [ ]
2.	Are all stairs with four or more steps equipped with handrails?	Yes [ ]　No [ ] NA [ ]

(*Continued*)

TABLE 15.2 (CONTINUED)

Example of a Teleworker Safety and Health Survey

3.	Is office equipment free of recognized electrical hazards?(frayed wires, bare conductors, sufficient number of permanent outlets, loose wires, etc.)	Yes [ ]　　No [ ]　NA [ ]
4.	Are surge protectors available to protect electrical office equipment from overload?	Yes [ ]　　No [ ]　NA [ ]
5.	Are aisles and passageways free and clear of obstructions and tripping hazards? (i.e. boxes, open file drawers, papers, books, supplies, wires phone lines frayed or worn carpet etc.)	Yes [ ]　　No [ ]　NA [ ]
6.	Does your work area have two exits to the outside? (for residential "basements," a minimum of one passage door leading to the outside).	Yes [ ]　　No [ ]　NA [ ]
7.	Are the phone lines, electrical cords, and extension wires secured under a desk or alongside a baseboard?	Yes [ ]　　No [ ]　NA [ ]
8.	Is the office space neat, clean, and free of an accumulation of combustibles?	Yes [ ]　　No [ ]　NA [ ]
9.	Are file cabinets secured from tipping and arranged to avoid tripping or obstructions?	Yes [ ]　　No [ ]　NA [ ]
10.	Are carpets secured to the floor and free of frayed or worn seams?	Yes [ ]　　No [ ]　NA [ ]
11.	Is the work area free from sources of accumulation of mold, mildew or musty smells?	Yes [ ]　　No [ ]　NA [ ]
12.	Is a smoke detector installed, working and suitably located near the work area?	Yes [ ]　　No [ ]　NA [ ]
13.	Is a fire extinguisher suitable for use on electronic equipment located near the work area (with a rating of ABC)?	Yes [ ]　　No [ ]　NA [ ]
14.	If radon is an issue, has it been tested? (basement offices)	Yes [ ]　　No [ ]　NA [ ]
15.	Is the room heating sufficient enough so that a local space heater is not used?	Yes [ ]　　No [ ]　NA [ ]
16.	Are carbon monoxide and gas detectors available?	Yes [ ]　　No [ ]　NA [ ]

B. Computer Workstation Ergonomics

17.	Is your chair adjustable?	Yes [ ]　　No [ ]　NA [ ]
18.	Do you know how to adjust your chair?	Yes [ ]　　No [ ]　NA [ ]
19.	Is your back well supported by a backrest?	Yes [ ]　　No [ ]　NA [ ]

2

(Continued)

TABLE 15.2 (CONTINUED)

Example of a Teleworker Safety and Health Survey

20. Are your feet flat on the floor or fully supported by a footrest?	Yes [ ] No [ ] NA [ ]
21. Are you satisfied with the placement of your monitor and keyboard?	Yes [ ] No [ ] NA [ ]
22. Is it easy to read the text on your screen?	Yes [ ] No [ ] NA [ ]
23. If you input data from printed documents, do you have a document holder?	Yes [ ] No [ ] NA [ ]
24. Are file cabinets and book cases arranged so you need to stand to access them? (Avoid sitting continuously throughout the day)	Yes [ ] No [ ] NA [ ]
25. Do you have enough leg room at your desk? (i.e. can you swivel your chair and or reach keyboards, phones and other equipment without having to lean forward?)	Yes [ ] No [ ] NA [ ]
26. Is the screen free from noticeable glare?	Yes [ ] No [ ] NA [ ]
27. Is the top of the screen at eye level? (bifocal users may prefer it lower)	Yes [ ] No [ ] NA [ ]
28. Is there space to rest the arms while not keying?	Yes [ ] No [ ] NA [ ]
29. When keying, are your forearms close to parallel with the floor?	Yes [ ] No [ ] NA [ ]
30. When keying or using other input devices do your elbows stay at your side?	Yes [ ] No [ ] NA [ ]
31. Are your wrists straight when keying?	Yes [ ] No [ ] NA [ ]
32. Have you received ergonomics training?	Yes [ ] No [ ] NA [ ]

C. Overall

18. Are you comfortable while working in your residential office? If no, please explain below.	Yes [ ] No [ ] NA [ ]
19. Do you believe your residential office is free from anything unsafe or that puts you at risk for an injury? If no, please explain below.	Yes [ ] No [ ] NA [ ]

NA = Not Applicable

D. Comments

_____

_____

_____

_____

3

(Continued)

TABLE 15.2 (CONTINUED)
Example of a Teleworker Safety and Health Survey

_____

_____

_____

_____

_____

Please return a copy of this form to your supervisor or manager. Optional-attach photos of your workstation if applicable.

Employee's Signature and Date: _____

Supervisor or Manager Signature and Date:_____

Action taken (to be completed by manager or supervisor)

_____

_____

_____

_____

_____

_____

_____

_____

REFERENCES

Apgar, M. 1998. The alternative workplace: Changing where and how people work. *Harvard Bus Rev.* 76(3):121–36.

Atkinson, W. 1985. Home/work. *Pers J.* 64(11):104–9.

Bailey, D. E., and N. B. Kurland. 2002. A review of telework research: Findings, new directions, and lessons for the study of modern work. *J Organiz Behav.* 23:383–400.

Baruch, Y., and N. Nicholson. 1997. Home, sweet work: Requirements for effective home working. *J Gen Manage.* 23:15–30.

Belanger, F., R. W. Collins, and P. H. Cheney. 2001. Technology requirements and work group communication for telecommuters. *Inform Syst Res.* 12(2):155–76.

Bergqvist, U., E. Wolgast, B. Nilsson et al. 1995. Musculoskeletal disorders among visual display terminal workers: Individual, ergonomic, and work organizational factors. *Ergonomics.* 38(4):763–76.

Carayon, P., and M. J. Smith. 2000. Work organization and ergonomics. *Appl Ergon.* 31(6):649–62.

Carayon, P., M. J., Smith, and M. C. Haims. 1999. Work organization, job stress, and work-related musculoskeletal disorders. *Human Factors.* 41(4):644–63.

Claes, R. 2000. Meaning of atypical working: The case of potential telecommuters. *Eur Rev Appl Psychol/Rev Euopeene Pscyhol Appl.* 50(1):27–37.

Dainoff, M., W. Maynard, M. M. Robertson, and J. Andersen. 2012. Office ergonomics. In Salvendy, G. (ed.) *Handbook of Human Factors and Ergonomics*, Fourth edn. Hoboken, NJ: John Wiley & Sons, 1550–73.

Davis, D. D., and K. A. Polonko. 2001. Telework in the United States: Telework American survey [online]. From http://www.workingfromanywhere.org/telework/twa2001.htm (accessed September 29, 2015).

Durbin, A. J. 1991. Comparison of the job satisfaction and productivity of telecommuters versus in-house employees: A research note on work in progress. *Psychol Rep.* 68(3):1223–34.

Dziak, M. J. 2001. Debunking telecommuting myths. *Chem Eng.* 108(11):149.

Ellsion, N. B. 1999. New perspectives on telework. *Soc Sci Comp Rev.* 17(3):338–56.

Gajendran, R. S., and D. A. Harrison. 2007. The good, the bad, and the unknown about telecommuting: Meta-analysis of psychological mediators and individual consequences. *J Appl Psychol.* 92(6):1524–41.

Galtiz, W. O. 1984. *The Office Environment: Automation's Impact on Tomorrow's Workplace.* Willow Grove, PA: American Management Society Foundation.

Global Workplace Analytics. 2015. Latest telecommuting statistics [online]. GlobalWorkplaceAnalytics.com. From http://globalworkplaceanalytics.com/telecommuting-statistics (accessed September 29, 2015).

Greengard, S. 1998. How technology will change the workplace. *Workforce.* 77(1):78–9.

Grzywacz, J. G., P. R. Casey, and F. A. Jones. 2007. The effects of workplace flexibility on health behaviors: A cross-sectional and longitudinal analysis. *J Occup Env Med.* 49:1302–9.

Hamilton, C. A. 1987. Telecommuting. *Pers J.* 66(4):90–101.

Harrington, S. S., and B. L. Walker. 2004. The effects of ergonomics training on the knowledge, attitudes, and practices of teleworkers. *J Safety Res.* 35:13–22.

Healy, M. L. 2000. Telecommuting: Occupational health considerations for employee health and safety. *AAOHN J.* 48(6):305–31.

Hendrick, H. W., and B. M. Kleiner (eds.). 2002. *Macroergonomics: Theory, Methods, and Applications.* Mahwah, NJ: Lawrence Erlbaum Associates.

Hequet, M. 1994. How telecommuting transforms work. *Training.* 31(11):56–61.

HFES. 2007. ANSI-HFES 100-2007: Human factors engineering of computer workstations. Santa Monica, CA: Human Factors and Ergonomics Society.

Ilozor, D., B. Ilozor, and J. Carr. 2001. Management communication strategies determine job satisfaction in telecommuting. *J Manage Develop.* 20(6):495–507.

Imada, A. S. 2002. A macroergonomics approach to reducing work-related injuries. In Hendrick, H. W., and B. M. Kleiner (eds.) *Macroergonomics: Theory, Methods, and Applications.* Mahwah, NJ: Erlbaum, 151–171.

International Telework Association and Council. 1999. U.S. Representative Frank Wolf announces five cities to be part of a federal telework pilot program [online]. From http://www.telecommute.org/policy/fed eral/wolf.shtml (accessed July 20, 2015).

Jaakson, K., and E. Kallaste. 2010. Beyond flexibility: Reallocation of responsibilities in the case of telework. *New Technol Work Employ.* 25(3):183–269.

Karnowski, S., and B. J. White. 2002. The role of facility managers in the diffusion of organizational telecommuting. *Environ Behav.* 34(3):323–34.

Karol, S., and M. M. Robertson. Implications of sit-stand and active workstations to counteract the adverse effects of sedentary work: A comprehensive review. *Work*. (In press).

Karsh, B.-T., P. Waterson, and R. J. Holden. 2014. Crossing levels in systems ergonomics: A framework to support 'mesoergonomic' inquiry. *Appl Ergon*. 45:45–54.

Lenckus, D. 1997. Home is where the risk is. *Bus Insur*. April 28:3.

Marcus, M., and F. Gerr. 1996. Upper extremity musculoskeletal symptoms among female office workers: Associations with video display terminal use and occupational psychosocial stressors. *Am J Indust Med*. 29:161–70.

McCloskey, D. W., and M. Igbaria. 1998. A review of the empir-icalempirical research on telecommuting and directions for future research. In *Virtual Workplace*. Hershey, PA: IGI Publishing, 338–58.

McGonegle, K. 1996. Taking work off-site—AEGON's telecommuting program. (AEGON Insurance Group). *Employ Relat Today*. 23(1):25–38.

Miller, H. 2009. Research Summary: The evolving nature of working at home [online]. Herman Miller, Inc. From http://www.hermanmiller.com/content/dam/hermanmiller/documents/research_summaries/wp _Working_at_Home.pdf (accessed September 25, 2015).

Mills, J. E., C. Wong-Ellison, W. Werner, and J. M. Clay. 2001. Employer liability for telecommuting employees. *Cornell Hotel Rest Admin Q*. 42(5):48–59.

Montreuil, S., and K. Lippel. 2003. Telework and occupational health: A Quebec empirical study and regulatory implications. *Safety Sci*. 41:339–58.

Murphy, L. A., M. M. Robertson, and P. Carayon. 2014. The next generation of macroergonomics: Integrating safety climate. *Accident Anal Prev*. 68:16–24.

NIOSH. 2002. *The Changing Organization of Work and the Safety and Health of Working People*. Atlanta, GA: DHHS (NIOSH) Publication No. 2002-116.

Olson, M. H., and S. B. Primps. 1984. Working at home with computers: Work and non-work issues. *J Soc Issues*. 40(3):97–112.

OPM (Office Program Management). 2015. Work-life: Telework [online]. U.S. Office of Personnel Management (OPM). From https://www.opm.gov/policy-data-oversight/worklife/telework/ (accessed September 29, 2015).

Piskurich, G. M. 1996. Making telecommuting work. *Train Dev* 50 (2):20–27.

Robertson, M. M., and P. Vink. 2012. Examining the macroergonomics and safety factors among teleworkers: Development of a conceptual model. *Work*. 41:2611–5.

Robertson, M. M., R. Henning, N. Warren et al. 2013. The intervention design and analysis scorecard. *J Occupl Environ Med*. 55(12, Supp. 125):S86–8.

Robertson, M. M., W. S. Maynard, Y. H. Huang, and J. R. McDevitt. 2004. Telecommuting: An overview of emerging macroergonomics issues. *Hum Fac Erg Soc P*. 46(15):1330–4.

Robertson, M. M., W. S. Maynard, and J. R. Mcdevitt. 2003. Telecommuting: Managing the safety of workers in home office environments. *Prof Safety*. 48(4):30–6.

Sznelwar, L. I., F. L. Mascia, M. Zilbovicius, and G. Arbix. 1999. Ergonomics and work organization: The relationship between tayloristic design and workers' health in banks and credit cards companies. *Int J Occup Safety Ergon*. 5(1):291–301.

TelCoa (The Telework Coalition). 2015. Telecommuting facts [online]. U.S. Department of Agriculture, Farm and Foreign Agriculture Services (FFAS) Human Resources Division. From http://www.fsa.usda.gov /FSA/hrdapp?area=home&subject=wpsv&topic=tel-tf (accessed September 29, 2015).

Trent, J. T., A. L. Smith, and D. L. Wood. 1994. Telecommuting: Stress and social support. *Psychol Rep*. 74(3):1312–4.

Unsworth, E. 1998. Telecommuting brings ergonomics risks. *Bus Insur*. 32(19):46.

Von Bergen, C. W. 2008. Safety and workers' compensation considerations in telework. *Reg Bus Rev*. 27:131–50.

Wilde, C. 2000. Telework programs are on the rise. *Info Week*. 189–90.

Youngblood, D. 1999. Bridgeworks tries to ease the generation gap at work: Baby boomer, generation Xer show companies how to survive the times *Star Tribune*. January 24.

Section III

Emerging Ergonomic Workplace
Design Issues

16 A Macroergonomics View of Transportation

Brian Peacock, Chui Yoon Ping, and Alan Hedge

CONTENTS

16.1 INTRODUCTION

This chapter will compare and contrast the micro- and macroperspectives of ergonomics as applied to transportation. Commonly, ergonomists use a microreductionist approach to both the research and the technology design, usually related to the human interfaces with technology. A macroapproach places these numerous microergonomics contributions in the context of human, technological, operational, and contextual variabilities, in which the various outcomes may result from particular combinations of these major factors. Governments, such as Singapore, attempt to address the challenge of transportation by a heavily managed approach in contrast with some other Southeast Asian countries, notably Vietnam, whose approach is more Darwinian (Bin and Ching 2013).

16.2 EVOLUTION AND LOCOMOTION

Humans are mobile bipedal animals. We crawl, walk, and run. Evolution may have favored our bipedal locomotion. Lieberman et al. (2009) provide a detailed description of the evolutionary development of humans as endurance runners. They suggest that, although quadrupeds exhibit considerable strength and speed, it is the biped human that, combined with our ability to efficiently thermoregulate through sweating, developed the greatest stamina capabilities. The evolution of the bipedal human locomotion may have given humans an advantage as a hunter. Early humans were also likely to have had a nomadic lifestyle. Even though only a small minority of humans still practice a nomadic lifestyle, the human desire to move and travel remains strong. Water transportation on lakes and along rivers, initially in human-powered vessels such as coracles and canoes, allowed humans to extend their geographical territories along the world's major inland waterways. Oar-powered transportation was energy intensive for humans, but the invention of sails allowed water transportation that required little human energy, and the evolution of sailing ships eventually allowed humans to circumnavigate the globe. On land, the domestication of animals, such as oxen, horses, camels, and elephants, extended the range that a human could travel in a given time, and together with the development of wheeled vehicles, such as chariots, carts, and wagons, the animal-powered technology allowed humans to transport themselves and greater loads at greater speeds and over longer distances, thereby also allowing for mass migrations of people over great distances. The industrial revolution was marked by the invention of rail travel, the iron horse, which further increased the speed, the size of the loads, and the distances that humans could travel per unit time, and the resulting railway networks further spawned the development of many towns and cities across the globe that were not connected by waterways. Just over 100 years ago, the introduction of the internal combustion engine marked the dawn of the age of the motor car and other motorized vehicles, and today, millions of people drive or ride in such vehicles on a daily basis, and by 2010 there were over one billion cars worldwide. An unfortunate downside of petrol-driven vehicular activity is the considerable air and environmental pollutions from the exhaust emissions. Another downside is the death of more than one million people a year on the roads (WHO 2015). To tackle these issues, there is a growing interest in electric vehicles of all shapes and sizes, and in driverless vehicles that should all but eliminate road accidents and will undoubtedly soon share our roadways.

Also, a little over 100 years ago, in 1903, the Wright brothers demonstrated the first powered flight, and in 2014, commercial airlines carried 3 billion passengers each year. The consequence of worldwide air travel is also atmospheric pollution and, although much safer than road transportation, each year, there are aviation accidents and fatalities (Aviation-Safety 2015). With the continuing demand for air travel, the sky too is rapidly becoming congested with both human-operated and, in the past few years, an increasing number of unmanned aerial vehicles (UAVs).

Most recently, the human desire to travel greater distances in shorter times was given a boost by the dawn of the space age, and visitors to the National Aeronautics and Space Administration (NASA) Johnson Space Center will see an unused Atlas rocket—another "giant leap for mankind." The most frequently flown space vehicle was the NASA space shuttle which flew 135 missions, two of which (Columbia and Challenger) catastrophically failed killing their crews. However, the quest for more power in transportation systems continues.

Vehicle manufacturers are driven by legislators to add substantial costs by the inclusion of devices that only come into effect when things go wrong. Cars have antilock brakes, seat belts, airbags, crush space, collision warning sensors, and a plethora of mechanisms to warn the driver of impending dangers and to keep the vehicle upright and stable in the event of loss of control. As the driver is often cited as the prime culprit in transportation accidents, an exciting, but challenging, prospect is the evolution of the driverless vehicle (Benenson et al. 2008). The shortcoming with most automation is when things go wrong, perhaps for contextual or technological reasons, and the driver (pilot) is unable to cope, as demonstrated by the Asiana Flight 214 crash in San Francisco.

A more down-to-earth experience is the increasing likelihood of being run down by a police officer on a Segway or a small child on a motorized scooter on a public sidewalk. These issues indicate the importance of a macroergonomics approach to the system design that includes the context and the operational traffic management as well as the microergonomics design of the driver and technology interfaces.

Another, more insidious, result of adding power to transportation is that people no longer have to perform as much physical work as the hunters of yesteryear. The evolution of sedentary lifestyles has resulted in obesity, diabetes, and other metabolic disorders. In the future, we may no longer walk to the bus stop; this last-mile chore will be aided by an electric scooter or even a moving walkway. Elevators and escalators replace the need for us to use our own energy to combat gravity. These issues raise doubts about the sometimes articulated objective of ergonomics to reduce the need for physical activity by making the tasks more convenient and comfortable. In the long term, powered transportation leads to illness as well as injury (Lakka et al. 2003).

16.3   TRANSPORTATION SYSTEM PURPOSES AND OUTCOMES

This evolutionary introduction sets the scene for the consideration of the role of ergonomics in the transportation system design. First, it is appropriate to consider the multiple purposes and outcomes of the processes and the systems that have some human involvement. Useful operational definitions are that systems themselves are, following Edward's software, hardware, environment, liveware (SHEL) model (Edwards 1988), things such as technology, people, contexts, and operational rules that combine as an interactive process to produce multiple measureable outcomes as articulated by the E4S4 model:

- Effectiveness
    - The process should fulfill its intended function—the quality requirement.
    - A transportation system uses vehicles to move people and goods along roads and rails, across the sea, and through the air, in all kinds of weather, according to many traffic rules. People are usually very competent and flexible in managing these systems in the face of technological, environmental, and regulatory complexities.
- Efficiency
    - The processes have measureable efficiency or productivity in terms of resource utilization, such as people, money, fuel, and time.
    - Mass transportation systems aspire to move many people quickly and inexpensively over long distances. Motor cycles, because of their weight and size, are much more efficient than cars in terms of fuel, footprint, and time resources with one or more occupants.
- Elegance
    - One expectation of systems and processes is that they should satisfy the affective requirements of their users. The affective dimensions have considerable impact on the peoples' purchase and operational decisions.
    - The affective response to "Daisy Bell" (Dacre 1892) reflects one instance of the affective domain:

        > "If you can't afford a carriage
        > There won't be any marriage
        > Cause I'll be switched if I'll get hitched
        > On a bicycle built for two"

- Ease of use
    - A familiar objective of ergonomics is the ease of use by the intended user and the difficulty of misuse by the intended or unintended users.
    - The unfamiliarity of the pilots in the Asiana Flight 214 with the autothrottle (automation) led to its crash on approach to the San Francisco airport. Many taxis on our streets have four or five aftermarket visual displays, which all compete for attention with the driver's primary task. Some of these systems are not very easy to use interfaces, which compound their distraction potential.
- Safety and health
    - An inevitable downside of all technology is that human health and safety will be compromised in some form and to some extent, in use, in manufacturing or in disposal.
    - Worldwide, there are more than one million fatalities on the roads each year. Interestingly, the transportation medium of these fatalities considerably varies among countries. In the United States, autoaccidents dominate, in much of Southeast Asia,

the culprit is the motorcycle. Pedestrian vulnerability also widely varies in different regions, due to varied engineering and administrative controls (WHO 2015).

- In the United States, over the last two decades, the number of fatalities in road accidents has declined, as the safety of cars has been improved. However, road accidents still account for a substantial number of accidents and fatalities. Since the records began from 1899 to 2013, 3,613,732 motor vehicle deaths have occurred in the United States. There were some 5,419,000 crashes on U.S. roads, of which 30,296 were fatal crashes killing 32,999 people and injuring 2,239,000 in 2012 (National Highway Traffic Safety Administration 2014).
- In 2014, there were 155 cases of fatal accidents on the road in Singapore; motorcyclists and pillion riders accounted for 76 out of these 155 cases. Likewise, there were 7791 cases of injuries in 2014, out of which 60% were suffered by motorcyclist and pillion riders.
- The ubiquitous health downside is air pollution, with cities in developing countries being particularly prominent; global warming is a universal unwanted outcome of transportation. The trends toward the central conversion of energy to electricity in fossil-fueled and nuclear power plants also present the potential for a catastrophic system failure. The Singapore government deals with the problems of pollution and congestion by introducing a quota limit to vehicles allowed on the roads by requiring car owners to bid, often a very high price, for a certificate to own a vehicle (known as certificate of entitlement). Road users also have to pay to enter certain parts of the city-state during peak hours.
- The regular daily chores of sitting in a car, sitting at a desk, and sitting in front of the television are not conducive to good physical health.
- Security
  - Security issues result from the malicious actions of some system user or third party. Aviation security is a familiar challenge as well as vehicle theft. These system security challenges impose considerable financial loads on manufacturers and operators. The rise in the electronic controls of cars and airplanes presents the opportunity for malicious hacking.
  - The ergonomics contribution is to design access systems to only permit use by the intended users, and this has become a major industry. How many passwords do we need, and remember? How reliable are the scans and the computer analysis of our faces, eyes, and thumbs?
- Satisfaction
  - There are many stakeholders associated with any system or process—the shareholder who wants to make a profit; the manufacturer who wants to cut costs; the user who may want function, elegance, or efficiency; the maintainer who spends most of his life under a vehicle; and the legislator who looks for tax revenue while shouldering the responsibility of traffic management. These satisfaction issues often pit one stakeholder against another with the result being unsatisfactory compromise.
  - Most vehicle buyers trade off effectiveness, efficiency (cost), elegance (styling), and safety. The solution to these satisfaction issues is for the manufacturer to offer a wide variation in styling and features, at various costs. But this strategy is expensive.
- Sustainability
  - Sustainability has two forms. First there is reliability—the system or the process should continue to act as intended over the expected life cycle. Resilience on the other hand requires that the system or the process performs well under unintended or extreme conditions.
  - Airplanes and cars do not perform well in bad weather or with bad pilots or drivers. Complex planes, trains, and automobiles often have a limited useful life with the maintenance costs rapidly increasing as the vehicle ages. On the other hand, a good bicycle, properly maintained, can last a lifetime!

The challenges for system design are the many stakeholders in the system life cycle and the trade-offs among their often-conflicting purposes and outcomes. In the transportation context, it is common to see conflicts between speed and safety; stationary vehicles do not crash, but do not fulfill their intended purpose. The landfills are full of vehicles that have passed their useful life.

16.4 HUMAN VARIABILITY

Most ergonomists do not get to deal with this big, macroergonomics picture. Rather, they deal with microproblems at the behest of some individual stakeholder, commonly the intended user. This produces a sometimes insurmountable problem. This user is usually not an individual; rather, s/he is a cohort or a population that exhibits considerable variability on many dimensions.

To illustrate this challenge, one should consider the design of the operator and passenger seats in public transport. The driver's seats are usually adjustable in two dimensions. In this way, they accommodate much, but not all, anthropometric variations of the driver for the purpose of vision and control actuations (Peacock and Roe 1988). The drivers in the tails of the anthropometric distribution may need special consideration by aftermarket add-ons, or simply not get selected for the job.

The passengers are not so lucky; usually one size fits all. In the case of an airliner or a bus seat, the requirements of the company to fit in as many paying customers as possible create knee room challenges for many. Some airlines provide reclining seats to allow for more comfortable sleep, but this may lead to altercations with the passenger behind. The seat heights may be a problem for shorter vulnerable passengers on long journeys (Vink 2011). The height of most airline and bus seats fails to match the ergonomics requirement (dogma) of accommodating the popliteal height of a 5th percentile female. Similarly, the seat widths also miserably fail to accommodate the ideal 95% hip or elbow widths. In practice, the actual design of the seat dimensions and the materials is the prerogative of an efficiency-driven policy maker, based on many other criteria, only occasionally with the help of an ergonomist.

An example of this one-size-fits-all seat design challenge is found in mass transit railway vehicle design (Peacock 1978). As the distances and the journey times are often quite short, it is common to have benches on either side of the carriage facing the center. This provides more space for the standing passengers who have a smaller footprint and can increase the payload at tolerable comfort costs. The ergonomics opportunity is the selection of the dimensions of these seats, including height, width, and depth. Again, because of the short journey time, the heights are usually such that many passengers' feet do not easily reach the floor, but are unlikely to cause undue circulatory stress. The widths, as with airplane seats, do not accommodate an increasing number of larger passengers. One solution adopted by the early Hong Kong mass transit seats was to allow variable widths by removing the scallops. Variants of this strategy could be adopted, perhaps with a pricing premium, in airplane seats. The once popular bench seat in large U.S. automobiles and trucks appears to have gone out of fashion.

Some ergonomists and customers for ergonomics advice consider that the familiar seat is the be-all and end-all of ergonomics, but, even here, as was illustrated in the previous paragraph, the advice based on this ergonomics dogma of pleasing 95% of the population is rarely implemented. Ergonomists will be quick to insist that the profession goes beyond the design of the seats, and note that the fact of variability coupled with the principles of accommodation on many other human dimensions sometimes stretch to the holy grail of universal design (Erlandson 2008). A little thought will show that such an aspiration is totally impractical unless we include the selection and the assignment of intended user populations to our ergonomics armory. But here again, our paying customers rarely allow such a comprehensive responsibility to be undertaken by the ergonomist.

Our target populations considerably vary on physical, sensory, cognitive, and affective dimensions. Furthermore, the contexts of use also considerably vary in terms of the physical, operational, and social environments. The example of the seat design is presented to illustrate the challenges of human variability in a familiar context. Some of the challenges of sensory, cognitive, behavioral, and contextual variabilities will be explored in the following sections.

16.5   HUMAN TEMPORAL AND OPERATIONAL FLEXIBILITY

In the one-size-fits-all case of public transport seats, human musculoskeletal flexibility allows the individuals to adapt to the spatial constraints of the seats. In the broader sense of transportation system design, especially public transport and commercial aviation, human flexibility is required to adapt to the temporal restrictions of the system design. The economies of scale require that airplanes, buses, and trains carry large passenger loads, and that full vehicles are needed to allow the providers to charge the minimum price. In practice, many providers try to meet the customer halfway by offering large-capacity long haul and small-capacity more frequent short haul options. Minibus and taxi services increase the level of ground system accommodation, using price as the way of offering more comfort, convenience, and shorter waiting and journey times. Even so, the slack in the service offerings is taken up by human temporal flexibility. The analysis of these server–customer systems is based on the queuing theory. With more complex systems, discrete event simulation is the technique of choice to search for optimal solutions—maximum profit for the provider and minimum waiting and travel times for the customer.

A more complex model of human requirements and behavior places the transportation system as one component of the larger system that includes housing, places of work and education, and other services, such as shops and recreation. Human temporal flexibility is required when their employment is in an organization that requires shift work to provide 24/7 service. In the long run, there is a human cost to this flexibility in terms of fatigue and greater incidence of metabolic illness.

This bigger picture implies a need for the adaptation of the transportation systems to accommodate the demands of the other services. Rush hour traffic is a common problem in all large cities. Some bus services offer greater frequencies during the morning and evening rush hours. Aviation services may offer greater capacity around public holidays when the demand for transportation increases. However, the costs of this adaptation in terms of capital equipment and variable manpower must be offset by the less tangible reductions in passenger waiting times—customer satisfaction.

A different opportunity for human operational flexibility in transportation system design is through operator training and assignment. It takes a trained person to steer a boat, drive a bus, or pilot an airplane, and as the technology grows, the training demands also expand. Greater system capacity and less reserve workforce cost can be achieved by cross-training, which allows flexibility in the assignment of drivers or pilots to the vehicle types. Such a strategy must be offset against the cost of training and the increased possibility of error due to cognitive interference and lack of practice. As airplanes, especially, become more complex, type rating becomes more important. Similar specialization and generalization issues also occur with vehicle maintenance. Eventually, the trade-offs have to be made between the cost of the waiting times for vehicles and the costs of training, hiring, and assigning flexible people.

16.6   THE COGNITIVE DIMENSION

Physical and cognitive microergonomics studies of cars address the matters of pedal and steering wheel placements, the instrument panel reach, and the number, size, shape, discrimination, direction, and range of motion of knobs and switches. Similarly, the design of the displays addresses physical issues such as size, color contrast, font and marking sizes, and scales and ranges of movement. One precomputer era issue was also the location and the number of information sources. Classical studies also investigated the interactions between the controls and the displays including the direction of motion stereotypes and control-display gain (Wierwille 1993). As vehicle information systems proliferate, debates flare regarding the value of analog or digital detail and the utility of simpler status lights. Talking cars address the visual workload problem by repeating messages such as "Please put on your seat belt." Contemporary car navigation systems use the auditory channel to relieve the visual channel for control activity. "Be prepared to exit on the left."

The modern information-dominated era offers a considerable increase of information types, forms, and details that are accessed by touch screen menus. The primary tasks of the car driver—speed and heading controls—remain paramount, while the secondary navigational, operational, and vehicle system information tasks continue to compete for attention, with ever more insistence. The tertiary tasks, such as entertainment, external communications, and climate management, also contribute to a sometimes impossibly cluttered informational context. Meanwhile, the driver's capacities remain the same, perhaps marginally improved with familiarity and the contextual demands of traffic density increase. An instance of these cognitive workload and distraction issues is found in taxis, which often sport four or five aftermarket attention-demanding displays that crowd the forward view and demand attention (global positioning system [GPS] navigation, hand phone, fare meter, forward video recorder, toll road meter).

These well-meaning opportunistic information system interventions, driven by technology, are even more prevalent in modern airplanes. The traffic density may not be as great as on the roads, but the cost of failure is infinitely greater as gravity is unforgiving. Airplanes are commonly equipped with information sources that allow the instrument-trained pilot to fly heads down. Modern GPS systems make navigation and collision avoidance relatively easy, although remote management by the air traffic control is still needed to achieve separation and route guidance. It should be noted that with the trend toward GPS-guided free flight, there is an increasing opportunity of conflicts between pilots and air traffic controllers (ATCs) (Avitabile et al. 2007). The human factor challenges of this information management and flight control problem can be easily demonstrated in a multiplayer flight simulation environment, without the significant costs of failure. In fact, the frequent occurrence of failure that is not feasible in a real flight environment can be very instructive to aviation students, hobbyists, and designers.

16.7 THE BEHAVIORAL DIMENSION

Behavioral analysis in ergonomics is usually through the methods of ethnography in which the observer classifies and counts the activities and their outcomes, which are usually reflected in terms of frequency, time, and error. One general observation is that different behaviors can sometimes lead to the same outcome, and that the same behaviors can result in different outcomes, depending on the context. Behavioral variability among different individuals adds to the analytic noise. An example related to the sitting behaviors of children in airplane and car seats demonstrates a plethora of alternatives, to some extent, controlled by restraints. Behavioral restraints in the cognitive domain are achieved by training. Drivers and pilots are trained to scan the outside scene and their instruments depending on the task at hand, such as collision avoidance or navigation.

These examples are conducive to the microergonomics investigation in which the contexts are controlled, the performance requirements are specified, and the appropriate and inappropriate behaviors are identified. An example in car control is the analysis of the foot behavior where, in automatic drive vehicles, either one- or two-foot behaviors may be acceptable. The anecdotal evidence is that the drivers of automatic transmission taxis in some big cities habitually use one foot on the accelerator and the other on the brake, similar to the toggle pedal on some fork trucks. Also in vehicles, eye movement and fixation research can clearly indicate the utility and the pitfalls of variable visual behaviors. Similar approaches are used in pilot training with regard to the instrument scanning and the transition to glass cockpits that require the deliberate selection of display content.

Whereas the human overt behavior, such as foot, hand, or eye movements, is generally easily accessible using ethnographic or video methods, these approaches do not necessarily tell the whole story of those more elusive cognitive factors that predicate the behavior and the outcomes. A complementary approach is to use verbal protocol or think-aloud techniques, in which the subject describes what he is attending to or thinking of or why he chose a particular action. These methods are sometimes contaminated by hindsight bias, depending on the performance outcome. They may also be somewhat intrusive. Another familiar example of limited cognitive capacity and attention is seen when the conversation between the driver and the passenger stops as the vehicle approaches an intersection.

16.8   USE AND MISUSE

A second model—the 6Us and 2Ms—will be used to explain the factors that must be considered in the human system design (Peacock and Resnick 2011).

- Utility—Is the system useful?
  - Most technology is developed with particular functions in mind, but in the long run, technology itself is subject to Darwinism—survival of the fittest. Notably, the likes of the Titanic were replaced by jumbo jets for transocean transportation, but survived as cruise ships.
  - The perception of utility is in the eye of the customer, well explained by the Kano model (http://www.kanomodel.com/). A car, a bus, a train, or an airplane may have considerable utility, but the choice among these modes will vary according to context and customer preference.
- Usage
  - What is the intended use of a system or process? Is the car intended to take the driver to work or the family on a camping trip? Such questions will give rise to alternative design solutions or compromise where the driver has only one vehicle.
- Misusage
  - Can the system be used for other than its intended purpose? Is a family car also used for transporting building materials or pets? In either case, the choice of the make and the model may change from a small sedan to the more resilient large utility vehicle.
- User
  - Who is the intended user? What are his or her intentions and capabilities? Might there be many users each with a different role? What kinds of variability does the user exhibit?
  - The family minivan has many users, including the driver and the family members of various sizes and requirements. Another source of user variation is their ability and interest in paying more for more features, services, comfort, and convenience. The commercial aviation industry exploits these differences by offering different classes of travel.
- Misuser
  - System security is a universal concern in transportation, from hijacked airplanes to stolen cars.
  - At a less sensational level, the misuser may be legitimate but incompetent or incapacitated. Should a teenager be given the keys to a Corvette? Perhaps with the fifth and sixth gears locked out? How can a fatigued or inebriated driver or pilot be detected and prevented from using the vehicle? At what age should an elderly pilot or driver hand over the keys?
- Utilization
  - How often will the vehicle be used and in what kinds of environments? These temporal questions of intensity, frequency, or duration of use will also influence the design of a product or a service. The demands for the maintenance of vehicles with high utilization, such as public transport, are such that the designs should be robust in order to reduce the incidence of component wear and failure, and conducive to rapid, frequent, or regular maintenance operations.
  - What should be the denominator in the transportation accident analysis, the number of journeys or the passenger miles?

- User error
  - Can normal or abnormal conditions be conducive to human error?
  - Can the driver hit the wrong pedal or switch at the wrong time? User errors occur all the time, but the consequences of these errors are only important in the context of a preexisting hazard.
- Usability
  - *Usability* is a widely used word to describe how easy it is for a user to interact with some hardware or software. However, as perception is in the eye of the beholder, usability is in the hands of the user in the context of use. An expensive golf club may be usable in the hands of a good golfer; conversely, a rear-wheeled muscle car will not be very usable in icy conditions, and a large passenger airplane will require careful pilot training and selection. Usability therefore begs the question of the user and the context, and the negotiated purposes and the trade-offs of the design.
  - It should be noted that the automobile is probably one of the most widely used and therefore by definition one of the most usable systems ever developed, given certain assumptions about the driver training and selection. As automobiles, airplanes, and computer technology evolve, user selection may have to become more discerning.

16.9   THE CONTEXTUAL DIMENSION AND RESILIENCE

Edwards' SHEL model (1988) emphasizes the interactions among people, technology, operational rules, and contexts in all processes. The process outcomes (E4S4) are dictated by the capabilities, the contributions, and the compatibilities of the contributing systems and their interactions. Resilience is achieved by the capabilities of one component compensating for the limitations of another.

A car driving example is the introduction of antilock brakes, which compensate for slippery roads and inadequate driver-braking behavior. Also in cars, the introduction of talking maps reduced the navigation workload and the distraction vulnerability of the drivers. The advent of the glass cockpit in airplanes opened the door for powerful computer capabilities to aid the pilot in aviation, navigation, and communication. Most of these devices are aimed at compensating for the driver/pilot information handling limitations; however, it is the predictive and problem-solving powers of the driver/pilot that saves the day when technology is unable to deal with extreme physical and operational contexts. Drivers in traffic look beyond the vehicle immediately in front of them in order to anticipate the need for foot movements between the pedals. This human capability places the attentive driver ahead of the technological developments in adaptive cruise control, which simply reacts to the car in front.

Technocrats predict the future of driverless cars; however, the consideration of resilience in demanding contexts will render this aspiration unlikely. For example, one paramount duty of a car driver and a pilot is to maintain the separation from other vehicles. The density and the two-dimensional nature of ground traffic make this separation management a greater challenge than the three-dimensional airplane separations, although the relative speed differences between cars and airplanes closes the temporal gap. The separation challenge is also important in shipping as they navigate congested waterways. In shipping and aviation, the use of computers to calculate the closure rates and suggest avoidance maneuvers has considerable promise, assuming all the participants have location sensors and responders. GPS-based automatic-dependent surveillance-broadcast (ADS-B) technology is vastly superior to radar in terms of time and resolution, and this technology enables pilots and ships captains to avoid collisions without the help of the traffic control organization (Avitabile et al. 2007). Such a transition, involving greater autonomy of the individuals, may fail in comparison with the traffic controller's ability to optimize the throughput by imposing non-selfish decision-making. Advanced systems may demand that all vehicles in a congested traffic context be fitted with the technology. Remote radar has this capability in air and sea contexts, but there are significant temporal lags. GPS-based technology such as ADS-B reduces the time delays and provides information directly to the pilot, but not all airplanes and small boats have this technology.

The density of road traffic presents a greater order of complexity. Road traffic also has to contend with large buses, trucks, motor cycles, bicycles, pedestrians, and a plethora of new electric transport aides. The example of a typical Southeast Asian city intersection will illustrate the problem. In some city intersections, there is no attempt to regulate the traffic. It is left up to the individual driver or the pedestrian to detect obstacles and carry out rapid collision avoidance maneuvers. The technology-aided progression goes from see and avoid through sense and avoid to predict and avoid. It is evident that training and experience go a long way to support effectiveness, efficiency, satisfaction, and resilience in this integration context. However, the accident statistics show that the reliance on individual behavior and performance is far from adequate. The introduction of technology to manage the separation is likely to do so at the cost of reduced efficiency—intersection throughput. Also, the cost and the widespread deployment difficulties of traffic control technology may be prohibitive.

The challenge for the proponents of driverless cars is to mimic the human ability to anticipate, perceive, understand, predict, decide, and control. If car separation algorithms simply relied on the closure rate with the car in front, then given normal speeds and variable following distances, there would still be rear-end collisions. Good drivers maintain sufficient headway and monitor the behavior of vehicles ahead of the immediately preceding car in order to buy time for anticipatory behavior.

16.10 TRANSPORTATION SAFETY

A macroview of the transportation safety is that accidents happen due to situational changes in the interactions among technology, people, and contexts, including both the physical and operational contexts. The accident may result from the changes in the individual systems over a long period or quite rapidly. A tire or a tired driver may gradually wear out and eventually fail; similarly, a driver may accumulate bad habits over time and eventually pay the price. It should be noted that accidents usually have information processing or cognitive causes, such as failures in anticipation, attention, memory, perception, decision-making, and control behavior. Some failures are due to deliberate risk taking or violations of the correct procedures, whereas most are simply ascribed to the ubiquitous human error.

The human factors analysis and classification system (HFACS) (Shappell and Weigmann 2000) describes accidents as a combination of the unsafe acts (errors or violations), the preconditions (in the technology, driver, or context), a failure of the supervisory system (tolerance or even incentives for bad behavior), and finally the organizational climate which may precipitate adverse conditions by, for example, cost cutting in traffic management. The HFACS is a perceptive approach to aviation accident analysis, and the concepts have spread to medical error analysis and beyond. The concepts are applicable to all forms of transportation and can apply to both accident analysis and preventive design. This HFACS analytic approach to accident reconstruction and design is greatly enhanced when amalgamated with the Edwards' SHEL model, the 6Us model, and the E4S4 model of design purposes and trade-offs.

At the microergonomics level, most vehicles are controlled by human operators (drivers or pilots) by interaction with a control panel, sometimes with reference to a head-up view outside the vehicle and sometimes with a visual focus on the internal displays. As the driver or the pilot becomes more experienced, s/he replaces visual with tactokinesthetic information sources for control selection and actuation, supplemented by sequential and procedural memories. For operational reasons, it is common for controls with sometimes opposite functions to be placed close together and differentiated by relative location, shape, size, movement, and labeling conventions. Given the frequency of operation, it is inevitable that errors in the control selection and input will occur from time to time, especially when the driver or the pilot is distracted. The classical problem of unintended acceleration in automobiles is described by Schmidt (1993) who indicated that, apart from the occasional engine control module and carpet issues, the major cause was driver error. An engineering solution to a common causal pathway was the introduction of the brake–transmission interlock, which prevented the movements between forward and reverse in automatic cars without the driver's foot being on the brake pedal. The reliance on procedural training and cosmetic control differentiation alone, without engineering intervention, will never be sufficient to prevent unintended actuation.

16.11 SIMULATION AND SIMULATORS

Simulation methods offer various degrees of realism, but usually without the stress caused by the threat of real damage to the individual, the technology, or the environment. The principal advantage of simulation for research and training is that the scenarios can be constructed to tax various human systems, frequently repeated for training purposes, and to examine and correct erroneous behavior. The recent crash of TransAsia GE325 in Taiwan was quickly followed by a requirement of all ATR 72-600 pilots to take remedial (simulator) training and testing regarding the actions following the burnout of one of the two engines. The simulators considerably vary in their level of sophistication. Low-fidelity simulators are freely and widely available on the Internet as games and for more serious uses. Such simulators test the same cognitive mechanisms as their high-fidelity cousins, but with less realism. The intermediate level of simulators such as Microsoft Flight Simulator, now offered as Prepar3D (http://www.prepar3d.com/), offer realistic cognitive challenges, again relatively inexpensively. More expensive simulators are motion based on robotic arms that accurately mimic the system movement responses to the pilot input and the external forces.

The Singapore Institute of Management University known as SIM University Virtual Airspace project requires large classes of aerospace students to form teams to navigate around various airspaces. The teams consist of a pilot (attitude and airspeed), a navigator (altitude and heading), a communicator and ATC (route planning and separation), and an observer (cockpit voice recorder and activities monitor). As the team members have minimal experience with flight, the occurrence of flight control, navigation, communication, and crew resource management errors is frequent and informative. Distraction is a common culprit. Lack of situation awareness regarding navigation, operation, and control status is commonplace. The sequel to this simulator experience is a real-world accident analysis, using the HFACS approach to identify the human factors causes.

Simulation-based research that addressed the introduction of head-up displays into cars, following their successful use in flying, offered two important conclusions. First, it is important to consider the context; airplanes do not usually rely on out-of-the-window detection of other airplanes, while this mode is usual and continuous in cars. The second observation was that it was not just the visual interference of head-up displays; rather it was the cognitive capture of the information contained in the display that is the greatest hazard. An example of such a hazard occurs where multiple menu steps are required to obtain the required information. Such sequential procedures may take many seconds to navigate, while the view outside is ignored. Simulation-based research, although sometimes criticized for lack of face, ecological, and outcome validities, is an invaluable medium for the demonstration of cognitive workload issues.

16.12 THE KANO APPROACH

In maritime and aviation contexts, the passengers may have choices among competing companies, which do not differ in their functional attributes—the same models and performances of airplanes are common to all the competing companies. The competition among companies is therefore dependent on cost, comfort, and service dimensions. Some airlines are low cost with minimum service levels in terms of interior design, meals, and baggage allowances. Other airlines offer greater creature comforts and luxury levels in first and business levels, but at a premium price. Similar trade-offs have long been exploited in many areas of public transport such as trains and passenger boats.

Private cars however demonstrate by far the greatest opportunity for nonessential feature exploitation. The human factors approach to the understanding of these function and form dimensions has historically been through the analysis of the purchase behavior after the event and through surveys and focus groups before and after the point of sale.

It is widely recognized that product choice, given the basic functional requirements, is often based on the affective dimensions of likeability. The Kano model (Kano et al. 1984) was developed for the purpose of product evaluation and has since been applied to services evaluation (Peacock et al. 2012). The model suggests that there are three types of features in a product or a service: must have, more the better, and excitement. The absence of must-have features will make the product or the service unacceptable or unusable, but an increase may not increase the vehicle's attractiveness. In the case of a car, the must-have features include four wheels, an engine and transmission, and a body containing seats and controls. More wheels, engines, or seats would not help. More (or less)-the-better features may include increased speed, more trunk space, better fuel consumption, lower price, and fewer service visits. The absence of excitement features does not necessarily reduce the car's utility, but its presence may increase the car's attractiveness. Examples of these excitement features could include leather seats, a navigation system, or an automatic driver recognition. Another concept in the Kano model is that over time, these excitement and more-the-better features may become must haves.

Running, driving, riding, sailing, and flying also bring another purpose—enjoyment—the affective domain. In the Kano classification, these are the excitement factors (Hartono et al. 2013). Consider the excitement waiting to start in the Boston marathon, driving at 100 mph or more on the open freeway, riding a mountain bicycle down a steep slope, fighting a gusty sea breeze in a sailing dinghy, or flying coast to coast at 10,000 ft. The transportation affective domain has its downsides too: "hitting the wall" at Heartbreak Hill, seeing the flashing lights in your rear view mirror, falling off your mountain bike, or seeing thunderclouds ahead. The technologies themselves contribute to the enjoyment. Consider a group of runners discussing their shoe styles, classic car owners extolling the virtues of past styling arts, the adrenaline rush of the mountain bikers, and dingy sailors and pilots drooling over the intricacies of their technology designs. These conversations are beautifully reflected in the discussions in *Harry Potter* of the Nimbus 2000 quidditch broomsticks and the Toad of toad Hall in the *Wind in the Willows*:

> "Whether a Ford or a Ferrari, whatever I can get to carry me near or far, just give me any car. I love to ride the Tar, an old Excalibar; yes, any motor car. And I'll be happy—ho-ho! Messing around in cars!"

Contrarily, one "upside" of the Boston marathon is the traffic—20,000 runners (arranged by qualifying time) who have trained for months or years to participate in this prestigious event—the world's greatest foot race, whereas traffic on the road or in the air definitely puts the brakes on the pleasures of powered transportation. The value of this Kano approach is that the principles can be applied to all levels of the vehicle or the transportations system. However, the ubiquitous challenge of human variability often clouds the investigations.

16.13　SEGREGATION OR INTEGRATION

Before the introduction of the internal combustion engine, walkers and riders of horse-drawn carriages shared the same roads. The cautious walker heard the horses' hooves and moved out of the way. Over time, various levels of segregation have developed. First, sidewalks and roadways, and now the roadways are sometimes segregated for bicycles, buses, heavy vehicles, and cars. In many cities, integration is the modus operandi. The mutual skills of the pedestrians and the motorcycles on Ho Chi Min City roads show integration at its highest level. The intersections can be facilitated by costly segregation facilities, bridges, and tunnels, or by temporal management—traffic signs and lights. Other traffic management strategies use priority systems. Despite these engineering and operational interventions, there remain many safety and temporal problems that result from mass and speed variabilities where spatial and temporal sharing are permitted. The temporal problems arise when the control system is not adaptive to the demand—batches of cars wait at the traffic lights even though there is no cross traffic. Light-controlled pedestrian crossings often unnecessarily delay vehicular traffic. The roundabout with various levels of complexity provides a priority-based continuous flow alternative, which requires compliance by the participants to the prioritization rules. This technological solution sometimes extends to multiple miniroundabouts at a single intersection where successful navigation requires considerable experience and confidence. Human adaptation to these temporal and spatial challenges creates situations where bicycles and now various electric vehicles use the sidewalks to escape from the cars. Another administrative aid involves the use of the car horn or the bicycle bell to create segregation where the general rule is integration. These surface situations are being mirrored in aviation, which although there is an additional dimension to play with, is experiencing increasing densities that demand administrative solutions. Should airplanes or UAVs be equipped with horns?

Trains, cars, airplanes, motorcycles, and scooters have operators and passengers. In the case of UAVs, the operators, but not necessarily the passengers, may be located away from the vehicle; when accidents occur, they are involved, but not committed. The operators must manage the control, the navigation, and the external communication; the passengers just want comfort and convenience, plus speed and safety. The microergonomics of human vehicle control involves attention, perception, decision-making, and control input. These human controllers vary in their expertise and situational capability, due to such things as fatigue or pressure to hurry. System failures occur due to the mismatch between operational and contextual demands and instantaneous operator capabilities and limitations. Such mismatches account for 70% of transportation accidents. The prevention strategy must deal with the operational demands, the operator capability, or both. Much is known in general terms about operator variability in terms of capability and situational vigilance. Similarly, vehicle and operational system research and designs continue to address the demand factors. But the intervention is inevitably about trade-offs—licensing restrictions (teenagers and elderly drivers), operational interventions (traffic lights, in the broadest sense), and compliance and enforcement. The segregation strategy is inefficient, whereas the integration approach places too much emphasis on the individual operator's skills. These segregation and integration strategies may be of theoretical interest; they may be subject to modeling and optimization approaches. However, the complexity in terms of spatial and temporal capacities, size and speed variations, costs of engineering solutions (sensing and segregation), and fallibility, in terms of safety, of administrative solutions is such that the tolerance of one million road fatalities a year may be a reality.

16.14   MANUFACTURING

The proliferation of manufactured technology, including bicycles, cars, airplanes, and computers over the past 100 years has greatly benefitted from the principles of the assembly line, as popularized by Henry Ford and Frederick Taylor (Taylor 1998). These principles had considerable success in enhancing product quality and improving productivity by the introduction of easily learnable and controllable short-cycle tasks. Although automation is a common aim, these repetitive manual tasks are still the norm in most manufacturing organizations. The downside of this short-cycle, repetitive work is that it causes cumulative trauma disorders to some degree in the majority of operators who have long-term exposure to this kind of occupation. Automation aside, engineering intervention by product, fastener, tool, and fixture designs can go some way in reducing the incidence of musculoskeletal disorders. However, it is the temporal dimension—duration, frequency, repetition, shift work, etc—that is the major culprit. These stressors can be reduced, but not totally eliminated, by administrative strategies such as job enlargement and rotation.

An organizational strategy that was also recommended by Ford and Taylor, although the detractors do not always acknowledge these sources, was the importance of the participation of the line operators in the task and job designs. These participation principles have been widely promoted by the ILO (Kogi 2012).

16.15 THE CAR AS AN OFFICE

Each day, millions of people worldwide use their car as their office for either a part or all of their workday. Professionals, such as a police officer, may spend their whole working shift in a car. Others, such as mobile insurance assessors, power company workers, and salespeople, may also use their vehicle as their mobile office. Millions of commuters spend several hours per week in their cars, and with modern mobile communications technologies, such as wireless phones and Wi-Fi-enabled tablets and laptops, many drivers can continue some of their work tasks from within their vehicles.

In 2010, 75% of the U.S. workforce were mobile workers (International Data Corporation 2011). Indeed, by 2015, the mobile worker population is predicted to reach 1.3 billion people worldwide, representing 37.2% of all workers. For many people, the car may be their office. However, driving and office tasks present a person with inherently conflicting demands. Ironically, the use of mobile phones while driving cars or riding motorcycles and even while walking is responsible for an increasing number of collisions, due to driver distraction. Research studies show that the drivers are distracted, and the driving performance is impaired when they are involved in performing tasks such as texting on their phone or holding a phone conversation, even in a hands-free arrangement. Distracted driving is a significant issue in the United States, and each day, approximately 660,000 drivers use their cell phones or manipulate electronic devices while driving, and in 2014, there were 3179 fatalities and 431,000 injuries distracted driving motor vehicle crashes (http://www.dis traction.gov/stats-research-laws/facts-and-statistics.html). An analysis of 4452 safety-critical events (i.e., crashes, near-crashes, crash-relevant conflicts, and unintentional lane deviations) and 19,888 uneventful, routine driving periods as baseline showed that nondriving-related tasks were involved in 71% of crashes, 46% of near crashes, and 60% of all safety-critical events (Olson et al. 2009). For specific tasks, such as texting, 5 seconds was the average time the driver's eyes were off the road, which is the equivalent of driving the length of a football field blindfolded when traveling at 55 mph (Olson et al. 2009). Fitch et al. (2013) found that the risk of getting into a crash is tripled if the driver is engaging in nondriving visual–manual subtasks, such as reaching for a phone, dialing, and texting. Consequently, the use of mobile communication technologies in a vehicle is best performed when the vehicle is stationary, although this is not always possible; for example, a laptop may need to be used when police officers are involved in chasing another vehicle.

Driving a vehicle for frequent and long durations can increase the risks of developing musculoskeletal symptoms, especially back and upper body injuries (Porter and Gyi 2002). Moreover, the interior layout of a car is not optimally designed for the driver to undertake computer-based tasks. While in the driver seat, the position of the steering wheel restricts the space in front of the driver, making it more difficult to use mobile technology. Consequently, while the vehicle is stationary, when a driver uses a tablet or a laptop in their vehicle, they often have to sit with their torso in a twisted position and with their arms in deviated postures to operate their device. A consequence of this is that the drivers can suffer from work-related musculoskeletal disorders; the onset of which is accelerated by the poor work postures that they adopt. Fortunately, studies on the use of a vehicle, such as a car, a van, or a truck, as a mobile office have been conducted, and design guidelines have been developed to minimize the injury risks. Saginus et al. (2011) tested how four different mobile computer locations for both keyboard and touch screen devices affected joint angles, muscle loadings, task performance, and subjective ease of use, comfort, and productivity. They found that the layouts shown in Figure 16.1 resulted in least postural deviations, lower muscle loadings, and 40% lower L4/5 spine compression.

A variety of ergonomic products are commercially available to support the driver's use of a tablet or a laptop in a vehicle, and for more intensive computer office tasks or for multiple workers,

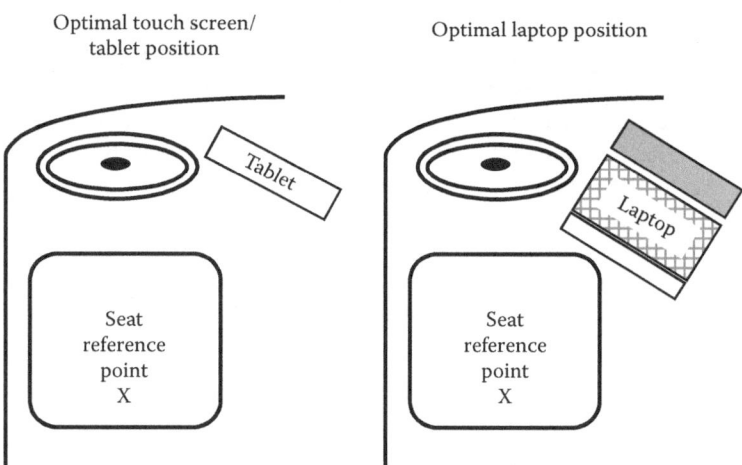

FIGURE 16.1 Optimal touch screen and laptop mobile computer locations. (From Saginus, K. A., R. W. Marklin, P. Seeley, G. G. Simoneau, and S. Freier, *Human Factors*, 53, 474–88, 2011. With permission.)

a vehicle, such as a minivan, can be converted into a mobile office. For high-mileage business driving, whether the car is being used as a mobile office or as a load transportation system and there is manual handling from the car, participatory ergonomics has been shown to be useful in raising both the awareness of the management and the drivers of the musculoskeletal risks associated with their driving, and also effective in affecting behavioral changes to reduce these risks (Gyi et al. 2013).

16.16 CONCLUSIONS—THE BIG PICTURE

Transportation is a complicated business that continually responds to technological innovation (push) and customer demand (pull). Managers and users seek optimal solutions in terms of effectiveness, efficiency, safety, comfort, and convenience. Transportation management becomes complex with the addition of human variability and flexibility; further complexity is introduced by environmental and operational forces. The system becomes chaotic due to the interactions among technology, people, and contexts, and the attempts by the regulators to achieve order and predictability. The affective domain greatly increases the challenges of traffic management.

The opportunities for microergonomics analyses and interventions are everywhere—from seats and safety through controls and communication to artistic design and individual choices. Microergonomics intervention in manufacturing systems is common in most large organizations, and checklists galore draw attention to posture, movement, force, and frequency. These opportunities have been met over the years by the piecemeal pursuit of topics that interest the researcher or are highlighted by the various customers. Many of these contributions have been extremely successful. Altimeters in airplanes were redesigned to remove the confusion between the orders of magnitude on the scales. The center high-mounted stoplight has reduced the incidence of rear-end collisions by shortening the driver response time. The brake transmission interlock addressed the human causes of many unwanted acceleration incidents. The addition of a shoulder strap to the lap belt and the addition of a head restraint have reduced many injuries in cars.

Three Japanese design processes have addressed the affective domain and the broader aspects of customer requirements. The Kansei approach (Nagamachi 2011) translates customer statements into design criteria. The Kano approach differentiates among must-have (must-not-have), more (less)-the-better, and excitement factors in product or service design, and notes that over time, some excitement features may become must have, as in the case of entertainment and navigation systems in cars. The third, Kaizen, approach emphasizes the importance of continuous improvement in product and service designs, based on customer articulations and other system outcome feedback. These three approaches have been prominent and successful in the Japanese automotive industry, both in product and production designs, and abroad over the past few decades. Collectively, these three approaches move the practice of microergonomics from reductionist laboratory-centered research into the broader field context of macroergonomics, which can be articulated as a high-level, comprehensive approach to addressing human-centered design constrained by technological, contextual, and operational complexities. Finally, macroergonomics accepts that there are many stakeholders and purposes associated with the design process, and these stakeholders may emphasize different outcomes in terms of effectiveness, efficiency, ease of use, elegance, safety, security, satisfaction, and sustainability. Macroergonomics is the process of making trade-offs in design.

REFERENCES

Aviation-Safety. 2015. Net/Statistics. From http://aviation-safety.net/statistics/period/stats.php?cat=A1.

Avitabile, J., G. Northam, B. Peacock, and J. Tank. 2007. Automatic ependence surveillance-broadcast utility for air traffic avoidance. *J Air Transport*. 12(3):6–21.

Benenson, R., S. Petti, T. Fraichard, and M. Parent. 2008. Towards urban driverless vehicles. *Int J Vehicle AutonSys*. 6(1/2):4–23.

Bin, S. B., and L. Ching. 2013. The evolution of public transport policies in Singapore. Le Kuan Yew School of Public Policy. From http://lkyspp.nus.edu.sg/wp-content/uploads/2014/01/Transport-Planning-for -Singapore.pdf.

Edwards, E. 1988. Overview. In Weiner, E. L., and D. C Nagel (eds.) *Human Factors in Aviation*. Houston, TX: Gulf Publishing.

Erlandson, R. F. 2008. *Universal and Accessible Design for Products, Services and Processes*. Boca Raton, FL: CRC Press.

Fitch, G. M., S. A. Soccolich, F. Guo, J. McClafferty, Y. Fang, R. L. Olson, M. A. Perez, R. J. Hanowski, J. M. Hankey, and T. A. Dingus. 2013. The impact of hand-held and hands-free cell phone use on driving performance and safety-critical event risk. Springfield, VA: National Highway Traffic Safety Administration-National Technical Information Service, U.S. Department of Transportation.

Gyi, D., K. Sang, and C. Haslam. 2013. Participatory ergonomics: Co-developing interventions to reduce the risk of musculoskeletal symptoms in business drivers. *Ergonomics*. 56(1):45–8.

Hartono, M., K. C. Tan, and B. Peacock. 2013. Applying Kansei engineering, the Kano model and QFD to services. *Int J Serv Econ Manage*. 5(3):256–74.

International Data Corporation. 2011. Worldwide mobile worker population 2011-2015 forecast. Doc #232073. Summary infographic at http://cdn.idc.asia/files/5a8911ab-4c6d-47b3-8a04-01147c3ce06d.pdf.

Kano, N., N. Seraku, F. Takahashi, and S. Tsuji. 1984. Attractive quality and must-be quality. *J Japan Soc Qual Control*. 14:39–48.

Kogi, K. 2012. Roles of participatory action-oriented programs in promoting safety and health at work. From http://www.ncbi.nlm.nih.gov/pmc/articles/PMC3443691/.

Lakka, T. A., D. E. Laaksonen, H.-M. Lakka et al. 2003. Sedentary lifestyle, poor cardiorespiratory fitness, and the metabolic syndrome. *Med Sci Sports Exer*. 35(8):1279–86.

Lieberman, D. E., D. A. Bramble, D. A. Raichlen, and J. J. Shea. 2009. Brains, brawn and evolution of human endurance running capabilities. In Grine, F. E., J. G. Fleagle, and R. E. Leakey (eds.) *The First Humans*. New York: Springer.

Nagamachi, M. 2011. *Kansei/Affective Engineering*. Boca Raton, FL: CRC Press.

National Highway Traffic Safety Administration. 2014. http://www.nhtsa.gov/About+NHTSA/Press+Releases /2014/traffic-deaths-decline-in-2013.

Olson, R. L., J. Hanowski, J. S. Hickman, and R. J. Bocanegra. 2009. Driver distraction in commercial operations. Washington, DC: Federal Motor Carrier Safety Administration, U.S. Department of Transportation.

Peacock, B. 1978. Ergonomics aspects of Hong Kong public transport. *Hong Kong Engineer*. 6(10):27–34.

Peacock, B., and M. Resnick. 2011. The six us: An ergonomic approach to enhancing product and process evaluations. *Ergon Design*. 9(2):25–9.

Peacock, B., and R. Roe. 1988. Human factors in vehicle design. *Body Eng: J Am Soc Body Eng*. 16(2).

Peacock, B., M. Hartono, K. C. Tan, and S. Ishihara. 2012. Incorporating Markov chain modelling and QFD into Kansei engineering applied to services. *Int J Human Factors Ergon*. 1(1):75–97.

Porter, J. M., and D. E. Gyi. 2002. The prevalence of musculoskeletal troubles among car drivers. *Occup Med.* 52(1):4–12.

Saginus, K. A., R. W. Marklin, P. Seeley, G. G. Simoneau, and S. Freier. 2011. Biomechanical effect of mobile computer location in a vehicle cab. *Human Factors.* 53(5):474–88.

Schmidt, R. A. 1993. Unintended acceleration. In Peacock, B., and W. Karwowski (eds.) *Automotive Ergonomics.* London, Washington, DC: Taylor & Francis.

Shappell, S. A., and D. A. Weigmann. 2000. The Human Factors Analysis and Classification System (HFACS) (Report Number DOT/FAA/AM-00/7) Washington, DC: Office of Aerospace Medicine.

Taylor, F. W. 1998. *The Principles of Scientific Management.* Norcross Ga: Engineering and Management Press.

Vink, P. K. 2011. *Aircraft Interior Comfort and Design.* Boca Raton, FL: CRC Press.

WHO. 2015. http://gamapserver.who.int/gho/interactive_charts/road_safety/road_traffic_deaths3/atlas.html.

Wierwille, W. W. 1993. Visual and Manual demands of in car controls and displays. In Peacock, B., and W. Karwowski (eds.) *Automotive Ergonomics.* Abingdon, Oxfordshire: Taylor & Francis.

17 Health and Productivity Effects of Hot Desks, Just-in-Time Work Spaces, and Other Flexible Workplace Arrangements

Jay L. Brand

CONTENTS

17.1 INTRODUCTION

Beginning in the late nineteenth century, modern office design permanently moved to brick and mortar and concrete construction early in the twentieth century (Hysom and Crawford 1997). Fueled by inventions such as the elevator, steel constructions, and a general optimism and enthusiasm for progress, the concept of office design as part of an overall business strategy became familiar to practitioners outside its specialty areas (e.g., architecture) with the Quickborner Team's development of the Bürolandschaft, or the landscaped office, in the 1950s (Hookway 2009). This relatively early framework actually included many of today's trends in the office workplace design, although their labels did not yet exist (e.g., occupant-centered design, personal control over the environment, the biophilia, the organic work spaces, the sustainable design, the activity-based planning, the biomimicry, the off-grid space planning, the flexible furniture and work spaces). Perhaps due partly to this groundbreaking concept (no pun intended), as of this writing, the Quickborner Team remains in practice outside of Hamburg, Darmstadt (Germany; http://www.quickborner-team.com).

More recently, the Quickborner Team has championed the combi-office idea, a design concept based on providing many diverse work spaces within an office workplace, differing in size, levels of privacy and available technology, and other tools to support various kinds of tasks and activities, reminiscent of the more recent term, *activity-based design* or *activity-based planning*. Landscaped offices began to reflect what has become more generally known as the open-plan office in the 1970s and 1980s (Pile 1986), with the open offices becoming the norm in the 1980s and 1990s particularly in North America, Europe, Australia, Japan, and China (Duffy 1999; Davis, Leach, and Clegg 2011).

Although the research literature has addressed many issues related to open-plan office design and space planning, new ways of working, and other recent workplace design strategies, an important distinction needs to be clarified regarding the quality of available knowledge relative to these developments. Some investigators merely document the trends in current practices in order to justify the proposed changes in the office design or the work styles (description); others also describe how employees use specific work spaces or work space tools (e.g., ethnographic observation, case studies—explanation). Finally, some studies additionally seek to understand the underlying factors that determine the impact of design in order to predict its current and future influences. These strategies return increasing research value, as illustrated in Figure 17.1.

In this regard, consider that when the field of human factors (North America) and ergonomics (Europe) began, none of the phenomena identified in this chapter's title existed. Perhaps a few of the concepts represented by these contemporary terms were understood, but they referred at most to functional adaptation or perhaps system flexibility, not to intentional variability in workplaces, workstations, desks, chairs, work schedules, or work styles. However, not only are these terms now familiar to most practitioners and researchers, but they are also becoming well-nigh ubiquitous in many office workplaces if not in production work environments. In fact, on June 30, 2014, legislation was implemented granting the right of every employee in the UK to request flexible work arrangements (New Ways of Working, www.newwow.net 2014*).

Because the human factors and ergonomics disciplines seek to improve not only the productivity but also the quality of work life, we must address these new developments in an effort to understand and mitigate any user-centered disadvantages they may involve. In this regard, we might also wonder whether these adaptations include any advantages for human occupants of the built environment (Oseland and Burton 2012; Creighton 2014*). An excellent recent review of this literature (Davis, Leachm, and Clegg 2011) aptly demonstrates the difference in perspective between what could be called a *physically-based* design approach to office workplaces and a *user-based*, a *user-centered*, or an *occupant-centered* approach. A focus on the physical design of office environments tends to view office facilities and corporate real estate as financial assets that benefit the enterprise mostly through cost cutting, primarily by putting more people into less space. A user-centered filter for design focuses instead on the perceptual, psychosocial, and behavioral experiences of the occupants, assessing the value to the enterprise in terms of psychological outcomes (e.g., enhanced work attitudes such as organizational loyalty or organizational citizenship behavior) or quality of knowledge work (e.g., job performance; productivity). Although the office design industry claims essential compatibility for these goals, the available evidence suggests that they may point in somewhat different directions (Brand 2010; Brady 2014; cf. Property Directors Forum 2015).

By using the study by Fisk (2000) (see Table 17.1), we can predict the potential impact of an increased understanding of these considerations. Fisk estimates that office workers (including teachers) make up approximately 50% of the work force in the United States, or 64 million workers. International Facility Management Association (IFMA) (2009) found that overall, 5% of corporate office real estate is devoted to unassigned or nondedicated space. Assuming conservatively that 5% of the workforce is involved in distributed or alternative forms of work, then 3.2 million workers could potentially be involved. However, using the direct estimates that 15% of marketing and 35%

---

* The paper can be obtained from James L. Creighton, 480 Kenolio, 15-104, Kihei, HI 96753.

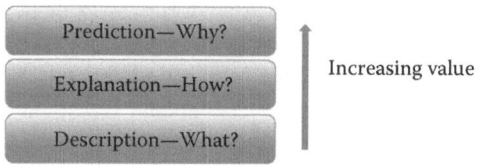

FIGURE 17.1 Types of design research and their knowledge value.

TABLE 17.1

Estimated Potential Productivity Gains from Improvements in Indoor Environments

Source of Productivity Gain	Potential Annual Health Benefits	Potential U.S. Annual Savings or Productivity Gain (96 $USD)
Reduced respiratory illness	16–37 million avoided cases of common cold or influenza	$6–$14 billion
Reduced allergies and asthma	8–25% decrease in symptoms for 53 million allergy sufferers and 16 million asthmatics	$1–$4 billion
Reduced SBS symptoms	20–50% reduction in SBS health symptoms frequently experienced at work by ~15 million workers	$10–$30 billion

Source: Fisk, W. J., Health and productivity gains from better indoor environments and their implications for the U.S. Department of Energy, LBNL 47458, E-Vision 2000 Conference, Washington, DC, 2000. With permission.

of accounting/consulting services employees work in a distributed or an alternative way yields a range from 9.6 million to 22.4 million workers potentially being affected.

In light of the estimates of the proportion of the workforce that are potentially involved in these new ways of working, we must next define the salient terms, as with any rational discipline. Hot desks or hot desking has nothing to do with popularity or preference, but refer to office work environments where chairs, desks, and other components of complete workstations are not dedicated or assigned to individual occupants, rather for affected groups of employees (e.g., a department); a collection of work spaces is provided in an office environment that anyone in that group can use at any time on a first-come, first-served basis. Usually, employee ID (card access) allows secure entry to these work areas. Although isomorphic space planning (i.e., one chair per person) may be used, often fewer chairs or work spaces are provided than the number of people supported under the assumption that everyone in the group will rarely be working in the office at the same time. Just-in-time work space strategies usually feature a flexible menu of different kinds of spaces that can adapt to various levels of occupancy and individual or group task needs. Well-designed just-in-time officing includes spaces that can support individual or group work as needed with both virtual and analog work tools and displays. They typically assume fewer seats than employees within the accommodated work groups. These space planning and work style strategies also typically assume an increasingly global and therefore transitional workforce (e.g., visitors from other office locations; full-time or part-time consultants).

Here are a few more related terms and their brief definitions. Some of these concepts began as early as the 1980s, and most have continued to influence office work environments throughout the ensuing two and a half decades (cf. IFMA 2007). Research questions have primarily addressed what the subset of the total office space is and what percentage of employees is affected by these nontraditional offices along with the trends in this regard (i.e., description). *Alternative officing*—a term that originally meant any approach to office design other than individual private offices—soon included any of a number of corporate real estate and facility management strategies that differ from providing a traditional, assigned (or dedicated) office for each employee; this concept has largely been supplanted by other options.

Coworking—office work environments provided for independent, satellite, or self-employed professionals who simply prefer to work around other people than alone in a home (or other) office arrangement—in design and appearance, differ very little from the more typical, traditional corporate offices; only the relationships among the occupants represent atypical circumstances; an update to this term, proworking, denotes coworkers affiliated with an organization that pays their membership in a coworking space. Distributed work primarily denotes work groups or teams whose members are not temporally or geographically colocated. Such groups may use quite sophisticated virtual, multimedia platforms that mimic the features of colocated, face-to-face interaction and collaboration; however, simple teleconferencing (voice-only) can also be used.

17.1.1 Hoteling

Hoteling is scheduling and providing office work space similar to securing reservations at a hotel that usually includes concierge services and management of secure, mobile storage and other work tools and technologies.

17.1.2 Mobile Workers

Mobile workers are employees who intentionally work anywhere, anytime, and usually have the tools and the technological savvy to do so; these workers may use three or more specific technologies (e.g., laptop, smart phone, tablet/iPad™); a distinction can be made between internally mobile and externally mobile workers—the former work anywhere within a corporate office environment (e.g., without an assigned seat or office; office design and corporate real estate strategies supporting these workers have been termed *activity-based working*), although the latter correspond with the familiar business term *road warrior*—representing employees who frequently travel and literally work from anywhere.

17.1.3 Satellite Office

A satellite office is a remote site that allows workers to access the company network and collaborate on various projects; it usually features videoconferencing, virtual presence, or other similar technology tools (e.g., AdobeConnect™, Bluescape™, FaceTime™, Skype™, Google™ hangout).

17.1.4 Telecommuting

Telecommuting primarily refers to working from home or at a semipermanent work space away from an office location (e.g., central corporate headquarters).

17.1.5 Telework

Telework is any of a number of alternative approaches to work and work space strategy that contrasts with dedicated one-worker, one-place strategies; new ways of working denotes a similarly broad category that acknowledges that some professional work may not involve technology (see Chapter 15).

17.1.6 Touchdown Spaces

Touchdown spaces are usually a fixed number of small, individual work spaces that accommodate mobile workers, such as external consultants or sales members, in corporate headquarters or satellite office locations. These spaces may or may not provide plug-and-play network access, and they invariably assume only transient use.

From a traditional human factors/ergonomics perspective, these kinds of office work strategies and their accompanying work environments offer a unique set of challenges. For example, the term *seat ratio* (number of employees for each chair) means that without adjustable chairs and adequate training for employees, the concept of a user-centered design of workstations does not make much sense; how should one adjust a chair to fit a dozen or more users (12:1 seat ratio)? However, some studies have found positive benefits from many of these nontraditional workspace strategies for the employees' work performance and work attitudes. Typically, personal control over work schedules and the work environment features prominently in lists of these relevant factors (Veitch and Gifford 1996; Leaman and Bordass 1999; Lee and Brand 2005; Thompson, Veitch, and Newsham 2014; Hellwig 2015).

The literature linking the characteristics of the physical environment to the psychological experience of workers (e.g., job satisfaction) (Newsham et al. 2009) and ultimately to job performance (Sundstrom, Burt, and Kamp 1980) if carefully and collectively considered, provides a clear theoretical framework for evaluating the impact of flexible and alternative ways of working among today's knowledge workers. The initial development of this framework has been usefully simplified in the following diagram, which links design with individual and organizational outcomes (Figure 17.2).

Although many practitioners fully recognize this as an oversimplification, they, particularly in the architecture and design communities, prefer to position the value of their services to their clients in terms of this diagram. They realize that most business leaders need a clear, direct rationale for investing in design, and this model provides just that. In this regard, some recent evidence suggests that at least some outcomes may indeed be directly related to office design (Danielsson et al. 2014). In this regard, we should pause here to note that corporate real estate strategies based on direct estimates of cost savings from placing more people into less space, invariably relying on various versions of open-plan office design, remain incomplete unless they include the potential savings in healthcare costs from enclosed (i.e., cellular) offices as predicted from the results of Danielsson et al.'s study.

But returning to the central issue at hand, as usual, reality may be more complicated than what the simple, direct model illustrated earlier implies. Instead, we might suggest the following modification, with psychosocial factors moderating/mediating the relationship between design and relevant outcomes (Figure 17.3).

Intriguingly, earlier research on workplace design anticipated much of the structure of these models without formalizing it (e.g., Hedge 1982; Evans and Johnson 2000). Accepting for the moment that this slightly more complicated version of a design outcomes model still represents a level of abstraction above or beyond many important details, it does convey an important implication—that any design change, element, component, or intent actually impacts individual and organizational outcomes by influencing the psychological experience of employees, teams, or work groups. The unit of analysis or the level of measurement may change, but such adaptations will not alter this fundamental relationship. For example, the design may provide better lighting or daylight in a work space, but this will improve the job performance only if it first improves the work attitudes. Organizational design may provide outcomes-related incentives for individual performers or work

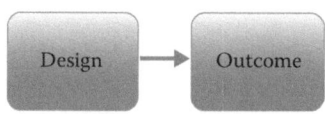

FIGURE 17.2 A simplified model for design research.

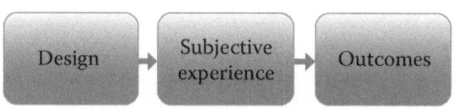

FIGURE 17.3 A more realistic model for design research.

teams, but these must first be perceived as fair (e.g., Colquitt et al. 2011) in order to improve the work attitudes or increase the quality or the quantity of work outcomes.

Depending on the particular outcome under consideration, several more detailed versions of this moderator/mediator model linking design with organizational performance have been proposed. Some modifications suggest possible design effects on the specific aspects of cognitive performance and even suggest that exposure to nature may attenuate these deficits (Jahncke et al. 2011; Kim and de Dear 2013). The former study defines the ubiquitous functional trade-off within open-plan offices—the balance between privacy and communication. However, regardless of such specific behavioral goals, the design invariably indirectly impacts the performance outcomes through influencing the subjective experience (e.g., perception). Illustrating this claim and defined at a somewhat broader level of conceptual abstraction, the following two versions (Figures 17.4 and 17.5) of a moderator/mediator model predict how the design influences creativity and innovation and job attitudes, respectively.

Figure 17.5 illustrates the empirical work highlighting more of the constituent mechanisms related to work attitudes such as job satisfaction (e.g., Newsham et al. 2009). Note also the additional mediation/moderation level in this model category termed *building performance*. This level of evaluation measures the objective, physical characteristics of the built environment thought to be relevant for human experience and performance (e.g., acoustics). The design of the physical environment can influence these performance characteristics, but the characteristics themselves ultimately determine the impact of the physical design on the quality of occupant experience.

Perhaps an example may clarify these distinctions. The surfaces in a work space may be highly reflective (i.e., reflect sound and incident light) or highly absorptive (i.e., deaden sound and scatter incident light) as a direct result of the design choices. However, in situ measures of the installed work space, e.g., the assessments related to speech articulation index or speech articulation class (Gover and Bradley 2009) or the assessments of glare for specific tasks or activities (Leder et al.

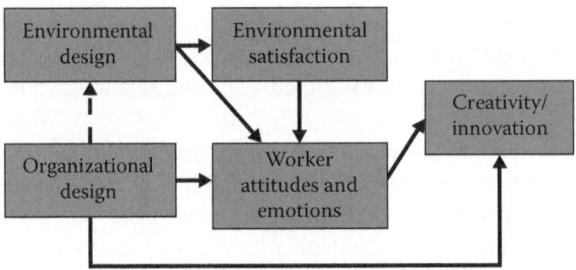

FIGURE 17.4 Research model linking design with creativity and innovation.

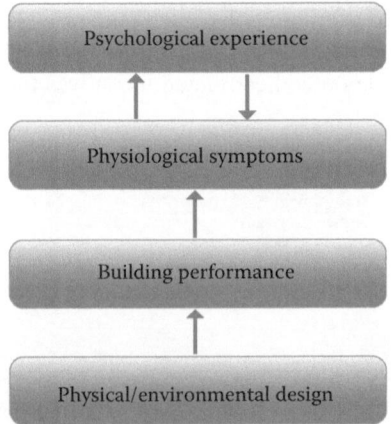

FIGURE 17.5 Design research model linking design to building performance and salient outcomes.

2016), would provide building performance indices rather than objective details of particular design elements. Such building performance evaluation has been related to human perception and performance (Frankel and Edelson 2015). However, according to this class of model, in order for even building performance factors to influence occupancy quality, they must first impact the physiological variables of the individual occupants. Only then, indirectly, do they change the psychological experience or the performance of the human occupants.

The important point suggested here is that flexible work strategies and alternative work design elements indirectly improve the well-being and ultimately the job performance by influencing the quality of psychosocial factors (e.g., employee perceptions of personal control over the work environment, work scheduling, and work–life balance). Other possibly salient moderator/mediator include contextual factors including leadership (cf. Kellerman 2012), feedback (Pritchard et al. 2008), and perhaps even organizational culture (Denison 1996; Denison et al. 2004). In any case, to understand the feasibility of alternative or flexible ways of working, corporations and practitioners must not only address the critical design characteristics necessary to implement these organizational strategies (both physical/environmental and organizational design) but also understand and measure perceptions, work attitudes, and organizational factors that ultimately determine the impact of the design within their particular context.

17.2 WORKPLACE DESIGN IN THE AGE OF REMOTE WORK

Telework, remote work, mobile workers, telecommuting, distributed work, and other similar alternatives to traditional, central office-based employees represent the growing trends for many—if not most—organizations of all sizes. Based on one current estimate, the number of telecommuters in the United States, including part-time workers and those who work from 1 to 4 days a week away from the office, is projected to increase from today's 34 million to 63 million, or 43% of the workforce, by 2016 (Schadler, Brown, and Burnes 2009). How best to encourage, motivate, measure, support, and integrate these employees constitute some of the primary challenges facing managers in the new millennium. Although the inexorable shift away from one-worker, one-place workspace strategies to much more flexible, adaptable definitions of corporate work strategies began in the 1980s and accelerated in the 1990s, several more recent events combined with certain economic realities have forced even cautious, conservative organizations to begin developing and adopting strategic remote worker policies for the new century.

For example, the globalization and internationally distributed teams, the threats of terrorism and communicable disease, the inevitable rise in fuel costs, the continuing departure from traditional family units, the traffic congestion and commuting times, the concern about global warming and the environmental impact of burning fossil fuels (needed for transportation), and the plethora of technology tools that allow workers to be productive outside a typical office building have all converged to produce a shift toward many different forms of mobile work. In order to secure the top talent for their interdisciplinary teams and to offset the high labor costs particularly in North America and Western Europe, multinational firms increasingly turn to globally distributed, self-managed work groups organized around specific projects (Robbins and Judge 2014). Such work groups do not easily lend themselves to being located in the same place at the same time.

Although little consensus exists regarding the likelihood of large corporations being the targets of terrorism or communicable disease, the prospect of a calamity wiping out much of an entire organization's workforce—as happened on September 11, 2001, and earlier in Oklahoma City—did not pass unnoticed by corporate leaders. In such extreme cases, the advantages of a distributed workforce linked by technology and common goals seem obvious. In situations where workforce strategies have embraced remote, distributed forms of work, Information Technology/Information Systems (IT/IS) concerns over data security, technology reliability, and suitable support for efficient collaboration tend to replace worries about car bombs, contamination from biohazards, and other forms of collective violence or pestilence.

The forecasts that world oil prices will continue to rise or at least stabilize at much higher prices than the already steep costs of today have also contributed to the trend toward work styles and strategies that do not require commuting—particularly in single-passenger vehicles over long distances. Savvy executives and managers have begun to realize that without lengthy commutes, telecommuters not only save money, but can also work longer and perhaps be more productive. Governments at the national, state, and local levels have begun to explore in earnest the advantages of the various forms of telework in terms of reduced fuel consumption, reduced emissions and pollution, reduced congestion, higher employee satisfaction, reduced absenteeism and turnover, and in some cases, even higher employee productivity (Lansons and Opinium 2015).

Wage earners in single-parent families and other nontraditional family units can often be reclaimed by organizations willing to provide the tools and the flexible schedules that let them work from home. Spouses who prefer to stay at home with children may also choose to earn a paycheck if that opportunity is available. In such cases, companies usually benefit from committed, loyal workers, and these employees save commuting and day care costs (e.g., Wesley 2012).

17.3 CHALLENGES FACED BY TELECOMMUTING, DISTRIBUTED WORK, AND OTHER REMOTE WORK STRATEGIES

Although relatively little systematic, rigorous research has been conducted comparing colocated work groups and teams with groups comprised of distributed team members, many corporate leaders have assumed numerous advantages for bringing the members of work groups together in one place at one time (cf. Springer 2011*). Traditional, centralized corporate headquarters in urban cores and suburban office parks continue to be designed and built around the world. The question is rarely "Should we build a corporate office or link remote employees with technology?" Instead, it is usually "How can we best balance the needs of colocated group collaboration with those of distributed, mobile workers?" or "How do we best integrate our fixed-in-place real estate portfolio with a distributed, mobile work program?" The following discussion focuses on the latter questions and similar issues.

17.4 TECHNOLOGY SUPPORT

Perhaps this is too obvious to mention, but ideally, distributed workers should enjoy all the access and the connectivity of their colocated counterparts, to the extent that the mobile workers experience technological or other barriers to being productive any time, any place, their performance, and value to their organizations will suffer. The more realistic and similar to real-time, face-to-face interaction videoconferencing, virtual presence, and other such technology tools are, the closer they will approximate the advantages of colocated groups of workers (Creighton 2011*). However, for many types of tasks, old-fashioned teleconferencing provides adequate support, particularly for groups of workers who have previously met face to face. Exactly which types of task benefit the most from realistic, full-scale, virtual presence technology platforms has not yet been confirmed by research.

17.5 GROUP IDENTITY

The relationships among team members—both interpersonal and task related—constitute an important determinant of group performance and effectiveness. An assumed maxim that has nonetheless not yet received adequate empirical scrutiny is that remote team members should first meet face to face to instill confidence and trust among the group participants, to determine optimal roles and leaders among self-managed teams, and to outline the scope and the timeline of the project to be completed. Certainly, strategies are needed to provide meaningful boundaries for group

---

* The paper can be obtained from James L. Creighton, 480 Kenolio, 15-104, Kihei, HI 96753.

membership and, as far as possible, a detailed description of successful task completion or the goal to be reached by the group (Creighton 2011*).

17.6 ORGANIZATIONAL CULTURE

Organizational leaders differ in terms of how integrated with the dominant, majority culture of their companies they would like the distributed team members to be and feel. Some prefer that mobile workers become oriented to their broader corporate culture, while others merely expect the members of the distributed teams to forge a common set of best practices, expectations, and interpersonal and task-related heuristics that can drive the group process toward shared, organizationally relevant goals (cf. Bezrukova et al. 2012). In any case, the psychosocial dimensions of the distributed work groups must be considered along with their task-defining constraints, because groups sharing a common purpose and a set of values tend to be more motivated throughout the duration of their projects (Raghuram 2013*).

17.6.1 SHARED MENTAL MODELS

Research has shown that groups whose members share a common conceptual framework for defining success and for addressing task-related problems outperform those with less similar mental models (McNeese, Salas, Endsley 2001). Thus, the efforts to provide group members with a common orientation, a common purpose, and a shared understanding of the group's mission and vision can be very productive. Whether such experiences are more effective if given in the same location or whether equivalent experiences can be provided to remotely located team members awaits future research to be established. In any case, intentional education and training techniques with the purpose of increasing the similarity of particularly task- or project-relevant cognitive schemata among team members—before the team begins a project—will pay many dividends throughout the life of the team.

17.7 INCENTIVE STRUCTURES

Many organizations remain somewhat naive in practice regarding how the members of teams in general—and of distributed teams in particular—should be reviewed, evaluated, and compensated. If adequate attention has been paid to defining a work group's mission, it should be possible to collectively assess and reward group performance, in addition to any individual incentive plans in operation. Of course, the initial definition of team member roles and the scope of responsibilities and contributions are extremely important, and any consequences of group productivity must be tied to these criteria. Often, distributed teams suffer from insufficient rewards and acknowledgment of their contributions to the larger organizational goals (Rock and Schwartz 2006). Although collective incentives should be provided, nonmonetary rewards—such as the acknowledgment of individual team members' contributions to chat rooms, e-rooms or other project-relevant repositories of collective, task-relevant wisdom and lore—should be embraced as well. Believe it or not, very few employees are motivated solely by money (Brand 2008).

In the end, the fact that distributed, mobile, remote, alternative forms of work and work space strategies imply transient/transitional tasks with accompanying postural changes and frequent movement(s) suggests that many of the traditional guidelines for ameliorating the risk factors for musculoskeletal disorders (e.g., Brand 2008) may not apply, at least not for interactive tasks. However, many contemporary office environments confront their occupants with a number of challenges. For example, the assumption that more open environments support interaction and collaboration better than traditional private offices has not enjoyed empirical support; in fact, private offices

---

* The paper can be obtained from James L. Creighton, 480 Kenolio, 15-104, Kihei, HI 96753.

remain the preferred form of work space on many if not most indoor environmental quality (IEQ) dimensions (Kim and de Dear 2012).

Fanger's (2000) call to turn from merely eliminating the negative aspects of indoor environmental quality (IEQ) in order to concentrate on optimizing the built environment for human health and well-being continues to find adherents (Oseland and Burton 2012; Thompson, Veitch, and Newsham 2014). We may continue to hope that more and more constituencies in both the public and private sectors will recognize the value of strategic investment in high-quality, high-performance work places—not only to provide work life and task supports but also to delight their human occupants (Creighton 2014*). A recent review of the voluminous research literature on office work spaces and workplace strategies continues in this tradition of defining and measuring successful outcomes from the construction or the renovation of office environments in terms of the quality of human experience (Augustin 2015).

17.8 IMPLICATIONS AND CONCLUSIONS

Irrespective of any additional pertinent considerations, new ways of working and flexible workplace environments designed to nurture the employees appear to be salient trends in need of the thoughtful application of human factors and ergonomics design principles. More targeted research is needed to fully understand how each manifestation of these alternative workplaces and work styles influence employee attitudes and performance. However, the available evidence suggests mostly positive organizational outcomes from the implementation of these techniques, provided that the relevant tasks and activities comprise knowledge work that can easily be conducted remotely or at various transient locations. From an organizational design and behavior perspective, the definition and the evaluation of such work must shift from the time-on-task to the quality of work outcomes, and the group or the department results should be consolidated and rewarded at the team level. In most instances, individuals within these teams will identify social loafers who do not contribute to team performance. This approach to incentives will ensure that individual performers maintain an identity with their group/team and continue to contribute to their shared goals through individual efforts. Finally, it remains important to apply traditional human factors and ergonomics design principles to work space design in this age of increasingly global, mobile work for employees who sit for longer than 3 or 4 hours at a time regardless of where they might be working.

REFERENCES

Augustin, S. 2015. Applying what scientists know about WHERE and HOW people work best. Houston, TX: IFMA. Available at http://www.ifma.org/marketplace/store/products.

Bezrukova, K., S. M. B. Thatcher, K. A. Jehn, and C. S. Spell. 2012. The effects of alignments: Examining group faultlines, organizational cultures, and performance. *J Appl Psychol.* 97(1):77–92.

Brady, J. 2014, August 6. The search for innovative workspaces is just a waste of time: Here's why. *Washington Post.* From https://www.washingtonpost.com/news/innovations/wp/2014/08/06/the-search-for-innovative-workspaces-is-just-a-waste-of-time-heres-why/.

Brand, J. L. 2008. Office ergonomics: A review of pertinent research and recent developments In Carswell, C. M. (ed.) *Reviews of Human Factors and Ergonomics.* Santa Monica, CA: Human Factors and Ergonomics Society, 4(1):245–282.

Brand, J. L. 2010. Can we rescue the open-plan office via evidence? In Martin, C. S. and D. A. Guerin (eds.) *The State of the Interior Design Profession.* New York: Fairchild Books, p. 143–9.

Colquitt, J. A., J. A. Lepine, R. F. Piccolo, C. P. Zapata, and B. L. Rich. 2011. Explaining the justice-performance relationship: Trust as exchange deepener or trust as uncertainty reducer? *J Appl Psychol.* 97(1):1–15.

Creighton, J. L. 2011. Team formation in remote virtual teams. New Ways of Working. Santa Clara, CA: From http://www.newwow.net.

---

* The paper can be obtained from James L. Creighton, 480 Kenolio, 15-104, Kihei, HI 96753.

Creighton, J. L. 2014, April. Employee wellbeing and new ways of working. New Ways of Working. Santa Clara, CA: From http://www.newwow.net.

Danielsson, C. B., H. S. Chungkham, C. Wulff, and H. Westerlund. 2014. Office design's impact on sick leave rates. *Ergonomics.* 57(2):139–47.

Davis, M. C., D. J. Leach, and C. W. Clegg. 2011. The physical environment of the office: Contemporary and emerging issues. In Hodgkinson, G. P. and J. K. Ford (eds.) *International Review of Industrial and Organizational Psychology.* Chichester: Wiley, vol. 26, p. 193–23.

Denison, D. R. 1996. What is the difference between organizational culture and organizational climate? A native's point of view on a decade of paradigm wars. *Acad Manage Rev.* 21(3):619–54.

Denison, D. R., S. Haaland, and P. Goelzer. 2004. Corporate culture and organizational effectiveness: Is Asia different from the rest of the world? *Organ Dyn.* 33(1):98–109.

Duffy, F. (1999). *The New Office*, Second edn. London: Conran-Octopus.

Evans, G. W., and D. Johnson. 2000. Stress and open-office noise. *J Appl Psychol.* 85:779–83.

Fanger, P. O. (2000). Indoor air quality in the 21st century: Search for excellence. *Indoor Air* 10(2):68–73.

Fisk, W. J. 2000. Health and productivity gains from better indoor environments and their implications for the U.S. Department of Energy. LBNL 47458. Washington, DC: E-Vision 2000 Conference.

Frankel, M., and J. Edelson. 2015, May. Getting to outcome-based building performance (report from a Seattle summit on performance outcomes). Washington, DC: National Institute of Building Sciences. From https://newbuildings.org/sites/default/files/Performance_Outcomes_Summit_Report_5-15.pdf.

Gover, B. N., and J. S. Bradley. 2009. Guide for assessment of the architectural speech privacy and speech security of closed offices and meeting rooms (RR-276). Ottawa, ON: National Research Council, Institute for Research in Construction.

Hedge, A. 1982. The open-plan office: A systematic investigation of employee reactions to their work environment. *Environ Behav.* 14:519–42.

Hellwig, R. T. 2015. Perceived control in indoor environments: A conceptual approach. *Build Res Inform.* 43(3):302–15.

Hookway, B. 2009. Rules of engagement: Architecture theory and the social sciences. Summer. in Frank Duffy's 1974 thesis on office planning. Working Paper Series #39. Princeton, NJ: Center for Arts and Cultural Policy Studies, Princeton University.

Hysom, J. L., and P. J. Crawford. 1997. The evolution of office building research. *J Real Estate Literat.* 5:45–157.

IFMA (International Facility Management Association). 2007. Space and Project Management Benchmarks, Research Report #28. Houston, TX: IFMA.

IFMA (International Facility Management Association). 2009. Distributed work. Research Report #31. Houston, TX: IFMA.

Jahncke, H., S. Hygge, N. Halin, A. M. Green, and K. Dimberg. 2011. Open-plan office noise: Cognitive performance and restoration. *J Env Psychol.* 31:373–82.

Kellerman, B. 2012. *The End of Leadership.* New York: HarperCollins.

Kim, J., and R. de Dear. 2012. Nonlinear relationships between individual IEQ factors and overall workspace satisfaction. *J Build Environ.* 39:33–40.

Kim, J., and R. de Dear. 2013. Workspace satisfaction: The privacy-communication trade-off in open-plan offices. *J Env Psychol.* 36:18–26.

Lansons and Opinium. 2015, July. Britain at work. 1–46. From http://www.lansons.com; http://www.opinium.co.uk.

Leaman, A., and W. Bordass. 1999. Productivity in buildings: The "killer" variables. *Build Res Inform.* 27(1):4–19.

Leder, S., G. R. Newsham, J. A. Veitch, S. Mancini, and K. E. Charles. 2016, February. Effects of office environment on employee satisfaction: A new analysis. *Build Res Inform.* 44(1):34–50.

Lee, S. Y., and J. L. Brand. 2005. Effects of control over office workspace on perceptions of the work environment and work outcomes. *J Env Psychol.* 25:323–33.

McNeese, M., E. Salas, and M. R. Endsley. 2001. *New Trends in Cooperative Activities: Understanding System Dynamics in Complex Environments.* Santa Monica, CA: Human Factors and Ergonomics Society.

Newsham, G. R., J. L. Brand, C. L. Donnelly, J. A. Veitch, M. B. C. Aries, and K. E. Charles. 2009. Linking indoor environment conditions to job satisfaction: A field study. *Build Res Inform.* 37(2):129–47.

Oseland, N., and A. Burton. 2012. Quantifying the impact of environmental conditions on worker performance for inputting to a business case to justify enhanced workplace design features. *J Build Survey, Appraisal Valuation.* 1:151–65.

Pile, J. F. 1986. *Open Office Planning: A Handbook for Interior Designers and Architects.* New York: Whitney Library of Design.

Pritchard, R. D., M. M. Harrell, D. Diaz Granados, and M. J. Guzman. 2008. The productivity measurement and enhancement system: A meta-analysis. *J Appl Psychol*. 93(3):540–67.

Property Directors Forum. 2015. Maximising your business's value through workplace strategy. www.property directorsforum.com.

Raghuram, S. 2013, March. Organization culture and distributed work. New Ways of Working. Santa Clara, CA: From http://www.newwow.net.

Robbins, S. P., and T. A. Judge. 2014. *Organizational Behavior*, 16th edn. Upper Saddle River, NJ: Pearson/Prentice-Hall.

Rock, D., and J. Schwartz. 2006, May 30. The neuroscience of leadership. *Strategy + Business*. Issue 43, Reprint No. 06207.

Schadler, T., M. Brown, and S. Burnes. 2009, March 11. US telecommuting forecast: 2009 to 2016. March 11. From http://www.forrester.com/US+Telecommuting+Forecast+2009+To+2016/fulltext/-E-RES46635 ?objectid=RES46635.

Springer, T. 2011, February. Measuring work and work performance. New Ways of Working. Santa Clara, CA: From http://www.newwow.net.

Sundstrom, E., R. E. Burt, and D. Kamp. 1980. Privacy at work: Architectural correlates of job satisfaction and job performance. *Acad Manage J*. 23(1):101–17.

Thompson, A. J. L., J. A. Veitch, and G. R. Newsham. 2014. Improving organizational productivity with building automation systems. Ottawa, ON: Continental Automated Buildings Association. From http://docs .caba.org/documents/IIBC-WP-Improving-Productivity-BAS.pdf.

Veitch, J., and Gifford, R. 1996. Choice, perceived control, and performance decrements in the physical environment. *J Env Psychol*. 16:269–76.

Wesley, K. 2012. Women in nontraditional occupations: A case study of worker motivation (Paper 118). Lincoln, NE: DigitalCommons@University of Nebraska-Lincoln.

18 Challenges and Future Research Opportunities with New Ways of Working

Lynn McAtamney, Christine Aickin, David Caple, Carlo Caponecchia, and Martin Mackey

CONTENTS

18.1 INTRODUCTION

Changes in technology have created new ways of working that were previously not possible. This has contributed to people choosing to work in places other than at their desk and for the commercial occupancy of buildings to drop to sometimes less than 50%. This, and research about the impact of sedentary work tasks on health, are drawing attention to cost savings achieved by changing the design of the workplaces and the way work is done. Research indicates that the changes can enhance health and well-being; however, there are some complex elements to consider before a successful outcome is achieved (Chau et al. 2013).

The traditional office ergonomics design criteria used for the single-user workstation still have some application; however, they no longer provide adequate solutions for managing the risks in many of the new ways of working. This creates new challenges in evaluating and advising on the person-centered ergonomic principles as they fit with new work environments, technologies, work patterns, collaborations, supervision, and performance measures.

   This chapter outlines the challenges in introducing ergonomic design to new ways of working and the related research opportunities in future workplace design.

18.2   WHAT ARE "NEW WAYS OF WORKING?"

New ways of working are encompassed in the alternative work strategies, which have become more popular since the early 1990s, and are presented as options such as telecommuting, flexible work hours or arrangements by which people can work from home or other locations. Other strategies include agile working where team-based arrangements are made for specific projects or development work and coworking, which is popular in the UK and with sales representatives where people from different businesses share an office facility, using any available desk. Also gaining popularity is real-time working where a combination of cloud technology and meeting rooms with connectivity are used for collaboration producing documents and plans together.

18.2.1   ACTIVITY-BASED WORKING

One popular new way of working is known as activity-based working (ABW), a concept developed by a Dutch consulting firm, Veldhoen, in 1997. ABW facilitates the freedom for people to organize their work style and work location. It is based on the three pillars of people (behavioral environment), place (physical environment), and technology (including knowledge sharing); so at first glance, it has strong foundations in ergonomic design (see Figure 18.1).

   The planning and design of ABW sees the creation of spaces for specific task needs so that people become mobile moving from one space to another (where people work), enabled by technology (how people work) and a management culture that empowers when they work and trusts them to work in this different way (when people work). Typically, people are provided with a laptop computer, a "follow-me" printing access, a fully mobile phone system, a wireless technology, and whatever personal items they require for the day. These changes result in a worker no longer needing their own workstation to access all their technology requirements.

   The allocation of workstations and storage for office workers has often been based on industrial agreements such as the amount of floor area per person or the size of the office depending on seniority. This entitlement-based design model has completely changed with the new workplaces. The

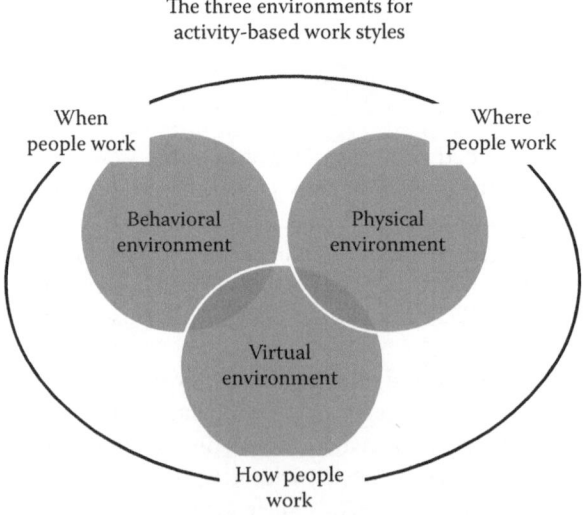

FIGURE 18.1   The three environments necessary for activity-based work styles.

focus is on the collective needs of all the workers to have a range of work points to suit the activities they need to undertake.

With the advancement of digitization technology, there is a significant reduction in the use of paper-based storage of information and an increased use of electronic storage. There are processes required for scanning important hard copy materials that need to be converted to a digital image prior to the commencement of the new ways of working. That workstations no longer need to be clustered close to the document storage areas results in greater flexibility of where people could go to work.

With more state, national, and international companies operating on organizational structures involving virtual teams, where people work together as members of a team but are working from different locations, ABW can enhance the work experience with communication technologies being incorporated into the office area. These include more opportunities for teleconferencing and video conferencing to enable the teams to effectively function.

Rather than communicating from a fixed desktop telephone, ABW requires the telephone systems to operate through the computer processes or to utilize mobile telephone technologies. Through the introduction of the new telephone systems, the opportunities to communicate as a virtual team can happen at multiple work locations.

There are groups of workers who have specific access requirements to their work environment. These include those in wheelchairs, with sight impairment, as well as those with temporary disabilities due to injury. Internationally, there are legal processes in place to ensure that the workplaces are designed for the access needs of persons with disabilities, for example, the Americans with Disabilities Act 1990 in the United States (U.S. Department of Labor 1990), the Australian accessibility and mobility standard (AS 1428.1 2001), or the UK Disability Discrimination Act (Her Majesty's Stationery Office 1995). Consideration is also needed for pregnant women and those with long-term physical or psychological illness. ABW provides opportunities for a wider range of work environments to be available. This enables the individuals to select specific workplaces that meet, not only their task requirements, but also their personal needs. ABW has been heavily promoted by corporate real estate, facilities management, and architects who are able to demonstrate the effective return on investment from space utilization. Although the cost is not the only factor in implementing ABW, it remains a key one. Ouye et al. (2012) conducted a benchmarking survey of alternative workplace practices and found that 76% of 143 organizations reported cost savings as the top driver, followed by employee work/life balance (66%) and productivity improvements (64%). The top three nominated barriers to implementing alternative workplace practices were found to be culture (65%), manager concerns (59%), and resistance to change (55%). The respondents to the survey, who were mostly from the corporate real estate sector, did not recognize funding as a major barrier, as it was only eight on the list (Ouye et al. 2012).

18.3 SUCCESSFUL IMPLEMENTATION OF ABW

The successful implementation of an ABW workplace requires coordination among IT, human resources (HR), and properties/real estate departments. Through the design and the development of ABW workplaces, a team of change managers are needed. Ergonomists are particularly useful additions to this team, as they provide a broader perspective on the risks that should be considered (see Figure 18.2).

Additionally, ergonomists with research skills are able to determine the return on investment through the evaluation of the impact on the people, the environment, and the productivity of the organization. Such research can contribute to measures called the triple bottom line—a way of determining the returns to and impact of an organization on its people, the planet, and their profits (Elkington 1994).

The culture of an ABW workplace is based on the assumption that workers are supportive of a work environment where they do not have an allocated base desk or workstation and work independently or with their team members to utilize different work settings. The ABW culture is based on an egalitarian management approach where all workers, including the managers, are in the same situation without an allocated space. Some companies make exceptions for individuals with special needs including managers who may be allocated specific areas of a floor to use.

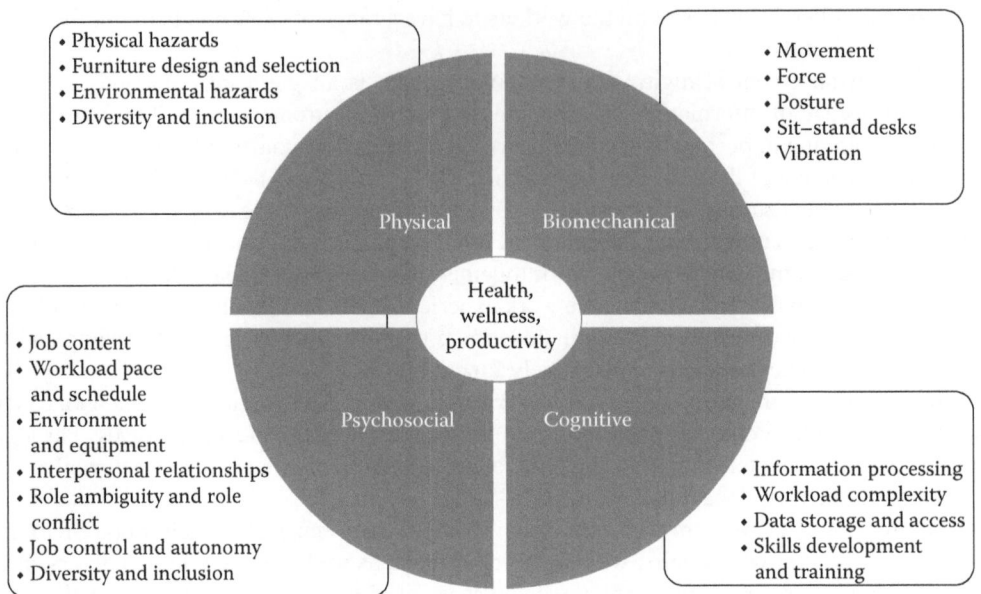

FIGURE 18.2 Elements considered by ergonomists in the design of alternative work spaces.

Mosselman et al. (2002) conducted a longitudinal study where after 6 years, workers reported a general improvement in the look, the layout, and the opportunities to communicate and socialize. However, they stated that attention was required regarding indoor climate, filing, storage, and privacy. There is an expectation that the workers will interact more and develop a stronger team-based culture with ABW. Further research is required to evaluate the relative advantages and disadvantages of these arrangements.

The basis of evaluating productivity with ABW is on measuring the work achievements or the outputs rather than the hours spent at the workplace. With this, there is greater accountability on the individual to organize their time and to select the most appropriate work location to complete their work duties. Productivity can be negatively impacted if the speed of logging onto computers is slow or if the worker is unable to quickly locate work colleagues when consultation and collaboration are required.

Successful implementation (Veldhoen + Company 2014) has established benefits that improve the following:

- Organization focus (e.g., improved collaboration)
- Worker focus (improved work satisfaction)
- Client focus (improved image)
- Financial focus (improved productivity and decreased costs)

For example, Armstrong (2013) found that ABW was influential in attracting new workers to an organization.

18.4 STAGES IN ESTABLISHING NEW WAYS OF WORKING

Taking the perspective of an ergonomist working within a company or as a consultant, let us look at the stages of engagement for a business establishing new ways of working. Understanding the business' capacity for change is a strong defining factor for which stage they are at and the different challenges and research opportunities that may be present.

18.4.1 STAGE 1—MINOR CHANGE OFTEN IN RESPONSE TO IMMEDIATE RISKS AND STRONG COST FOCUS

Have you been in the situation where there is little or no budget, however, the organization wants to explore changes in response to incident data, absenteeism, or to explore flexible work arrangements? Essentially, such organizations have not yet embarked on the road to new ways of working, yet this is an important opportunity to educate them while addressing the risks and the issues within the cost constraints.

Preparing practical, evidence-based workplace policy and ergonomic guidelines is an important basis on which to build new ways of working. These must embrace the psychosocial and cultural features of healthy work environments and provide education and early intervention components as well as reference to the injury management processes in place. This may also include equipment, furniture, and fit-out standards that define what the acceptable ranges are for purchasing personnel. Standards, such as workstation standards in AS/NZS 4443 (1997) for office desks and AS/NZS 4438 (1997) for height adjustable chairs, may provide for specific work environments (home, call center, etc.) or more flexible work environments, recognizing and including the physical workplace, how best to use the spaces to reduce physical and mental demands/risks along with supervisory and work organization components. National and international standards may provide technical information; however, as technology advances so rapidly, the information may not be up to date. In a 2014 review, the Canadian Standards provided the most useful information (CSA-Z412-00 [R2011] 2011). However, there is certainly a need for further research in the area of integrated design and ergonomics.

Do you remember the fixed-work posture diagrams of the 1980s and the 1990s with all joints at the 90° correct angle? Now, research (e.g., Dunstan et al. 2012; Katzmarzyk and Lee 2012) has demonstrated the benefits for light and frequent activities including more standing and walking instead of sitting. Research is moving from correct postures to movement. However, this brings some challenges in the work environments seeking minor changes such as the introduction of some sit–stand desks or standing meeting rooms. Even these changes require the endorsement from the leadership, the integration into the culture of the workplace, and the education in how to benefit. From a musculoskeletal health perspective, prolonged standing is no better than prolonged sitting.

As yet, there are no evidence-based guidelines on the time before changes in posture are required, so that workers can still sit or stand for long periods. Given that the workers can choose to use a laptop for long periods in work spaces not set to enable a good posture, a further risk of musculoskeletal fatigue may be introduced. Some guidelines are provided by regulators such as Comcare.* Professional associations such as the Human Factors and Ergonomics Society of Australia are also developing guidelines.

Another research challenge is in defining the triggers to encourage movement and activity along with the education that encourages movement while working for extended periods in the same work location. Healy et al. (2012) undertook a comprehensive review of the research regarding work breaks. They reported that break times varied (all within the hour); however, the data were short term with no follow-up. The prompting of breaks is also a consideration for further research, because they may be more disruptive than assistive.

Some organizations are purchasing sit-to-stand units placed either on existing fixed-height desks or as a new work desk, along with appropriate work chairs and equipment, such as laptop or monitor arms/stands. There is limited success with this strategy unless workers are engaged and motivated to use the equipment and the culture of the office encourages sit to stand movement as part of their daily work (Wilks et al. 2006). Other low-cost initiatives include ensuring people take lunch breaks away from their desk. In the UK, a Bupa survey found that less than 30% of workers take an hour of lunch break despite the benefits for health and productivity from having frequent breaks (http://www.bupa.com/media-centre/press-releases/uk/take-a-break/). Walking meetings similarly have some success where technology is not required (Oppezzo and Schwartz 2014).

---

* Comcare is the workplace health and safety regulator for Australian government organizations (Comcare 2014).

The advice provided by Comcare indicates that prolonged sitting poses significant risks to the health of workers and to the organizations for which they work (Chau et al. 2011; Thorp et al. 2012). Sedentary behaviors involve prolonged sitting or reclining, resulting in little or no physical activity energy expenditure (Biswas et al. 2015). As working Australians spend most of their adult life in the workplace, it is important to measure and implement changes to reduce sedentary behavior. As with other first-world countries, many Australian workers spend 76% of their waking time per day seated. This can increase the blood sugar levels and the risks of cardiovascular disease, and T2D (Chau et al. 2013). Ciccarelli et al. (2014) found that office workers using electronic-based tasks had more neutral postures, with less postural variability, which increases the risk of musculoskeletal injury risks. To address the risk factors of sedentary work, several strategies have been recommended (Comcare 2014), including the following:

- Vary work tasks throughout the day so that there is a change in posture and different types of muscles are used or alternate between sitting and standing, e.g., find a reading area that requires you to stand.
- Ensure a standing-friendly culture is promoted and supported, e.g., have a regular standing agenda item and encourage workers to stand during meetings.
- Use a height-adjustable desk, so the workers can work either standing or sitting.
- Encourage managers to be role models of standing behaviors and regular movements.
- Use *I-mails* instead of an e-mail, where the person walks to their colleague and talks to them ("I deliver").
- Encourage light activity such as walking to a printer further away.
- Encourage the workers to move during breaks, e.g., eat lunch away from their desk.

18.4.1.1 Steps to Recommending Office Furniture and Equipment

1. Expert review of office furniture and equipment against standards

    Over the last decade, the variety and design of office furniture and equipment including workstations, desks, chairs, and screen raisers has significantly increased in response to the ABW. However, the Australian design standards for workstations, desks, and chairs, which were most recently updated in 1997, have not kept pace with these design improvements and remain prescriptive in approach (AS 1428.2 1992; AS/NZS 4438 1997; AS/NZS 4442 1997; AS/NZS 4443 1997).

    When assessing the suitability of workstations, desks, and office chairs, these standards do not provide sufficient guidance, and there is no Australian design standard for liquid crystal display (LCD) monitor arm or laptop risers. These standards are based on the use of cathode ray tube monitors and do not address the specific requirements for the use of laptops or LCD monitors at workstations. The same ergonomics principles relating to posture and sightlines are the same regardless of the technology, but the means of achieving this using monitor arms and adjustable laptop stands are not covered in the existing Australian Standards. Ergonomists working in this field need to have a good working knowledge of the office furniture and equipment that are available and the assessment tools to assist with choosing between the different brands, in line with the organizational needs.

    With many workers now spending time in a number of different work spaces, it is also important to have input into the selection of equipment and the design of different work areas, such as lounges, occasional chairs, and kitchen furniture, for example. Any non-workstation furniture also needs to have an ergonomic assessment to ensure that it suits the requirements of that workplace. The design of a couch or an armchair in a workplace is very different in a home application. Such work settings need to enable the workers to easily sit down and get up from short-term use compared to a home setting which may be used for longer periods and where people may be in more reclined postures on casual furniture. Generally, furniture retailers offer domestic seating that is too low and deep and

lacks adjustability, which is in conflict with the needs for a workplace application. Hedge et al. (2002) found that home working increases the risk of musculoskeletal disorders.
2. Office furniture and equipment user trials

Along with an ergonomic expert assessment, it is necessary to involve the worker in user assessments to assist in the decision-making around the furniture and equipment selections. This can involve a small group of workers who are representative of the workplace (say male and female in ranging stature and build, through short to tall and slim to obese, respectively). The consideration of disabled users is also necessary. The workers should be asked a range of questions around function, support, and comfort. Asking open questions such as "What do you like about this chair?" and "What do you dislike about this chair?" provides a wealth of information to assist in making the final decision about which brand to recommend for purchase. An example of criteria that might be used is provided in Figure 18.3.

Seated workstation evaluation

Date	Click here to enter a date.	Lab ID	
Occupation		Workstation #	

Function and support

1. How comfortable is the workstation for all the tasks you do?	Click here to select
2. How functional is the workstation for all your tasks?	Click here to select

Body comfort

1. Shoulders	Click here to select
2. Arms and hands	Click here to select
3. Upper back	Click here to select
4. Lower back	Click here to select
5. Buttock	Click here to select
6. Thighs	Click here to select
7. Lower legs and feet	Click here to select
8. Overall comfort	Click here to select

What do you like about this workstation?

What do you dislike about this workstation?

FIGURE 18.3   Example of the questions in a user trial of seating (administered after using the seat for 3 days).

18.4.2　Stage 2—Moderate Change to the Space, the Technology, the Work Process, or the Culture Perhaps with One Alternative Work Strategy

Moderate changes, as opposed to Stage 3, which encompasses a significant redesign, may occur in one team or building. For example, a new executive may want to introduce collaborative spaces and standing meeting rooms in his area where there was a significant amount of teamwork undertaken. Having an accommodation policy and ergonomic design standards in place to provide ergonomic criteria assists in projects and initiatives at this stage.

A clear understanding of what the organization wants to achieve with the changes is required. For example, they may want to achieve the following:

- Enable distributed teams or cloud workers.
- Communicate the culture of the organization by including social and wellness spaces.
- Provide more collaboration spaces.
- Use IT and wireless platforms for better communication and/or "bring your own device."
- Ensure that multigenerational workers are more comfortable.
- Accommodate diversity in user's physical needs, anthropometry, health, and well-being.
- Enable specific task purposes.
- Enable frequent movement.
- Enhance cognitive and sensory function in users.
- Offer workers flexible work options and trust them to be responsible when working away from the desk (e.g., café or home).

The need for a stable IT platform to provide any changes to communications technology will take precedence for the workers over the physical environment because the technology is crucial for their work. Working with senior management, HR, IT, and properties/facilities stakeholders to understand the expectations and the necessary budget is critical in the success of introducing elements of an ABW environment.

People are naturally unsure of change until they are clear about "what is in it for me," so the consultation process in establishing change and education on how to use the new work space is also critical for success. Not everyone will see the change as a normal process and be on board with the new environment. For those with specific physical or sensory needs, an assessment and support process needs to be in place for the managers to refer their people to when these needs are identified. These workers can often be well accommodated, as the ABW environment is flexible by nature. For example, a worker with back pain has more options to stand and sit, while a person with light or noise sensitivity may be able to move to a different area of the office.

However, the change can be a particular struggle for some workers, and preparation is needed to assist them and avoid the following coping behaviors observed in some ABW environments where such issues are not addressed early enough in the transition. Such behaviors include the following:

- Reserving: Reserving is putting post-it notes on an equipment to say that it is broken—a way of reserving the desk.
- Bagsing: Bagsing is putting a jacket on a chair or a bag on a neighboring desk to save it for a friend.
- Being an early bird: Being an early bird is getting to work even earlier to get a desk by the window or other favorite areas.
- Squirrelling: Squirrelling is when the workers find spaces to leave or store items, sometimes colonizing the team storage.

- Pulling rank: Pulling rank is using seniority or other leverages to move the workers and secure the preferred spaces.
- Hiding: Hiding is being in the building but dialing into meetings on the pretense of being off site—thus enabling multitasking such as doing e-mails while listening to the meeting.

Further consultation and research on how to transition people through the change and avoid these coping behaviors would assist the implementation of new ways of working.

One area requiring more definition and guidance is in using laptops in casual work settings such as the home or the cafe environment. The cumulative risks caused to the neck, the shoulders, the arms, and the wrists from prolonged use of laptops in these locations can be estimated by tools such as the RULA developed by McAtamney and Corlett (1993). The "Manual Handling Codes of Practice" provided by the Australian occupational health and safety regulators (Safework 2011) recommends 30 min as a guide to limit the use of equipment that does not provide the appropriate ergonomics postures. Hence, using these compact technologies, including laptops, iPads, and tablets, should be limited in duration unless a docking station or similar is provided.

18.4.3 STAGE 3—SIGNIFICANT CHANGES OR A COMPLETE REDESIGN OF THE BUILDING, THE WORK SPACES, THE WORK PROCESS, AND THE TECHNOLOGY

The redesign of organizations to new ways of working is a significant business decision. Apart from defining the future for that organization, success requires significant numbers of processes and cross-business collaboration. To define the vision requires the implementation of the following steps:

- Researching other organizations' success in new ways of working and defining what has relevance for the business under change (learning from others using consistent criteria provides valuable insights)
- Building an understanding of the culture, the technology needs, the work patterns, and the vision of the business through a broad consultation process at the different levels of the organization
- Understanding the people and how they use the work environment and what their fears and aspirations are through a range of workshops
- Building a mock-up or a pop-up of a space that allows people to experience the proposed changes, along with some suggested new furniture, which will enable their further understanding and feedback
- Consulting with people with specific workplace requirements such as those needing wheelchair access or vision-impaired facilities
- Recognizing subgroups who need specific consultation such as senior leaders or their personal assistants who may feel that they are "losing" an office during the change
- Building a relationship with the manufacturers and the suppliers to provide furniture and equipment that meet the specific needs and the best practice for success
- Developing a workplace strategy based on a well-researched understanding of the functional design needs of the organization
- Building and using a "living lab" where people come and work in a fully functioning area for several weeks, providing an opportunity to test the technology, the work environment, and the induction needs, and modifying the design from the user's responses
- Informing senior leadership of the outcomes and the recommendations for their endorsement
- Ensuring that the transition into the work space is enabled by engaging education and information about the use and the adjustability of the space to meet personal needs
- Creating a living lab that is an opportunity to research the physical and psychosocial responses of people as they are immersed in the new technology, work environment, and ways of working

18.5 ASSESSING PHYSICAL ACTIVITY RESPONSES TO ABW ENVIRONMENTS

ABW-designed environments may provide opportunities for health-enhancing reductions in sitting time and complementary increases in light-intensity activity (standing and walking) at work (Appel-Meulenbroek et al. 2011), but such laudable goals need to be evaluated. A recommended methodical approach is to first measure the time-dependent work postures and the activity in the workers' standard office environment prior to transitioning to the ABW office environment to establish a baseline. A second measure should be conducted after 2–3 weeks of accommodating to the new way of working in the living lab. Further measures 6–12 months after occupying the new ABW environment will determine if any initial health-enhancing activity changes have been sustained (Radas et al. 2013; Chau et al. 2014).

Several validated self-report and objective tools are available to measure the work posture behavior. However, it should be acknowledged that while self-report tools are easier and less costly to administer and evaluate than some objective tools, human nature is such that we tend to underestimate how much time we sit and overestimate our participation in light–moderate physical activity (Sallis and Saelens 2000; Prince et al. 2008; Marshall et al. 2010). Nonetheless, self-reported sitting time can be established using either the occupational sitting and physical activity questionnaire (OSPAQ) (Chau et al. 2012) or the workforce sitting questionnaire (WSQ) (Chau et al. 2011). Both instruments are acceptable measures for the assessment of the sitting time at work with high test–retest reliability, and they correlate well with objective measures of sedentary behavior. OSPAQ has the advantage that it measures standing and walking times in addition to sitting time, while WSQ measures sitting time throughout the working day in multiple domains, including in transport, in leisure time, and in workplace. Including sitting time in extended domains may be important because modern mobile technology permits work activity to continue beyond the standard office hours. In addition, there is some evidence that workers may compensate for less sitting at work by more sitting in leisure time (Alkhajah et al. 2012), potentially negating any ABW-associated health benefits.

The broad availability of powerful portable devices (e.g., smartphones, tablets, wearable sensors) offers great opportunities for IT-enhanced solutions aimed at promoting incidental physical activity in sedentary work environments and linking these benefits to nonwork hours. These technologies vary in cost, availability, and expertise required for evaluating the data outputs. Currently, research is underway to validate these IT activity measurement tools for use in the workplace.

Evidence on the health benefits of ABW-related office chair-sitting reduction is scant and inconsistent and is mostly derived from a small number of short-term laboratory-based studies (Dunstan et al. 2012; Bailey and Locke 2014). Little attention has been paid to potential unintended or adverse effects of working in an ABW-type environment, for example, the potential effects of prolonged standing on fatigue, on musculoskeletal discomfort, and on sleep quality—all of which may influence the adherence to incidental physical activity in an ABW environment. Therefore, it is also important to evaluate the frequency and the extent of any musculoskeletal complaints when transitioning to a new way of working such as ABW. The evaluation can be achieved by reference to work injury data such as sick leave or workers compensation claims. However, there is a lag time in the access to such hard data and sole reliance on it may underestimate the true extent of musculoskeletal complaints. Several self-report measures are readily available for use including a discomfort visual analog scale and the more extensive Nordic Musculoskeletal Questionnaire (Kuorinka et al. 1987), which measures the prevalence of discomfort over both the last 7-day and 12-month periods and has been shown to have acceptable reliability and validity in workplace studies (Palmer et al. 1999; Descatha et al. 2007; Griffiths et al. 2012). The measurement of how well the workers perceive their health, skills, and experience match the demands of their job in a new ABW can be measured by instruments such as the work ability index questionnaire (Tuomi and Oja 1999), which has been validated for use in a number of workplace studies (Mackey et al. 2007, 2011; Gould et al. 2008).

18.6   ASSESSING THE PSYCHOSOCIAL IMPACTS OF ABW

Changes to the physical work environment, through activity-based work and other new ways of working, inevitably change the social work environment and the ways in which work is performed, monitored, shared, and evaluated. Together, this can affect the antecedents to psychosocial risks.

Psychosocial hazards are defined as

> aspects of job content, work organization and management, and environmental, organizational conditions that have the potential for psychological and physical harm. (Cox 1993; cf. McKay et al. 2004)

Exactly what events and experiences constitute psychosocial hazards (often just termed *psychosocial issues* at work) varies depending on the level of description and context in which the term is used, but Leka and Cox (2008) note that typically this includes issues such as

- Job content—variety, use of skills
- Workload, pace, and schedule
- Control and autonomy
- Environment and equipment
- Relationships and interpersonal interactions
- Role ambiguity and role conflict

The issues that can affect such hazards, relevant to ABW programs, include the changes to the perceived levels of privacy (cf. Rashid and Zimring 2005); the notions of space ownership or territoriality (Brown et al. 2005; Brown 2009), identity, and personalisation; the availability and the accessibility of equipment or desirable situations (e.g., lighting, views) at critical times; and noise (cf. Vischer 2007 for a review).

How these factors affect psychosocial issues, positively or negatively, in any particular case, likely depends on wider organizational variables, independent of the activity-based work program, for example, the culture and the relationships present in the work group over time. The existing job design, supervision, and performance management arrangements will contribute to the nature of any psychosocial ill effect. This not only makes isolating the consequences of an ABW program more difficult, but also reflects how the success of any ABW program (or indeed any safety or ergonomic intervention) also depends on wider systemic factors.

The following example highlights how ABW-related changes could have a range of different effects for different stakeholders in particular contexts.

In an ABW scenario, the workers can have greater access to senior managers because those managers are not behind closed doors nor accessed through the physical and social barriers of professional and technical assistants. At the same time, the workers can be more easily monitored by the senior staff, who may stand next to them as they perform particular tasks. Depending on the situation, for the worker, this could evoke feelings of being excessively monitored, or conversely, of being better supported. For the manager, this could lead to them feeling overworked with tasks with which they would not normally consider, just because they are now more accessible to others. Alternatively, they could feel that they are more collaboratively involved in the work tasks. It is conceivable that the situation portrayed earlier, if extended over time, could affect workloads, relationships, and role ambiguity thus potentially building to wider psychosocial outcomes.

In either case, more than just a rearrangement of furniture has occurred: the work tasks and the organizational climate have fundamentally changed. Consequent effects on psychosocial variables, negative or positive, can be expected in the medium and longer term.

While it is generally noted that most studies comparing open-plan offices to traditional workspaces find negative effects on employees in terms of noise exposure, stress, perceived productivity, and satisfaction, and no real benefits in terms of cooperation (cf. Brennan et al. 2002; Oommen et al. 2008;

Kaarlela-Tuomaala et al. 2009), there are some contradictory findings in this area (cf. Evans and Johnson 2000; Smith-Jackson and Klien 2009). ABW is arguably more complex, however, with the manner in which the work is performed being transformed, rather than just having the same work performed without walls. The possibilities of positive and negative effects on productivity and satisfaction exist, and are mediated by the quality of program implementation, consultation, training, equipment, and resources. Longer-term in-depth evaluations of ABW, similar to those that have occurred for open-plan office spaces, will be needed across industries. This will help establish whether activity-based work generally results in better work, from psychosocial, ergonomic, and productivity points of view, while controlling for potential confounds of organizational culture, work tasks, and job design, and accounting for quality of program implementation (e.g., consultation, training, resourcing, etc.).

Individual organizations are likely to want shorter-term evaluations which are specific to their own activity-based work program, and which help check for any risks going forward.

Choosing a suite of measures appropriate for an organizational evaluation as opposed to those for ongoing research purposes is not easy. Which indices are included will depend on the research questions and the project scope. The time taken to provide data at a sampling point (e.g., through survey responses, physiological samples, recording activity or location, etc.), as well as the number of sampling points at which data will be collected, is a major constraint for the research and the evaluation performed within organizations. The methods and the measures need to be selected and planned with this practical constraint in mind. Providing nonwork time for employees to participate in data collection may need to be considered. In general, the integrity of these data is improved with a degree of independence between those collecting the data and the organization. A lack of negative effects could just reflect a response bias due to not wanting to report being unproductive, or negatively affected, to one's organization.

One strategy would be to measure the outcomes of psychosocial hazards (e.g., levels of stress, depression, anxiety, using the depression anxiety stress scale described by Lovibond and Lovibond [1995], the state-trait anxiety inventory developed by Spielberger et al. [1983], and the general health questionnaire produced by Goldberg [1978]), along with organizational data that could be consistent with those outcomes (such as sickness and absenteeism, use of employee assistance programs or support services). It can take some time for these outcomes to become apparent, and they may need to be quite severe or extreme in number before they appear to identify a new problem associated with the activity-based work program. The outcome measures over a short duration may identify a halo effect, where a novel, bright new work space makes people feel good about their jobs. The job affect well-being scale may also be useful, as it is specific to the emotional effects at work rather than measuring how people are feeling in general (Van Katwyk et al. 2000). Survey-based measures are likely to be the most frequently used for these purposes, but wider evaluations interested in stress and health effects may consider the physiological indices (such as blood pressure, heart rate, salivary cortisol, or immunoglobulins; cf. Danna and Griffin 1999; Evans and Johnson 2000; Maina et al. 2008; Richardson and Rothstein 2008).

Collecting other non-health-related variables may be relevant and desirable, but these are not always specific to work environment changes. Job satisfaction, for example, can be measured before and after moving to an ABW arrangement, but even if it is measured over a long duration, it may be affected by other aspects of work, including opportunities for career progression, rewards, and employees' own values (cf. Paton et al. 2003), in addition to the physical environment and how this might influence the social aspects of work. It may be more useful to include the measurement of organization- and job-related variables (e.g., job satisfaction, job commitment, perceived organizational support, intentions to leave, etc.) in a wider evaluation of an ABW program, rather than in the evaluation of an individual organization's program.

A lack of standardized, well-validated scales for assessing the ratings of the work environment has been noted as a problem in the field (Veitch et al. 2007). Accordingly, planned evaluations of ABW programs need to be informed of what specific research instruments are available. Ongoing assessment of the scales should also be attempted. The examples of scales which index the satisfaction with the environmental elements of the workspace include the 37-item physical work environment satisfaction questionnaire (Carlopio 1996) and a 16-item questionnaire developed by Kim and de Dear (2013) based on indoor environment quality dimensions from the postoccupancy questionnaire by the Center for Built Environment at the University of California Berkeley (cf. Zagreus et al. 2004).

More local issues can be assessed through the ratings or the perceptions of traditional compared to ABW workspaces, in terms of the issues outlined earlier (noise, privacy, lighting, effects on communication, and social interactions). These may be particularly useful for individual organizations seeking to evaluate their particular ABW program in its early stages, or as an adjunct to risk identification. However, plans should be made for these kinds of evaluations to be performed again, later, when the space is no longer new, and when people have relaxed into the modes of using the space, which may not be consistent with the design intentions. Knowing how people use the space, for example, what tasks tend to be performed in particular zones, whether particular zones are more used than others, and the times at which the zones are used, is important for assessing the problems that may occur, and for designing and implementing appropriate interventions (e.g., training, consultation, workflow redesign). Accepting that work is fundamentally different, and changing, in ABW scenarios may lead to an ongoing and updated analyses of task and space matching. In such an analysis process, the manner in which the tasks are performed, where they are performed, and the frequency with which they are done in certain areas would be observed, in addition to regular physical ergonomic assessments of equipment use. Because work is more dynamically performed in ABW, with more movement and interaction between people and spaces, the more holistic assessment of how the work tasks may affect the whole work group may need to be considered. ABW thus requires not only a change in how work is performed, but also a consequent tailoring of the methods used to identify the risks. This kind of analysis informs risk assessments, ensures appropriate provision of resources to facilitate productivity, and reduces the potential for any negative psychosocial impacts.

18.7 HOW TO EMBED ERGONOMICS INTO FUTURE WORKPLACE DESIGN?

Ergonomics has been successfully integrated into design when the environment and the work organization meet the workers' fit and functional needs without them being conscious of it. How is this sustained in new ways of working? We return to the notion of people, place, and purpose and that of "change is constant." Where there is adequate flexibility in the workplace design and ways the organization can evolve through collaborative processes, the adaption to new purposes is possible for the people involved. There are internal and external drivers that enable people to continue to thrive in changing workplaces. Success is not measured in the worker who is immediately comfortable with change and the challenges of new work environments. It is measured in enabling all the workers, particularly those who do not like change, to be engaged and productive with the new ways of working.

18.7.1 Internal Organizational Drivers

As with the rest of the business, success is driven by the culture and the leadership that enable the workers to embrace the change and feel supported in the new systems and ways of working. If managers cling to their private office space for the prestige rather than the purpose, the success of an ABW program will be compromised. If the business fails to adequately plan the investment needed, particularly with respect to the IT requirements, then damage can occur to the concept both now and in the future. It is better to have a successful, staged approach than to rush into a full implementation that may be at risk of introducing behaviors and attitudes that are negative. If the managers are not upskilled in managing virtual teams or using performance measures that reflect ABW principles, the effectiveness is also likely to be compromised. How an organization anticipates and manages the processes to incorporate all workers into the change process will greatly influence the success of new ways of working. Given that change is a continuum in many organizations, this is an ongoing program of review and refinement, especially when there are changes to the structure or the function of teams.

18.7.2 External Drivers Aiming to Embed Ergonomics into the Workplace Design

In recent years, a system of Green Stars has been introduced that enables ergonomics to be embedded in the design process. The Green Building Council of Australia (GBCA) promotes a national, voluntary sustainability rating system for interior fit-outs, buildings, and communities through a rating system known as Green Star (GBCA 2015a). One of the rating tools within the Green Star rating system is Green Star—Interiors (GBCA 2015b), which focuses on the interior design of spaces within buildings. The original version of this rating tool only addressed the sustainability of the office interior fit outs, yet was significantly revised in 2012 to cover all types of interior fit-out projects as well as a number of new credit categories. A "credit" within this rating system is a set of criteria and requirements that outline the best practice benchmarks. Project teams that demonstrate that they achieved these benchmarks are awarded a point, which contributes to an overall Green Star rating. One of the newer credit categories is the option to address ergonomics and occupant comfort through an ergonomics credit (GBCA 2015c), developed in consultation with the industry and led by the GBCA (Aickin and Pollard 2013). Outside of Green Star—Interiors, this credit may be used in other Green Star rating tools via the innovation category, as an innovation challenge (GBCA 2015c). The existence of this credit has begun to increase the use of ergonomists by organizations wishing to achieve the kudos of a Green Star rating.

The ergonomics credit requires the furniture and the equipment to be reviewed in areas where it is planned that employees may work for more than 1 to 2 hours per day. These areas can include quiet areas, collaborative work areas, meeting spaces, and more traditional computer-based workstations along with task-specific workstations such as reception/concierge areas.

At this stage, the ergonomics credit is limited in its application, as it is most easily applied to office areas or other areas where there is a well-defined work setting. It should be possible to additionally apply the credit to laboratories, workshops, healthcare situations, and industrial projects in the future with some further development of the ergonomic credit criteria.

For fit-outs below 2000 m², the project team can use the guidance provided in "Performance Oriented Checklist for Computer VDT Workstations" produced by Cornell University (2015) (http://ergo.human.cornell.edu/cutools.html) or a recognized standard in nonoffice fit outs.

In fit-outs of over 2000 m², the GBCA has identified that only ergonomists with the professional grading of Certified Professional Ergonomist (CPE) of the Human Factors and Ergonomics Society of Australia are able to do this work (http://www.gbca.org.au/green-star/green-star-interiors/the-rating-tool/).

Engaging a CPE must occur at the schematic design stage, and when the fit out is constructed, the ergonomist will verify that the advice they provided was delivered in the fit out. Additionally, the project must provide ergonomics information to the users. This can be done in a variety of ways including information guide, worker training, and induction where it must address the furniture adjustment to enable acceptable working postures.

The GBCA requires a submission template to be filled out, along with the provision of supporting evidence, describing the process undertaken in the fit-out design, the selection of furniture, and the confirmation that the ergonomics information has been provided to the users to enable it to assess the ergonomic credit as part of a Green Star—Interiors rating. This information is then used to assess whether the applicant is successful or not in gaining the ergonomic credit as part of their overall rating.

It is hoped that over time, this external driver will further encourage the integration of ergonomics into the design process of not only offices but also other more specialized work areas.

18.8 CONCLUSIONS

New ways of working have opened up great opportunities for flexible work, organizational benefits in costs, and productivity, as well as potential for increased movement and reduced sedentary time at work. The key challenges in implementing effective ABW systems include investing in infrastructure and equipment, managing and consulting on change, and providing well-tailored ergonomic solutions in the work space. Ongoing multidisciplinary research needs to monitor the possible benefits of ABW in encouraging the movement at work, as well as the medium- and longer-term psychosocial impacts of these redesigned work systems. Ergonomists are in a strong position to influence new ways of work, from design to consultation, implementation, and ongoing evaluation of their health effects.

REFERENCES

Aickin, C., and B. Pollard. 2013, November. Green star interiors rating tool—Ergonomic credit: What should CPE's know? *Proc 49th Human Factors and Ergonomics Soc Australia Conf.* Perth.

Alkhajah, T. A., M. M. Reeves, E. G. Eakin, E. A. Winkler, N. Owen, and G. N. Healy. 2012. Sit-stand workstations: A pilot intervention to reduce office sitting time. *Am J Prev Med.* 43:298–303.

Appel-Meulenbroek R., P. Groenen, and I. Janssen. 2011. An end-user's perspective on activity-based office concepts. *J Corp Real Estate.* 13(2):122–35.

Armstrong, T. 2013. Using Activity based work to attract and retain the best talent. *Keeping Good Companies.* 65(7):439–42.

AS 1428.1. 2001. Design for access and mobility: Part 1: General requirements for access—New building work. From http://www.saiglobal.com/pdftemp/previews/osh/as/as1000/1400/nn14281.pdf.

AS 1428.2. 1992. Design for access and mobility: Part 2: Enhanced and additional requirements—Buildings and facilities. Sydney: Standards Australia.

AS/NZS 4438. 1997. Height adjustable swivel chairs. Sydney: Standards Australia and Wellington: Standards New Zealand.

AS/NZS 4442. 1997. Office desks. Sydney: Standards Australia and Wellington: Standards New Zealand.

AS/NZS 4443. 1997. Office panel systems—Workstations. Sydney: Standards Australia and Wellington: Standards New Zealand.

Bailey, D. P., and C. D. Locke. 2014. Breaking up prolonged sitting with light-intensity walking improves postprandial glycemia, but breaking up sitting with standing does not. *J Sci Med Sport.* 18(3):294–98.

Biswas, A., P. I. Oh, G. E. Faulkner, R. R. Bajaj, M. A. Silver, M. S. Mitchell, and D. A. Alter. 2015. Sedentary time and its association with risk for disease incidence, mortality, and hospitalization in adults: A systematic review and meta-analysis. *Ann Intern Med.* 162(2):123–32.

Brennan, A., J. S. Chugh, and T. Kiline. 2002. Traditional versus open office design: A longitudinal field study. *Env Behav.* 34:279–99.

Brown, G. 2009. Claiming a corner at work: Measuring employee territoriality in their workspaces. *J Env Psychol.* 29:44–52.

Brown, G., T. B. Lawrence, and S. L. Robinson. 2005. Territoriality in organisations. *Acad Manage Rev.* 30:577–94.

Carlopio, J. R. 1996. Construct validity of a physical work environment satisfaction questionnaire. *J Occup Health Psychol.* 1:330–44.

Chau, J. Y., M. Daley., A. Srinivasan, S. Dunn, A. E. Bauman, and H. P. van der Ploeg. 2014. The effectiveness of sit-stand workstations for changing office workers' sitting time: Results from the Stand@Work randomized controlled trial pilot. *Int J Behav Nutr Phys Act.* 11:127.

Chau, J. Y., A. C. Grunseit, T. Chey, E. Stamatakis, W. J. Brown, C. E. Matthews, A. E. Bauman, and H. P. van Der Ploeg. 2013. Daily sitting time and all-cause mortality: A meta-analysis. *PLoS One.* 8(11):e80000.

Chau, J. Y., H. P. van der Ploeg, S. Dunn, J. Kurko, and A. E. Bauman. 2011. A tool for measuring workers' sitting time by domain: The workforce sitting questionnaire. *Br J Sports Med.* 45:1216–22.

Chau, J. Y., H. P. van der Ploeg, S. Dunn, J. Kurko, and A. E. Bauman. 2012. Validity of the occupational sitting and physical activity questionnaire. *Med Sci Sports Exer.* 44(1):118–25.

Ciccarelli, M., L. Straker, S. E. Mathiassen, and C. Pollock. 2014. Posture variation among office workers when using different information and communication technologies at work and away from work. *Ergonomics.* 57(11):1–9.

Comcare. 2014. Strategies to help you stand up, sit less, move more. From http://www.comcare.gov.au/Forms _and_Publications/publications/services/fact_sheets/fact_sheets/strategies_to_help_you_stand_up,_sit _less,_move_more (accessed August 18, 2015).

Cornell University. 2015, August 18. Performance oriented ergonomic checklist for computer (VDT) workstations. From http://ergo.human.cornell.edu/cutools.html.

Cox, T. 1993. *Stress Research and Stress Management: Putting Theory to Work*. Sudbury: HSE Books.

CSA-Z412-00 (R2011). 2011. Guideline on office ergonomics. Ottawa, ON: Standards Council of Canada.

Danna, K., and R. W. Griffin. 1999. Health and well-being in the workplace: A review and synthesis of the literature. *J Manage*. 25:357–84.

Descatha, A., Y. Roquelaure, J. Chastang, B. Evanoff, M. Melchior, C. Mariot, C. Ha, E. Imbernon, M. Goldberg, and A. Leclerc. 2007. Validity of Nordic-style questionnaires in the surveillance of upper-limb work-related musculoskeletal disorders. *Scand J Work Environ Health*. 33:58–65.

Dunstan, D., B. Kingwell, R. Larsen, G. Healy, E. Cerin, M. Hamilton, J. Shaw et al. 2012. Breaking up prolonged sitting reduces postrandial glucose and insulin responses. *Diabetes Care*. 35:976–83.

Elkington, J. 1994. Towards the sustainable corporation: Win-win-win business strategies for sustainable development. *Calif Manage Rev*. 36(2):90–100.

Evans, G. W., and D. Johnson. 2000. Stress and open office noise. *J Appl Psychol*. 85:779–83.

GBCA (Green Building Council of Australia). 2015a. Green star. From http://www.gbca.org.au/green-star/ (accessed August 18, 2015).

GBCA. 2015b, August 18. Green star—Interiors. From http://www.gbca.org.au/green-star/green-star-interiors /the-rating-tool/ (accessed August 18, 2015).

GBCA. 2015c. Green star—Innovation in green star. From https://www.gbca.org.au/green-star/technical -support/innovation-in-green-star/ (accessed August 18, 2015).

Goldberg, D. 1978. *Manual of the General Health Questionnaire*. Windsor: NFER Publishing.

Gould, R., J. Ilmarinen, J. Järvisalo, and S. Koskinen. 2008. *Dimensions of Work Ability: Results of the Health 2000 Survey*. Helsinki: Finnish Institute of Occupational Health.

Griffiths, K. L., M. G. Mackey, B. J. Adamson, and K. L. Pepper. 2012. Prevalence and risk factors for musculoskeletal symptoms with computer based work across occupations. *Work*. 42:533–41. http://www.ncbi .nlm.nih.gov/pubmed/22523044.

Healy, G., S. Lawler, A. Thorp, M. Neuhaus, E. Robson, N. Owen, and D. Dunstan. 2012. *Reducing prolonged sitting in the workplace: An evidence review*. Victorian Health Promotion Foundation. Publication Number: P-031-GEN_B.

Hedge, A., M. Rudakewych, and L. Valent-Weitz. 2002. Investigating total exposure to WMSD risks: The roles of occupational and non-occupational factors. *Hum Fac Erg Soc P*. 46(15):1325–9.

Her Majesty's Stationery Office. 1995. Disability Discrimination Act. From http://www.legislation.gov.uk /ukpga/1995/50/contents.

Kaarlela-Tuomaala, A., R. Helenius, E. Keskinen, and V. Hongisto. 2009. Effects of acoustic environment on work in private office rooms and open-plan office—Longitudinal study during relocation. *Ergonomics*. 52:1423–44.

Katzmarzyk, P., and I. M. Lee. 2012. Sedentary behaviour and life expectancy in the USA: A cause-deleted life table analysis. *BMJ Open*. 2:e000828.

Kim, J., and R. de Dear. 2013. Workspace satisfaction: The privacy-communication trade-off in open-plan offices. *J Environ Psychol*. 36:18–26.

Kuorinka, I., B. Jonsson, A. Kilbom, H. Vinterberg, F. Biering-Sørensen, G. Andersson, and K. Jørgensen. 1987. Standardised Nordic questionnaires for the analysis of musculoskeletal symptoms. *Appl Ergon*. 18:233–7.

Leka, S., and T. Cox. 2008. The future of psychosocial risk management and the promotion of well-being at work in the EU: A PRIMA time for action. In Leka, S. and T. Cox (eds.), *The European Framework for Psychosocial Risk Management*. Nottingham: I-WHO Publications, p. 174–84.

Lovibond, S. H. and P. F. Lovibond. 1995. *Manual for the Depression Anxiety Stress Scales*, Second edn. Sydney: Psychology Foundation. From http://www2.psy.unsw.edu.au/groups/dass/.

Mackey, M., C. M. Maher, T. Wong, and K. Collins. 2007. Study protocol: The effects of work-site exercise on the physical fitness and workability of older workers. *BMC Musculoskel Dis*. 8:9.

Mackey, M. G., P. Bohle, P. Taylor, T. Di Biase, C. McLoughlin, and K. Purnell. 2011. Walking to wellness in an ageing sedentary university community: Design, methods and protocol. *Contemp Clin Trials*. 32:273–9.

Maina, G., A. Palmas, and F. L. Filon. 2008. Relationships between self-reported mental stressors at the workplace and salivary cortisol. *Int Arch Occup Env Health*. 81:391–400.

Marshall, A. L., Y. D. Miller, N. W. Burton, and W. J. Brown. 2010. Measuring total and domain-specific sitting: A study of reliability and validity. *Med Sci Sports Exerc*. 42:1094–102.

McAtamney, L., and E. N. Corlett. 1993. RULA: A survey method for the investigation of work-related upper limb disorders *Appl Ergon*. 24(2):91–9.

McKay, C. J., R. Cousins, P. J. Kelly, S. Lee, and R. H. McCaig. 2004. Management standards and work-related stress in the UK: Policy background and science. *Work Stress: Intl J Work Health Organ*. 18(2):91–112.

Mosselman, N., A. Gosselink, and M. Beijer. 2002. Long-term effects of activity-based working. From http://www.cfpb.nl/fileadmin/cfpb/images/publicaties/artikelen/2010/Long_term_effects_activity_based_working.pdf (accessed August 18, 2015).

Oommen, V. G., M. Knowles, and I. Zhao. 2008. Should health service mangers embrace open plan work environments? A review. *Asia Pac J Health Manage*. 3:37–43.

Oppezzo, M., and D. L. Schwartz. 2014. Give your ideas some legs: The positive effect of walking on creative thinking. *J Expl Psychol: Learn Mem Cog*. 40(4):1142–52.

Ouye, J., G. Nagy, and J. Langhoff. 2012. New ways of working in the post recession economy: Results from New Ways of Working's 2011 Benchmarking study. In: State of Art Report. New ways of Working. 2012. p. 60. http://www.vtt.fi/inf/pdf/technology/2012/T17.pdf.

Palmer, K., G. Smith, S. Kellingray, and C. Cooper. 1999. Repeatability and validity of an upper limb and neck discomfort questionnaire: The utility of the standardized Nordic questionnaire. *Occup Med*. 49:171–5.

Paton, D., D. Jackson, and P. Johnson. 2003. Work attitudes and motivations. In O'Driscoll, M., P. J. Taylor, and T. Kaliath (eds.) *Organisational Psychology in Australia and New Zealand*. Oxford: Oxford University Press.

Prince, S. A., K. B. Adamo, M. E. Hamel, J. Hardt, S. C. Gorber, and M. Tremblay. 2008. A comparison of direct versus self-report measures for assessing physical activity in adults: A systematic review. *Int J Behav Nutr Phys Activity*. 5:56.

Radas, A., M. Mackey, A. Leaver, A.-L. Bouvier, J. Y. Chau, D. Shirley, and A. Bauman. 2013. Evaluation of ergonomic and education interventions to reduce occupational sitting in office-based university workers: Study protocol for a randomized controlled trial. *Trials* 14:330.

Rashid, M., and C. Zimring. 2005. On psychosocial constructs in office settings: A review of the empirical literature. Paper presented at the 36th Annual Conference of the Environmental Design Research Association, Vancouver, BC. From http://edra.org/sites/default/files/publications/EDRA36-Rashid_0.pdf.

Richardson, K. M., and H. R. Rothstein. (2008). Occupational stress management intervention programs: A meta-analysis. *J Occup Health Psychol*. 13:69–93.

Safework. 2011. Hazardous manual tasks: Code of practice. From http://www.safeworkaustralia.gov.au/sites/SWA/about/Publications/Documents/640/COP_Hazardous_Manual_Tasks.pdf.

Sallis, J. F., and B. E. Saelens. 2000. Assessment of physical activity by self-report: Status, limitations, and future directions. *Res Q Exerc Sport*. 71(Suppl. 2):1–14.

Smith-Jackson, T. L., and Klein, K. W. 2009. Open-plan offices: Task performance and mental workload. *J Environ Psychol*. 29:279–89.

Spielberger, C. D., Gorsuch, R. L., R. Lushene, P. R. Vagg, and G. A. Jacobs. 1983. *Manual for the State-Trait Anxiety Inventory*. Palo Alto, CA: Consulting Psychologists Press.

Thorp, A., G. Healy, E. Winkler, B. Clark, P. Gardiner, N. Owen, and D. Dunstan. 2012. Prolonged sedentary time and physical activity in workplace and non-work contexts: A cross-sectional study of office, customer service and call centre employees. *Int J Behav Nutr Phys Activity*. 9(1):128.

Tuomi, K., and G. Oja. 1998. *Work Ability Index*. Helsinki: Finnish Institute of Occupational Health.

U.S. Department of Labor. 1990. Americans with Disabilities Act. From http://www.dol.gov/dol/topic/disability/ada.htm.

Van Katwyk, P. T., S. Fox, P. E. Spector, and E. K. Kelloway. 2000. Using the job-related affective well-being scale (JAWS) to investigate affective responses to work stressors. *J Occup Health Psychol*. 5:219–30.

Veitch, J. A., K. E. Charles, K. M. J. Farley, and G. R. Newsham. 2007. A model of satisfaction with open-plan office conditions: COPE field findings. *J Environ Psychol*. 27(3):177–89.

Veldhoen + Company. 2014. Workplace trends 2012, activity based working in the Netherlands louis lhoest. From http://www.slideshare.net/maggieprocopi/workplace-trends-2012-activity-based-working-in-the-netherlandslouis-lhoest?related=2.

Vischer, J. C. 2007. The effects of the physical environment on job performance: Towards a theoretical model of workspace stress. *Stress Health*. 23:175–84.

Wilks, S., M. Mortimer, and P. Nylén. 2006. The introduction of sit–stand worktables: Aspects of attitudes, compliance and satisfaction. *Appl Ergon*. 37:359–65.

Zagreus, L., C. Huizenga, E. Arens, and D. Lehrer. 2004. Listening to the occupants: A web-based indoor environmental quality survey. *Indoor Air*. 14:65–74.

19 Sustainable Design in the Workplace

Julie Dorsey

CONTENTS

19.1 INTRODUCTION: SUSTAINABILITY AND SUSTAINABLE DEVELOPMENT

The global ecosystem currently faces significant threats including climate change, resource depletion, soil degradation, and loss of biodiversity (United Nations [UN] 1987; World Federation of Occupational Therapists 2012). According to the Millennium Ecosystem Assessment (2005, p. 2), "over the past 50 years, humans have changed ecosystems more rapidly and extensively than in any comparable period of time in human history, largely to meet rapidly growing demands for food, fresh water, timber, fiber, and fuel." Although the actions causing this ecological destruction have advanced economic development and human well-being, the trade-off has rendered significant ecosystem changes and growing social and economic inequities (Millennium Ecosystem Assessment 2005). The direct and indirect effects of these environmental issues are being experienced by people and communities worldwide, from an increase in the frequency and the severity of natural disasters (McMichael and Lindgren 2011), to decreased and/or unstable crop yields causing food insecurity and changing water supplies (Centers for Disease Control and Prevention 2015), to many others.

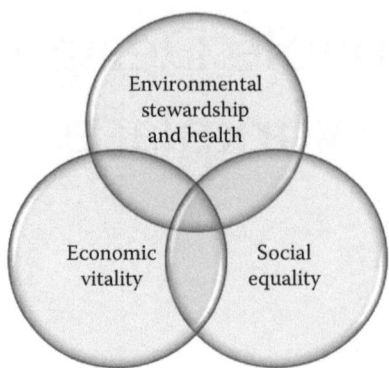

FIGURE 19.1   Tripod balance of sustainable development.

How are we addressing these issues? The term *sustainability* has become very popular in the last few decades, although sustainability efforts have been documented for much longer. Sustainability consists of a balance of social, economic, environmental, and cultural components (Werbach 2009), and can be conceptualized as the intersection of environmental stewardship and health, economic vitality, and social justice (Yanarella, Levine, and Lancaster 2009; Pearsall and Pierce 2010). Sustainable practices must address the interrelationships between economic, environmental, and social factors and strive for an equilibrium between these areas (Figure 19.1).

The United States Environmental Protection Agency (U.S. EPA) states that

> …sustainability creates and maintains the conditions under which humans and nature can exist in productive harmony, that permit fulfilling the social, economic and other requirements of present and future generations. Sustainability is important to making sure that we have, and will continue to have, the water, materials, and resources to protect human health and our environment.

The Brundtland Report, which coined the term *sustainable development* and is the most commonly cited reference, defines sustainable development as meeting the needs of the present without impeding the ability of future generations to meet their needs (UN 1987). Sustainable development is described further as "…not a fixed state of harmony, but rather a process of change in which the exploitation of resources, the direction of investments, the orientation of technological development, and the institutional change are made consistent with future as well as present needs" (UN 1987, p. 17).

In consideration of the significant threats faced by our global ecosystem, there have been many efforts to facilitate both adaptation and mitigation strategies, and more recently to facilitate resilience among people and communities. Sustainability efforts became popular in the late 1960s and 1970s through grassroots advocacy and the formation of environmental groups such as Friends of the Earth, Greenpeace, and the National Resources Defense Council (International Institute for Sustainable Development 2012). These efforts have significantly expanded over the past four decades with continued focus on sustainable development initiatives. In 2012, the UN hosted Rio + 20, an international conference on sustainable development that resulted in world leaders approving the document "The Future We Want," and pledging more than $513 billion toward sustainable development initiatives. The UN has an ongoing commitment to environmental sustainability and sustainable development, as recently evidenced with the adoption of the 2015 Millennium Development Goals, with goal 7 to ensure environmental sustainability (UN 2015). Examples of popular sustainable development include green building and maintenance practices, creation and expansion of jobs in the green sector, and design of sustainable communities. The primary focus of many recent sustainable development practices has been related to the environmental pillar, with some attention to the economic and social pillars (Institute of Medicine [IOM] 2011; US EPA 2015).

It has been suggested (Hedge, Rollings, and Robinson 2010; Miller 2010; IOM 2011) that this is of concern for various reasons that will be discussed later in the chapter.

19.2 RELATIONSHIP TO HUMAN HEALTH AND OCCUPATIONAL ENGAGEMENT

While the effects of the global crisis have the potential to disrupt how humans engage in occupations, it is also important to consider the effect of human occupations on the environment. Capon (2014, p. 10) opines that "...it all comes down to the things we do—our occupations...the scale of human enterprise, and the technologies we have, are enabling us to disrupt planetary systems." Balancing consumption with conservation in all aspects of development would promote sustainability, yet there are inherent flaws in this goal. The definition of sustainable development is vague, and it becomes near impossible to measure the true needs of both current and future generations. Developed countries consume disproportionate amounts of the world's natural resources to meet their needs, and this has a direct effect on those in developing countries. The very nature of how developed countries are structured yields more needs by more people, tipping the scale away from global social equity. A more in-depth discussion of these inherent issues in sustainable development is out of the scope of this chapter, but is important to be recognized. The approaches relevant to human factors/ergonomics (HF/E) professionals presented in this chapter are important steps to contribute to a larger global issue of protecting the earth's natural resources in order to promote a more sustainable future for all. A first step is to gain an understanding of our relationship with the world around us, one that Thatcher (2013, p. 391) describes as bidirectional in which "...humans influence the health of their natural environment and the health of the natural environment, in turn, impacts on the health and wellbeing of humans" (see Figure 19.2). To this end, it is imperative that we consider the relationship between human factors and sustainability from a systems perspective.

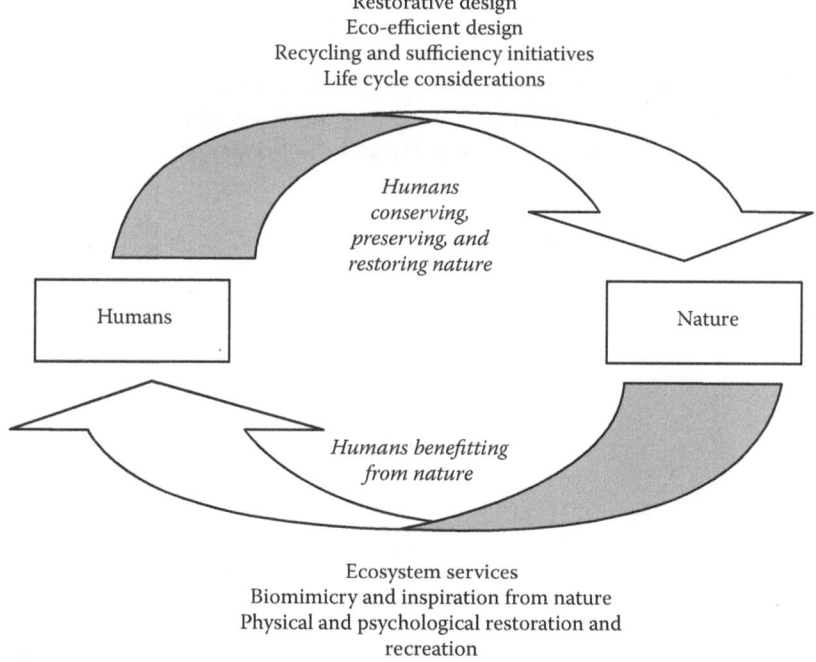

FIGURE 19.2 Bidirectional relationships for green ergonomics. (Reprinted from Thatcher, A., *Work*, 49, 381–93, 2013. With permission.)

19.3 GREEN ERGONOMICS

The term *green ergonomics* (Hedge 2008; Hedge, Rollings, and Robinson 2010) has been appearing in the literature over the past few years with increased attention as of late, and its definition is currently under discussion in the HF/E literature. Hedge, Rollings, and Robinson (2010) see green ergonomics as "the role that ergonomics can play in improving the quality of the built environment for occupants." Hanson (2013, p. 400) suggests that green ergonomics is "the application of ergonomics/human factors to the design of products, jobs, environments, and systems that have an intentional objective of reducing a negative impact on the environment." Thatcher (2013, p. 391) defines green ergonomics as "ergonomics interventions that have a pro-nature focus; specifically, ergonomics that focuses on human affinity within the natural world." Yet another perspective is that green ergonomics is the integration of ergonomics into sustainable development to enhance human performance, productivity, health, and well-being, thereby promoting sustainability at both the individual and the systems level (Dorsey and Miller 2013).

Common to all these definitions is the application of ergonomics to sustainable development in order to—ideally—positively affect the environment, the economy, and the people involved in all aspects of the development on both a micro- and a macrolevel. If we consider a systems perspective, in which everything we do has an impact on the broader system, and vice versa, it becomes impossible to separate the human from the environment. This concept is integral to the theory and the practice of ergonomics, and HF/E professionals are well positioned to make significant contributions to sustainability efforts by integrating green ergonomics into practice (see Figure 19.3).

19.4 INTRODUCTION TO GREEN BUILDING DESIGN

It has been well established that due to the large portion of time that individuals spend within the built environment, especially in North America, the IEQ should be of concern (Choi, Loftness, and Aziz 2011). There have been numerous studies and recommendations made for IEQ standards and practices mainly related to conventional buildings (Choi, Loftness, and Aziz 2011; Peretti and Stefano 2011). With the environmental crisis and the resultant sustainability efforts, there has been tremendous growth in green building practices (United States Green Building Council [USGBC] 2015a). This necessitates specific attention, from an ergonomics perspective, that must be paid to all aspects of green buildings, including design, construction, operation, maintenance, renovation, and demolition.

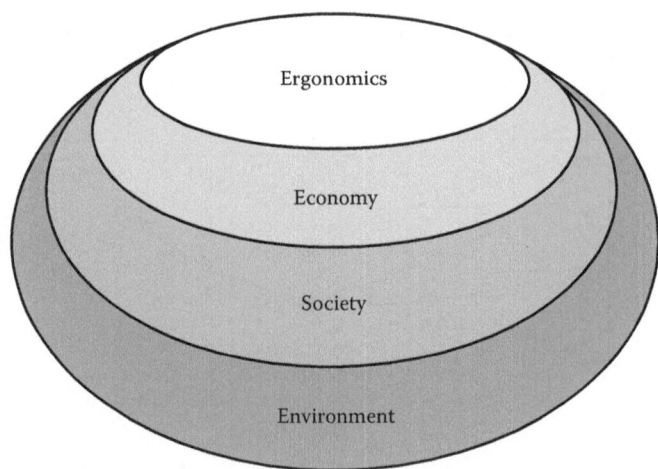

FIGURE 19.3 Relationship of ergonomics to sustainability. (Reprinted from Pavlovic-Veselinovic, S., *Work*, 49, 395–9, 2014. With permission.)

19.4.1 CERTIFICATION PROGRAMS

There are many certification programs worldwide, some voluntary and some mandated, that seek to ensure that green buildings meet certain criteria as related to sustainable development principles and practices. Some programs such as Energy Star (Energy Star 2015) are government sponsored and focus on a single attribute such as energy efficiency, and others are multi-faceted and strive to focus on the varied attributes of green buildings such as design, operation, and maintenance (Whole Building Design Guide 2015). The programs that fall into the latter category include Leadership in Energy and Environmental Design (LEED; international), Green Star (South Africa, Australia), Green Globes (Canada), and Building Research Establishment Environmental Assessment Method (BREEAM; international), among many others. BREEAM has been used since 1990 and serves as the basis for many other certification programs. However, it can be argued that LEED is the most widely used program in that it has certified over 10 billion ft^2 of commercial and residential building spaces, spanning more than 54,000 projects in over 135 countries worldwide (USGBC 2015a) (see Figure 19.4).

Many of the certification programs function similarly with a scorecard outlining a certain amount of credits or points to be awarded for particularly strategies, with various rating systems depending on the project. Indoor air quality (IAQ) and IEQ are categories in most of these rating systems which speaks to the need for ergonomics professionals to be involved in green building projects. Green buildings employ many different design, construction, and operation strategies and utilize different materials than conventional buildings. Ergonomics professionals must identify those strategies that are related to human factors in order to assist the design teams and the construction professionals with the green building process to ensure that attention is paid to human needs.

19.4.2 LEED RATING SYSTEM

The USGBC created the LEED certification program in 2000 with the intent of certifying green building performance (USGBC 2015a). Since that time, the rating system has undergone numerous revisions with the most current system implemented in November 2013. The LEED v4 rating system certifies projects based on the total number of points achieved: certified, 40–49 points; silver, 50–59 points; gold, 60–79 points; and platinum, 80+ points (USGBC 2015b). Five rating systems exist within LEED v4 for various project types: building design and construction, interior design and construction, building operation and maintenance, neighborhood development, and homes.

FIGURE 19.4 LEED platinum building. (Courtesy of Alan Hedge.)

LEED for New Construction and Major Renovations (v4)

		POSSIBLE: 1
Credit	Integrative process	1

LOCATION & TRANSPORTATION   POSSIBLE: 16

Credit	LEED for Neighborhood Development location	16
Credit	Sensitive land protection	1
Credit	High priority site	2
Credit	Surrounding density and diverse uses	5
Credit	Access to quality transit	5
Credit	Bicycle facilities	1
Credit	Reduced parking footprint	1
Credit	Green vehicles	1

SUSTAINABLE SITES   POSSIBLE: 10

Prereq	Construction activity pollution prevention	REQUIRED
Credit	Site assessment	1
Credit	Site development - protect or restore habitat	2
Credit	Open space	1
Credit	Rainwater management	3
Credit	Heat island reduction	2
Credit	Light pollution reduction	1

WATER EFFICIENCY   POSSIBLE: 11

Prereq	Outdoor water use reduction	REQUIRED
Prereq	Indoor water use reduction	REQUIRED
Prereq	Building-level water metering	REQUIRED
Credit	Outdoor water use reduction	2
Credit	Indoor water use reduction	6
Credit	Cooling tower water use	2
Credit	Water metering	1

ENERGY & ATMOSPHERE   POSSIBLE: 33

Prereq	Fundamental commissioning and verification	REQUIRED
Prereq	Minimum energy performance	REQUIRED
Prereq	Building-level energy metering	REQUIRED
Prereq	Fundamental refrigerant management	REQUIRED
Credit	Enhanced commissioning	6
Credit	Optimize energy performance	18
Credit	Advanced energy metering	1
Credit	Demand response	2
Credit	Renewable energy production	3
Credit	Enhanced refrigerant management	1
Credit	Green power and carbon offsets	2

MATERIAL & RESOURCES   POSSIBLE: 13

Prereq	Storage and collection of recyclables	REQUIRED
Prereq	Construction and demolition waste management planning	REQUIRED
Credit	Building life-cycle impact reduction	5
Credit	Building product disclosure and optimization - environmental product declarations	2
Credit	Building product disclosure and optimization - sourcing of raw materials	2
Credit	Building product disclosure and optimization - material ingredients	2
Credit	Construction and demolition waste management	2

INDOOR ENVIRONMENTAL QUALITY   POSSIBLE: 16

Prereq	Minimum IAQ performance	REQUIRED
Prereq	Environmental tobacco smoke control	REQUIRED
Credit	Enhanced IAQ strategies	2
Credit	Low-emitting materials	3
Credit	Construction IAQ management plan	1
Credit	IAQ assessment	2
Credit	Thermal comfort	1
Credit	Interior lighting	2
Credit	Daylight	3
Credit	Quality views	1
Credit	Acoustic performance	1

INNOVATION   POSSIBLE: 6

Credit	Innovation	5
Credit	LEED Accredited Professional	1

REGIONAL PRIORITY   POSSIBLE: 4

Credit	Regional priority	4

TOTAL		110

40-49 Points	50-59 Points	60-79 Points	80+ Points
CERTIFIED	SILVER	GOLD	PLATINUM

FIGURE 19.5 Sample LEED scorecard. (Reprinted from USGBC, LEED credit library, http://www.usgbc .org/credits/new-construction/v4, 2016. With permission.)

Sustainable site design, water quality and conservation, energy and environment, IEQ, and materials and resources are the five main areas of sustainability and are used to guide the green building design process and the area reflected as such on the scorecards (see Figure 19.5).

In addition to the items listed under each of the main scorecard areas, there are up to five points that can be awarded in the innovation category. A pilot credit library is maintained and can be accessed to provide successful past examples. In past LEED versions, there was a pilot credit for a comprehensive ergonomics strategy, and someone on the project team would have to know enough to search for the strategy in the pilot credit library in order to find the criteria and apply for the credit. In LEED v4, the ergonomics strategy is listed in a more readily accessible location, which will hopefully lead to more project applications in the future.

The intent of the one credit for an ergonomics strategy is to "improve occupant well-being (human health, sustainability and performance) through integration of ergonomics principles, specifically in the design of work spaces for all computer users" (USGBC 2015c). The pilot credit was updated in March 2016 and the requirements of the credit are presented in Table 19.1.

19.4.3 STRATEGIES USED IN GREEN BUILDINGS

As mentioned earlier, there are five areas of sustainability that are used to guide the green building design process: sustainable site design, water quality and conservation, energy and environment,

TABLE 19.1
LEED Ergonomics Strategy Requirements

Requirement	Criteria (If Indicated)
During the conceptual or schematic design phase: • Engage an ergonomist or health and safety specialist to assist in the development of the ergonomics strategy. • Make a commitment to integrate ergonomics principles into the overall design.	Confirmation that an ergonomist or health and safety specialist was utilized for the project Written commitment from the owner or owner's representative that the ergonomics strategy will be implemented

The ergonomist or health and safety specialist, in conjunction with the client, must develop a description of the ergonomics strategy that will be implemented and include the following:
• Statement identifying the goals of the ergonomics strategy
• Description of occupant needs, including occupant characteristics and/or demographics, tasks, and machines, equipment, tools, and work aids (METWA's) used to perform these tasks
• Process for selecting workstation layouts and furnishings based on relevant standards or guidelines; for computer workstations, reference to one or more of the following (or their most up-to-date versions):
  • BIFMA G1-2013
  • ANSI/HFES 100-2007
  • CSA Z412-00 (R2011)
  • ISO 9241-5:1998
• Education program for move-in and during ongoing operations
• Process for evaluating and maintaining occupant well-being upon move-in and during ongoing operations to ensure that the ergonomics strategy goals are being met; selection of appropriate metrics and/or measurements for this evaluation, the evaluation frequency, and how soon after implementation the process will begin
	Detailed description of the ergonomics strategy, including the following:
	• Goals of the ergonomics strategy
	• Process to define occupant needs
	• Standard/guideline(s) that will be used to inform design options/selections
	• Design layouts, options, and cut sheets for the product selections
	• Education program
	• Ergonomics process that will be used for maintaining occupant well-being

Source: USGBC, Ergonomics strategy, http://www.usgbc.org/node/4631863?return=/credits/new-construction/v4/pilot-credits, 2015. With permission.

IEQ, and materials and resources. Each of these areas can impact the end users of the buildings as related to the design and/or the operation, and many of these areas also impact those people involved in the construction, renovation and/or demolition phases (Miller, Dorsey, and Jacobs 2012). HF/E professionals should gain an understanding of green building strategies in order to ensure that attention is given to the human element. Loftness, Lam, and Hartkopf (2005) present a framework for considering the human needs in relation to green, high-performance building needs (see Table 19.2). It is of value to note that "even modest improvements in productivity, absenteeism, and/ or employee retention can substantially outweigh the traditionally sought-after efficiency benefits such as energy savings" (Pyke, McMahon, and Dietsche 2010). Additionally, there are many other potential health benefits due to green building design such as improved physical activity through supports, including project siting, and reduced illnesses through lowered exposure to known toxins (USGBC 2015b). Human resources are by far the most expensive considerations in any building project, and they need to be protected (Romm 1994; Fuller 2010). However, when final decisions need to be made regarding design and layout, budget and environmental concerns (i.e., the credit being sought) will take precedence over the human element as it is not as easy to see the impact on humans especially in the planning and construction stages.

TABLE 19.2

Understanding Environmental Performance for the Design of High-Performance Buildings

	Physiological Needs	Psychological Needs	Sociological Needs	Economic Needs
Performance Criteria Specific to Certain Human Senses, in the Integrated System				
1. Spatial	Ergonomic comfort, handicap access, functional servicing	Habitability, beauty, calm, excitement, view	Wayfinding, functional adjacencies	Space conservation
2. Thermal	No numbness, frostbite, no drowsiness, heat stroke	Healthy plants, sense of warmth, individual control	Flexibility to dress the custom with...	Energy conservation
3. Air quality	Air purity no lung problems, no rashes, cancers	Healthy plants, not closed in, stuffy, no synthetics	No irritation from neighbors, smoke, smells	Energy conservation
4. Acoustic	No hearing damage, music enjoyment, speech clarity	Quiet, soothing, activity, excitement, "alive"	Privacy, communication	
5. Visual	No glare, good task, illumination, wayfinding, no fatigue	Orientation, cheerfulness, calm, intimate, spacious, alive	Status of window, daylit office, "sense of territory"	Energy conservation
6. Building integrity	Fire safety, structural strength plus stability, weather lightness, no out-gassing	Durability, sense of stability, image	Status/appearance, quality of constant "craftsmanship"	Material/labor conservation
Performance Criteria General to All Human Senses, in the Integrated System				
	Physical comfort	Psychological comfort	Privacy	Space conservation
	Health	Mental health	Security	Material conservation
	Safety	Psychological safety	Community	Time conservation
	Functional	Aesthetics	Image/status	Energy conservation
	Appropriateness	Delight		Money/investment conservation

Source: Loftness, V., K. P. Lam, and V. Hartkopf, *Build Res Inform.*, 33, 196–203, 2005. With permission.

To this end, IEQ will be discussed as a way to understand how the principles of sustainability are put into practice in green buildings. The strategies presented are based on the LEED v4 rating system for new construction and major renovation; however, they have varied applications across certification programs and green building projects. It is important to note that not all aspects of each IEQ category are required, and project managers can choose which strategies to incorporate. This yields high variability in IEQ across green building projects. Following the strategies are the considerations for HF/E professionals regarding how the green building strategies may impact the building occupants. If we are more clearly able to articulate the potential impact of green design strategies on the end user, we can advocate for the inclusion of human-centered, ergonomic design strategies during the planning and design stages to yield more positive occupant outcomes.

19.4.3.1 Lighting

The importance of lighting to comfort, productivity, and well-being is recognized in green building design. The focus of lighting is multifaceted, addressing lighting control, lighting quality, and day-lighting. In regards to lighting control, it is presumed that providing a large percentage of building occupants with individual lighting controls to meet their preferences will positively contribute to the desired outcomes of comfort, productivity, and well-being. There should also be easy access to adequate lighting control in shared work spaces, with multiple zones to meet the needs of the groups. Lighting quality takes into account illuminance, surface reflectance, and energy efficiency based on set standards. Increasing the daylight in green buildings provides occupants with a connection to the outdoors and can have physiological benefits, and this is presumed to have a positive impact on occupant experiences. In addition, increased daylighting can reduce the overall electricity, thereby saving organizational and environmental resources. Improving daylight within the built environment is achieved through such strategies as the use of large windows, central atria with skylights, and interior windows from offices, and building siting that takes into account natural sunlight exposure.

While each of these green building strategies has the potential to positively impact the occupants, it is imperative to consider all the potential impacts. For example, low-energy bulbs may provide a different quality of illumination from a user perspective that could take some adjustment. In addition, the natural variations in daylighting over the course of the day can create significant fluctuations at a given work space. The quality of natural light may vary as well, and the sheer number of windows or the location of workstations in relation to the windows does not guarantee adequate lighting (Lee and Guerin 2009). Large windows and skylights can contribute to excess glare and surface reflectance. Additionally, there have been a significant number of injuries including overexertion and falls, related to the construction of atria and the installation of skylights and large windows (Fortunato et al. 2012).

Motion sensors are often used to account for periods of inactivity in work spaces, and while this is useful from an energy perspective, it can be perceived as an annoyance by the occupants. A common scenario is when employees are working later than coworkers in a shared space, deeply involved in a computer task when the lights suddenly turn off. The employee must then make large movements to trigger the sensors, affecting his/her concentration with the task, which can in turn affect productivity. While HF/E professionals may see this as a natural consequence of being too sedentary in essence forcing the employee to take a break, an unanticipated break in concentration can be detrimental to employee productivity and satisfaction with the work environment.

19.4.3.2 Acoustics

The acoustics within a work space are important for overall well-being, effective communication, and productivity. Within green buildings, an effective acoustic design is based on industry standards and takes into account such factors as background noise from heating, ventilation and air conditioning (HVAC) systems, sound isolation, reverberation time, sound reinforcement, and masking systems. Other systems may have an impact on the acoustics, such as the natural/displacement-type ventilation systems that typically produce lower background noise and lower noise isolation (Hodgson et al. 2008). This can be positive but can also be negative in that masking systems may need to be used instead to address privacy issues. Additionally, the increased number of operable windows that are included as part of the ventilation system can have an impact on the acoustics, as this can introduce more noise from external sources (e.g., traffic). Other factors that can influence the acoustics include the usage of partition walls in shared spaces and properties of interior furnishings. Green buildings may use the strategy of large shared work spaces with open layouts or with varied height partition walls for socialization/collaboration benefits as well as for improved air flow and thermal comfort. This can affect the experience of background noise and perceived sound privacy, which can be a challenge for workers, especially if they were used to private offices or different layouts. When masking systems are used, occupants may require an adjustment period especially if they have never encountered this before. If less carpet is used in an effort to lower VOCs, the designers must take into account the reverberation from the different, perhaps more eco-friendly, finishes such as concrete or other natural materials.

19.4.3.3 Thermal Comfort

It is important for occupants to experience thermal comfort within the work environment to enhance productivity and well-being. In green building design, thermal comfort addresses both the design of the HVAC systems and the building envelope as well as the provision of individual thermal comfort controls. Attention to the design of the HVAC systems and the building envelope are also important from an energy perspective. Providing occupants with control over the temperature, the air speed and/or the humidity is seen as a good practice for facilitating comfort and productivity, and must be balanced with the overall energy picture of a building.

As mentioned earlier, utilizing large windows is one of the green building design strategies that accomplishes many goals, such as thermal benefits and reduced energy costs. While the use of these windows is certainly beneficial for certain aspects of sustainable development, they can cause occupants to experience temperature fluctuations over the course of the day as well as in varied seasons. These fluctuations may negatively impact occupant comfort and productivity. Additionally, with the increased use of common shared spaces, it can be difficult for all occupants to achieve thermal comfort due to personal preferences, and the control systems may not truly meet the needs of all users.

19.4.3.4 Air Quality

IAQ assessment and management is an important part of the green building process. It is hoped that through strategies to ensure adequate IAQ, the occupants will experience enhanced comfort, well-being, and productivity. Green building practices include strategies for mechanically ventilated spaces, naturally ventilated spaces, and mixed-mode systems, and the use of low-emitting materials is encouraged. Consideration should also be given to what workers are exposed to during the construction phase.

One of the strategies that may be used in green buildings under IAQ is to provide occupants with increased access to operable windows. While this can give occupants more satisfaction over their air quality, the bigger picture must be considered. IAQ monitoring systems are in place to ensure that the standards are continually met, and this can result in periods in which window use is restricted. If occupants do not follow the window use guidelines, then the IAQ can be compromised, reducing the overall building performance. With temperature fluctuations noted earlier, occupants may rely on windows to regulate their own comfort. They may also open their private office doors for increased airflow for thermal comfort. Open windows and open doors can affect the acoustical quality for those in surrounding areas and must be considered as a significant issue for occupants (Hodgson et al. 2008). Another issue that should be considered as related to IAQ is during the construction phase. Construction workers may not be familiar with the properties of the low VOC materials, and some studies have shown that workers perceive the materials to have poorer performance, resulting in rework of projects (Fortunato et al. 2012). Chapter 3 in this book gives more details on IAQ issues.

19.4.3.5 Layout

There are several layout design trends that are being seen in green buildings to promote social, economic, and environmental benefits. LEED v4 has recognized the importance of increased access to quality views to help occupants have a connection to the outdoors during their workday. This is achieved by a design that places workstations in a direct line of sight with exterior windows (preferable) or interior atria windows (less desirable). This can create issues with glare and visual comfort for employees depending on the layout of the computer and the work space in relation to the windows and the path of the sun, as well as the glazing or other control methods used. Another trend is for smaller, narrower building footprints in general, which results in smaller personal space footprints within the building and more occupant workstations located along the perimeter. This can take adjustment on the part of the occupant to reorganize his/her space, utilize more common meeting spaces versus private, and adapt to shared resources such as shared printing. Additionally, appropriate emergency plans must be put into place that take into account the decreased utilization of private spaces and location of workers, and adequate shelter in place and emergency exit routes must be planned.

19.4.3.6   Office Furnishings and Materials

It is important to consider the office furnishings and materials that are being used in green buildings. Certain materials may have different properties that can affect work, such as the use of concrete as mentioned earlier and its potential impact on the acoustics. These impacts must be considered, and the occupant needs must be addressed in relation to the issues that arise or that are predicted. Additionally, HF/E professionals are often involved in selecting and/or recommending products and furnishings and must take into account the principles of sustainability during this process. Through the thoughtful incorporation of proactive ergonomics, the need for retrofits can be reduced. Retrofits are not only costly from an economic standpoint, but also from an environmental standpoint in terms of product waste, and from a social standpoint, in terms of occupant dissatisfaction and lost productivity during renovations. HF/E professionals can assist designers and project managers in creating flexible layouts that will have minimal waste if renovations are needed. HF/E professionals are encouraged to reuse and repair products instead of purchasing new whenever possible. When recommending products, HF/E professionals should consider the environmental impacts of these recommendations and attempt to locally purchase when possible to decrease the economic and environmental costs of shipping. Products that utilize sustainable materials and have low VOC or other pollutants are preferred whenever possible. Additionally, there may be guidelines set by the organization for purchasing that must be considered in order to consider all aspects on sustainable development (see Table 19.3).

19.4.4   Occupant Experiences in Green Buildings

With the significant growth in the green building industry, more and more people worldwide inhabit these green spaces, and therefore they deserve our attention. It has been suggested by numerous sources that there is a need for increased attention to the impact of green buildings on the health of occupants, since people spend the majority of their time indoors (Newsham 2010; IOM 2011). The IOM reports that there is limited research on the intersection of the "effects of climate change on human health in the indoor environment" (IOM 2011, S-3). It has been assumed that if buildings employ sustainable design principles, they will naturally improve the indoor environments for occupants. However, the extent to which this is true has not yet been fully explored. Additionally, some issues have arisen in the construction of green buildings wherein workers face significant safety issues. We must consider the following question: can a building be considered sustainable if it addresses the environmental and economic dimensions of sustainability, yet fails to meet the needs of the people?

During this time of growth in the green building industry, there have been postoccupancy evaluations conducted to investigate the human element of sustainable design. While this area of research is still being developed and is in its infancy, the initial results have been mixed. There have been many benefits of green design strategies from a user perspective, and there have also been a significant number of concerns raised. The most recent literature will be discussed to highlight both the successes and the concerns, and to highlight design suggestions that are based on the available evidence. The following studies discussed include international perspectives and investigate various green building rating systems including LEED (majority of studies), Green Star (Australia, South Africa), and the Korean Green Building Council.

19.4.4.1   LEED Certifications

Some attention has been given to investigating the differences among the various levels of LEED certifications. This is important because projects that seek higher levels of certification may apply for more credits in IEQ, thereby influencing the design and potentially influencing the occupants to a greater extent. LEED platinum building occupants have been found to have higher satisfaction

TABLE 19.3

Potential Role for HF/E Professionals Related to Green Building Strategies

Green Building Strategy	Potential Impact on Occupants	Potential Role for HF/E
Lighting—control, quality, daylighting	Breaks in concentration from motion lights deactivating Adjustment to fluctuations, different lighting Construction risks	Establishment of healthy work routines—build in timed movement breaks Coordination with designers to ensure user friendly lighting controls Training occupants in lighting controls including the proper use of shades etc. Placement of appropriate, energy-efficient task lighting Proactive ergonomics—construction workers
Acoustics—HVAC background noise, reverberation, masking, sound reinforcement	Adjustment to masking systems Less sound privacy More background noise in shared spaces	Training and education for occupants in regards to masking systems—for personal spaces as well as adjusting to common systems Assistance with individual workstation setup in regards to seat positioning
Thermal comfort—design and control	Less individual control over temperature Temperature fluctuations	Education regarding appropriate temperatures to support productivity Training workers to adapt to changes
Air quality—strategies, materials, construction, assessment	Restricted window use Construction risks	Education for occupants Proactive ergonomics—construction workers
Layout—view	Glare Decreased visual comfort More common spaces and shared resources	Proactive consultation on ergonomic workstation design Training and education for workers in adapting to smaller spaces, shared resources, and safety plans
Office furnishings and materials		Reduce need for costly retrofits through proactive ergonomics and flexible designs Reuse and recycle products whenever possible, repair instead of purchasing new when possible Recommend ergonomic products that are made locally, utilize sustainable materials and have low VOC or other pollutants whenever possible Consider corporate policies

and perceived job performance attributed to acoustic and lighting qualities than their LEED gold counterparts; however, there were no overall positive relationships found between the level of certification and the levels of satisfaction and performance (Lee 2011a). Hedge and Dorsey (2013) found issues related to IAQ and noise impacting the perceived health and productivity of the occupants in a LEED platinum building. This suggests that a higher level of certification does not necessarily mean improved occupant experiences, and individual design decisions may be more important. This speaks to the importance of involving HF/E professionals in the planning stages of green buildings to ensure that specific attention is paid to balancing the occupant needs with the certification criteria. It is not a given that a higher level of certification will yield more positive occupant experiences and outcomes, and there must be deliberate planning.

It is also not a given that occupants in LEED buildings will be more satisfied with IEQ than those in conventional buildings, and in fact they may be more dissatisfied with certain features upon post-occupancy investigations (Altomonte and Schiavon 2013). One study conducted a secondary analysis of postoccupancy evaluation data from a subset of the Center for the Built Environment (CBE's) IEQ survey and compared 144 buildings (65 LEED certified) with 21,477 occupant responses

(10,129 in LEED buildings) (Altomonte and Schiavon 2013). There was no significant difference for the overall IEQ satisfaction between the two groups; however, the LEED building occupants were slightly more satisfied with the air quality and slightly less satisfied with the amount of light (Altomonte and Schiavon 2013). These same authors analyzed these data again in a later study to investigate other non-IEQ factors that may impact the occupant satisfaction in LEED buildings (Schiavon and Altomonte 2014). Layout, office type, distance to window, building size, and occupant demographics were investigated, and while some significant differences were found, the effect size was generally negligible (Schiavon and Altomonte 2014). There was some suggestion that smaller buildings, shorter time occupying the building, and open work spaces can contribute to improved IEQ satisfaction; however, this needs to be further investigated (Schiavon and Altomonte 2014).

Thatcher (2014) conducted a one-year longitudinal study that compared a group of employees that moved to a green building with counterparts in a conventional building, with three data collection points (time 1, before move; time 2, six months after; time 3, 12 months after). The green building group had significantly higher self-reported productivity at time 3 as well as significant improvements in the physical well-being (Thatcher 2014). Singh et al. (2010) conducted two longitudinal survey design case studies investigating the effects of green buildings on employee health and productivity. The mean numbers of the negative health symptoms reported (asthma, allergies, depression, and stress) decreased after the move to the green buildings (Singh et al. 2010). Additionally, absenteeism was reduced, perceived productivity improved, as did perceptions of health and well-being. In their postoccupancy evaluation of two LEED platinum buildings, Hedge and Dorsey (2013) found that the respondents generally rated their LEED workplace experiences as higher than their prior conventional buildings, specifically related to satisfaction, comfort, stress, health, and productivity. Symptom prevalence was generally lower in LEED-certified buildings, indicating higher perceptions of health; however, the frequency of musculoskeletal symptoms were similar in green and conventional buildings (Hedge, Miller, and Dorsey 2014). These findings are an important step in helping to understand the relationship between green buildings and health. However, it is quite clear that more research must be done in this area in order to draw conclusions and have a clearer picture of the nature of the relationship.

Newsham et al. (2012) conducted a comprehensive multipart study, which investigated the effects of green buildings on occupants. Postoccupancy evaluations conducted in the United States and Canada were analyzed examining both green and conventional office buildings. Physical data (i.e., layout and design features, thermal conditions, air quality, acoustics, and lighting) were collected from and supplemented with user survey data. The overall findings indicate that while the impact on occupant outcomes in green buildings is improved, consideration is still needed in the area of acoustic performance and IAQ, and the enhancement of the interdisciplinary design process and the postoccupancy evaluation (POE) protocols (Newsham et al. 2012). Specifically, green buildings had improved outcomes related to overall environmental satisfaction, views to the outside, satisfaction with thermal conditions, aesthetics, less disturbance from HVAC noise, and some ratings of personal health (Newsham et al. 2012). An early study by Abbaszadeh et al. (2006) investigated occupant experiences in green and conventional buildings. In general, those in green buildings were more satisfied with thermal comfort and air quality and had similar ratings of lighting and acoustics (Abbaszadeh et al. 2006). When these findings were further investigated, it was revealed that those in green buildings were dissatisfied with light levels (lighting controls and window coverings) and sound privacy (people talking nearby, private conversations being overhead, phones ringing, and hearing others talk on the phone) (Abbaszadeh et al. 2006). Turner's (2006) postoccupancy evaluation of 11 LEED-certified buildings found that occupants were generally satisfied with green buildings and their workplaces, and that they rated the lighting and air qualities as being good. Occupants were somewhat satisfied with temperature and dissatisfied with noise and sound privacy (Turner 2006). Heerwagen and Zagreus (2005) also found occupants in one LEED platinum building to be highly satisfied with the building in general, and also with air quality, daylighting, amount of light, and access to views. The LEED building occupants reported a sense of pride in their workplace; however, they did report acoustical concerns related to lack of speech privacy and distractions from coworkers talking (Heerwagen and Zagreus 2005). In contrast, Paul and Taylor (2008) found no significant differences between occupants in green and conventional buildings, as related to various measurements of satisfaction.

19.4.4.2 Lighting

Building siting and location of workstations in relation to windows are important considerations in green building design. A large-scale study was conducted to investigate the effects of indoor lighting on 2744 occupants in a green office building (Hwang and Kim 2011). Measurements were taken throughout the building, and the occupants were surveyed via questionnaire to ask their perceptions on the impact of the lighting on their work. Issues with lighting levels were related to the direction of the windows, with those workers near north-facing windows having more evenly distributed illuminance (Hwang and Kim 2011). Workstations with other exposures may require additional lighting controls and employee training in such controls (Hwang and Kim 2011). Occupants rated higher levels of visual annoyance due to such factors as glare, darkness, and shade materials in the winter months due to less sunlight penetration (Hwang and Kim 2011).

Specific interior design strategies can affect lighting quality as evidenced by the results of post-occupancy evaluations. The height of partition walls in offices with cubicles can affect lighting quality, with walls over 5 ft (1.5 m) causing more issues with satisfaction and job performance than other types of green offices (Lee and Guerin 2010). Using different layouts such as bullpen offices, utilizing different materials at the top of partition walls to allow for transparency, providing more lighting control, and task lighting may improve the lighting quality when high partition wall cubicle design must be used (Lee and Guerin 2010). For energy purposes, green buildings typically rely less on artificial lighting and more on daylighting; however, this can result in the occupants perceiving the space as dimmer (Thatcher 2014). A postoccupancy evaluation of two LEED office buildings in Hong Kong revealed more natural light, more glare from sunlight and skylights, and mixed results for occupant satisfaction with lighting as compared to surrounding conventional buildings (Gou, Lau, and Shen 2011). Hedge, Miller, and Dorsey (2014) found occupants to be generally satisfied with daylight and found varied results related to the influence of artificial lighting on headaches and eyestrain. Lee and Kim (2008) found issues in a LEED-certified building related to low illuminance and lack of task lighting in personal work spaces. These findings may suggest the need for more individualized lighting controls, appropriate and adjustable window coverings, and addition of task-based lighting, although these strategies must also take into account the overall energy picture of the building.

19.4.4.3 Acoustics

Lee (2010) investigated the influence of various types of office layouts on occupant perceptions of privacy, interaction, and acoustic quality in LEED buildings by using secondary data from the CBE's IEQ Survey. Open, bullpen (no partition) layout offices were found to have higher satisfaction with noise level and acoustic quality versus both high and low partitions. High cubicles had lower satisfaction and job performance related to visual privacy and interaction than both private and enclosed shared offices, while lower partitions contributed to perceptions of improved coworker interactions. These findings suggest that higher partition walls may not contribute to acoustical quality, and that bullpen style layouts can be a good option for small shared spaces in relation to balancing acoustical needs with environmental and economic needs.

Hodgson et al. (2008) took measurements within six green office buildings and found persistent acoustical issues in green office buildings including excessively high reverberation times, inadequate speech privacy in shared and open-plan offices, and high noise levels near exterior walls. Lee and Guerin (2009) found low occupant ratings for performance and satisfaction related to acoustic quality, with the main reasons for dissatisfaction identified as overhearing colleagues talking in surrounding areas, overhearing private conversations, and hearing others talk on the phone, and these findings support an earlier study by Lee and Kim (2008). Gou, Lau, and Zhang (2012) found that one of the green buildings investigated in their study had perceptions of more noise from nearby colleagues, more ambient noise, and more frequent unwanted interruptions. Hedge and Dorsey (2013) also found persistent issues with the occupants' perceptions of noise levels impacting perceived work performance of the occupants of a LEED platinum building. It is clear that the acoustics within green buildings are an ongoing issue and more work needs to be done in this area.

19.4.4.4   Thermal Comfort

Persistent issues with thermal comfort have been identified by occupants in green buildings, specifically related to heating/cooling load distribution, inaccessible thermostats, and control of thermostats (Lee and Guerin 2009). Enclosed private offices were found to have the lowest mean scores for thermal comfort as compared to other office types, while low cubicles had the highest mean scores (Lee and Guerin 2010). Enclosed private offices also had the lowest mean score for job performance enhanced by thermal quality, while enclosed shared offices were the highest (Lee and Guerin 2010). However, across all office types, there were low mean scores in general indicating persistent issues with thermal comfort regardless of layout (Lee and Guerin 2010). Another study conducted by Lee (2011b) analyzed the data from the CBE's IEQ survey and solely focused on IAQ and thermal comfort as related to occupant satisfaction and perceived job performance. The results were analyzed overall and by level of LEED certification. Those in platinum buildings had the highest ratings of thermal comfort quality satisfaction, while those in gold buildings had the highest ratings of thermal comfort quality enhancing job performance (Lee 2011b). In the lowest rating groups, the concerns were inaccessible thermostats, thermostats controlled by other people, and temperature of personal work space versus other areas (Lee 2011b). In their postoccupancy evaluation of two LEED platinum buildings, Hedge and Dorsey (2013) found that occupants rated air temperature as negatively impacting work performance, and this finding was replicated in a later study as well (Hedge, Miller, and Dorsey 2014). These collective findings point to common issues that must be addressed within green buildings in order to have a positive impact on the occupants.

19.4.4.5   Air Quality

Air quality can have a direct relationship with the health of occupants and demands specific attention, especially in green buildings. IAQ was found to have a significant positive relationship with the occupants' perceived performances in overall work space (Lee and Guerin 2009). Enclosed private offices were found to have the highest mean scores related to IAQ measures, while bullpen offices had the lowest satisfaction with IAQ (Lee and Guerin 2010). Enclosed private offices also had the highest mean scores for job performance enhanced by IAQ with high cubicles rating the lowest (Lee and Guerin 2010). In Thatcher's (2014) longitudinal comparison study, the green building group reported significantly improved air quality, including reduced stale air, less drafty conditions, improved airflow, improved ventilation, and decreased humidity. In Lee's (2011b) previously mentioned study, those in platinum buildings had the highest ratings of satisfaction with IAQ and IAQ-enhancing job performance, and silver buildings had the lowest (Lee 2011b). Those in the silver buildings most frequently reported stuffy and stale air, unclean air, and bad air smell (Lee 2011b). Issues were also found by Hedge and Dorsey (2013) wherein occupants in one LEED platinum building reported low air quality. In contrast, satisfaction with ventilation and air quality were found to be significantly higher in green buildings than in conventional buildings (Hedge, Miller, and Dorsey 2014). These mixed findings related to IAQ point to the need for future studies, especially from the perspective of the impact on occupant health.

19.4.4.6   Office Furnishings and Materials

Lee and Guerin (2009) found that occupants reported greater satisfaction with office furnishings and IEQ in LEED-certified buildings as opposed to those in non-LEED-certified buildings. In this study, the quality of office furnishing was the only IEQ category that was found to significantly affect both satisfaction and performance (Lee and Guerin 2009). Comfort, adjustability, and colors and textures of flooring, furniture, and surface finishes were the measures of office furnishings quality (Lee and Guerin 2009). Additional studies have found that LEED building occupants are generally more satisfied with office furnishings quality (Lee and Kim 2009), and that the furnishings benefit comfort, performance, and health (Hedge and Dorsey 2013). While the quality of office furnishings is not directly addressed in LEED, it was perceived to be significant by occupants and therefore should be of concern by designers and HF/E professionals as related to green buildings.

19.5   IMPLICATIONS AND DIRECTIONS FOR THE FUTURE

While it is evident that there are benefits of green buildings to the environment and the economics of organizations, the full impact on building occupants is not yet clear. There is a continued need to study occupant experiences within green buildings in order to inform the design process and ensure that the occupants' needs are being met. The research in the area of green buildings and their influence on human health is in its infancy. Humans are an organization's biggest asset and comprise the majority of the total expenses in both green and conventional buildings. Improvements in IEQ can positively impact the occupant experience leading to improved health, well-bring, and productivity (Newsham et al. 2012; Thatcher 2014). However, energy-related research typically receives the majority of funding, while IEQ research receives a negligible amount in comparison (Baum 2007). This line of research needs to become a priority if we are to truly understand the human element of sustainable development. HF/E professionals have a clear role in advancing the agenda of green ergonomics, and there are exciting opportunities on the horizon.

One of the areas that require focused attention by HF/E professionals is assisting green building projects in seeking the LEED credit for an ergonomics strategy. If more buildings are proactively considering ergonomics during the design phase in new construction or major renovations, then the needs of the occupants can be more fully met. Since the credit is not on the LEED scorecard, HF/E professionals must work toward educating those involved in green building projects on its existence and on the social, economic, and environmental benefits of pursuing the credit. Lynch (2014) presents a useful case study to highlight successful strategies in seeking and achieving the LEED ergonomics credit. Additional work needs to be done in this area both inside and outside the profession to raise awareness of the credit as well as increase the number of projects seeking it.

A relatively new certification system, the WELL Building Standard, evaluates buildings in regards to human wellness within the built environment (Delos 2015). The certification system works in alignment with LEED and other global programs and has ongoing recertification requirements, and it incorporates an ergonomics requirement (Delos 2015). There are seven rating categories, with the comfort category directly addressing ergonomics. This is different from the LEED rating system in that it is an essential part of the rating system in the WELL Building Standard versus an option within LEED (Delos 2015; USGBC 2015a). The WELL Building Standard is gaining momentum internationally, and it poses new considerations for ergonomics professionals. See Chapter 21 for more information on the WELL Building Standard.

Another exciting development is the new focus on human health and green building that was articulated by the USGBC. The "2013–2015 Strategic Plan Charge" from the 2013 Summit on Green Building and Human Health states that (p. 20)

> While green building has strong roots in economic and environmental performance, it offers vitally important benefits for human health and well-being. Designing for these benefits, "human-based design" can promote health and well-being in many ways, from reduced exposure to toxic chemicals to enhanced access to natural light, from routine physical activity to improved indoor and ambient air quality, from injury prevention to increased mobility. USGBC will pursue a range of strategies

to understand, document, implement and disseminate the health benefits of green building including healthier climate, healthier environment and healthier individuals… it's now time to raise the profile of human health concerns as part of our measures of sustainability.

While the ergonomics strategy credit is certainly an important piece of LEED and green buildings, it is one piece of the larger puzzle. As outlined in this chapter, there is a need to apply ergonomics throughout the entire green building process and rating systems, and there are many more possibilities that have yet to be defined. One area that is beginning to get connected to green building design is that of layout to enhance socialization and encourage physical activity. Our jobs are sedentary, our workers have chronic conditions, and we spend the majority of our time indoors including within our workplaces. Design that encourages socialization and physical activity must become an essential component of our buildings. As HF/E professionals, we need to make this happen and be on the cutting edge of this development. We must take our collective expertise in the varied applications of HF/E and apply them to sustainable development in order to ensure a healthy and sustainable future for all.

REFERENCES

Abbaszadeh, S., L. Zagreus, D. Lehrer, and C. Huizenga. 2006. Occupant satisfaction in indoor environmental quality in green buildings. *Proc Healthy Build*. 3:365–70. Lisbon, Portugal.

Altomonte, S., and S. Schiavon. 2013. Occupant satisfaction in LEED and non-LEED certified buildings. Center for the Built Environment. From http://escholarship.org/uc/item/4j61p7k5 (accessed November 2, 2015).

Baum, M. 2007. Green building research funding: An assessment of current activity in the United States. Washington, DC: U.S. Green Building Council.

Capon, A. G. 2014. Human occupations as determinants of population health: Linking perspectives on people, places and planet. *J Occup Sci*. 21(1):8–11.

Centers for Disease Control and Prevention. 2015. Climate effects on health. From http://www.cdc.gov/climateandhealth/effects/default.htm.

Choi, J., V. Loftness, and A. Aziz. 2011. Post occupancy evaluation of 20 office buildings as a basis for future IEQ standards and guidelines. *Energy Build*. 46(2012):167–75.

Delos. 2015. Well building standard. From http://delos.com/about/well-building-standard.

Dorsey, J., and L. Miller. 2013. Green ergonomics: Occupational therapy's role in the sustainability movement. *OT Practice*. 18(16):9–14.

Energy Star. 2015. The simple choice for energy efficiency. From http://www.energystar.gov (accessed November 2, 2015).

Fortunato, B. R., A. M. Hallowell, M. Behm, and K. Dewlaney. 2012. Identification of safety risks for high-performance sustainable construction projects. *J Constr Eng M ASCE*. 138(4):499–508.

Fuller, S. 2010. Life-cycle cost analysis (LCCA). From http://www.wbdg.org/resources/lcca.php.

Gou, Z., S. Lau, and J. Shen. 2011. Indoor environmental satisfaction in two LEED offices and its implications in green interior design. *Indoor Built Env*. 21:503–14.

Gou, Z., S. Lau, and Z. Zhang. 2012. A comparison of indoor environmental satisfaction between two green buildings and a conventional building in China. *J Green Build*. 7(2):89–104.

Hanson, M. 2013. Green ergonomics: Challenges and opportunities. *Ergonomics*. 56:399–408.

Hedge, A. 2008. The sprouting of "green" ergonomics. *Bull HFES*. 51(12):1–3.

Hedge, A., and J. Dorsey. 2013. Green buildings need good ergonomics. *Ergonomics*. 56:492–506.

Hedge, A., L. Miller, and J. Dorsey. 2014. Occupant comfort and health and green and conventional university buildings. *Work*. 49(3):363–72.

Hedge, A., K. A. Rollings, and J. Robinson. 2010. "Green" ergonomics: Advocating for the human element in buildings. *Hum Fac Erg Soc P*. 54(9):693–7.

Heerwagen, J., and L. Zagreus. 2005. *The Human Factors of Sustainable Building Design: Post occupancy Evaluation of the Phillip Merrill Environmental Center, Annapolis, MD*. University of California, Berkley, CA: Center for the Built Environment.

Hodgson, M., R. Hyde, B. Fulton, and C. Taylor-Hell. 2008. Acoustical evaluation of six "green" office buildings. *Can Acoustics*. 36(3):72–3.

Hwang, T., and J. Kim. 2011. Effects of indoor lighting on occupants' visual comfort and eye health in a green building. *Indoor and Built Env*. 20(1):75–90.

International Institute for Sustainable Development. 2012. Sustainable development timeline. From http://www.iisd.org/pdf/2012/sd_timeline_2012.pdf (accessed November 2, 2015).

IOM (Institute of Medicine). 2011. *Climate Change the Indoor Environment, and Health*. Washington, DC: The National Academies Press.

Lee, Y. S. 2010. Office layout affecting privacy, interaction, and acoustic quality in LEED-certified buildings. *Build Env*. 45:1594–600.

Lee, Y. S. 2011a. Lighting quality and acoustic quality in LEED-certified buildings using occupant evaluation. *J Green Build*. 6(2):139–55.

Lee, Y. S. 2011b. Comparisons of indoor air quality and thermal comfort quality between certification levels of LEED-certified buildings in USA. *Indoor and Built Env*. 20(5): 564–76.

Lee, Y., and D. Guerin. 2009. Indoor environmental quality related to occupant satisfaction and performance in LEED-certified buildings. *Indoor and Built Env*. 18(4):293–300.

Lee, Y. S., and D. Guerin. 2010. Indoor environmental quality differences between office types in LEED-certified buildings in the US. *Build Env*. 45:1104–12.

Lee, Y. S., and S. Kim. 2008. Indoor environmental quality in LEED-certified buildings in the US. *J Asian Arch Build Eng*. 7(2):293–300.

Loftness, V., K. P. Lam, and V. Hartkopf. 2005. Education and environmental performance-based design: A Carnegie Mellon perspective. *Build Res Inform*. 33(2):196–203.

Lynch, M. 2014. Achieving LEED credit for ergonomics: Laying the foundation. *Work*. 49(3):401–10.

McMichael, A. J., and E. E. Lindgren. 2011. Climate change: Present and future risks to health, and necessary responses. *J Int Med*. 270(5):401–13.

Millennium Ecosystem Assessment. 2005. *Ecosystems and human well-being: Synthesis*. From http://www.maweb.org/documents/document.356.aspx.pdf.

Miller, A. 2010. The relationship between usability and sustainable environments. Paper presented at the Association of Canadian Ergonomists Annual Conference 2010: Kelowna, BC.

Miller, L., J. Dorsey, and K. Jacobs. 2012. The importance of ergonomics to sustainability throughout a building's life cycle. *Work*. 41:2129–32.

Newsham, G. 2010. Research matters post-occupancy evaluation of green buildings. National Research Council Canada. From http://www.nrccnrc.gc.ca/obj/irc/doc/pubs/nrcc53568.pdf.

Newsham, G., B. Birt, C. Arsenault, L. Thompson, J. Veitch, S. Mancini, A. Galasiu, B. Gover, I. Macdonald, and G. Burns. 2012. *Do Green Buildings Outperform Conventional Buildings? Indoor Environment and Energy Performance in North American Offices*. Ottawa, ON: National Research Council Canada.

Paul, W., and P. Taylor. 2008. A comparison of occupant comfort and satisfaction between a green building and conventional building. *Build Env*. 43:1858–70.

Pavlovic-Veselinovic, S. 2014. Ergonomics as a missing part of sustainability. *Work*. 49(3):395–9.

Pearsall, H., and J. Pierce. 2010. Urban sustainability and environmental justice: Evaluating the linkages in public planning/policy discourse. *Local Env*. 15(6):569–80.

Peretti, C., and S. Stefano. 2011. Indoor environmental quality surveys: A brief literature review. Indoor Air 2011, Dallas. June 5–10.

Pyke, C., S. McMahon, and T. Dietsche. 2010. Green building and human experience. USGBC. From http://www.usgbc.org/ShowFile.aspx?DocumentID=7383.

Romm, J. J. 1994. *Lean and Clean Management: How to Boost Profits and Productivity by Reducing Pollution*. New York: Kodansha International.

Schiavon, S., and S. Altomonte. 2014. Influence of factors unrelated to environmental quality on occupant satisfaction in LEED and non-LEED certified buildings. Building and Environment. From http://www.solaripedia.com/files/1174.

Singh, A., M. Syal, S. C. Grady, and S. Korkmaz. 2010. Effects of green buildings on employee health and productivity. *Am J Pub Health*. 100:1665–8.

Thatcher, A. 2013. Green ergonomics: Definition and scope. *Ergonomics*. 56(3):389–98.

Thatcher, A. 2014. Changes in productivity, psychological well-being and physical well-being from working in a "green" building. *Work*. 49(3):381–93.

Turner, C. 2006. LEED building performance in the Cascadia Region: A post occupancy evaluation report. Cascadia Region Green Building Council.

UN (United Nations). 1987. Report of the World Commission on Environment and Development: Our common future. From http://www.un-documents.net/our-common-future.pdf.

UN. 2012. The future we want. From http://www.uncsd2012.org/content/documents/727The%20Future%20 We%20Want%2019%20June%201230pm.pdf.

UN. 2015. Goal 7: Ensure environmental sustainability. From http://www.un.org/millenniumgoals/environ .shtml.

US EPA (United States Environmental Protection Agency). 2015. What is sustainability? From http://www.epa .gov/sustainability/basicinfo.htm#sustainability.

USGBC (United States Green Building Council). 2013, January. Health is a human right: Green buildings can help. A report from The Summit on Green Building and Human Health. From http://www.usgbc.org /sites/default/files/GBHH_Final_1.pdf.

USGBC. 2015a. About USGBC. From http://www.usgbc.org/about/.

USGBC. 2015b. We take our rating systems seriously. From http://www.usgbc.org/about/leed.

USGBC. 2015c. Ergonomics strategy. From http://www.usgbc.org/node/4631863?return=/credits/new-construction /v4/pilot-credits.

USGBC. 2016. LEED credit library. From http://www.usgbc.org/credits/new-construction/v4.

Werbach, A. 2009. When sustainability means more than green. *McKinsey Quarterly*. 4(2009):74–9.

Whole Building Design Guide. 2015. Green building standards and certification systems. From http://www .wbdg.org/resources/gbs.php.

World Federation of Occupational Therapists. 2012. *Position Statement: Environmental Sustainability Sustainable Practice within Occupational Therapy*. From http://www.wfot.org/ResourceCentre.aspx.

Yanarella, E. J., R. S. Levine, and R. W. Lancaster. 2009. Green versus sustainability: From semantics to enlightenment. *Sustainability*. 2(5):296–302.

20 Exploring the 3C Workplace
For Connectedness, Collaboration, and Creativity

So-Yeon Yoon and Susan S. E. Chung

CONTENTS

20.1 INTRODUCTION

In today's rapidly changing and seamlessly interconnected global economy, connectedness, collaboration, and creativity (3C) have become increasingly important workforce qualities to cultivate for any business organization. These 3Cs ensure that a business is agile, responsive, capable of adapting to predictable or unpredictable challenges, and prepared to successfully manage change and to grow in the competitive market. Organizations have gradually recognized the value of the physical environment in shaping human and organizational performances; it is the second most expensive resource after human resources (McCoy 2005). With increasing frequency, organizations are transforming their physical spaces and integrating emerging technologies to better facilitate collaboration, as collaboration across disciplines, departments, and silos is recognized as the core of knowledge workers' creativity and innovation. The effectiveness of innovation labs is a result of escaping from conventional workplace designs to work spaces that facilitate the creative process (Magadley and Birdi 2009). Enhanced creativity is fostered by spaces that enhance the flow of knowledge and the exchange of ideas (Martens 2008) and generate movement and spatial patterns that facilitate serendipitous creative encounters (Wineman, Kabo, and Davis 2009). Organizational "whitespace" (unoccupied territory) can also facilitate organizational innovation (Maletz and

Nohria 2001; Jakobsson and Stiernstedt 2010; Johnson 2010). Other important considerations when designing to facilitate creativity include providing separate spaces for different types of activities (Meusburger 2009), accommodating personal preferences for physical environmental conditions, and offering personal control over the task environment (Martens 2011). The presence of creativity-supporting elements (e.g., furniture, indoor plants/flowers, calming colors, inspiring colors, privacy, window view to nature, any window view, quantity of light, daylight, indoor physical climate, positive sound, and positive smell) is related to higher creative performance in knowledge workers (Dul, Ceylan, and Jaspers 2011). The evaluation of a studio environment revealed that the personalization of a studio space enriches the creative process and outcomes (Hasirci and Demirkan 2007). Research also shows that spaces designed to encourage physical activity and movement enhance creativity (Singh-Manoux et al. 2005).

Previous studies have addressed the following: the critical role played by the interior designer in creating physical environments for organizational creativity, the design of spaces as innovation incubators (Mumford 2001; Hatch 2006) and spaces that enhance organizational performance (Gregson and Rose 2000; Lewis 2008), and the design of spaces that facilitate new organizational forms (Girard and Stark 2008). Unfortunately, empirical studies elucidating this relationship are scarce, and most research studying environmental influences on creativity and innovation has focused on the social environment (Hunter, Bedell, and Mumford 2007). Investigating the environmental design factors conducive to creative knowledge environments is becoming a priority (Hemlin, Allwood, and Martin 2008). Case studies of innovation labs—work spaces built to enhance and support creativity and innovation in organizations—suggest that structures should be designed with flexibility, and that the infrastructure should complement the physical design through the use of computer and other technology-based tools, through multiple open work spaces, and through designs and furnishings that eliminate hierarchy (Lewis and Moultrie 2005). Mobile and social sensing technologies offer new design research opportunities for capturing photo/voice/video data directly from the participants' phones, phablets, and tablets, and the resulting big data offers greater fidelity for postoccupancy evaluations. Social sensing devices (Olguín-Olguín et al. 2009) can also capture movement, face-to-face interaction, speech quality, and location information. The elements of the environment can be captured using environmentally aware computing (Olguín-Olguín and Pentland 2010; Tripathi and Burleson 2012). In this chapter, a case study is described that was conducted to explore how the emerging social sensing technologies can be combined with narrative feedback and strategic storytelling to identify specific design attributes in the workplace that deliver a measurable impact on process and output.

20.2   WORKPLACE CONNECTEDNESS, COLLABORATION, AND CREATIVITY

The physical environment is a multifaceted construct composed of many attributes that interact in creating a holistic environment. Understanding these attributes of the physical environment is essential for creating work spaces for creativity. To date, several classifications of the physical environment exist. Davis (1984) categorized the physical environment attributes in office settings as physical structure, physical stimuli, and symbolic artifacts, those that are observable and can be described somewhat accurately. The physical structures include the architectural design and the physical placement of furnishings in relation to social interaction; the physical stimuli are the aspects of the physical setting that impact the user's behavior; and symbolic artifacts are objects that individually or collectively guide the interpretation of the social setting. McCoy (2005) also identified five physical environment components related to creativity: spatial organization, architectonic details, views, resources, and ambient conditions. The spatial organization consists of many characteristics, which include size, shape, allocation, or division of space through furniture configuration and circulation routes, level of enclosure, proxemics, territoriality, flexibility, visual access, and so on. Architectonic details are fixed or stationary aesthetics, ornaments, or materials that communicate a sense of identity or purpose through how those elements are used. The views are observable features that can be of the natural or the built environment and can be intimate or panoramic. Resources are related to the accessibility and the functionality of the environment and can be scarce, finite, nonrenewable resources (i.e., money, time) or expandable, renewable resources (e.g., motivation). Finally, ambient conditions are environmental stimuli, such as illumination, heating, ventilation, and acoustics. From these two frameworks, the workplace physical environment for creativity can be organized into ambient, spatial, and symbolic factors. Ambient factors are atmospheric elements of the indoor built environment, spatial factors describe the structural and interior spatial elements, and symbolic factors are features that communicate messages and add meaning to the work space. These factors together create behavioral, emotional, and technological connectedness in the workplace, can also provide opportunities for collaboration, and lead to creative performance and satisfaction.

20.2.1 Connectedness

Connectedness defined here is positive behavioral, emotional, and technological interactions with interior attributes in the physical setting directly or indirectly contributing to creativity. Connectedness relates to the creative interior attributes from a review of evidence that focuses on three factors: ambient factors (e.g., acoustics, luminous, air quality, thermal), spatial factors (e.g., form, shape, layout, view), and symbolic factors (e.g., plants, color, furniture, artifacts, material). The sources of this evidence include scholarly publications across disciplines, industry white papers, and findings from questionnaires from the 3C design case study.

20.2.1.1 Ambient Factors

Ambient factors include acoustics, lighting, temperature, and air quality. Acoustics, in particular, have become a pressing issue in offices, especially in an open layout. Indoor ambient noise can be a distraction or interruption that inhibits creativity (Martens 2011). More specifically, noise affects arousal and can serve as a distractor that inhibits attentional focus and impairs cognitive performance; however, it can also be an arousing stimulus that provides cues to aid the task performance. For example, Mehta, Zhu, and Cheema (2013) found an inverted U relationship between noise level and creative performance in novelty and divergent thinking as measured by the remote associates test and in idea-generation tasks (judged), in which the creative performance was highest in 70 dB of various background noise recorded from real-life venues, compared to 50 and 85 dB. This optimal sound level was associated with higher arousal, more processing difficulty, and higher construal level. Toplyn and Maguire (1991) also suggested this inverted U-shaped relationship, but found that the creative performance was best at 80 dB of white noise than in 60 or 100 dB only among high creative potential participants. The type of noise also affects creativity: unpredictable noise has a more damaging effect on judged creativity than predictable noise or no noise (Kasof 1997). Noise unpredictability and intelligibility impaired creativity more for individuals who could not screen environmental stimuli. Although these studies allude to a moderate level of noise (78–80 dB) leading to creative performance, this may depend on the individual's creativity potential and screening ability, and also on the particular creativity task.

Lighting in general has many psychological and behavioral effects including mood and cognitive performance. From interviews with employees of Brazilian organizations, de Alencar and Bruno-Faria (1997) identified adequate light as a stimulant to creativity and insufficient illumination as an obstacle to creativity. Similar to acoustics, different lighting characteristics such as illuminance levels, color temperature, and color rendering index (CRI) affect the creative performance in different ways. Dim illuminance (150 lux; 14 fc) was found to enable global processing and facilitate freedom from constraints and hence support the creative performance for insight problem solving and a structured imagination task compared to bright illuminance (1500 lux; 140 fc) (Steidle and Werth 2013). Warm white light at 300 lux (28 fc) illuminance and cool white light at 1500 lux illuminance in low CRI was found to be optimal for problem-solving (Knez 1995).

High temperature was unanimously mentioned as a physical attribute that inhibits creativity according to interviews of 10 creative leaders in a study conducted by Martens (2011). However, the interviewees did not provide specific temperature values, and the geographical characteristics of the interviewees' environment (i.e., UK and the Netherlands) may have an impact on their perception of temperature levels. Air ventilation was identified as a characteristic of the physical environment that stimulated creativity (de Alencar and Bruno-Faria 1997).

20.2.1.2 Spatial Factors

Physical space affects the well-being of people, the channels of information, and the availability of knowledge tools, and sets the stage for coherence and continuity, which may contribute to competitive advantages in creativity (Kristensen 2004). The spatial form of the interior space, defined by the structural size and shape, can be determined by rectilinearity and complexity. Although the size and rectilinearity of space was found to have no relation to the creative potential of interior spaces, higher visual complexity was perceived to have higher creative potential among college students (McCoy and Evans 2002), whereas the same was associated with low creative potential among office managers in a similar study (Celyan, Dul, and Aytac 2008).

The physical layout of workplaces can enhance the social process of creativity through interaction patterns (Martens 2011; Sailer 2011). Creativity, in particular, takes place in a physical context where the configuration, the design, and the management of space can restrict and support the flow of knowledge and the exchange of ideas, while inducing emotions that facilitate or reduce the enhancement of creativity (Kristensen 2004; Martens 2008). Layouts also facilitate different benefits for different activities: long corridors facilitate a hierarchical organization with people in separated rooms, flat structures afford open space where people can interact at many levels, narrow paths that only allow sequential passages reduce interaction, and linear spaces make it difficult for a group to assemble and have discussions, whereas centralized or radial shapes are appropriate for communal space and creativity (Kristensen 2004). The type of office arrangement (i.e., closed or open-plan) has also a great impact on employee behavior and communication (Ornstein 1989). Open-plan offices increase communication, rate higher in aesthetics, and have more group sociability among employees than conventional designs (Maher and von Hippel 2005). Relationships may form according to the spatial proximity of employees, having an effect on communication and employee satisfaction (Oldham and Rotchford 1983). Conversely, open-plan offices have been linked to increased workplace noise, disturbances, distractions, feelings of crowding, and reduced privacy (Maher and von Hippel 2005). The increase in distractions may offer more cues to utilize for creative ideation; however, when attention or concentration is required, the overstimulation from the open environment may hinder cognitive processes and performance.

Workplaces that provide opportunities for physical change (i.e., flexibility) through design also support creativity. Csikszentmihalyi (1996) observed that creative individuals are most likely to change their environment according to the rhythm of their thoughts or habits of their actions. The interaction between people and the environment played out by flexibility in spatial arrangement can enable many of the dimensions for a creative climate. When users have the control to design an environment to accommodate their needs for the given task and goal, the process supports dynamism, provides opportunities for idea support, gives a sense of freedom over the environment, and challenges the exploration of the adequate arrangement for the given task and situation (Vithayathawornwong, Danko, and Tolbert 2003). Similarly, higher levels of perceived control can influence the employees' ability to effectively use their work space and its adjustable features, and can lead to higher environmental satisfaction and communication (Huang, Robertson, and Chang 2004). More personal control over the physical work space and easy access to team or communal places lead to higher perceived group cohesiveness and job satisfaction (Lee and Brand 2005). Particularly, the perceived control symbolizes freedom, trust, and organizational support, which may lead to the improvement in creative tasks in the long run (Veitch and Gifford 1996).

20.2.1.3 Symbolic Factors

Symbolic attributes focus on the communicative messages in the workplace. Windows or access to an exterior view is highly preferred in offices, especially due to the positive and restorative effects of natural views on performance (e.g., Kaplan 1995). Natural elements in the indoor environment have been found to convey messages that may not impact the performance in tasks, but rather impact the emotional perception of the environment. The amount of wood grain texture and the overall use of natural materials, preferred over synthetic and composite materials, were positively associated with the creative potential of the physical environment (McCoy and Evans 2002) and an overall favorable first impression on interpersonal perception (including creativeness) in offices (Ridoutt, Ball, and Killerby 2002). In addition, the presence of indoor plants has been evaluated to be calmer and less distracting and found to positively affect association tasks (Shibata and Suzuki 2002).

Messages are also conveyed through the furnishings and the overall style of the workplace. The mood states of arousal and psychological safety can be conveyed through the combination of color, lighting, materials, sitting arrangement, and furniture type. For example, a study investigating mood and creativity designed a room with green colors, armchairs placed in a circle with a low round table, wooden materials, a poster of a natural environment, objects to create a domestic atmosphere, and dim lighting (de Korte, Kuijt, and van der Kleij 2011). The overall impression of this room was rated to be high in pleasantness and social status. On the other hand, the room designed for arousal had warm, red colors, stools placed in a circle with no table, a poster with complex figures of fractals and bright lighting. This room was rated to be high in originality and complexity. No direct effects were found between the rooms on creativity measures; however, the arousal tested by heart rate variability was found to be a possible mediating factor, and an interaction effect between the interior and the task type indicated that the optimal pairing of these two could enhance creative performance.

The social aspects of the environment such as the creative climate dimensions (Ekvall 1996) can be expressed through the physical environment and provide a workplace that is conducive to creativity. For example, some companies include playfulness in the design of their workplace from the belief that a fun workplace, which intentionally encourages, initiates, and supports a variety of activities that positively impact the attitude and the productivity of individuals and groups, may improve employee morale, communication, performance, recruitment, and retention (Ford, McLaughlin, and Newstrom 2003). Environmental distractions and poor social climate at work can restrict the employees' experiences of creativity by interfering with their concentration on job-related tasks or by increasing unpredictability and uncontrollability, resulting in the belief that the workplace does not support their efforts to be creative (Stokols, Clitheroe, and Zmuidzinas 2002). Work environment satisfaction was also associated with the extent of the organization's personalization policy (i.e., amount of control over the individual work space), with the amount of personalization found to positively correlate with creativity (Wells 2000). Personalization allows for the individual expression and a very subtle interaction between the people and the environment.

20.2.2 COLLABORATION

Industry leaders including Google, Apple, Facebook, and recently Yahoo have been promoting collaboration and creativity by mixing workers from different departments in shared spaces (Miller and Rampell 2013) and overlapping movement paths (Lehrer 2011). The spatial layout or the configuration influences the social interactions that are necessary for both effective task performance and satisfaction of the social needs in organizations. Layouts that create pathways for people to get around to different areas of the workplace can be planned to induce spontaneous interaction that is believed to encourage creative idea exchange (Leonard and Swap 1999). Collaboration tasks may benefit from working in a layout that is more open for free-flowing communication or having work spaces (i.e., desks) that are close to each other for simultaneous idea exchange or having a large team table or tools to facilitate team discussions. While our understanding of what factors in organizational and environmental designs will most accelerate creative innovation is still forming, it is established that various types of interactions and collaborations serve as catalysts for creativity and innovation in knowledge workplaces.

Different spatial settings, activities and cognitive intensity, and personal preferences can have an impact on how the physical workplace supports creativity (Martens 2011). Case studies of innovation laboratories—structures purposely designed to enhance and support creativity—have also suggested different spatial designs according to the needs of each phase of the creative process (i.e., preparation, incubation, insight/illumination, and elaboration/evaluation). While open plan offices can facilitate the collaborative phase of the creative process, places that provide moments of relaxation are preferred for creative thinking (Haner 2005). Penn, Desyllas, and Vaughan (1999) conducted spatial syntax analyses of innovative organizations and identified that centrally located spaces afford unplanned interaction and rapid transfer of ideas, whereas segregated spaces afford better execution of tasks. In addition to the functional support of the physical environment for creativity, Lewis and Moultrie (2005) also suggested the importance of a dislocation effect in which users diverge from traditional norms and are surrounded with design elements that communicate opportunities that are different from day-to-day activities (e.g., writing on walls, doors utilizing high technology, etc.). Magadley and Birdi (2009) describe this as "getting away from the workplace" in order to differently think, and suggest that creative behavior can be affected depending on the opportunities provided by the physical environment.

20.2.3 CREATIVITY

Creativity is a complex and multifaceted phenomenon that has become an essential part of work in various disciplines. Creativity has been defined as "the interaction among *aptitude, process, and environment* by which an individual or group produces a *perceptible product* that is both *novel and useful* as defined within a *social context*" (Plucker, Beghetto, and Dow 2004, p. 90, emphasis in original). Businesses from all sectors consider creativity as a competitive advantage and seek to hire creative individuals and implement creative processes in both management and problem-solving tasks, as it is recognized to be fundamental for positive change (Hennessey and Amabile 2010). The degree of creativity (i.e., incremental creative solutions to monumental breakthroughs) expected in organizations may depend on the field of work and its task; however, in general, individual creativity has been linked to organizational creativity, its performance, growth, and survival (Woodman, Sawyer, and Griffin 1993).

As creativity outcomes, objective performance, and subjective satisfaction data will be gathered from the participants and the managers in each workplace. The KEYS scales, a widely recognized instrument and the current standard for measuring team creativity and innovation in workplaces (Amabile et al. 2005), will be adopted to measure a variety of factors such as affect, rewards, and motivation. It asks participants to self-report their creativity on a Likert scale (Amabile 1983; Amabile et al. 1996). The KEYS survey also assesses the perceived stimulants and obstacles to creativity in organizational and physical work environments, i.e., negative and positive aspects of the environment. In collaboration with the managers of the participating workplaces, we will develop a workplace performance measurement rubric derived from the business metrics to assess the behavior, the quality outcome, and the process efficiency of the workplace. Additionally, we will adopt the job demand control model (Karasek 1979) to assess the employees' views of workplace performance.

20.3 DEVELOPMENT OF THE 3C DESIGN FRAMEWORK

As illustrated in Figure 20.1, the 3C framework is an integrated approach using social sensing technologies and strategic self-reported data to identify critical evidence-based design parameters and

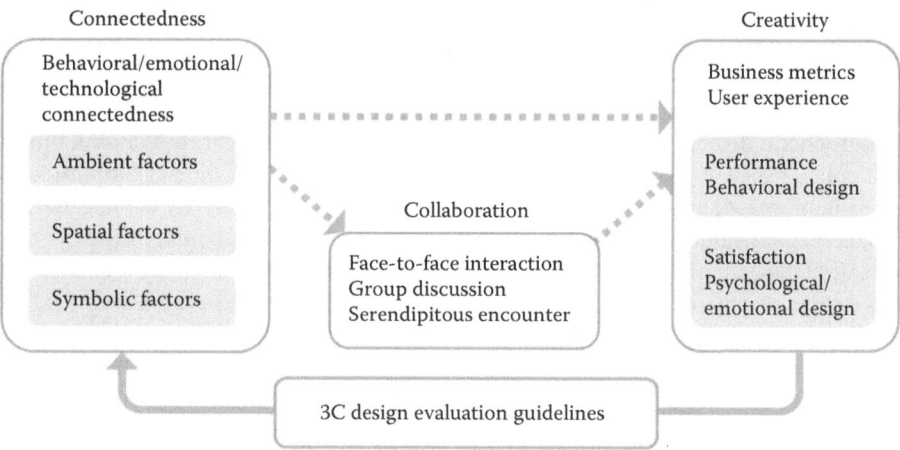

FIGURE 20.1   3C framework.

to develop guidelines for strategic decision-makers who want to create 3C workplaces that support an organizational culture of innovation. Connectedness, the positive behavioral, emotional, and technological interactions with the interior attributes in the physical setting directly or indirectly contributing to creativity, creativity outcomes, objective performance, and subjective satisfaction data, were gathered from the participants and the managers in each workplace from self-reported surveys. The self-reported survey scales were developed to measure the subjective satisfaction on perceived connectedness, collaboration support, productivity, and creativity on a Likert scale. It allows us to examine the relationships between the 3Cs. In collaboration with the managers of the participating workplaces, a workplace performance measurement rubric derived from the business metrics to assess the behavior, the quality outcome, and the process efficiency of the workplace can be also developed and incorporated into the analysis.

Acknowledging that various types of interactions and collaborations serve as catalysts for creativity and innovation in knowledge workplaces, the interactions were measured through sociometric badges that gather data on spatial movement patterns, track the location, and analyze the elements of the participants' social interactions. It is a recently developed social sensing technology that is often used in organizational studies, and this is the first time that it is used for design research purposes.

The social sensing badges use four types of measurement technologies (infrared, Bluetooth, microphone, and accelerometer) to capture a different dimension of the social interaction. The badges consist of a wearable sensor worn as a pendant around the neck that will record the network data at 17 Hz and the body movements (two-dimensional accelerometer) at 50 Hz, and also include an embedded speaker at 8 kHz. They communicate data through a bidirectional infrared transceiver, an accelerometer, and by low-resolution microphone analysis. The badges detect and time face-to-face interactions, proximity to other workers wearing badges and to the base stations, speech and nonverbal features, and physical activity. To ensure the privacy of the participants, no personally identifiable data are recorded. A sociometric software analyzes the daily adjacency patterns of face-to-face interactions between the participants. Since their recent debut at the Massachusetts Institute of Technology Media Lab, sociometers have been extensively validated in several studies (Olguín-Olguín et al. 2009; Jayagopi et al. 2010; Dong et al. 2012). The present study specifically looked at body movement activity, consistency of body movement, speech activity, face-to-face interaction time, cohesion, and centrality. The accelerometer data were collected with a sampling frequency of 50 Hz in three dimensions, which estimates the degree of the body movement by averaging the signal power over the three axes (Lepri et al. 2012). It measures the body movement by averaging all the samples' acceleration signal magnitudes for each 1 min interval. The consistency of the body movement is measured by calculating the standard deviation (SD) of the accelerometer signal magnitude for each 1 min interval for all samples, and subtracting that value from a constant representing 100% consistency (Olguín-Olguín and Pentland 2010).

The microphone detects speech data without recording the content, thus protecting the user's privacy. It collects data at a sampling rate of 8 kHz. Every 50 ms, the signal's amplitude, SD, minimum, maximum, mean, and variance values are taken. The speech activity is measured by averaging all sample signal values in 1 min intervals (Olguín-Olguín and Pentland 2010).

The infrared data are collected when two badges are in the direct line of sight of each other. Each badge has a cone-shaped detection zone with a height (h) up to 1 m and a radius of $r \leq h \tan \theta$, where $\theta = \pm 15°$ (Lepri et al. 2012). Face-to-face interactions are recorded with this measure only when the two badges are facing each other and are in each other's detection zone. Cohesion and centrality are also measured with infrared by looking at which badges are interacting with each other to see if the communication is equally distributed among the badges or if a few badges are the center of interactions.

The Bluetooth data are collected in an omnidirectional fashion within a 10 m radius (Olguín-Olguín et al. 2009). Radio signal strength indicators (RSSIs) are used to measure the signal strength between transmitting and receiving badges. When the RSSI is over a predetermined threshold, then the two devices are recorded as being in close proximity.

20.4 CASE STUDY

20.4.1 DESIGN CHARRETTE: COLLABORATIVE DESIGN PROBLEM-SOLVING

A design charrette was conducted to explore short and intensive collaboration toward creative solutions. The design charrette is a specific form of engaging people in the planning process (Kelbaugh 2014) often used in design education and practice. Since the end of the nineteenth century, the design charrette has been used in architectural design education to collect students' drawings for jury critiques using a pushcart (*charrette* in French) as the deadline approaches. Design charrettes are mostly for complex and controversial problems and considered as an effective way for creative problem-solving. While some define the design charrette as "two or more day intensive design workshops in which mixed group of participants work collaboratively toward designing future visions for a certain area" (Roggema 2014), it may also refer to any collaboration session with mixed design teams to provide a creative environment to think and draft design solutions. Today, charrettes take place in workplaces of many fields beyond design and has become a technique for intensive group workshops and meetings to engage all stakeholder groups for joint ownership in planning and solving problems.

Problem-solving has been a long-standing topic in many fields, generally understood as the process involved in seeking solutions to problems (Yoon and D'Souza 2009). In the context of interior and architectural design studios as well as professional workplaces, problems are authentic and peer collaboration is a common practice. Design problems are nonroutine and often ambiguous; thus, it needs to be creative. MacKinnon (1978) viewed creativity as a process characterized by a novel response to a problem and defined the creative process as a problem-solving process. Gelfand (1988) also noted that creativity is a special form of problem-solving thinking needed for an unusual or a unique situation. Especially for design problems, creativity and problem-solving are seamlessly intertwined concepts.

In design charrettes, the four most significant rules that are acknowledged (Condon 2008) regard the participants to do as follows:

1. Design with everyone in the team in the design process during the charrette to be integrative and to generate a variety of possible solutions.

2. Start with a blank sheet, so the challenge is to fill in the future and fill up the empty paper.
3. Provide just enough information to avoid decision paralysis or inappropriate proposals.
4. Consider the drawing as a contract to have a well-understood agreement on submittals at the end of the charrette.

20.4.2 Research Methodology

One full day of a collaborative design charrette with college student teams majoring in interior design and architecture was conducted as a case study to test the 3C framework. After invitation e-mails and flyers were sent out to recruit participants for the charrette and case study, 18 students volunteered to participate in the design charrette from 9:00 a.m. to 6:00 p.m. in a 1050 ft^2 (~98 m^2) classroom on a Saturday. The challenge was to design an innovative office space based on the provided materials including site and building layout information and specific submission requirements. Their academic levels in design programs and computer skills were comparable, and the instructor obtained signed consent forms from each participant. Then, the participants were asked to wear the social sensing badges and keep them on during the entire time (Figure 20.2).

Five teams, each consisting of three to four members, were formed. At the end of the charrette, each team was asked to propose a novel and applicable design solution. After the submission of the work, each participant completed a set of survey questionnaires on their subjective evaluation about the physical environment and the experience of collaboration and creativity. Connectedness in the study was operationalized by the degree of flexibility in the group seating arrangement. All other physical factors were controlled to keep the design of the study simple. Four teams were randomly assigned to their desks. Low-back chairs on casters providing high mobility were used for all participants. Trapezoidal desks on casters were provided to four teams and one team was assigned to rectangular desks stationed without casters (Figure 20.3).

In the postcharrette survey, the participants rated their subjective satisfaction of the general environment and the team work area layout in supporting their tasks. Collaboration was measured with subjective ratings as well as social sensing data collected from the badges. Perceived quality and satisfaction of the collaboration experience were asked in the survey. In an enclosed space, all the teams were relatively close to each other, so the audio sensors and the Bluetooth sensors of social sensing devices were also technically detecting signals from adjacent participants outside the team. Therefore, speech and proximity interaction data between members outside their own teams were not considered for analysis. The interaction across teams can be one of the important types of

FIGURE 20.2 Social sensing badges.

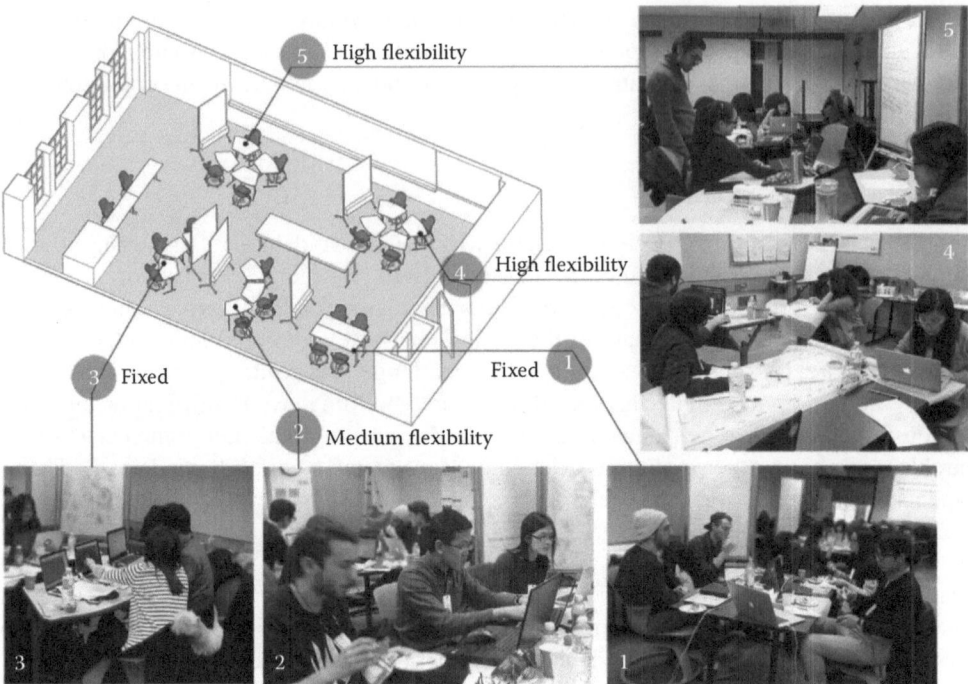

FIGURE 20.3   Five teams with three levels of seating flexibility.

creative activities for many workplaces. However, there was no cross-team interaction observed as the teams were competing with each other in this case. Among a number of social sensing data features from five categories for analysis, i.e., body movement, posture analysis, speech analysis, social network, and turn-taking analysis, four features were selected based on the balance between relevance and convenience: overall activity, body movement consistency, face-to-face and proximity combined interaction, turn-taking, and speech duration.

Creativity was measured based on the participants' subjective evaluations on the perceived creativity they experienced during the collaborative problem-solving charrette in addition to the review scores from the professional designers who served as reviewers. Two professional architects and

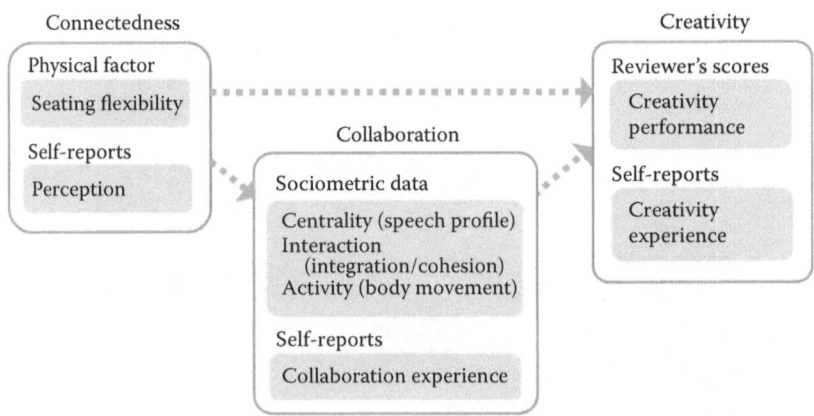

FIGURE 20.4   The case study model following the 3C framework.

two design professors were invited, and they reviewed the five teams' design proposals together. While the reviewers were exchanging opinions, each assessed the proposals individually using three assessment criteria: creativity in the concept and design development, quality and completeness of presentation and applicability of the solution. The reviewers agreed on the ranking of the teams. Both ranking and mean scores were considered as creative performance outcome measures. The following diagram summarizes the measures used in the case study following the proposed 3C framework (Figure 20.4).

20.4.3 Analysis and Results

20.4.3.1 Connectedness

Connectedness was manipulated with high, medium, and low levels of seating flexibility. In addition, the participants' responses to the survey questions were examined in the correlation analysis to explore how they are related to sociometric dimensions and how the subjective perceptions of the workplace and the layout of the team's desks are related to perceived collaboration and creativity. As seen in Table 20.1, both overall space connectedness and layout connectedness were strongly correlated with each other and perceived collaboration satisfaction and support. It was also observed that there is a negative correlation between perceived connectedness and body movement/activity.

20.4.3.2 Collaboration

Face-to-face interaction via infrared detections and proximity via Bluetooth detections determine the number of seconds that a badge was within the interaction distance of other badges. No significant difference in the average detection numbers between teams was found from a one-way analysis of variance. However, strong significant differences were found in speech levels, speech centrality, and energy level changes over time. Each badge at every timestamp records the amount of time spent speaking, listening, and in silence. By comparing the total speaking and the silence time between badges within a team, it is possible to examine the sound balance. Figure 20.5 shows the SD values across the teams to compare the centrality within each team. The higher the SD, the stronger the centrality the team has. In other words, team 1 (fixed)'s exceedingly higher SD value indicates that there were members dominating their discussions, while team 2 (medium flexibility) members' verbal interactions were noticeably more balanced than the other teams. The energy level changes over time (in hours) were explored with body movement/activity data. The activity data can be analyzed in any intervals; the average scores by hour between teams were compared for this study.

The activity levels can also be interpreted as energy levels. As shown in Figure 20.6, participants with higher levels of flexibility obviously moved more throughout the charrette. The two fixed seating teams demonstrated significantly lower activity levels compared to the two high flexibility teams. It was interesting to notice that only the two fixed teams' energy levels continued to decrease near the end of the charrette, while the other teams showed their activities picking up when gathering materials by the time for wrapping up and putting together the submittals.

20.4.3.3 Creativity

Creative performance outcomes of the collaborative design problem-solving charrette were analyzed with review scores based on the criteria to judge originality (creative concept and approach), applicability (functionality and human impact), and elaboration (fulfillment of requirements, presentation quality) using a nine-point rating scale. The two fixed seating teams scored the lowest; the high flexibility teams were ranked second and third. Team 3 with medium flexibility received the highest scores.

In the centrality analysis (Figure 20.7), team 2 had the most balanced speech proportions among the members, whereas team 1 had significantly strong centrality. The activity data demonstrated overall lower activity and energy levels for the two fixed seating teams compared to the other teams.

TABLE 20.1

Correlations between Survey Responses and Sociometric Data

	1	2	3	4	5	6	7	8	9
1. Body movement consistency	1.00								
2. Activity	−1.00***	1.00							
3. Speech	0.12	−0.13	1.00						
4. Interaction	0.60**	−0.59**	0.71***	1.00					
5. Perceived creativity	0.41	−0.46*	0.83***	0.64**	1.00				
6. Collaboration support	0.82***	−0.83***	0.05	0.64**	0.37	1.00			
7. Perceived collaboration	0.19	−0.17	−0.03	0.51*	−0.06	0.66	1.00		
8. Space connectedness	0.54*	−0.55**	−0.12	0.50*	0.14	0.92***	0.86***	1.00	
9. Layout connectedness	0.49*	−0.51*	−0.28	0.35	0.02	0.88***	0.83***	0.99***	1.00

$*p < 0.05$; $**p < 0.01$; $***p < 0.001$.

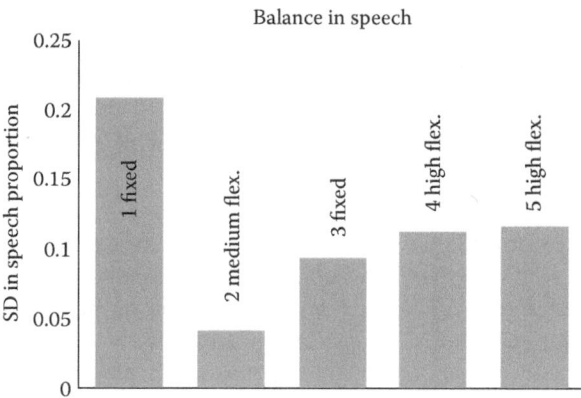

FIGURE 20.5 SD in speech proportion.

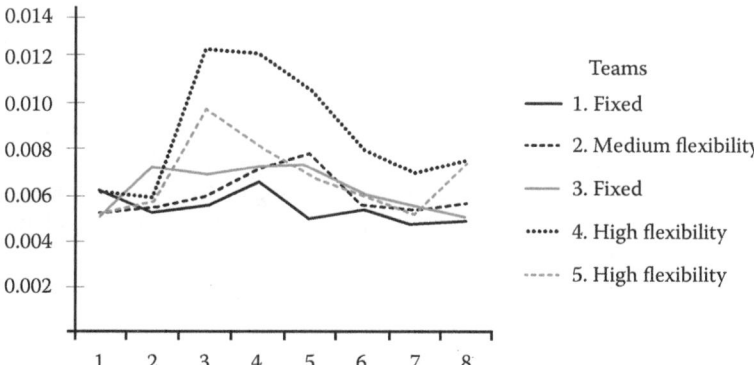

FIGURE 20.6 Activity levels over time.

Teams	1 Fixed	2 Medium	3 Fixed	4 High	5 High
Review criteria and scores (9 pt ratings)				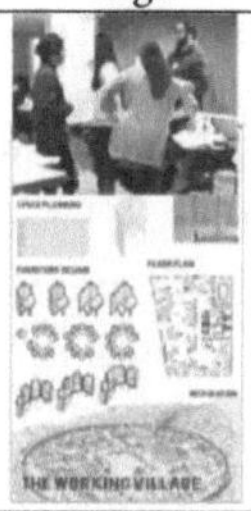	
Creativity	4.67	7.38	6.25	7.63	6.5
Applicability	4.86	6.88	5.25	6.38	4.67
Elaboration	4.67	6	2.75	5.75	5.33
Average	4.47	6.4	4.45	5.65	4.93
Ranking [ ]	[4]	[1]	[5]	[2]	[3]

FIGURE 20.7 Creative performance outcomes, review scores, and rankings.

In this case study, it was also observed that high flexibility is strongly related to body movement and activity but not necessarily to higher creative performance compared to the team with medium flexibility. In the context of a design charrette, perhaps high flexibility may lead to more distraction than control.

20.5   FUTURE DIRECTIONS

The connectedness among knowledge workers is essential for creativity and innovation. Although the change in the nature of work has knowledge workers relying on technology to stay remotely connected, the significance of face-to-face interactions has never been undermined. The connectedness among knowledge workers in the physical workplace intensifies the creative potential of the organization and is a necessary component to consider when designing the physical work environment. In order to make evidence-based design decisions for connectedness, collaboration, and creativity in the workplace across multiple disciplines and for organizations of multiple sizes, future research in this area is needed.

Technology gives us the advantage to capture big data in a short time. In the case of sociometric badges, one small device can track multiple data points, especially when interacting with other devices. When the badge data is exported in the DataLab software as a spreadsheet, many data forms are generated for each category, i.e., body movement, turn-taking, speech, interactions, and miscellaneous. There are many ways to select data types to analyze. From the series of case studies, we found that centrality, integration/cohesion, and activity/energy have been the most useful sociometric dimensions linked to the creativity outcomes and the employees' experience of the workplace. However, those dimensions can be interpreted differently from one workplace to another; thus, different data types can be used for the same dimensions depending on the context. For example, centrality can be measured with face-to-face and proximity data in a workplace, but speech balance data were more appropriate for the design charrette case study. The numerous capabilities of the device and the social sensing technology allow for research to quickly advance in the investigation for connected, collaborative, and creative workplace designs. It should be noted that this case study was conducted for the purpose of providing proof of concept, and that multiple case studies in various disciplines of work is being conducted for this project to gather further data. The capabilities of the device and the social sensing technology allow for research to quickly advance in the investigation for connected, collaborative, and creative workplace designs. We will develop a website for the 3C design knowledge base explaining the ambient, spatial, and symbolic factors for connectedness and how they relate to collaboration and creativity in the workplace. Design guidelines will also be generated based on the case studies from this project and other future studies to provide designers and organization leaders with the most cutting edge information for creating innovative knowledge workplaces.

The project schedule, particularly the data collection, was significantly affected due to unforeseen issues in conducting the field studies with the partner offices. The primary issue came from the conflict between the company policies' strong emphasis on information security and the built-in audio recording features of the wearable social sensing devices. While the researchers demonstrated the technology's capability to turn on or off the audio recording feature during initial site visits to global marketing and design companies, the concern was not resolved due to intensified sensitivity from the increase in intellectual property cases in the field. The interest in big data and new technology is great, yet participating in it still seems to make both organizations and individuals uncomfortable and hesitant.

Research using social sensing technology should take precautionary steps when introducing the study to individuals and organizations. Demonstrating the capabilities of the sociometric badges can help participants understand the device and how it captures data. The major concern for speech privacy is not subdued even though researchers emphasize that audio measures only include speech volume, speed, duration, and pitch. Showing the audio data output in numeric form and not linguistic form may help ease the participants with this sensitivity. Surveillance is another issue with social tracking for individuals in organizations, as participants in company case studies have commented on the "fine line between Big Data and Big Brother," and felt uncomfortable that managers may be able to track their every move (Silverman 2013). Researchers should continue to reassure the participants that individual data are not reported to the managers, and only aggregated data for the organization are disclosed. Sharing sample results from a pilot study or another study can also help participants understand how the data are analyzed and translated.

Social sensing badges may be best when the wearers become unaware of their presence. The design of the badges is very similar to keycard holders worn around the neck, commonly worn by employees. However, some study participants have commented on the weight of the device, and some female employees found it to be unattractive to their attire. When the presence of the badge is conscious to the wearer, the reactive effects of the experimental arrangement and the experimenter effects seem to take place. Knowing that they are in a research study, some participants intentionally interact more with others to ensure that they are providing data that are conducive with the study. Also, some participants have also commented that they felt obliged to interact with others more during the study so that their data would not show negatively to managers. Longer studies using this technology may consider to discard the data from the first few days or to check the full data for normality. Observations of the organizational dynamics prior to data collection may also be a way to ensure that the data capture the normal work environment. Shorter studies, such as the charrette case study, allow for participants to focus on the task and not the experiment. Finding the optimal study duration and creating an experimental environment that does not distract from the normal work environment is necessary for future studies.

ACKNOWLEDGMENT

This project was supported by the American Society of Interior Designers Foundation.

REFERENCES

Amabile, T. M. 1983. The social psychology of creativity: A componential conceptualization. *J Person Soc Psychol.* 45(2):357–76.

Amabile, T. M., S. G. Barsade, J. S. Mueller, and B. M. Staw. 2005. Affect and creativity at work. *Admin Sci Q.* 50(3):367–403.

Amabile, T. M., R. Conti, H. Coon, J. Lazenby, and M. Herron. 1996. Assessing the work environment for creativity. *Acad Manage J.* 39(5):1154–84.

Celyan, C., J. Dul, and S. Aytac. 2008. Can the office environment stimulate a manager's creativity? *Human Factors Ergon Manufact.* 18(6):589–602.

Condon, P. M. 2008. *Design Charrettes for Sustainable Communities.* Washington, DC; Covelo, CA; London: Island Press.

Csikszentmihalyi, M. 1996. *Creativity: Flow and the Psychology of Discovery and Invention.* New York: Harper Collins.

Davis, T. R. V. 1984. The influence of the physical environment in offices. *Acad Manage Rev.* 9(2):271–83.

de Alencar, E. M. S., and M. F. Bruno-Faria. 1997. Characteristics of an organizational environment which stimulate and inhibit creativity. *J Creativ Behav.* 31(4):271–81.

de Korte, E., L. Kuijt, and R. van der Kleij. 2011. Effects of meeting room interior design on team performance in a creativity task. *Ergon Health Aspects Work Comp.* 6779:59–67.

Dong, W., B. Lepri, T. Kim, F. Pianesi, and A. S. Pentland. 2012. Modeling conversational dynamics and performance in a social dilemma task. In *2012 5th Int Symp Comm Control Signal Proc.* 1–4. Rome: Institute of Electrical and Electronics Engineers. From http://ieeexplore.ieee.org/lpdocs/epic03/wrap per.htm?arnumber=6217775.

Dul, J. A. N., C. Ceylan, and F. Jaspers. 2011. Work environments for employee creativity. *Hum Res Manage.* 50(6):715–34.

Ekvall, G. 1996. Organizational climate for creativity and innovation. *Eur J Work Org Psychol.* 5(1):105–23.

Ford, C., F. McLaughlin, and J. Newstrom. 2003. Questions and answers about fun at work. *Hum Res Plan.* 26(4):18–33.

Gelfand, B. 1988. *The Creative Practitioner.* New York: The Free Press.

Girard, M., and D. Stark. 2008. Heterarchies of value: Distributing intelligence and organizing diversity in a new media startup. In Ong, A., and S. J. Collier (eds.) *Global Assemblages: Technology, Politics, and Ethics as Anthropological Problems.* Oxford: Blackwell Publishing Ltd.

Gregson, N., and G. Rose. 2000. Taking butler elsewhere: Performativities, spatialities and subjectivities. *EnvPlan D: Soc Space.* 18(4):433–52.

Haner, U.-E. 2005. Spaces for creativity and innovation in two established organizations. *Creativ Innov Manage.* 14(3):288–98.

Hasirci, D., and H. Demirkan. 2007. Understanding the effects of cognition in creative decision making: A creativity model for enhancing the design studio process. *Creativ Res J.* 19:259–71.

Hatch, J. M. 2006. *Organization Theory: Modern, Symbolic and Postmodern Perspectives.* New York: Oxford University Press.

Hemlin, S., C. M. Allwood, and B. R. Martin. 2008. Creative knowledge environments. *Creativ Res J.* 20:196–210.

Hennessey, B., and T. Amabile. 2010. Creativity. *Ann Rev Psychol.* 61:569–98.

Huang, Y., M. Robertson, and K. Chang. 2004. The role of environmental control on environmental satisfaction, communication, and psychological stress: Effects of office ergonomic training. *Environ Behav.* 36(5):617–37.

Hunter, S. T., K. E. Bedell, and M. D. Mumford. 2007. Climate for creativity: A quantitative review. *Creativ Res J.* 19:69–90.

Jakobsson, P., and A. Stiernstedt. 2010. Googleplex and informational culture. In Ericson, S., and K. Riegert (eds.) *Media Houses: Architecture, Media and the Culture of Centrality.* 111–32. New York: Peter Lang Publishing.

Jayagopi, D. B., T. Kim, A. Pentland, and D. Gatica-Perez. 2010. Recognizing conversational context in group interaction using privacy-sensitive mobile sensors. *Proc 9th Int Conf Mobile and Ubiq Multimedia.* 1–4. Limassol: ACM Press. From http://portal.acm.org/citation.cfm?doid=1899475.1899483.

Johnson, M. W. 2010. *Seizing the White Space: Business Model Innovation for Growth and Renewal.* Cambridge, MA: Harvard Business Press.

Kaplan, S. 1995. The restorative benefits of nature: Toward an integrative framework. *J Environ Psychol.* 15(3):169–82.

Karasek, R. A. 1979. Job demands, job decision latitude, and mental strain: Implications for job redesign. *Admin Sci Q.* 24(2):285–308.

Kasof, J. 1997. Creativity and breadth of attention. *Creativ Res J.* 10(4):303–15.

Kelbaugh, D. S. 2014. The design charrette. In Roggenma, R. (ed.) *Companion to Urban Design.* Dordrecht; Heidelberg; New York; London: Springer.

Knez, I. 1995. Effects of indoor lighting on mood and cognition. *J Environ Psychol.* 15:39–51.

Kristensen, T. 2004. The physical context of creativity. *Creativ Innov Manage.* 13(2):89–96.

Lee, S. Y., and J. L. Brand. 2005. Effects of control over office workspace on perceptions of the work environment and work outcomes. *J Environ Psychol.* 25:323–33.

Lehrer, J. 2011. The Steve Jobs approach to teamwork. *Wired Magazine.* From http://www.wired.com/2011/10/the-steve-jobs-approach-to-teamwork/.

Leonard, D., and W. Swap. 1999. *When Sparks Fly.* Boston: Harvard Business School Press.

Lepri, B., J. Staiano, G. Rigato, K. Kalimeri, A. Finnerty, F. Pianesi, N. Sebe, and A. Pentland. 2012. The sociometric badges corpus: A multilevel behavioral dataset for social behavior in complex organizations. *E 2012 Int Conf Privacy, Security, Risk and Trust (PASSAT) and 2012 Int Conf Soc Comp.* 623–28. Piscataway, NJ: IEEE Press.

Lewis, P. 2008. Emotion work and emotion space: Using a spatial perspective to explore the challenging of masculine emotion management practices. *Brit J Manage.* 19(s1):S130–40.

Lewis, M., and J. Moultrie. 2005. The organizational innovation laboratory. *Creativ Innov Manage.* 14:73–83.

MacKinnon, D. W. 1978. *In Search of Human Effectiveness: Identifying and Developing Creativity.* Buffalo, NY: Creative Education Foundation, Inc.

Magadley, W., and K. Birdi. 2009. Innovation labs: An examination into the use of physical spaces to enhance organizational creativity. *Creativ Innov Manage.* 18:315–25.

Maher, A., and C. von Hippel. 2005. Individual differences in employee reactions to open-plan offices. *J Environ Psychol.* 25:219–29.

Maletz, M. C., and N. Nohria. 2001. Managing in the whitespace. *Harvard Bus Rev.* 79(2):102–11.

Martens, Y. 2008. Unlocking creativity with the physical workplace. *Proc CIB W070 Conf Facil Manage.* Edinburgh: Heriot Watt University.

Martens, Y. 2011. Creative workplace: Instrumental and symbolic support for creativity. *Facilities.* 29(1/2):63–79.

McCoy, J. M. 2005. Linking the physical work environment to creative context. *J Creative Behav.* 39(3):167–89.

McCoy, J. M., and G. W. Evans. 2002. The potential role of the physical environment in fostering creativity the potential role of the physical environment in fostering creativity. *The Creativ Res J.* 14(3 and 4):409–26.

Mehta, R., R. Zhu, and A. Cheema. 2013. Is noise always bad? Exploring the effects of ambient noise on creative cognition. *J Consum Res.* 39(4):784–99.

Meusburger, P. 2009. Milieus of creativity: The role of places, environments, and spatial contexts. *Milieus of Creativ.* 97–153.

Miller, C., and C. Rampell. 2013. Yahoo orders home workers back to the office. *The New York Times.* February.

Mumford, L. 2001. Space and time to connect. In Cohen, D., and L. Prusak (eds.) *In Good Company: How Social Capital Makes Organizations Work.* 81–102. Boston: Harvard Business School Press.

Oldham, G. R., and N. L. Rotchford. 1983. Relationships between office characteristics and employee reactions: A study of the physical environment. *Admin Sci Q.* 28:542–56.

Olguín-Olguín, D., and A. Pentland. 2010. Sensor-based organisational design and engineering. *Int J Org Design Engineer.* 1(1/2):69–97.

Olguín-Olguín, D., B. N. Waber, T. Kim, A. Mohan, K. Ara, and A. Pentland. 2009. Sensible organizations: Technology and methodology for automatically measuring organizational behavior. *IEEE Trans Syst Man Cybern: Part B: Cybern: Publ IEEE Syst Man Cybern Soc.* 39(1):43–55.

Ornstein, S. 1989. The hidden influences of office design. *Acad Manage Exec.* 3(2):144–47.

Penn, A., J. Desyllas, and L. Vaughan. 1999. The space of innovation: Interaction and communication in the work environment. *Env Plan B: Plan Design.* 26:193–218.

Plucker, J. A., R. A. Beghetto, and G. T. Dow. 2004. Why isn't creativity more important to educational psychologists? Potentials, pitfalls, and future directions in creativity research. *Educ Psychol.* 39(2):83–96.

Ridoutt, B. G., R. D. Ball, and S. K. Killerby. 2002. Wood in the interior office environment: Effects on interpersonal perception. *Forest Prod J.* 52(9):23–30.

Roggema, R. 2014. The design charrette. In Roggema, R. (ed.) *The Design Charrette: Ways to Envision Sustainable Futures.* 15–32. Dordrecht: Springer.

Sailer, K. 2011. Creativity as social and spatial process. *Facilities.* 29:6–18.

Shibata, S., and N. Suzuki. 2002. Effects of the foliage plant on task performance and mood. *J Environ Psychol.* 22:265–72.

Silverman, R. E. 2013. Tracking sensors invade the workplace. *The Wall Street Journal.* From http://online .wsj.com/article/SB10001424127887324034804578344303429080678.html?mod=WSJ_mgmt_Lead StoryCollection.

Singh-Manoux, A., M. Hillsdon, E. Brunner, and M. Marmot. 2005. Effects of physical activity on cognitive functioning in middle age: Evidence from the Whitehall II prospective cohort study. *Amer J Pub Health.* 95(12):2252.

Steidle, A., and L. Werth. 2013. Freedom from constraints: Darkness and dim illumination promote creativity. *J Environ Psychol.* 35:67–80.

Stokols, D., C. Clitheroe, and M. Zmuidzinas. 2002. Qualities of work environments that promote perceived support for creativity. *Creativ Res J.* 14(2):137–47.

Toplyn, G., and W. Maguire. 1991. The differential effect of noise on creative task performance. *Creativ Res J.* 4(4):337–47.

Tripathi, P., and W. Burleson. 2012. Predicting creativity in the wild: Experience sample and sociometric modeling of teams. *Proc ACM Conf Comp Supported Coop Work.* Seattle, WA.

Veitch, J. A., and R. Gifford. 1996. Choice, perceived control, and performance decrements in the physical environment. *J Environ Psychol.* 16:269–76.

Vithayathawornwong, S., S. Danko, and P. Tolbert. 2003. The role of the physical environment in supporting organizational creativity. *J Interior Design.* 29:1–16.

Wells, M. 2000. Office clutter or meaningful personal displays: The role of office personalization in employee and organizational well-being. *J Environ Psychol.* 20:239–55.

Wineman, J. D, F. W. Kabo, and G. F. Davis. 2009. Spatial and social networks in organizational innovation. *Env Behav.* 41(3):427–42.

Woodman, R. W., J. E. Sawyer, and R. W. Griffin. 1993. Toward a theory of organizational creativity. *Acad Manage Rev.* 18(2):293–321. From http://amr.aom.org/cgi/doi/10.5465/AMR.1993.3997517.

Yoon, S.-Y., and N. D'Souza. 2009. Different visual cognitive styles, different problem-solving styles? *Int Assoc Soc Design Res.* 18–22. Seoul.

21 Ergonomics and Wellness in Workplaces

Alan Hedge and Sara Pazell

CONTENTS

21.1 INTRODUCTION

As you have seen from the other chapters in this book, in most workplaces, the ergonomics focus is on implementing physical ergonomic solutions to prevent injuries, accommodate those with injuries or different abilities, and improve work performance. Although physical ergonomics is only one area of the discipline, this focus has led to an influx of practitioners from different though related disciplines, such as physical therapists, occupational therapists, chiropractors, kinesiologists, etc., who practice as ergonomists in organizations. Perhaps inevitably, this physical focus on accidents, injury prevention, and accommodation of injured employees has led many to adopt a narrow vision of the role of ergonomics in relation to the health, the wellness, and the productivity of workers as only being concerned with changing the design of the tools and the work spaces for people. Yet, in addition to physical and environmental design aspects of workplaces, ergonomics has a systems perspective that also embraces cognitive, social, and organizational considerations and how these interact to establish a healthful and productive work system.

There is evidence to suggest that ergonomics programs go well beyond the link to safety performance and occupational health (Dul and Neumann 2009). In fact, the design of work for health is part of a continuum of the design for safety (e.g., Laitinen et al. 1998; Dul and Neumann 2009; Punnet et al. 2009; Pazell 2015). Laitinen et al. (1998) demonstrated a significant link between the psychosocial and physical work conditions through an organic analysis of an organization and the assessment of the workers' perceptions of the company, their jobs, their future, and their work. They concluded that the technical improvements in equipment, work, and work systems through a participative ergonomic process improved psychological health profiles at work. Postintervention, the subjects perceived improved prospects for their future and viewed their company more favorably.

Participatory ergonomics, as a component of human-systems integration, may improve organizational work climate, positively affect safety culture, enhance learning opportunities, improve communication, and improve comfort while reducing musculoskeletal disorders (Laitinen et al. 1998; Laing 2007; Burgess-Limerick 2011; Lallemand 2012; Pazell 2015). Human factors ergonomics employs a systems approach, is design driven, and focuses on performance and well-being (Dul et al. 2012).

Participatory ergonomics, human factors engineering, and human-centered design help mobilize a workforce beyond engagement, wherein workers become architects of a well workplace (Pazell and Burgess-Limerick 2015a). They help redesign their work systems, procedures, and equipment. The workers become coauthors of superior work design (Cantley et al. 2014; Burgess-Limerick 2011; Pazell and Burgess-Limerick 2015a,b). These methods go beyond co-designing and user experience, as they draw upon evidence-based findings related to optimum human work conditions: the design process may be facilitated by skilled professionals that draw upon a large body of science and validated research regarding human performance technology (U.S. Air Force 2009; Pazell 2015).

Ergonomics is a process, not a product, and this process addresses not merely the physical needs of people but also their cognitive and emotional needs. The process of ergonomics provides for robust, dynamic, participative, and engaged activity among people. Design provides for a creative outlet, and design that is facilitated by a framework of participative ergonomics leads to solutions that are richly meaningful to daily living in work, play, or social activity (Daniellou and Rabardel 2005; Barcellini et al. 2015). In this respect, ergonomics is an essential component of any wellness program in an organization.

Wellness is a broad concept that is derived from the WHO definition that "health is a state of complete physical, mental and social well-being and not merely the absence of disease or infirmity" (WHO 1948). Founded in 1977, the National Wellness Institute (NWI) defines wellness as "an active process through which people become aware of and make choices toward a more successful existence" (http://www.nationalwellness.org/?page=Six_Dimensions). The NWI defines six interconnected dimensions of wellness (Hettler 1976). Note that these are not presented in order of importance, but all six dimensions contribute to the overall well-being:

1. Occupational—Occupational wellness is the importance of work in enriching one's life and providing personal satisfaction. It is enhanced by choosing a career that is consistent with a person's values, interests, and beliefs, and is rewarding and affords the development of functional, transferable skills.
2. Physical—Physical wellness is the need for physical activity and regular exercise, and a healthy diet while avoiding smoking, recreational drugs, and excessive alcohol consumption.
3. Social—Social wellness is the importance of engaging in actions that preserve the beauty and the balance of nature, enhance personal relationships, strengthen friendships, and build a better living space and community.
4. Intellectual—Intellectual wellness is sharing knowledge and skills with others, engaging in intellectual, creative, stimulating mental activities, and avoiding worry and procrastination.
5. Spiritual—Spiritual wellness is the importance of developing tolerance, finding meaning, and leading a purposeful human existence.
6. Emotional—Emotional is the importance of feeling positive and enthusiastic about one's self and life and the value of optimism over pessimism.

The University of California, Davis (2015) proposes two additional dimensions of wellness:

7. Financial—Since money plays a critical role in our lives, it is important to learn how to successfully manage financial expenses to avoid financial stress.
8. Environmental—Environmental wellness is an extension of the social wellness, and it encourages taking action to protect the planet and the lifestyles that are in harmony with the earth.

Michaels and Greene (2013) propose that worksite wellness programs incorporating the dimensions should aim to protect and promote employee health and improve the quality of work life and morale while reducing direct and indirect health care costs, absenteeism, and presenteeism (present at work but not working). However, they also note that comprehensive worksite health promotion strategies have yet to be adopted by a majority of U.S. employers.

Organizations that have wellness programs often limit the scope of these strategies to focusing on nutrition and exercise, and in most companies, the wellness program is independent from any ergonomics program. Good ergonomic workplace design is essential for the promotion of wellness, and such designs also improve the comfort and the satisfaction of the employees. Yet, all too often, ergonomics programs are subsumed under a safety and health function, and they have to accommodate cases where people have already been injured and/or are unhealthy, such as those who are unfit and overweight. Wellness programs typically promote exercise and healthy diet, yet frequently the employees participating in these programs spend much of their workday working at workstations that are not ergonomically designed and that place them in unhealthy postures that can increase injury risks and stress.

Wellness programs often include activities that appeal to the fittest and most active workers. These workers may be least likely to derive the greatest benefit from workplace intervention if they are already intrinsically motivated and engaged in healthful activities such as regular sport, sound nutritional plans, and mindfulness practices or meditation. The key is to design work for health that will likely nudge workers toward healthful behavior while assuring safe work performance. Design teams must target strategic workplace intervention for safety, occupational health, health, and wellness according to the critical activity demands of the workers. As C. Badke (personal communication, 2015) explains, safety managers will embrace the design of wellness interventions that result in improved work performance for critical decision-making. The evolution of job roles in industrialized nations is such that the work demands have evolved in terms of significance for perception, cognition, distributed situation awareness, critical decision-making, and rational execution of actions (Hassall and Sanderson 2012; Hedge 2013).

There is an opportunity to strategically link evidence-based findings from health, exercise science, and wellness models to safety systems. For example, Buckley et al. (2015) describe evidential findings of ill-health risk associated with sedentary behaviors. They provide a prescription for the workplace design to support at least 2 hours of low-impact physical activity embedded in a workday. These findings can be placed on a safety risk register in terms of the need to address work design and encourage active movement. Once integrated with traditional safety management systems, the public health provisos become part of business accountability and are communicated throughout many levels of management.

It is clear that, to achieve well-being, we must be preemptive and not merely focused on ill-being (Robert Wood Johnson Foundation 2010; Pazell 2015). Health begins where we live, learn, work, and play (WHO 2012). Health must start long before the illness, and our jobs may be one area where health intervention and promotion may begin (Robert Wood Johnson Foundation 2010). As such, work provides a medium to positively affect and promote health and well-being (Pazell 2015). There is, in turn, a caveat that work conditions should be safe and accommodating (Waddell and Burton 2006).

Pazell and Burgess-Limerick (2015a,b) argue that the effective design of work for health must adopt a human-centered approach. With this approach, safety may be connected with health and wellness initiatives, leadership is developed, complementary business practices may be unified, and systems thinking may be applied (Carayon 2006; Karanika-Murray and Weyman 2013). The process of ergonomic design enables work improvement along a continuum of safety, occupational health, health, well-being, and organizational resilience (Pazell 2015). Under this design model and consistent with an occupational perspective of health (Wilcock 2006), humans may experience life beyond doing: they may do, be, belong, and become (Wilcock 2006; Pazell 2015). The fundamental work that they perform may occur in safe conditions with reduced likelihood of fatality, disablement, or discomfort; they may achieve comfort; they may be physically and mentally conditioned; they may connect socially and contribute new learnings or be inspired in the creative process; and they may contribute in a fundamental and meaningful way to robust organizational performance (Pazell 2015).

Organizational interventionists posit that to achieve promotion of health and organizational high performance, changing the nature of work and work environment is required. Changing work and work systems or conditions prevents the stressors and optimizes the performance. This is a preventive strategy. The effort to change people and their behavior is latent and represents a corrective strategy (Burke 2014). Herein lies the tension between traditional workplace public health initiatives: the adoption of an individual behavioral change model may divert the resources and the attention that should otherwise be put to positive work design that recognizes fundamental workplace influence on employee well-being (Munz et al. 2001; Kohler and Munz 2006; Mellor et al. 2012; Karanika-Murray and Weyman 2013; Pazell 2015). Targeted intervention with changes to the work system or the equipment design may be far more constructive than the less evidential approach afforded by behavioral change models (Carayon 2006; Henning et al. 2009; Karanika-Murray and Weyman 2013; Pazell 2015). Fundamentally, employee well-being is rooted in the design of work (Munz et al. 2001; Kohler and Munz 2006; Mellor et al. 2012; Karanika-Murray and Weyman 2013; Burke 2014; Pazell 2015).

Henning et al. (2009) describe a study in which traditional workplace safety and health protection was integrated with health promotion. The foundation of the program involved a participative model for engaging the workers that facilitated innovation through an iterative design process. The program was modeled after a traditional participatory ergonomics program with targeted intervention to address the perceived health needs. The program was referred to as PE × HP or participatory ergonomics × health promotion. With the application of a participatory ergonomics model, the changes were made at work: the end users of a system were involved in priority task identification, analysis of need, solutions finding, and strategy development. Initially, the change was made to those issues most readily or easily addressed, such as changes to a vending machine to offer more healthful snack options under the guidance of a nutritionist. Plans were also developed to introduce a new walking path to encourage healthful walking activity at a nursing home. More complicated issues, such as a change in rostering to address fatigue, were scheduled for address in small and manageable steps owing to the complexities of industrial relations and union involvement. This began with targeted efforts to review the coping strategies of management and staff associated with roster demands. The program provided training to the staff and the management as to how to develop a business case to address the broader industrial relations issues of fatigue and rostering.

Workplace health programs offer residual, effective, positive changes when the programming is integrated into core organizational strategies and aligns with business need, engages all levels of the organization, is segmented and targeted to populations most in need of intervention, and the messages and programs are simple (Karanika-Murray and Weyman 2013; Kickbush 2013). The models for the design improvement in safety and occupational health in the workplace share similar philosophy with health promotion initiatives when approached from an organizational design perspective (e.g., Haslam 2002; Vink et al. 2006; Karanika-Murray and Weyman 2013; Randall and Nielson 2014). The methods to improve occupational health and safety initiatives in the workplace share similarities to the methods employed to address community health promotion. If we recognize and integrate these program methods, the platform of the design for safety through the continuum of health and wellness may be better articulated (Urlings et al. 1990; Haslam 2002).

An optimum approach is to combine the power of wellness and ergonomics, and as we look to the future, there are several recent initiatives that attempt to do this, at least in some part.

21.2 THE LEED ERGONOMICS CREDIT

As discussed in Chapter 19 by Dorsey in this book, since 2008, the USGBC has offered a credit point in their highly successful LEED certification system for a good ergonomics program (LEED pilot credit #44). The goal of this credit is to encourage the implementation of an ergonomics program to promote healthy, comfortable, and productive work by designing the workplace to accommodate its users. The credit requirements focuses on the development and the implementation of a strategic

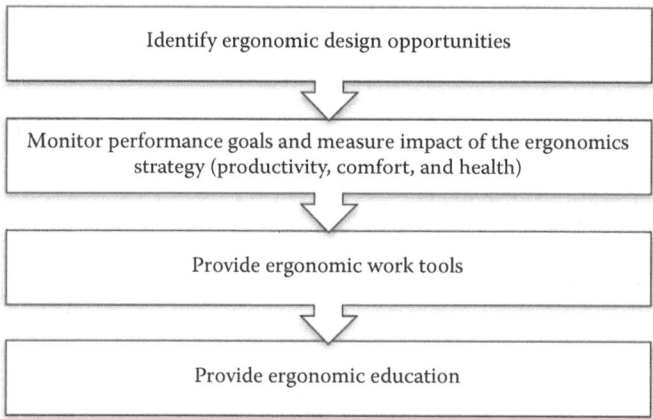

FIGURE 21.1   Summary of the main requirements for the LEED ergonomics credit.

plan for a comprehensive ergonomics strategy that, at the time of planning, the workspace identifies "activities and building functions that would benefit from the application of ergonomics in the selection of appropriate furnishings, equipment and education" (U.S. Green Building Council 2011). The credit encourages architects and designers to consult "current ergonomics standards and guidelines relevant to the tasks that will be performed in the building." The pilot credit requirements list three standards for the ergonomic design of computer workstations (ANSI/HFES 100 2007; BIFMA G1 2011; CSA Z412-00 [R2011] 2011) that should be consulted as appropriate. It also gives three standards for noncomputer workstations (OSHA 3192-05N 2004; OSHA 3182 [revised] 2009; Z1004-12 2012).

Architects and designers can work with ergonomists to analyze all the opportunities for improving the ergonomic design of the work space. This involves conducting an ergonomics user survey of the occupants to analyze their characteristics (age, size, shape, weight, ability/disability, gender) and consulting with them about the relative importance and the priority of the tasks they perform and their frequency, duration, plus the equipment and the materials used to perform the tasks along with their storage requirements, and any other needs. The key ergonomic design considerations are those that facilitate occupant well-being (health, performance, and satisfaction), and these include versatility and flexibility, fit, postural change, and maintainability and adaptability. All this work should be undertaken prior to designing the interior, which includes designing the lighting, the thermal environment, the office layout, the individual workstation design, and the furnishings and equipment. When design alternatives are available, these options should be reviewed with the occupants. To obtain the credit, the design team has to demonstrate that they incorporated key interrelated ergonomic principles into their interior design and their plans for worker education and training. After the installation of the furniture and the equipment and once the space is occupied, all the users should be provided with the appropriate ergonomics education and training in both face-to-face and electronic formats, and a follow-up survey should be conducted to evaluate the success of the design and identify any issues requiring further attention. Figure 21.1 summarizes the LEED ergonomics credit requirements.

21.3   WELL BUILDING CERTIFICATION

In 2013, the International WELL Building Institute was launched in the United States, and a WELL Building Standard was introduced. The focus of this standard is that the built environment needs to be more than just an energy-efficient enclosure built from sustainable materials to being one that also beneficially affects the health, the well-being, the happiness, and the productivity of the occupants. Given that we spend the majority of our lives indoors (Klepeis et al. 2001; Schweizer et al. 2007), the design of the built environment impacts our behavior, work performance, and health both directly and indirectly. The WELL Building Standard advocates a human-centered approach to building design and identifies seven areas of health and well-being that can be beneficially impacted by the design of the environment:

1. Air quality—Design requirements for optimal IAQ for the occupants
2. Water quality—Safe, clean water for the occupants
3. Nourishment—Access to healthy and nutritious foods
4. Lighting—Lighting that enhances productivity, provides for appropriate visual acuity, increases alertness, and minimizes the disruptions to the circadian rhythms and sleep
5. Fitness—Designs that integrate exercise and fitness into everyday life
6. Comfort—Physical and visual ergonomics, olfactory, thermal and acoustic comfort, noise levels
7. Mind—Environmental requirements for optimal cognitive and emotional health

Several pilot projects are underway that have applied the WELL certification requirements to their buildings. Starting in 2013, Coldwell, Banker Richard Ellis (CBRE), a commercial real estate company, renovated part of their downtown headquarters in Los Angeles to meet WELL certification requirements as well as implementing a "Workplace360" design to meet changing needs and preferences throughout the day and promoting choice, flexibility, and mobility by using a free-address desking strategy and through paperless working with digitized files. The employees were surveyed 1 year after moving into the Workplace360 offices. Also, more than 50 wellness features were integrated into the workplace, and this was the first WELL-certified office in the world. Sit-to-stand desks are used by 51% of employees, and stairs rather than elevators are used by 74% of employees. A postoccupancy evaluation of 191 employees in the new office space showed the following:

- 83% feel more productive.
- 86% have ample access to focused or private space when needed.
- 88% say it has helped generate business.
- 90% can work from anywhere, anytime.
- 90% would recommend it to colleagues and friends.
- more than 90% of employees prefer it to their previous way of working.
- 92% have seen a positive effect on their health and well-being.
- 93% are able to more easily collaborate with others.
- 93% report a positive impact on their business performance.
- 93% can easily access their digitized files off-site or from handheld devices such as a tablet or a smartphone.
- 100% say clients are interested in the new design.

Overall, the employees say that they are happier, more productive, and feel healthier than before, and that they would not choose to go back to a traditional way of working. Over 10,000 clients, business, government, and community leaders have toured the CBRE office space. The design has won numerous awards and received considerable media coverage. It has been transformative for CBRE, and it is being introduced in 11 other U.S. sites and 10 additional sites globally.

21.4 TOTAL WORKER HEALTH

The Centers for Disease Control and National Institute for Occupational Safety and Health (NIOSH) launched a Total Worker Health (TWH) approach in 2011. TWH is defined as "policies, programs, and practices that integrate protection from work-related safety and health hazards with promotion of injury and illness prevention efforts to advance worker well-being" (Centers for Disease Control 2013). The TWH approach protects and promotes the total health, safety, and well-being of workers by integrating occupational safety and health protection with health promotion in an organization (Schill and Chosewood 2013). The TWH approach includes strategies to undertake risk assessments; control hazards and exposures; prevent injuries, illness, and fatalities; and promote safe and healthy working. It combines these actions with the development of healthier behaviors, such as improved nutrition, tobacco use cessation, and increased physical activity, through employee engagement and support, and it also focuses on improvements in work/life balance (Schill and Chosewood 2013). The types of risks and factors embraced by the TWH approach are shown in Figure 21.2.

A meta-analysis of 17 TWH interventions that addressed both injuries and chronic diseases showed that these can effectively and more rapidly improve workforce health than employing several more narrowly focused programs to achieve the same outcomes in serial fashion, and the return on investment can be $4.61 returned for every $1.00 invested (Anger et al. 2015).

21.5 ACTIVITY-BASED WORKING AND AGILE WORKING

Activity-based working and *agile working* refer to flexible real estate strategies that reflect an emerging trend in the workplace design. The design provides people with a choice of settings in which to perform the activities that support their job roles. The workers can physically locate themselves according to the type of work effort required or desired: concentration, social connection and collaboration, standing work, sitting work, or interface with nature, for example. The space is designed to provide opportunity for self-directed work within an environment most conducive to the nature of that work and personal needs (LaSalle 2012; Hosking 2015). It is not hot desking, which requires shared work pods, and it is not hoteling, which requires prebooking of work areas.

Activity in the context of activity-based-work refers to functional job roles. A design brief objective may include provision of nonassigned desks for individuals to promote different objectives (work team interaction, solo concentrated time with little noise distraction, nutrition and meal space, team meetings, recreational areas, or multipurpose areas). It must be contextualized to suit the right population, their readiness for change, and the task need (LaSalle 2012). A change management process is often critical to support the workers in their adoption of this evolution in the work space design (Hosking 2015).

For business efficiency, the implications are significant. A flexible real estate strategy enables the expansion or the contraction to readily accommodate the changing workforce demand. Predictive design enables the operations to assign a weighting according to a job role or a person assigned to a critical work activity. Consideration is given to the percentage of time a worker may likely spend at a dedicated workstation or a work area at any given time. This probability weighting is described as a percentage, and a number of work space options are designed to be available at any time, rather than a single dedicated desk. For example, an accounting team may be considered to be using some form of desk and computer workstation 70% of the time in their day. As such, a weighting of 0.70 is provided. Rather than allocating the building capacity of a single dedicated desk and supportive capacity (electrical, egress strategies, air conditioning, or water supply) for one worker 100% of the time, consideration is made to accommodate work activity in a number of work space options, e.g., 4–6, for at least 70% of that workforce population at any time.

The predictive design may be married with workforce strategy and extended to include calculations for anticipated work team expansion or contraction (R. Hosking, personal communication, November 12, 2015). Cost-savings calculations by Marsden (2015) notes that office workstations are often unoccupied, so companies are paying for empty space, and this is a hidden cost.

In addition to real estate cost savings, flexible work space designs can also improve the productivity of employees, especially when combined with ergonomics training. Robertson et al. (2008) conducted a pre-postevaluation study of the effects of a macroergonomics intervention (flexible work space design or flexible work space design plus ergonomics training) in a computer-based office

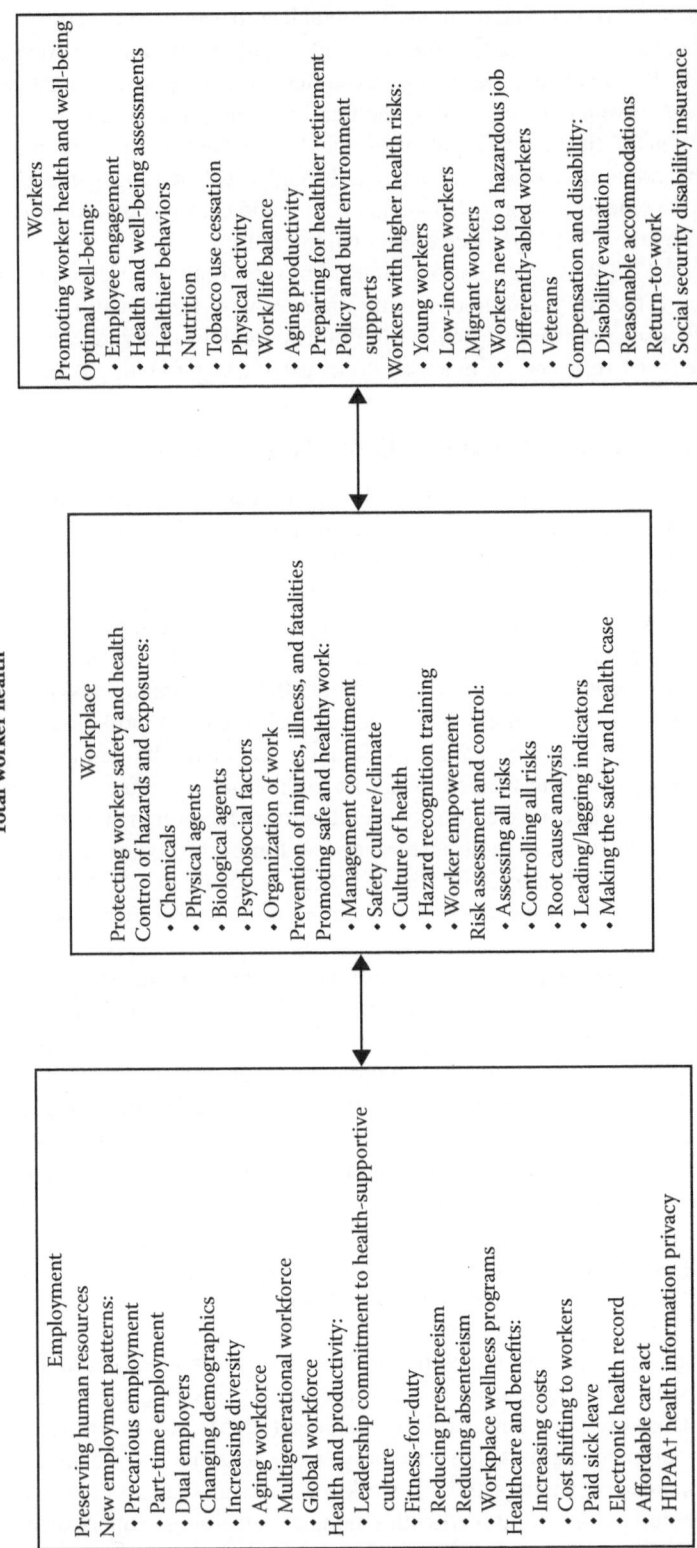

FIGURE 21.2 Examples of the issues embraced by a TWH strategy. †HIPAA—Health Insurance Portability and Accountability Act of 1996.

on the psychosocial work environment, the musculoskeletal health, and the work effectiveness. The intervention showed significant improvements in work-related musculoskeletal discomfort, job control, environmental satisfaction, sense of community, ergonomic climate, communication and collaboration, and business process efficiency (time and costs). The business process improved 5% for those in the flexible work space but 10% for those who also received ergonomics training in this work space. Ergonomic considerations for new ways of working are further discussed in detail in Chapter 18 of this book.

21.6 HEALTH-BASED WORKING

Although activity-based and agile working can improve the efficiency of space utilization and lower the real estate costs, there is no guarantee that such designs will also increase the at-work activity levels of the employees and promote their health and wellness. To achieve this, the integration of the benefits of good ergonomic design and wellness programs requires consideration of the total work environment as a system. In the 1980s, a new building style emerged from Europe, the "groundscraper," in which offices were built as horizontal low- to medium-rise structures rather than vertical as a skyscraper. This design allowed for the incorporation of an interior space that served as a street, allowing employees to walk between locations. Built in 1985, the Nederlandsche Middenstandsbank (NMB) building (now Internationale Nederlanden Groep [ING] Bank) in Amsterdam is an extremely energy-efficient building that houses 2400 employees in 538,000 ft^2 (50,000 m^2) of office space that consists of 10 medium-rise, slanted towers connected by an undulating irregular S-shaped street (see Figures 21.3 through 21.5). A detailed evaluation of the energy performance of this building shows that it performs at a very efficient level (Browning 1992). Mumtaz (1993) evaluated the impact of this building on employees compared with an adjacent more conventional office building of comparable age. Although the objective measures of activity levels were not taken, he found that the employees surveyed in the NMB building generally rated it as more satisfactory and spent more time interacting with others in the public street level.

FIGURE 21.3 Exterior view of the NMB (ING) Bank building.

FIGURE 21.4   Interior view of the NMB (ING) Bank building indoor street.

FIGURE 21.5   Interior view of the NMB (ING) Bank building stairs.

21.6.1   Medibank, 720 Bourke Street Building, Melbourne

Medibank is Australia's largest private health insurer. The building at 720 Bourke Street is a 16-level, 500,000 ft² (46,500 m²) mixed-use building in the Docklands area of Melbourne, Australia. Medibank occupies seven floors of the building (Figure 21.6). The floors have been designed as fully integrated activity-based work spaces by up to four different architectural firms working together under the master plan of one primary firm to integrate activity-based work space themes (Hosking 2015; Rebecca Hosking, Workplace Transition Manager, personal communication, November 12, 2015). The core of the building comprises prominent staircases designed to promote more walking, and office employees take an average of 1400 steps per day more compared with those in other Medibank facilities which accumulates to over 62 miles (100 km) per year (Olly Bridge, Head of Corporate Health and Wellbeing at Medibank, personal communication). The office floors are consistently color coded when viewed either from below or above, and a variety of activity hubs are provided (Figure 21.7). The Medibank office space is organized into 26 different work zones, varying from private to collaborative spaces. The interior provides circadian lighting that changes the light spectrum through the day, and there are some 3500 plants interspersed on the office floors.

    The employees do not have assigned desks, but they do have an assigned locker with a photo sleeve located at a "launch pad" at either end of each floor for them to retrieve what they may need from their locker for their scheduled work (Figure 21.8), and they can return to replace and retrieve

FIGURE 21.6   Exterior view of the Medibank 720 Bourke Street building.

FIGURE 21.7   Interior stairs, floor views, and activity hubs.

FIGURE 21.8   Launch pad and locker space.

FIGURE 21.9   Interior Medibank office space for the 720 Bourke Street building.

items to their locker as often as is required in the day. The employees can then choose the desk location where they wish to work (Figure 21.9).

The building provides for incubation or think-tank space, work booths, concentration pods, sit-or-stand stations that are traditional and atypical, collaborative team planning, social play, shared meals and kitchens, and garden space (Bridge 2015; Hosking 2015; R. Hosking, personal communication, November 12, 2015).

A significant change management process was implemented with award-winning gamification to aid the transition of workers to the new location, building style, and work demand. Wellness, in this regard, was supported by the design process and the design of facilities (Hosking 2015).

The results from a 12-month post-ccupancy survey (Bridge 2015; Hosking 2015; R. Hosking, personal communication, November 12, 2015) show a majority of employees in agreement with the following statements:

- 79%: "I am more collaborative with others at Medibank Place."
- 71%: "I am more connected to our 'for better health' purpose working at Medibank Place."
- 70%: "I am healthier working at Medibank Place."
- 66%: "I am more productive at Medibank Place."

To date, the business has also seen 5% reduction in call center absenteeism.

21.7 CONCLUSIONS: FROM WORKSTATION TO WORKSPACE— ERGONOMICS EVERYWHERE!

For a traditional computer-based office, the focus of ergonomics has typically been solely on the design of the workstation where the employee spends much of their day, but with modern portable computing devices and Wi-Fi networks, an employee can now work almost anywhere in a building. This changes the focus of ergonomics from simply addressing the design of an employee's workstation to considering the ergonomic design of the total workplace—inside and out. It means thinking about the building as an enclosure in which an employee can work in a variety of different spaces, and it means determining every space where an employee can potentially interact with technology (at their desk, in a cafeteria, in a library, in a meeting room, in a coffee area, etc.) and applying good ergonomic design principles to each of these locations, a process we call *Everywhere Ergonomics*. In addition, there is also a need to train employees in appropriate ergonomic working techniques for each of the locations where they may be working. With the mobility of computing technologies, the employees can now move from place to place throughout their workday, which should boost their activity level. As summarized by Buckley et al. (2015), multiple health risks, e.g., diabetes, cancer, heart disease, are associated with an overly sedentary lifestyle, and in the workplace, many computer users spend much of their day seated. Hedge (2006) calculated that if a person consumed an extra 34 cal daily, this alone would increase their bodyweight by ~3.5 lb (1.6 kg) per year. Conversely, if a person's activity level can be increased by 34 cal daily, they would lose this weight each year. Behavioral changes may achieve this target, but this is most likely to occur in an enduring manner when linked to design changes. Compared with sitting, standing consumes an extra 0.7 cal per minute, so the activity pattern shown in Chapter 1, Figure 1.6, which recommends 8 min of standing per 30 min of work period, results in the burning of an additional 12 cal per hour, or 96 cal per day. This can be facilitated by encouraging the use of a sit-to-stand workstation or allowing employees to move between sitting and standing workstations. Using a staircase at a moderate pace burns around 0.2 cal per step for going up the stairs and 0.05 cal per step for going down the stairs, so taking the stairs for 160 steps rather than an elevator will burn ~35 cal. Building designs that make stairs more prominent and easier to use can facilitate this behavior, as shown in the Medibank case study. Compared to standing, every 20 steps walked will burn 1 cal, so walking 680 steps will burn 34 cal. The strategies to encourage such movement include replacing desktop printers with a networked printer that an employee has to walk to, placing copy machines in more distant locations and relocating the vending machines, so employees have to take the stairs to reach them, and replacing their contents with healthy options. Buildings in which people can move around from place to place to do their work throughout the day, in combination with work design that has involved the users to design for health, and workplace strategies that nudge the adoption of public health initiatives, such as healthful nutrition, over time should lead to improvements in employee health, wellness, and productivity.

REFERENCES

Anger, W. K., D. L. Elliot, T. Bodner, R. Olson, D. S. Rohlman, D. M. Truxillo, K. S. Kuehl, L. B. Hammer, and D. Montgomery. 2015. Effectiveness of Total Worker Health interventions. *J Occup Health Psychol.* 20(2):226–47.
ANSI/HFES 100. 2007. Human Factors Engineering of Computer Workstations. From http://www.hfes.org /publications/ProductDetail.aspx?ProductID=69.

Barcellini, F., L. Van Belleghem, and F. Daniellou. 2015. Chapter 13: Design projects as opportunities for the development of activities. In Falzon, P. (ed.) *Constructive Ergonomics*. Boca Raton, FL: Taylor & Francis Group, LLC.

BIFMA G1. 2011. Ergonomics Guideline for Furniture Used in Office Work Spaces Designed for Computer Use. From https://www.bifma.org/store/ViewProduct.aspx?id=1375341.

Bridge, O. 2015. For better health… Medibank's thrive journey. The Integration of Emerging Trends: Creating a Healthy Workforce, National Workplace Health State Half Day Conferences 2015. Workplace Health Association Australia.

Browning, W. 1992. NMB bank headquarters: The impressive performance of a green building. *Urban Land*. June 23–25. From http://library.uniteddiversity.coop/Ecological_Building/NMB_Bank_Headquarters _The_Impressive_Performance_of_a_Green_Building.pdf (accessed November 13, 2015).

Buckley, J. P., Hedge, A., Yates, T., R. J. Copeland, M. Loosemore, M. Hamer, G. Bradley, and D. W. Dunstan. 2015. The sedentary office: A growing case for change towards better health and productivity. Expert statement commissioned by Public Health England and the Active Working Community Interest Company. *Brit J Sports Med*. 49(21):1357–62.

Burgess-Limerick, R. 2011. Ergonomics for Manual Tasks. In CCH Australia Ltd. *Australian Master OHS and Environment Guide*. Maryborough: McPherson's Printing Group, 261–78.

Burke, R. J. 2014. Improving individual and organisational health: Implementing and learning from interventions. In Biron, C., R. J. Burke, and C. L. Cooper (eds.) *Creating Healthy Workplaces: Stress Reduction, Improved Well-Being, and Organisational Effectiveness*. Surrey: Gower Ashgate.

Cantley, L. F., O. A. Taiwo, D. Galusha, R. Barbour, M. D. Slade, B. Tessier-Sherman, and M. R. Cullen. 2014. Effect of systemic ergonomic hazard identification and control implementation on musculoskeletal disorder and injury risk. *Scan J Work Environ and Health*. 40(1):57–65.

Carayon, P. 2006. Human factors of complex sociotechnical systems. *Appl Ergon*. 37:525–35.

Centers for Disease Control. 2013. Total worker health. From http://www.cdc.gov/niosh/TWH/ (accessed November 1, 2015).

CSA Z412-00 (R2011). 2011. Guideline on Office Ergonomics. From http://www.ccohs.ca/products/csa /27011972000pubs.

Daniellou, F., and P. Rabardel. 2005. Activity-oriented approaches to ergonomics: Some traditions and communities. *Theor Issues Ergons Sci*. 6(5):353–7.

Dul, J., and W. P. Neumann. 2009. Ergonomic contributions to company strategies. *Appl Ergon*. 40:745–52.

Dul, J., R. Bruder, P. Buckle, P. Carayon, P. Falzone, W. S. Marras, J. R. Wilson, and B. van de Doelen. 2012. A strategy for human factors/ergonomics: Developing the discipline and profession. *Ergon*. 55(4):377–95.

Haslam, R. A. 2002. Targeting ergonomics interventions—Learning from health promotion. *Appl Ergon*. 33:241–9.

Hassall, M. E., and P. M. Sanderson. 2012. A formative approach to the strategies analysis phase of cognitive work analysis. *Theor Issues Ergons Sci*. 15(3):1–47.

Hedge, A. 2006, July. Macroergonomics and the obesity epidemic. *Proc. IEA2006:16th World Congress on Ergonomics*. Maastricht. 6 pages. (CD).

Hedge, A. 2013. Ergonomics and U.S. Public Policy. *Assoc Comp Machine (ACM)*. From http://interactions .acm.org/archive/view/january-february-2013/ergonomics-and-u.s.-public-policy (accessed November 10, 2015).

Henning, R., N. Warren, M. Robertson, P. Faghri, M. Cherniack, and CPH-NEW Research Team. 2009. Workplace health protection and promotion through participatory ergonomics: An integrated approach. *Pub Health Rep*. 124(Suppl 1):26–35.

Hettler, B. 1976. Six dimensions of wellness model. National Wellness Institute. From http://c.ymcdn.com /sites/www.nationalwellness.org/resource/resmgr/docs/sixdimensionsfactsheet.pdf (accessed March 22, 2016).

Hosking, R. 2015, November 12. For better health: Medibank's thrive journey. Presented for Healthy and Active Living Day Webinar. Green Building Council of Australia, Sydney.

Karanika-Murray, M., and A. K. Weyman. 2013. Optimising workplace interventions for health and well-being. *Int J Workplace Health Manage*. 2(6):104–17.

Kickbush, I. 2013. *Health Literacy: The Solid Facts*. Copenhagen, Denmark: World Health Organisation.

Klepeis, N. E., W. C. Nelson, W. R. Ott, J. P. Robinson, A. M. Tsang, P. Switzer, J. V. Behar, S. C. Hern, and W. H. Engelmann. 2001. The national human activity pattern survey (NHAPS): A resource for assessing exposure to environmental pollutants. *J Expo Anal Environ Epidemiol*. 11:231–52.

Kohler, J. M., and D. C. Munz. 2006. Combining individual and organisational stress interventions: An organisational development approach. *Consult Psychol J: Prac Res*. 58(1):1–12.

Laing, A. C., D. C. Cole, N. Theberge, R. P. Wells, M. S. Kerr, and M. B. Frazer. 2007. Effectiveness of a participatory ergonomics intervention in improving communication and psychological exposures. *Ergonomics*. 50(7):1092–109.

Laitinen, H., J. Saari, M. Kivisto, and P.-L. Rasa. 1998. Improving physical and psychological working conditions through a participatory ergonomic process: A before-after study at an engineering workshop. *Int J Indust Ergon*. 21(1):35–45.

Lallemand, C. 2012. Contributions of participatory ergonomics to the improvement of safety culture in an industrial context. *Work*. 41(Suppl 1):3284–90.

LaSalle, J. L. 2012. Activity based working. From http://www.jll.com.au/australia/en-au/Documents/jll-au -activity-based-working-2012.pdf (accessed June 30, 2015).

Marsden, M. 2015. Hot desking, ABW, flexible working… Not for you? From http://marsdencollective.com .au/hot-desking-abw-flexible-working-not/ (accessed September 4, 2015).

Mellor, N., M. Karanika-Murray, and K. Waite. 2012. Taking a multi-faceted, multi-level, and integrated perspective for addressing psychosocial issues at the workplace. In Biron, C., M. Karanika-Murray, and C. L. Cooper (eds.) *Improving Organizational Interventions for Stress and Well-Being: Addressing Process and Context*. East Sussex: Routledge, pp. 39–59.

Michaels, C. N., and A. M. Greene. 2013. Worksite wellness: Increasing adoption of workplace health promotion programs. *Health Promot Pract*. 14(4):473–9.

Mumtaz, A. 1993, January. *Comparative post occupancy evaluation of work spaces and public areas in corporate headquarters of ABN-AMRO BANK and NMB POST BANK at Amsterdam, Netherlands*. MS Thesis. Cornell University, Ithaca, NY.

Munz, J. M., D. C. Kohler, and C. I. Greenberg. 2001. Effectiveness of a comprehensive worksite stress management programme: Combining organizational and individual interventions. *Int J Stress Manage*. 8(1):49–61.

OSHA 3192-05N. 2004. Guidelines for Retail Grocery Stores. From https://www.osha.gov/Publications/osha 3192.pdf.

OSHA 3182 (revised). 2009. Guidelines for Nursing Homes Ergonomics for the Prevention of Musculoskeletal Disorders. From https://www.osha.gov/ergonomics/guidelines/nursinghome/final_nh_guidelines.html.

Pazell, S. 2015. *Design of work for health: Strategies to embed participatory ergonomics and human-centred design in organisational systems* (Unpublished doctoral thesis). Brisbane: University of Queensland.

Pazell, S., and R. Burgess-Limerick. 2015a. Design of work for health: A human-centred design perspective: Part I. *Quarry Mag*. 23(10):74–6.

Pazell, S., and R. Burgess-Limerick. 2015b. Human-centred design: Integration of health and safety: Part II. *Quarry Mag*. 23(11):35–41.

Punnet, L., M. Cherniack, R. Henning, T. Morse, P. Faghri, and CPH-NEW Research Team. 2009. A conceptual framework for integrating work health promotion and occupational ergonomics programs. *Pub Health Rep*. 124(Suppl 1):16–25.

Randall, R., and K. Nielsen. 2012. Does the intervention fit? An explanatory model of intervention success and failure in complex organizational environments. In Biron, C., M. Karanika-Murray, and C. L. Cooper (eds.) *Improving Organizational Interventions for Stress and Well-Being*. East Sussex: Routledge, pp. 120–34.

Robert Wood Johnson Foundation. 2010. A new way to talk about the social determinants of health. From http://www.rwjf.org/content/dam/farm/reports/reports/2010/rwjf63023 (accessed October 13, 2015).

Robertson, M. M., Y.-H. Huang, M. J. O'Neill, and L. M. Schleifer. 2008. Flexible workspace design and ergonomics training: Impacts on the psychosocial work environment, musculoskeletal health, and work effectiveness among knowledge workers. *Appl Ergon*. 39(4):482–94.

Schill, A. L., and L. C. Chosewood. 2013. The NIOSH Total Worker Health™ Program: An overview. *J Occup Environ Med*. 55(Supplement 12S):S1–S88.

Schweizer, C., R. D. Edwards, L. Bayer-Oglesby, W. J. Gauderman, and M. J. Jantunen. 2007. Indoor time-microenvironment-activity patterns in seven regions of Europe. *J Expo Sci Environ Epidemiol*. 17(2):170–81.

University of California, Davis. 2015. Environmental wellness. From https://shcs.ucdavis.edu/wellness/environ mental/#.VjaJH_mrSM8 (accessed November 1, 2015).

Urlings, I., I. Nuboer, and J. Dul. 1990. A method for changing the attitudes and behaviour of management and employees to stimulate the implementation of ergonomics improvements. *Ergonomics*. 33(5):629–37.

U.S. Air Force. 2009. *Air Force Human Systems Integration Handbook*. Brooks City-Base, TX: Directorate of Human Performance Integration. From http://www.wpafb.af.mil/shared/media/document/AFD-090121 -054.pdf.

U.S. Green Building Council. 2011. LEED Pilot Credit 44: Ergonomics Strategy. From http://www.usgbc.org /Docs/Archive/General/Docs10097.pdf (accessed March 22, 2016).

Vink, P., E. A. P. Koningsveld, and J. F. Molenbroek. 2006. Positive outcomes of participatory ergonomics in terms of greater comfort and higher productivity. *Appl Ergon.* 37:537–46.

Waddell, G., and K. A. Burton. 2006. *Is Work Good for Your Health and Well-Being?* London: TSO.

WHO. 1948. WHO definition of Health. From http://www.who.int/about/definition/en/print.html (accessed November 1, 2015).

WHO. 2012. Social Determinants of Health. Geneva. WHO. From http://www.who.int/social_determinants/en/ (accessed October 13, 2015).

Wilcock, A. 2006. *An Occupational Perspective of Health*, Second edn. Thorofare, NJ: SLACK Incorporated.

Z1004-12. 2012. Workplace ergonomics—A management and implementation standard. From http://shop.csa.ca/en/canada/general-workplace-ergonomics/z1004-12/invt/27032732012.

Index

Page numbers followed by f and t indicate figures and tables, respectively.